STATISTICS AND CAUSALITY

A complete list of the titles in this series appears at the end of this volume.

STATISTICS AND CAUSALITY

Methods for Applied Empirical Research

Edited by

WOLFGANG WIEDERMANN
ALEXANDER VON EYE

WILEY

Published by John Wiley & Sons, Inc., Hoboken, New Jersey

Published simultaneously in Canada

Library of Congress Cataloging-in-Publication Data:

Names: Wiedermann, Wolfgang, 1981- editor of compilation. | Eye, Alexander von, editor of compilation.
Title: Statistics and causality : methods for applied empirical research / edited by Wolfgang Wiedermann, Alexander von Eye.
Other titles: Wiley series in probability and statistics.
Description: Hoboken, New Jersey : John Wiley & Sons, 2016. | Series: Wiley series in probability and statistics | Includes bibliographical references and index.
Identifiers: LCCN 2015047424 (print) | LCCN 2015050865 (ebook) | ISBN 9781118947043 (cloth) | ISBN 9781118947050 (pdf) | ISBN 9781118947067 (epub)
Subjects: LCSH: Statistics–Methodology. | Causation. | Quantitative research–Methodology.
Classification: LCC QA276.A2 S73 2016 (print) | LCC QA276.A2 (ebook) | DDC 001.4/22–dc23
LC record available at http://lccn.loc.gov/2015047424

Typeset in 10/12pt TimesLTStd by SPi Global, Chennai, India

Printed in the United States of America

10 9 8 7 6 5 4 3 2 1

CONTENTS

LIST OF CONTRIBUTORS

Bethany C. Bray The Methodology Center and The College of Health and Human Development, The Pennsylvania State University, University Park, PA, USA

Claus H. Carstensen Psychology and Methods of Educational Research, University of Bamberg, Bamberg, Germany

Yadolah Dodge Institute of Statistics, University of Neuchâtel, Neuchâtel, Switzerland

Ulrich Frick Department of Applied Psychology, HSD University of Applied Sciences, Cologne, Germany and Swiss Research Institute on Public Health and Addiction, University of Zurich, Zurich, Switzerland and Psychiatric University Hospital, University of Regensburg, Regensburg, Germany

Ned Hall Department of Philosophy, Harvard University, Cambridge, MA, USA

Kateřina Hlaváčková-Schindler Department of Adaptive Systems, Institute of Information Theory and Automation, Academy of Sciences of the Czech Republic, Prague, Czech Republic

Aapo Hyvärinen Department of Computer Science, University of Helsinki, Helsinki, Finland

Daeyoung Kim Department of Mathematics and Statistics, University of Massachusetts, Amherst, MA, USA

Seongyong Kim Department of Applied Statistics, Hoseo University, Asan-si, Republic of Korea

Ingrid Koller Institute for Psychology, Department of Developmental and Educational Psychology, Alpen-Adria-Universität Klagenfurt, Klagenfurt, Austria

Stephanie T. Lanza Department of Biobehavioral Health and The Methodology Center, The College of Health and Human Development, The Pennsylvania State University, University Park, PA, USA

Lawrence L. Lo Quantitative Developmental Systems Methodology, Department of Human Development and Family Studies, The Pennsylvania State University, University Park, PA, USA

Peter C. M. Molenaar Quantitative Developmental Systems Methodology, Department of Human Development and Family Studies, The Pennsylvania State University, University Park, PA, USA

Valeriya Naumova Center for Biomedical Computing, Simula Research Laboratory, Lysaker, Norway

Sergiy Pereverzyev Jr. Applied Mathematics Group, Department of Mathematics, University of Innsbruck, Innsbruck, Austria

Steffi Pohl Department of Education and Psychology, Methods and Evaluation / Quality Management, Freie Universität Berlin, Berlin, Germany

Jürgen Rehm Social and Epidemiological Research (SER) Department, Centre for Addiction and Mental Health, Toronto, Canada and Addiction Policy, Dalla Lana School of Public Health, University of Toronto, Toronto, Canada and Department of Psychiatry, Faculty of Medicine, University of Toronto, Toronto, Canada and PAHO/WHO Collaborating Centre for Mental Health & Addiction, Toronto, Canada and Institute of Medical Science, University of Toronto, Toronto, Canada and Epidemiological Research Unit, Technische Universität Dresden, Klinische Psychologie & Psychotherapie, Dresden, Germany

Valentin Rousson Division of Biostatistics, Institute for Social and Preventive Medicine, University Hospital Lausanne, Lausanne, Switzerland

Megan S. Schuler Department of Health Care Policy, Harvard Medical School, Boston, MA, USA

Marie-Ann Sengewald Methodology and Evaluation Research, Institute of Psychology, Friedrich-Schiller-University Jena, Jena, Germany

Shohei Shimizu Institute of Scientific and Industrial Research, Osaka University, Osaka, Japan

Peter M. Steiner Department of Educational Psychology, School of Education, University of Wisconsin-Madison, Madison, WI, USA

Rolf Steyer Methodology and Evaluation Research, Institute of Psychology, Friedrich-Schiller-University Jena, Germany

Alexander von Eye Department of Psychology, Michigan State University, East Lansing, MI, USA

Wolfgang Wiedermann Department of Educational, School & Counseling Psychology, College of Education, University of Missouri, Columbia, MO, USA

Michael Wilde Department of Philosophy, School of European Culture and Languages, University of Kent, Kent, UK

Jon Williamson Department of Philosophy, School of European Culture and Languages, University of Kent, Kent, UK

Phillip K. Wood Department of Psychological Sciences, University of Missouri, Columbia, MO, USA

Kazuo Yamaguchi Department of Sociology, University of Chicago, Chicago, IL, USA

Kun Zhang Max Planck Institute for Intelligent Systems, Tübingen, Germany and Carnegie Mellon University, Pittsburgh, PA, USA

PREFACE

The discussion of concepts of causality has been a staple of philosophical discourse since at least Aristotle. Very well known are Aristotle's four types of causes: the material cause, the formal cause, the efficient cause, and the final cause. Having been introduced into scholarly thinking slightly later, statistics took a moment to make a contribution to causal thinking. Early efforts put forth by statistics reside in two domains. First, in the domain of design, it was discussed whether only experimental data are needed for researchers to make conclusions about causal processes (Fisher, 1926, 1935), or whether observational data can also lead to trustworthy conclusions (see, e.g., Cochran and Chambers, 1965). Second, in the theoretical domain, concepts were developed that would allow one to derive testable hypotheses. Examples of such concepts include counterfactual statistical theory (for discussions, see Holland, 1986; Neyman, 1923/1990; Rubin, 1974, 2005) and causal structural modeling (e.g., Sobel, 1994).

These efforts were needed and important because it is well known that, with standard methods of statistics, that is, with methods from the family of Generalized Linear Models (GLM; Nelder and Wedderburn, 1972) one of the key characteristics of causal effects, direction, cannot be ascertained (for an illustration, see von Eye and DeShon, 2012). For example, the standardized slope parameter for the linear regression of a variable Y on a variable X is exactly the same as the standardized slope parameter for the regression of X on Y and the correlation between X and Y. Thus, conclusions concerning the direction of effects have to be guided by a priori theoretical considerations.

Both, the philosophical and the statistical lines of research have made most impressive progress. In philosophy, various theories of causality have been elaborated, and Hume's classical causality theory (Hume, 1777/1975) now is just one among a number of others. An overview of philosophical theories and discussion can be found in Beebee *et al.* (2009). In statistics, known approaches have been further developed, in particular, in the domain of models for nonexperimental research, and novel and most promising ideas have been presented, in particular, in the domain of methods of analysis. The links between philosophical theories, design, and statistical data processing have been discussed. Methods of analysis are available that match particular philosophical theories.

This book is concerned with novel statistical approaches to causal analysis, in the context of the continuing development of philosophical theories. This book presents original work, in five modules. In the first module, *Bases of Causality*, Hall presents an account of causal structures from a foundationalist perspective and explicitly connects it to the aims of any scientific inquiry (Chapter 1). Causal structures are seen as ways in which states of localized bits of the world depend on states of other localized bits. The author discusses why unpacking and rendering this localized dependence account (which is, in essence, a version of the well-established counterfactual or "interventionist" account; e.g., Holland, 1986; Pearl, 2009; Rubin, 1974) may lead to several problems, which so far lack adequate solutions. Further, the author explains why treating causal structures as localized dependences may lead to an abandonment of a core feature of causation, that is, the idea that causes need to be connected to their effects via mediating processes. In Chapter 2, Wilde and Williamson discuss issues associated with standard mechanistic and difference-making theories of causality. Both lines of causality theories are often discussed in the face of counterexamples, and may struggle to explain the evidential practice of establishing causal claims. Similarly, common lines of response to the issue of counterexamples (such as simply dismissing the counterexamples or moving to pluralism) suffer from difficulties in accounting for the practice of establishing causal claims. The authors present an epistemic theory of causality as a valuable alternative. Here, causality is perceived as being purely epistemic in the sense that causal claims are not claims about causal relations that exist independent of humans. Instead, these causal claims enable humans to reason and interact with the environment.

In the empirical sciences, the Pearson correlation coefficient is one of the most widely used statistics to measure the linear association of two variables. Covariances/correlations constitute the essential source of data information used in countless statistical models, such as Factor, Path, and Structural Equation Models (e.g., Bollen, 1989), which are nowadays indispensable for both theorists and applied researchers. A very important (as well as thorny) feature of covariances and correlations is that both do not depend on the order of the variables (i.e., $cov(X, Y) = cov(Y, X)$ and $cor(X, Y) = cor(Y, X)$). Thus, in particular, in observational data setting, one has to sharply distinguish between correlation and causation. However, in recent years, tremendous theoretical progress has been made, which led to the development of so-called asymmetric facets of the Pearson correlation, that is, situations in which the status of a variable (in terms of "response" or "predictor") is no longer exchangeable. Dodge and Rousson (2000, 2001) proposed the first asymmetric facet of the correlation coefficient through considering the third moments (i.e., the skewness) of two nonnormally distributed variables. The second module, *Directionality of Effects*, presents novel generalizations of the asymmetric characteristics of the correlation coefficient. All methods presented in this module share that information beyond the second moments of variables (skewness and kurtosis) is considered being informative. In Chapter 3, Dodge and Rousson present new empirical evidence on the adequacy of methods for statistical inference for determining the direction of dependence in linear regression models. The authors present a modified approach to identify the direction of effects in the bivariate setting. Further, direction of dependence approaches in case of lurking/confounding variables, sampling from subpopulations, and in the presence of outliers are discussed. In Chapter 4, Wiedermann and von Eye extend approaches to determine the direction of effects to cases of mediational hypotheses, that is, situations in which a third intervening variable is assumed to affect a predictor–outcome relation. Significance tests are proposed designed to empirically test a putative mediation model against a plausible alternative model (i.e., a model in which the reverse flow of causality is considered). Results from a Monte Carlo simulation study as well as practical applications are presented. In Chapter 5, von Eye and Wiedermann then discuss potential application of direction of dependence methods in the categorical variable setting. The authors present the "generalized direction dependence principle" and propose log-linear model specifications that allow directional statements in terms of both univariate probability distributions and structural elements of observed associations. Early theoretical results of Dodge and Rousson (2000) have also been discussed from a Copula perspective (Sungur, 2005) that led to the development of directional Copula regression methods (Kim and Kim, 2014). In Chapter 6, Kim and Kim discuss recent advances in making directional statements based on Copula regression techniques. The authors present skew-normal Copula-based regression models to analyze directional dependence based on the joint distributional behavior of variables. An empirical demonstration of this new model is given using data from adolescent aggression research. The last two chapters of this module give an excellent overview of recently proposed causal discovery algorithms for nonnormal data. In Chapter 7, Shimizu introduces the so-called linear acyclic non-Gaussian model (LiNGAM; Shimizu *et al.*, 2006) and discusses extensions to various data analytic domains including time series analysis and models in case of latent common causes. Chapter 8 is devoted to causal discovery algorithms for nonlinear data problems. Starting with a summary of linear non-Gaussian causal models, Zhang and Hyvärinen review nonlinear additive noise models, propose a likelihood ratio to decide between two directional candidate models, and embed the approach within an information-theoretic framework. Further, the authors generalize the approach to the postnonlinear causal model (which contains the linear non-Gaussian model and additive noise model as special cases). The performance of these causal discovery approaches is discussed using 77 cause–effect data sets from various scientific disciplines.

The aspect of temporality became a widely accepted requirement to distinguish between association and causation (implicitly following Hume's proposition that the "cause must precede the effect"). In time series analysis, the majority of methods for causal inference use temporal precedence as an essential element to deriving causal statements. However, at least since Yule's seminal papers on 'nonsense' correlations among time-variables (Yule, 1921, 1926), statisticians are well aware that temporal priority cannot per se be regarded as a causal factor. One of the most prominent attempts to incorporate the time factor in elucidating causation was introduced by Granger (1969). In essence, testing "Granger causality" relies on a prediction error approach. A variable X is said to "Granger-cause" a variable Y if the prediction error variance of Y_t given a universal set of information up to time point t (i.e., Ω_t) is smaller than the prediction error variance of Y_t given Ω_t without the information of X. The third module, *Granger Causality and Longitudinal Data Modeling*, is devoted to novel advances in Granger causality testing and related issues. In Chapter 9, Molenaar and Lo discuss important theoretical ambiguities associated with Granger causality testing, discuss Granger causality testing in the light of standard vector autoregressive models (VAR), structural VARs, and hybrid VARs, and propose a new approach to empirically determine which VAR best describes the dynamic stochastic process underlying observed time series. This new approach is promising in correctly recovering the underlying true model and, thus, yielding correct results concerning lagged Granger causality. In Chapter 10, Koller, Carstensen, Wiedermann, and von Eye link the Granger causality principle to Item Response Theory (IRT). The authors discuss formulations of multidimensional longitudinal item response models to test hypotheses compatible with Granger causality hypotheses, which enables researchers to estimate an underlying measurement model while simultaneously assessing the predictability of latent person abilities over time *sensu* Granger. Chapter 11 by Hlaváčková-Schindler, Naumova, and Pereverzyev Jr. is devoted to applications of the Granger causality principle in the case of high-dimensional data. The authors consider Granger causality as a special case of an inverse ill-posed problem and discuss novel regularization techniques to uncover causal relations in the case of high-dimensional data and evaluate these approaches in a case study on gene regulatory networks reconstruction. Chapter 12 by Wood is then devoted to reciprocal causal models. Such models often involve estimation of effects in longitudinal data settings, where earlier assessments have effects on subsequent measurement occasions. However, some processes can be modeled using path diagrams containing instantaneous feedback loops uniquely associated with a measurement point that may involve reciprocal effects between two constructs, circular effects of three or more constructs, or autocausal effects associated with the dissipation/acceleration of levels of a variable within the system. The author first starts with discussing how these models differ from more commonly applied cross-lagged correlation and Granger causality models. Then it is shown that autocausal effects are equivalent to models in which variables involved with reciprocal or circular effects are omitted. These unconsidered reciprocal effects can lead to biased parameters estimates. Further, it is shown that for some research designs and research questions, it is possible to distinguish between nonrecursive and recursive models. Empirical examples from alcohol research as well as results from a Monte Carlo simulation experiment are presented, which show that multiwave assessments have sufficient power to identify autocausal effects.

Over the decades, various statistical frameworks have been developed that outline necessary assumptions under which statistical results can be endowed with causal interpretation. One of the most widely recognized conceptualizations is Rubin's potential outcome representation (Rubin, 1974), which, in essence, can be regarded a generalization of Fisher's principles of experimentation (Fisher, 1926, 1935). The first appearance of potential outcome representations can be traced back to Neyman (1923/1990) who explicitly linked the results from randomized experiments to the logic of counterfactuals (for a detailed historical account see Barringer *et al.*, 2013). The potential outcome framework is deeply rooted in the philosophical foundation of counterfactual causal analysis (e.g. Lewis, 1973), that is, causal statements can only be derived if one additionally considers what would have happened had a person experienced something different than she/he did experience. This conceptualization of causal analysis inevitably leads to what has been called the fundamental problem of causal inference (Holland, 1986), that is, the fact that only one condition-outcome pair can be observed. The fourth module is devoted to *Counterfactual Approaches and Propensity Score Analysis*. In Chapter 13, Yamaguchi discusses causal analysis of categorical variables. The author proposes a solution to the so-called lack of collapsibility issue associated with models with a logit-link function (Gail *et al.*, 1984). This novel approach enables researchers to accurately estimate causal effects of cross-classified variables. Propensity score (PS) techniques, such as PS matching, PS stratification, or inverse-propensity weighting are routinely used to estimate causal treatment effects from purely observational data. However, in practice, it is rarely recognized that PS designs can be analyzed according to design- or model-based formulations. In Chapter 14, Steiner provides an excellent overview of PS approaches under design- and model-based formulations, which highlights that the type of formulation used affects estimators of average treatment effects and the generalizability of the results. Chapter 15 contributed by Pohl, Sengewald, and Steyer is devoted to covariate adjustment to obtain unbiased treatment effects. The authors discuss the impact of measurement error of covariates on estimated treatment effects. The authors specify conditions under which latent or manifest (fallible) covariate adjustment should be used to avoid biased causal effect estimates. Theoretical and empirical evidence is provided on the impact of measurement error in covariates for causal effect estimation and various adjustment methods based on latent covariates are discussed. In the last chapter of this module (Chapter 16), Lanza, Schuler, and Bray discuss extensions of causal inference methods to the domain of latent class analysis. The authors discuss the application of inverse propensity weighting to estimate causal effects of explanatory variables to predict latent class memberships and demonstrate this new approach using data of adolescent depression and adult substance use. Their empirical example reveals that this novel modeling technique enables researchers to (i) identify latent patterns based on a series of manifest indicators, (ii) consider potential moderator effects, and (iii) arrive at causal statements concerning additional explanatory variables within a single modeling framework.

In the final module of the volume, *Designs for Causal Inference*, Frick and Rehm (Chapter 17) provide an excellent overview of research designs commonly used in the field of Epidemiology (such as cohort and case-control designs). Starting with a discussion of epidemiological theories of causality, the authors use various examples from recent epidemiological research to vividly remind the readers that even the most elaborated and complex statistical tools cannot compensate potential weaknesses in the process of data collection, such as, ill-designed questionnaires, failing to adequately standardize interview situations, and low measurement quality.

WOLFGANG WIEDERMANN
University of Missouri
Columbia

ALEXANDER VON EYE
Michigan State University
East Lansing

REFERENCES

Barringer, S.N., Eliason, S.R., and Leahey, E. (2013) A history of causal analysis in the social sciences, in *Handbook of Causal Analysis for Social Research* (ed. S.L. Morgan), Springer-Verlag, Dordrecht, pp. 9–26.

Beebee, H., Hitchcock, C., and Menzies, P. (2009) *The Oxford Handbook of Causation*, Oxford University Press, Oxford.

Bollen, K.A. (1989) *Structural Equations with Latent Variables*, John Wiley & Sons, Inc., New York.

Cochran, W.G. and Chambers, S.P. (1965) The planning of observational studies of human populations. *Journal of the Royal Statistical Society. Series A (General)*, **128**, 234–266.

Dodge, Y. and Rousson, V. (2000) Direction dependence in a regression line. *Communications in Statistics: Theory and Methods*, **29** (9-10), 1957–1972.

Dodge, Y. and Rousson, V. (2001) On asymmetric properties of the correlation coefficient in the regression setting. *American Statistician*, **55** (1), 51–54.

von Eye, A. and DeShon, R.P. (2012) Directional dependence in developmental research. *International Journal of Behavioral Development*, **36** (4), 303–312.

Fisher, R.A. (1926) The arrangement of field experiments. *Journal of the Ministry of Agriculture of Great Britain*, **33**, 503–513.

Fisher, R.A. (1935) *The Design of Experiments*, Oliver & Boyd, Edinburgh.

Gail, M.H., Wieand, S., and Piantadosi, S. (1984) Biased estimates of treatment effect in randomized experiments with nonlinear regressions and omitted covariates. *Biometrika*, **71** (3), 431–444.

Granger, C.W. (1969) Investigating causal relations by econometric models and cross-spectral methods. *Econometrica*, **37**, 424–438.

Holland, P.W. (1986) Statistics and causal inference. *Journal of the American Statistical Association*, **81** (396), 945–960.

Hume, D. (1777/1975) *Enquiries Concerning Human Understanding and Concerning the Principles of Morals*, Clarendon Press, Oxford.

Kim, D. and Kim, J.M. (2014) Analysis of directional dependence using asymmetric Copula-based regression models. *Journal of Statistical Computation and Simulation*, **84** (9), 1990–2010.

Lewis, D. (1973) Causation. *Journal of Philosophy*, **70** (17), 556–567.

Nelder, J.A. and Wedderburn, R.W.M. (1972) Generalized linear models. *Journal of the Royal Statistical Society. Series A (General)*, **135** (3), 370–384.

Neyman, J. (1923/1990) Sur les applications de la theorie des probabilites aux experiences agricoles [On the application of probability theory to agricultural experiments; D. Dabrowska and T. P. Speed, translators]. *Excerpts reprinted in Statistical Science*, **5**, 463–472.

Pearl, J. (2009) *Causality: Models, Reasoning, and Inference*, 2nd edn, Cambridge University Press, Cambridge.

Rubin, D.B. (1974) Estimating causal effects of treatments in randomized and nonrandomized studies. *Journal of Educational Psychology*, **66** (5), 688–701.

Rubin, D.B. (2005) Causal inference using potential outcomes. *Journal of the American Statistical Association*, **100** (469), 322–331.

Shimizu, S., Hoyer, P.O., Hyvärinen, A., and Kerminen, A. (2006) A linear non-Gaussian acyclic model for causal discovery. *Journal of Machine Learning Research*, **7**, 2003–2030.

Sobel, M.E. (1994) Causal inference in latent variable models, in *Latent Variable Analysis: Applications for Developmental Research* (eds A. von Eye and C.C. Clogg), Sage Publications, Thousand Oaks, CA, pp. 3–35.

Sungur, E.A. (2005) A note on directional dependence in regression setting. *Communications in Statistics: Theory and Methods*, **34** (9-10), 1957–1965.

Yule, G.U. (1921) On the time-correlation problem, with especial reference to the variate-difference correlation method. *Journal of the Royal Statistical Society*, **84** (4), 497–537.

Yule, G.U. (1926) Why do we sometimes get nonsense-correlations between time-series? A study in sampling and the nature of time-series. *Journal of the Royal Statistical Society*, **89** (1), 1–63.

ACKNOWLEDGMENTS

In the year 2014, a conference took place in Vienna, Austria, on the topic of *Statistics and Causality*. We are grateful for the support and sponsoring of the Austrian Research Foundation and the University of Vienna. A number of the papers presented on the occasion of this conference can be found in the present volume. To cover additional important topics, the main protagonists of these topics were invited to contribute chapters. We are indebted to all of the authors for sharing their wonderful work.

The organization of this conference was in the powerful hands of Martina Edl and Abla Marie-José Bedi, and they were supported by our graduate assistants Felix Deichmann, Karin Futschek, Francie Mißbach, Vanessa Mitschke, Nele Motzki, and Sandra Peer. Without these capable experts, the conference would not have run half as smoothly. In addition, we extend our thanks to Philipp Gewessler. He translated the chapter manuscripts into LaTeX. His timely and professional work is most appreciated.

We are also grateful to Wiley, publishers, for their interest in this topic and their support. This applies in particular to Stephen Quigley, Sari Friedman, Bhargavi Natarajan, Divya Narayanan, and Sumalini Vivekanandan who have gracefully guided our efforts from the very first contact through the completion of this volume. It was a great pleasure to work with them, and we benefited enormously from their experience and professional style. Thank you all!

Most important, we are happy to emphasize that we consider ourselves lucky for being able to work in a social environment of love and support. Wolfgang

Wiedermann is grateful to be allowed to experience a rare and truly unbiased (and reciprocal, as he hopes) process–the unconditional and tireless support of Anna. Alexander von Eye knows that the happiness he enjoys is caused by Donata, in a nonstochastic manner.

Finally, we welcome and we are deeply indebted to Linus A. Wiedermann who decided it was time to enter this world … on the second day of the 2014 conference, in Vienna. All our best to Linus!

PART I

BASES OF CAUSALITY

1

CAUSATION AND THE AIMS OF INQUIRY

NED HALL

Department of Philosophy, Harvard University, Cambridge, MA, USA

1.1 INTRODUCTION

I often like to ask my students how many of them have heard the advice not to confuse correlation with causation. Most raise their hands. I then ask how many of them have taken a class that explained, precisely, what correlation is. Not as many raise their hands–but still, plenty do, and some of them can even go on to articulate, quite lucidly, some of the different ways in which statistical correlation can be precisely defined. Then I ask a *mean* question. "Have your statistics classes also explained to you, with similar precision, what causation is?" Nope. Never.

That is too bad: after all, the advice not to confuse X with Y falls a little flat, if we do not really know what Y is. But it is not just the value of this (excellent) advice that is at stake. As I will try to explain, clarity about what causal structure is–or better, clarity about one thing we can *usefully mean* by "causal structure"–promises to bring clarity about the very aims of scientific inquiry. The first goal of this essay is to sketch one especially simple, attractive account of causal structure and to advertise its virtues, by connecting it explicitly to the most general aims of inquiry. The account I will lay out is, more or less, a version of a kind of counterfactual or "interventionist" account that has recently gained a good deal of popularity in the literature; see especially Woodward (2005) and Pearl (2009). But I find the contemporary literature on interventionism, structural equations, causal modeling, and so on unnervingly quietist

Statistics and Causality: Methods for Applied Empirical Research, First Edition.
Wolfgang Wiedermann and Alexander von Eye.

about the metaphysical *foundations* of the account; accordingly, I will be focusing much more attention than most other authors do on those foundations.[1]

That is the good news. But there is also bad–or, at least, *challenging*–news. For when we turn to the foundations of interventionist approaches to causation, and try to construct those foundations in the most straightforward and plausible manner, we quickly encounter several deep problems, for which we so far lack adequate solutions. I do not think these problems are intractable in a way that threatens the viability of these approaches. But they are serious and deserve more scrutiny than the current literature is giving them. The second goal of this essay is to present them as clearly as possible.

Finally, there is *perplexing* news. In brief, the kind of account of causal structure that we will be focusing on sees that structure as constituted by the ways in which states of localized bits of the world *depend on* states of other localized bits. As you might have guessed, the crucial work comes in unpacking and rendering precise this relation of dependence. But what may come as a surprise is that in the course of this unpacking and rendering precise, a certain core feature of causation–at least, as we *ordinarily* conceive of it–gets abandoned, in a quite blatant and striking manner: we give up the idea that there is any special sense in which causes need to be connected to their effects via mediating *processes*. The third and final goal of this essay is to assess this intuitive cost and to offer some reasons for thinking it considerable enough to amend the very foundations of our interventionist account.

We will march through the good news, the challenging news, and the perplexing news in order. But we will be wise to cover some important methodological precepts, first.

1.2 THE AIM OF AN ACCOUNT OF CAUSATION

Just what is a philosophical account of causation? This will have to do: it is a comprehensive answer, pitched at a very high level of generality and abstraction, to the question, "What is causation?" (It is the generality and abstraction that make it "philosophical"; nothing more.) Then it might seem quite easy to say what counts as "success," in giving such an account: the account should be correct; it should tell the *truth* about causation. But that is naïve–not because we must capitulate to the postmodernist forces of darkness and eschew talk of "truth," but for two better reasons.

1.2.1 The Possible Utility of a False Account

First, we should not ignore a standard of success that is quite compatible with falsehood: namely, limning a conception of causation that is useful for some well-defined

[1]For those unfamiliar with the terminology, by "metaphysical foundations," I just mean those very basic theses about how our world works that we need to draw upon in order to give the account a precise and noncircular content.

and important purpose, even though it is false. Or, if you prefer, limning a conception of some *other* relation–very similar to, but not to be confused with, causation itself–which, once we have got it in our conceptual toolkit, can help us solve some important problem. Or, to say it in yet another way: success might come in the form of a revisionary account, which is allowed to reject certain home truths about causation on account of the utility thereby earned. In short, we might want an account of some "causal-like" relation and, provided it earns its keep, not care too much that it fails to fit perfectly what we have in mind when we ordinarily talk about causation.

Here are some examples. It matters a great deal, in legal settings, that we have some way of assigning responsibility for harm. An account of causation might succeed by enabling a simple and exact method for demarcating such legal responsibility. Again, cognitive psychologists have begun to make real progress in understanding how our capacity to reason about our environment (especially, in ways that allow us to successfully navigate and manipulate it) develops, in infancy and early childhood; it seems clear, by this point, that part of this capacity consists in the ability to represent the world around us as causally structured. So, an account of causation might try to elucidate that structure–that is, whatever structure it is that we learn to represent to ourselves, in learning how to navigate and manipulate our environment. Or, yet again, we might want our account to help clarify how, precisely, statistical data may be used to draw inferences about causal structure. In each of these cases, we should not blink if an account succeeds at the task we have set it, but at the cost of denying certain causal claims that strike us as obviously correct.

1.2.2 Inquiry's Aim

Here is a final example, one that I am particularly interested in, and that will help bring that last point into sharper focus. Suppose we endorse the following sweeping claim about science:

Inquiry's Aim: Scientific inquiry aims to discover and describe the causal structure of the world.

I think **Inquiry's Aim** is almost certainly correct. But it is quite another question whether there is any way to unpack it that will make it at all *illuminating*. (To see the problem, suppose that when asked what we mean by the "causal structure of the world," all we can say is that it is the kind of structure that scientific inquiry aims to discover and describe.) So, there is another task we might set ourselves, which is to answer this question: what philosophical account of causation will make **Inquiry's Aim** a true and illuminating thing to say about science?

Section 1.3 sketches an account that answers this question, in a rather elegant and attractive manner. But we will see that it forcefully denies certain perfectly obvious claims about causation. Consider, for example, the following scenario:

Suzy First: Suzy and Billy, two young vandals, throw rocks at a particularly choice window. Both throw with deadly accuracy, but Suzy is a bit quicker: her rock hits the window first, breaking it. Billy's rock flies through a now empty window pane.

Ask what causes the window to break, in **Suzy First**, and the answer can seem blindingly obvious: it is Suzy's throw, and *not* Billy's. Alas, according to the elegant and attractive account we are about to see, the two throws are *on a par*: each counts as just as much of a cause of the window's breaking as the other.[2] Why isn't that a fatal defect? In part, because the account can prove its worth by making **Inquiry's Aim** a true and illuminating thing to say about science. But in part, too, because the background methodology that gives a powerful role to firm intuitions about cases is itself deeply suspect.

1.2.3 The Role of "Intuitions"

Earlier, I said that there were two good reasons for rejecting the naïve, "we just want the truth about causation" standard of success for an account. As I have just argued, the first reason is that falsehood can be useful. The second is that the truth in this domain might turn out to be not very useful or interesting at all.

Let us suppose we go about finding this truth–at, remember, the appropriately high level of generality and abstraction–in the usual way. That is, we propose an analysis of "event X is a cause of event Y" in other terms and systematically test our analysis against claims that we already know to be true (drawn, as we philosophers like to say, from "intuition"). In doing so, we are following a methodology neatly laid out by Lewis (1986b):

> When common sense delivers a firm and uncontroversial answer about a not-too-far-fetched case, theory had better agree. If an analysis of causation does not deliver the common-sense answer that is bad trouble. But when common sense falls into indecision or controversy, or when it is reasonable to suspect that far-fetched cases are being judged by false analogy to commonplace ones, then theory may safely say what it likes. (Lewis 1986b, p. 194)

As an example of this method in action, suppose we start by proposing a very simple counterfactual analysis: event X is a cause of event Y if and only if X and Y both occur, but if X had not occurred, Y would not have occurred. (However dated, this is not all that far from currently popular interventionist accounts. Thus, some now like to say that "variable" X is a cause of "variable" Y if and only if an "intervention" on the value of X would have led to a change in the value of Y (again, see Woodward, 2005 and Pearl, 2009)). This account fails–right?–since it delivers the incorrect result that in **Suzy First**, Suzy's throw is not a cause of the window's breaking ("bad trouble").

This method has driven decades of philosophical work on causation, but is rightly falling out of fashion. For a good question never got a good answer: suppose we succeed. We come up with an account of causation that rigorously passes the sorts of tests that treat our firm judgments about causation as nonnegotiable data, in just

[2]On some versions of the account, both are causes; whereas on others, neither is. But the important point is that basic features of the account guarantee that the two throws be treated the same way. While I will not defend that claim here, see, for example, Halpern and Pearl (2005) for a popular attempt to distinguish the two throws from within the interventionist framework and Hall (2006) for a rebuttal. For in-depth discussion of the challenges posed by this kind of example, see Paul and Hall (2013, Chapter 3, Section 4).

the way Lewis recommends. Why should we care? What value thereby attaches to the resulting account? Perhaps it will be useful: it will help us clarify the nature of scientific inquiry, or the standards for statistical inference, or the way responsibility for harm ought to be assessed, and so on. But then it is that very utility that serves as the mark of success.

Here is an important upshot. Our causal judgments–even ones as firm as that, in **Suzy First**, Suzy's throw and Suzy's throw alone is a cause of the breaking–cannot serve as nonnegotiable data. An account may permissibly controvert them, or at any rate, some of them.[3]

Myself, I prefer to stop there and not simply set aside our causal judgments entirely. For it seems to me a modest and sensible revision to Lewis's methodology to view our intuitive causal judgments not as *data*, but as *clues*: clues, specifically, to where a potentially useful causal-like concept or concepts might be found. They might be misleading clues. But one should not just assume so, at the outset. I draw on this modest methodological orientation, in Section 1.5. First we will turn, though, to the forthrightly revisionary account of causation I have been hinting at, an account that answers our question about **Inquiry's Aim** rather neatly (albeit in a way that treats the "clues" contained in our intuitions about cases such as **Suzy First** as wholly misleading).

1.3 THE GOOD NEWS

1.3.1 The Core Idea

A very simple idea lies at the heart of the account: the world possesses a *localized dependence structure*, constituted by the totality of facts about how conditions at different spatially and temporally bounded places and times depend on conditions at other such places and times. Here, at a certain location, at a certain time, lie some shards of glass on the ground. On what does that fact depend? That is, of what other conditions, characterizing what other places and times, is it the case that, had those conditions not obtained (or, had they obtained in a somewhat different manner), then the condition in question–that there are, at this place and time, shards of glass lying on the ground, just so–would not have obtained (or, would have obtained in a somewhat different manner)? The answer to that question will tell you how the target condition, itself localized, depends on other localized conditions. (e.g., *part* of the answer might be that, at a certain earlier place and time, a girl threw a rock in the direction of the window.)

Now, generalize. Imagine that you can consult an Oracle. Given any two distinct regions of space and time, she can tell you how conditions in one depend on conditions in the other–again, in the sense captured by counterfactuals of the form "Had conditions in this region been different in such-and-such a way, then conditions in this

[3]Controvert too many of them, and your account will look like it is changing the subject–which is fine, it is just that at that point, you should not advertise it as having anything to do with causation.

other region would have been different in such-and-such a way." What she is thereby in a position to convey to you is the world's localized dependence structure.

Lewis, in a much later work, captured this idea (under the label "influence") with characteristic pithiness:

> Think of influence this way. First, you come upon a complicated machine, and you want to find out which bits are connected to which others. So you wiggle first one bit and then another, and each time you see what else wiggles. Next, you come upon a complicated arrangement of events in space and time. You can't wiggle an event: it is where it is in space and time, there's nothing you can do about that. But if you had an oracle to tell you which counterfactuals were true, you could in a sense 'wiggle' the events; it's just that you have different counterfactual situations rather than different successive actual locations. But again, seeing what else 'wiggles' when you 'wiggle' one or another event tells you which ones are causally connected to which. (2004, p. 91)

Lewis intended this concept of "influence" to provide the foundation for an account of causation that he hoped would succeed on *his* terms–that is, the terms that demand that an account recover the "firm and uncontroversial" opinions of common sense; the terms we rejected in the last section. But we may use influence for our own ends, as a helpfully evocative explication of the notion of localized dependence structure. Where Lewis speaks of "events," we may substitute "conditions that obtain in some localized region of space and time." Then the relations of influence between the events that obtain in our world collectively constitute its localized dependence structure.

Now we may go a step further and offer a simple proposal about **Inquiry's Aim**: the structure that it is the aim of the sciences to discover and describe is precisely its localized dependence structure. This is just what "causal structure" amounts to, in the sense needed to make **Inquiry's Aim** not only true but also illuminating.

We should forestall an immediate worry, which is that this structure, grand though it may be in scope, is nonetheless too specific and concrete to serve as the target of inquiry of any mature science. After all, don't the sciences traffic primarily in explanatory *generalizations*? Of course, they do. But there is no conflict here: in highlighting the localized dependence structure of the world, we are simply drawing attention to the subject *about which* the sciences try to generalize. Over here, in this corner of the world, these conditions depend in such-and-such a way on those conditions; so far, we may not have anything of much scientific interest, since we have narrowed our attention to just one little part of reality. But if that very structure of dependence gets *repeated* in other corners of the world, and better yet, if it can be seen as a particular instance of yet more abstract structures of dependence that are even more widely instantiated–then we have the stuff out of which a proper science can be made. That does not show that the sciences investigate something other than the world's localized dependence structure; it merely shows that they investigate it at a certain level of generality and abstraction and that in order for them to succeed, we must hope that our world's dependence structure has enough of the right kind of order and systematicity to it. (So far so good, on that score.)

Our proposal about the content of Inquiry's Aim strikes me as quite plausible (even though in the end, I am going to suggest that it is crucially incomplete). But in order to unpack it properly, and to appreciate just how illuminating it is, we need to look a bit more closely at the notions of "condition" and "dependence." That is the business of the next two subsections.

1.3.2 Taxonomizing "Conditions"

Suppose we wish to map out how conditions at one place and time depend on conditions at some other place and time. Then, as with the construction of any map, we need to make several choices. First, we need to choose a *scale*, and since the structure we are mapping has both spatial and temporal dimensions, this needs to be a choice of both spatial and temporal scale. It is unsurprising, then, that one of the most important distinguishing features of any mature science is the characteristic scale at which it operates.

But scale is not enough. We must also decide which aspects of the world's structure, visible at that scale, to focus attention upon. The social sciences provide ready examples of how differently this choice can be made; think, for example, of the different ways in which a sociologist and an economist might analyze the very same institution. In fact, the importance of such choices of aspect to focus upon shows up even in the most mundane examples. Suppose, for example, that we wished to study–scientifically!–the way in which the shattering of windows depends on the projectiles thrown at them. There is a fairly obvious choice of scale (within rough boundaries), but also some fairly obvious choices about what to attend to and what to ignore: for example, the volume, mass, and shape of the projectiles are worth modeling, but not their color. (And why is that? Precisely because there *are* interesting generalizations about how window health depends on the volume, mass, and shape of thrown projectiles, but no such interesting generalizations about dependence on *color*.) Next, we need to make a choice about precision: how fine are the discriminations between possible conditions that we wish to be able to represent? Here we may observe that it is far from the case that more precision is always desirable; to the contrary, it may serve only to obscure the dependence structure we are seeking to capture.

So far, our choice of scale, salient aspects, and degree of precision will yield some kind of flexible taxonomic scheme, in terms of which we can look at the two regions whose dependence relations we wish to capture, divide each region into salient parts, assign each such part one of a range of possible states, and thereby capture the "conditions" that obtain in each region in a way that will allow us to track facts about how the condition in one region would have varied, had the condition in the other region been different in some specified way. But we will also want these taxonomic schemes to come equipped with *similarity* metrics–that is, ways of systematically assessing which differences in possible states of some part count as *larger* and *smaller* and by how much. The way we reason about dependence structure is, even in the most mundane cases, shot through with a reliance on such similarity metrics. Suppose Suzy and Billy throw rocks at separate windows, breaking both. We might comment that Billy's throw was just hard enough to break his window, whereas Suzy's was more than hard enough. Notice what that means and how what it means relies on a background similarity metric: had Billy's rock been thrown with just slightly less velocity, his window would not have broken, whereas the same is not true of Suzy. And in more serious cases–namely, in any scientific domain in which the use of mathematics is essential to capturing explanatory generalizations–we rely on similarity metrics so automatically that it is easy to forget that that is what we are doing.

What emerges, thus far, is a picture of the world as possessing a localized dependence structure that is not only richly detailed but also layered, so that different patterns of dependence will come into focus with different choices of scale, aspect, degree of precision, and similarity metric. Those choices give each science its tools for constructing generalizations about localized dependence structure, and since they can be reasonably made in so many different ways, it is no surprise that we find ourselves with so many branches of science. By contrast, though, I think that the notion of dependence itself is quite univocal across the sciences. Let us see why.

1.3.3 Unpacking "Dependence"

In order to appreciate this univocality, it is crucial to recognize that one science–fundamental physics–is special. Why? To set up the answer, I am going to endorse a certain grand metaphysical picture, broadly (though mildly) reductionist in spirit. Here it is:

Two distinct fundamental features characterize reality as a whole. First, there is a total history of complete physical states. Second, there are fundamental laws that dictate exactly how earlier states generate later states. All other natural facts are ultimately explicable in terms of these.

What makes fundamental physics special is that it is its job, and its job alone, to map this basic structure. Thus, physics will succeed exactly if it does three things: (i) provide a taxonomy of fundamental physical magnitudes and kinds; (ii) accurately characterize, in terms of this taxonomy, the range of physically possible states the world can exhibit; (iii) accurately characterize the laws that govern how these physical states evolve in time.

Will physics succeed? Hard to say. (The prospects seem a little dim. Superconducting supercolliders are so very expensive.) But it does not really matter, for our purposes, whether physics will ever achieve the aims I have attributed to it. What matters is the grand view of the world that motivates and makes intelligible this conception of its aims: that is a view of the world as being wholly constituted by a complete history of fundamental physical states, whose evolution is governed by exact and universal fundamental laws.[4]

[4]Arguably, this view is under some threat from within fundamental physics itself–namely, from spacetime theories such as Einstein's general relativity, which seem to lay down global constraints on how spacetime as a whole can be, and so seem *not* to have the characteristically dynamical form just described. I think appearances here may be misleading, but at any rate, it does not matter. While I will not try to argue the point in detail here, all that is *really* needed, in order to give sharp content to the notion of localized dependence structure, is a distinction at the level of fundamental physical reality between those total histories of the world that are *physically possible* and those that are not. It is easiest to see how counterfactuals function

Augmented by a modest amount of reductionism, this view gives us the resources to say fairly precisely what the relations of localized dependence are whose patternings it is the business of the other sciences to uncover, and, as we see in the next section, to expose some important open questions. The tool to use to capture these dependence relations is just a certain kind of counterfactual conditional, with the following regimented form:

If conditions C1 had obtained in region R1, then conditions C2 would have obtained in region R2.

I will now sketch truth conditions for these conditionals. (Here I am closely following the excellent discussion of counterfactuals in Maudlin, 2007.) Bear in mind that I am *not* trying to capture the truth conditions such conditionals have *in everyday discourse*. I am, rather, aiming to elucidate how they should be understood, *given* their own role in elucidating the notion of localized dependence structure.

We will start with the easiest case (saving complications for the upcoming discussion of the challenging news): the region R1 is a localized region of space at a single instant of time, *t*; in addition, the condition C1 specifies a single fundamental physical state for that region, at that time. Similarly, condition C2 picks out a unique fundamental physical state for region R2 (although we will not need to assume that R2 is instantaneous in the way that R1 is). We will also assume that R2 lies to the future of R1.[5] Then consider a possible complete physical state of our world at *t* that is exactly like its actual state, save that C1 obtains in R1. Assuming determinism, the actual laws of nature will, when applied to this complete "counterfactual" state, yield a unique forward evolution. If this forward evolution makes R2 instantiate C2, then the conditional is true; otherwise, false.[6]

Think of what these conditionals are doing, in such easy cases, as capturing the outcomes of *perfectly* controlled experiments. Imagine that you have a god's-eye view of the world, past, present, and future. Here is R1; over there is R2. You are curious about how conditions in R2 depend *specifically on the state of* R1. So, using your god-like powers, you reach in and intervene *just* on the state of R1, changing it to some specific alternative; you leave the time-*t* state of the rest of the world *unchanged*. (That is why this is a *perfect* controlled experiment.) You then let the fundamental dynamical laws do their work, watching to see what sort of altered history unfolds from time *t*, and in particular, how conditions in R2 have changed. And the results of this "experiment"

if we think of these histories as generated by a choice of initial state, plus dynamical laws that prescribe an exact evolution therefrom. But it is not required.

[5]In so doing, we gloss over deep and important questions about why the relations of localized dependence the sciences are in the business of tracking almost always have a past-to-future direction. For good book-length treatments of these issues, see Price (1996) and Albert (2000).

[6]There are a couple of deep issues I am bracketing here. One is obvious: if the fundamental laws are stochastic, then we will really need to replace the form of conditional given in the text with conditionals whose consequents specify probabilities for R2 to instantiate some or other condition. The other is more subtle and is connected to the phenomenon of quantum mechanical entanglement: very roughly, quantum mechanics give us good reason to believe that, when we specify a complete physical state for one part of the world at a given time and then specify a complete physical state for the remainder of the world at that time, we nevertheless *underspecify* the complete physical state for the world as a whole, at that time.

tell you–unambiguously if, inevitably, only partially–how conditions in R2 depend on conditions in R1.

Of course, we do not have such god-like powers. But the point is that if we knew the truth value for any of these "easy" counterfactuals concerning R1 and R2, we would thereby know *precisely* what the use of such powers would reveal.

These easy cases put very clearly on display the central and significant role that fundamental laws play, in determining dependence structure. Mind you, they are also highly idealized, since we are taking R1 to be instantaneous and C1 and C2 to be as specific as they can possibly be. For now, let us naively assume that relaxing these idealizations will introduce no new difficulties. The crucial conditionals, we will pretend, work in just the same way: *in general*, for it to be the case that if C1 had obtained in R1, then C2 would have obtained in R2, is just for it to be the case that a state of the world at the time of R1 that differed only in that C1 obtained in R1 evolves, under the fundamental physical laws, in such a way as to make R2 obtain in C2. Then we get the unity and diversity of the sciences in a simple and attractive package deal: unity, because there is–as far as the structure of the underlying counterfactuals is concerned–just one sort of localized dependence structure for the sciences to study; diversity, because they can (and should) vary widely in their selection of shape and size of regions and in their taxonomies of conditions.

1.3.4 The Good News, Amplified

There is much more to say in favor of this way of understanding causal structure as localized dependence structure. But I will be brief, partly for reasons of space, but also because the literature already contains excellent and detailed treatments of several of the virtues I will point out (see especially Woodward, 2005).

Our proposal about the *nature* of causal structure forges an immediate connection to our best understanding about how to *study* causal structure–namely, by conducting controlled experiments where possible and watching out for confounding variables. No surprise, for causal structure *just* is what would be revealed by *perfectly* controlled experiments. This tight link between the metaphysics and epistemology of causation is no small virtue, for one can easily find in the philosophical literature accounts of causation that leave it wholly obscure why, for example, controlled experiment should be a remotely effective means for finding out about it (Armstrong, 2004 is a prime example).

The proposal can also help enforce a welcome kind of explanatory hygiene. After all, explanatory talk comes cheap: it is all too easy to invent hypotheses about what explains what, in a given domain. Popper famously tried to separate wheat from chaff by demanding that a properly scientific hypothesis be *falsifiable* (see, e.g., Popper, 1959). But so far, no one has been able to articulate a precise standard for falsifiability–or more generally, testability–that is neither too weak to exclude obvious pseudoscientific quackery, nor too strong to permit perfectly respectable conjectures. Our proposal can come to the rescue, at least partially, for if we link explanation with causation–as we probably should, at least for the most part[7]–then we can demand of any putatively explanatory hypothesis that it be recast explicitly as an hypothesis about what depends, locally, on what. If that cannot be done, at least in a sufficiently clear and coherent manner, then so much the worse for that hypothesis. (But see the next section for some reservations.)

These hygienic effects extend into philosophy as well. Specifically, we are now in a position to debunk philosophical worries about "causal exclusion," arising from the thought that if a complete set of causes of some condition can be found at the fundamental, microphysical level, then there will not be any *higher* level causes of that condition, since there will not be any more causal "work" for such higher level causes to do. Here is a crisp statement of the most famous version of this worry, which has arisen in discussions of mental causation (Robb and Heil, 2014; Section 6.2):

> If mental properties are not physical, how *could* they make a causal difference? Whenever any mental (functional) property M is instantiated, it will be realized by some particular physical property P. This physical property is unproblematically relevant to producing various behavioral effects. But then what causal work is left for M to do? It seems to be causally idle, "excluded" by the work of P.

The background picture seems to be that causation is some kind of special, metaphysical "juice," so that if a given effect gets all it needs from one source, then there is nothing left for any other source to contribute. But this picture simply makes no sense, if causal structure is constituted by localized dependence structure. Suppose we have some downstream condition C. Given some earlier time t, we have identified, say, all the microphysical t-conditions P1, P2, … , on which C counterfactually depends. That does nothing to prevent there being a host of more macrolevel conditions obtaining at t and on which C *also* depends. There is simply no sense in which counterfactual dependence on one source "excludes" such dependence on another source. It is an illusion that there is any exclusion problem to be solved.

Now for two final bits of good news, but ones that are bound to be more controversial. The first concerns the regrettable quietism about foundations that infects the current literature on interventionist approaches to causation. Those approaches draw on the "causal modeling" framework, in which we represent the causal structure of any situation by means of a set of *variables*, connected to one another via *structural equations* that are meant to capture patterns of dependence between them. But if we ask for a noncircular account of what must be the case for a given structural equation to be *correct*, we get radio silence. Instead, we are typically offered an explanation in terms of potential "interventions," where this notion is itself cashed out by appeal to (different) causal models.

[7]The reason for the qualification is that not all explanation is causal explanation. Explanation in mathematics provides a counterexample. But even in the empirical sciences, it is not hard to find examples of explanations that do not count as such because they cite causes. But the link between explanation and causation is certainly tight enough, I think, to support the present point.

But the foregoing discussion of counterfactuals and their truth conditions shows that there was never any reason for such timidity. Once we have the basic framework in place for evaluating counterfactuals by means of fundamental physical laws, it is not that hard to extend it in such a way as to give systematic, noncircular correctness conditions for structural equations (see Hall, 2006). Doing so does not only enhance our understanding of causal models but also improves our ability to construct them. For in some notable cases, practitioners have written down *incorrect* structural equations, drawing on their intuitive feel for what the relations of dependence are, without rigorously evaluating the underlying counterfactuals. To my mind, the most egregious example is the typical treatment of cases with the structure of **Suzy First** (as in Halpern and Pearl, 2005). Without going into the gory details (for which see, again, Hall, 2006), suffice it to say that the structural equations typically offered to model such cases fail badly, in a way that is hard to catch unless you focus explicit attention on their (noncircular!) correctness conditions.

A final observation–this time, about how our proposal can foster greater clarity about a certain kind of causal inference. It is a good idea, when asking what sorts of conclusions about causal structure you can draw from a certain set of data, to have clearly in mind the *scale* that you are interested in. For, when the data are statistical in nature, there is danger of a very specific kind of confusion. I will illustrate by example.

Suppose there is a large population of people in their 50s. Some have lung cancer, some do not; some smoked heavily in their youths, some not at all. Suppose we have studied this population over time with sufficient care (e.g., in spotting and controlling for confounds) that we can now confidently make the following claim: had *none* of these people smoked, the present incidence of cancer would have been much lower. Then we can truthfully say that, within this population at least, *smoking causes lung cancer.*

But this claim is ambiguous. On the one hand, it might be construed as a claim about the dependency structure of *the population as a whole*, considered as a kind of unit. It is a truth about this very population–this thing–that if the incidence of early smoking in *it* had been zero, then the incidence of later lung cancer in *it* would have been much lower than in fact it is. Thinking of the population as an entity in its own right, there are dependency facts that pertain to it, and this is one of them.

But there are also, of course, dependency facts pertaining to each of its *members.* Consider Randolph: he smoked a lot in his youth and now has lung cancer. It might in addition be true of Randolph that, had he not smoked in his youth, he would not now have lung cancer. On the other hand, for all we have said so far, it might be true of Randolph that had he not smoked in his youth, he would *still* have lung cancer. Suppose that people come in two types: there are those who are highly sensitive to the toxic effects of smoking and who almost invariably contract lung cancer, after a history of heavy smoking; and then there are the lucky ones, who can smoke as much as they want without the slightest increase in the risk of lung cancer. Suppose that half the people in our population are of the first type, half of the second. Then our dependency claim about the population as a whole will still be true. But for all that, it may be that *Randolph* belongs to the immune half.

Or, consider a different hypothesis: this time, everyone in the population is *alike* with respect to how smoking affects them. But heavy smoking in one's youth does not guarantee lung cancer; it merely raises its probably pretty substantially. So, what is true of *every* smoker in the population is that, had they not smoked, their chance of contracting lung cancer would have been much lower than in fact it was.[8] Then it is not true of *any* smoker in the population that, had she not smoked, she would not have contracted lung cancer. What is true is rather that, had she not smoked, her probability of contracting lung cancer would have been much lower than in fact it was. That still counts as a kind of localized dependence, all right. But it is much weaker than the nearly sure-fire dependence that obtains at the level of the population as a whole.

Now return to our claim that, in this population, smoking causes lung cancer. We noted that this could be construed as a kind of singular claim about the dependency structure of the population itself. But, of course, it could also be construed as a *generalization* about the dependency facts pertaining to each individual member. Understood this second way, it lacks any sort of precise content, but should probably be taken to mean something roughly like this: in a reasonably significant percentage of the population, one's probability for contracting cancer depends on smoking to a reasonably high degree.

Here is the upshot: the claim that smoking causes lung cancer is ambiguous. It can mean something relatively sharp, when it is understood as a claim about a population as a whole; or, it can mean something rather squishy, when it is understood as some sort of generalization about the members of the population. Unfortunately, this ambiguity between individual-scale and population-scale readings of a causal claim can breed confusion, in philosophical, practical, and moral domains.

Philosophical: Many philosophers working on causation hold that while causation can certainly relate token events–as when Randolph's youthful smoking causes his later contraction of lung cancer–it can also relate event *types*–as when *smoking causes lung cancer*. Not so; there is no such thing as causation as a relation between event types. But we can now see why this mistake tempts. Suppose you take it to be clearly true that smoking causes lung cancer. But you know that this cannot be construed as the *universal* claim that everyone who smokes will get lung cancer as a result. Still, it seems to you that the truth you are asserting when you say that smoking causes lung cancer is *not* just some vague, ill-defined generalization about the members of the population. So, what else could it be? Answer: a nice, sharp claim about a causal relation between event types! But that answer overlooks another, clearly better alternative: the construal of this claim that makes it both true and sharp is the one that

[8]Of course, that means that we slightly overstated the claim about the dependency structure of the population as a whole: what is really the case is that if the incidence of smoking had been zero, then it would have been highly probable that the incidence of cancer would have been much lower. But if the population is large, "highly probable" can be quite close to 1. Here, for example, are some numbers: suppose our population has 1 million people in it. Half of them smoked heavily in their youths; half smoked not at all. Suppose that the probability of contracting lung cancer if you do not smoke is 5%; if you do, 50%. The incidence of lung cancer in the population is in fact about 27.5%. But if no one had smoked, the incidence would, with probability greater than 0.99999, have been less than .051.

takes it to be a *singular* causal claim, not about any person, but about a *collection* of people.

Practical: Suppose, to use another example, that you know that in any sufficiently large population of drivers, the death rate from car accidents depends on the rate of seat belt use: the higher the latter, the lower the former. What you know is a population-scale dependency fact. Now, if you are a public policy maker, concerned to design rules that will affect the characteristics of entire populations of drivers, then that causal information might be enough to convince you to pass a seatbelt law. But if you are an individual driver, that information may not be relevant *at all* to your decision about whether to fasten your seatbelt. Maybe you are an eccentric and rich safety nut and have had your car custom-built with all sorts of safety measures that render wearing a seatbelt irrelevant to *your* prospects of surviving an accident. There is absolutely no contradiction here. One kind of decision is, in the abstract, a decision about how to intervene on an entire population. The other kind of decision is a decision about how to intervene on an individual member of that population. Information about the dependency structure of the population as a whole is clearly going to bear on the first kind of decision in a more direct and powerful way than it will on the second.

Moral: There is a puzzling phenomenon that the shift in scale from a population to the members thereof gives rise to. It was first pointed out to me by Johann Frick (personal communication; but see Frick, 2015). Here is an example. A chemical company has been dumping toxic waste into the river upstream from a large city. Studies have established that, thanks to this practice, the instance of a certain disease in the city has risen from 10,000 cases per year to 10,500 cases per year. Let us take this dependency fact to be unambiguous: had the chemical company not been dumping, there would have been (or, would almost certainly have been) 500 fewer cases of the disease, per year. But for all that, it may not be the case, of any individual sufferer from this disease, that *her* contraction of the disease depended on the company's behavior: the most that may be true is that if the company had not dumped, she would have had a somewhat lower *chance* of contracting the disease.

That raises an interesting moral question (and no doubt legal ones, as well): it seems perfectly clear that in some sense, the company's actions have *caused harm*, so the company is thereby morally culpable. Indeed, if the disease is fatal, we would blame the company for *killing 500 people*. But for all that, there might be no one person of whom we can truly say: the company's actions killed *them*. The puzzle–which I will not try to solve–is simply this: what sort of principles should we use to evaluate a circumstance in which an agent's actions indisputably cause harm to people, even though there is no person of whom it is true to say that the agent's action caused *them* harm?

It is a powerful argument in favor of our proposed identification of causal structure with localized dependence structure that it adds so much clarity, to so many issues. All the same, some powerful challenges remain to making this proposal *itself* adequately clear. The next session canvasses a few of the most important of these challenges.

1.4 THE CHALLENGING NEWS

Recall our canonical counterfactual conditional:

If conditions C1 had obtained in region R1, then conditions C2 would have obtained in region R2.

We started, in Section 1.3.3, with the easiest case, taking R1 to span but a single instant of time *t* and C1 to specify a single fundamental physical state. We counted the conditional *true* just in case a complete physical *t*-state of our world that is exactly like its actual state, save that C1 obtains in R1, evolves forward under the laws so as to make R2 instantiate C2. We will now explore three significant challenges. The first arises when we relax the assumption that C1 specifies a *single* fundamental physical state. The second arises when we relax the assumption that R1 spans but a single instant of time. And the third arises when we allow the conditions C1 and C2 to be specified by means of taxonomies other than that of fundamental physics.

1.4.1 Multiple Realizability

We might specify, in the antecedent and consequent of our counterfactual conditional, not a single fundamental physical state, but a range of them: If R1 had been in one of the following states: …, then R2 would have been in one of the following states: … (As we might put it, the conditions C1 and C2 are "multiply realizable," by any of a range of fundamental physical states.) On the end of the consequent, this adds no complications: we must simply assess whether a forward evolution makes it the case that one of the specified states obtains, in R2. But on the end of the antecedent, matters are trickier. The simplest idea is to consider every way of constructing a t-state for the world that is exactly like its actual state, save that C1 obtains in R1. For each such way, evolve it forward under the (actual) laws and determine whether C2 results. (You use your god-like powers to run multiple experiments, exploring the consequences of every way of altering R1 so that C1 obtains.) If so, the conditional is true; if not–if even one such forward evolution makes it the case that C2 does not obtain–then the conditional is false. Understood this way, the conditional might be more perspicuously rendered like this: no matter which way of realizing C1 in R1 had obtained, C2 would have obtained in R2.

But that is really *too* demanding. Consider, for example, a case where there are uncountably many states in the range covered by C1, and all but one of them yields a forward evolution in which C2 obtains.[9] More realistically, suppose we ask what would have happened to a certain ice cube, had it been placed in a glass of very hot water. Now, we have very good assurances from statistical physics that there are exact physical states that realize the placing of an ice cube in a glass of very hot water and that evolve forward in startlingly antientropic ways–so that, for example, the ice cube *grows*, and the surrounding water *heats up*. It is fair to conclude, then,

[9]It is entirely realistic that C1 should cover an infinite range. Consider, for example, a conditional that begins "If particle p had been located in the following region of space …."

that if our ice cube had been placed in very hot water, it *might* have grown.[10] But, intuitively, it is also correct to insist that the cube *almost certainly* would have melted. Probably, the cleanest way to get these results is to impose a *measure* over the range of C1-states. We can then introduce more subtle conditionals (really, quantifications over them), such as this one: it is true of *most* ways of realizing C1 in R1 that, had that way obtained, then C2 would have obtained in R2. We can thus treat the dependence between C1 and C2 as coming in degrees: the greater the proportion of C1-states that would yield C2-evolutions, the greater the dependence. Thus, the overwhelming majority of ice-cube-placed-in-hot-water states yield ice-cube-melted states; that is why it is okay, for example, to treat it as a robust higher level causal generalization that placing ice cubes in hot water causes them to melt.

We will not pursue this matter further, save to point out that the methodological points discussed in Section 1.3 apply equally, here. We should judge an account of causation on the basis of its utility; so too should we judge an account of counterfactuals–where, in this case, the usefulness we seek is in providing conceptual tools for capturing the relations of localized dependence that it is the business of the sciences to track. What we have just seen is that we will need to refine our tools, when dealing with counterfactuals whose antecedents are multiply realizable. Some of those tools have begun to be developed within philosophy of physics; see, for example, Albert (2000). But the exact form they should take remains controversial (see Weslake, 2014), and more to the point, the literature on interventionist approaches to causation has yet to properly incorporate them.

1.4.2 Protracted Causes

The next complication is much more challenging. For we will, of course, often want to assign a condition to a region that is *extended in time*. (*e.g.,* consider any medical investigation of the consequences of some long-term health condition.) But an issue of consistency now arises. As far as we can tell, the fundamental laws of our world impose few if any *synchronic* constraints: focusing on a specific time, any way of assigning a fundamental physical state to one region of space is compatible with any way of assigning a state to a distinct region of space.[11] As an example, consider Newtonian particle mechanics (treated, for present purposes, as a candidate for a fundamental physical theory). According to that theory, we specify the complete physical state of the world at any moment of time by giving the positions, velocities, and intrinsic properties (mass and charge, say) of all the particles. The theory leaves it open *how many* particles there are; even a conservative version of the theory will allow as possible *any* finite number of particles. What is more, the theory places almost no constraints whatsoever on how positions and velocities of different particles may be *combined*. We should probably understand it to forbid particles from

[10] See Hájek (2015) for an excellent discussion this and a range of related challenges to the idea that ordinary "would" counterfactuals are ever true.

[11] Perhaps within broad limits–the Pauli exclusion principle, for example, seems to be an instance of a synchronic constraint.

being located at the exact same point in space; but that is about it.[12] That is what it comes to, to say that Newtonian particle mechanics imposes almost no synchronic constraints. Notice that the *fact* that it is so permissive in this respect is crucial, for the viability of our recipe for evaluating counterfactuals. For if, as a matter of fundamental physical law, an exact state for one part of the world, at a moment in time, could only be combined with a certain limited range of states for the *rest* of the world, then a counterfactual situation in which R1 manifests condition C1 might *be incompatible with the fundamental laws*–in which case there is just no point to asking what forward evolution those laws would prescribe for it.

By contrast, the laws impose severe *diachronic* constraints, particularly under our working assumption of determinism. So, if region R1 is extended in time, and we wish to assign it some fundamental physical state (or better, some sequence of fundamental states, one for each moment that it contains), then two questions arise. Is the given sequence of states even *possible*, given the laws? And, assuming that it is, what must the world outside of region R1 be like, in order for the given conditions within R1 to lawfully obtain?

If the answer to the first question is "no," then it may be that we have no well-defined way to assess what would have happened, had the impossible condition obtained; and if so, there simply will not be any localized dependence structure to be captured. But we will see next that in an important range of cases, there is a (reasonably!) well-defined way to construct the nomologically impossible counterfactual situation. We will see most clearly how that emerges, though, by first tackling the second question, which turns out to raise a very important and quite subtle issue.

To bring this issue into focus, we need to remember what makes localized dependence structure *localized*. It is not just that the regions R1 and R2 are bounded in space and time; it is also that we are trying to assess how R2 would have differed, if R1 had differed in some specified way, *but everything else contemporaneous with R1 had been the same*. In other words, how would a change in the state of the world *localized* to R1 have made a difference to R2? *That* is the kind of question whose answer reveals causal structure. The question we face–quite a tricky one, as we will shortly see–is how to make sense of this "everything else is the same" clause, in the case where R1 is extended in time.

Couldn't we avoid this difficulty, by asking simpler questions? For example, these two: what lawfully follows, concerning the state of R2, just from the premise that condition C1 obtains in R1? In particular, does it follow that C2 obtains in R2? Assuming as we are that the laws *allow* condition C1 to obtain in R1, it will presumably be nontrivial what lawfully *follows* from the claim that it does.

Nontrivial, perhaps–but still, almost certainly wholly uninteresting. We have a range of possible worlds, all alike in these respects: R1 is in the condition C1; and the actual laws hold. But other than that, they differ in *every way possible*, compatible with the laws. So, they will differ in ways that can make a great deal of difference to

[12]Well, maybe we should also understand it to forbid configurations of particles that would *lead*, under its dynamics, to such colocation.

the conditions in R2 (except in exceptional circumstances, where R1 is sufficiently comprehensive and temporally proximate to R2 that the state of the former almost entirely determines, given the laws, the state of the latter). Example: What follows, from the claim that in a certain region at a certain time, Suzy stands a certain distance from a window and throws a rock with such-and-such features in such-and-such a manner? Very little–certainly *not* that the window breaks, a few moments later. For it is perfectly compatible with our two premises (that conditions in that region are as described and that subsequent events unfold in accordance with the actual laws) that all sorts of *interfering processes* are poised to block her rock, preventing it from breaking the window.

So, if we wish to analyze the causal relations between R1 and R2 by considering, as it were, the *lawful significance* of the hypothesis that C1 obtains in R1, we obviously need to fill in more details–in particular, details about the state of the *rest* of the world at the time of R1. We solved this problem, when considering our "easy" cases, by explicitly stipulating that the state of the rest of the world was to be *exactly as it actually is*. That stipulation was not ad hoc; rather, it reflected a deep point about what it means to attribute to the world a localized dependence structure. To conceptualize the world in such terms is precisely to conceptualize it by means of a contrast between what actually happens, and what would have happened, had matters differed *only* in a localized respect.

At this point, you may be impatiently wondering what, exactly, the problem is supposed to be. Fine; we have a region R1 that is extended in time, and we wish to consider what would have happened, if it had been in some nonactual condition C1. Why *can't* we just construct a counterfactual possible world in which, over the course of time that R1 spans, it is "put" in condition C1, while the rest of the world (over that stretch of time) remains as it actually is? Then we just proceed as before: evolve forward from this stretch of history, and check to see whether C2 ends up obtaining in R2.

Why can't we? Lots of reasons. Let us start with a simple one. R1 has, let us suppose, a first moment. In the counterfactual situation we are trying to construct, we are altering the state inside of R1, at this first moment, from what it *actually* is, precisely to conform to the requirement that C1 obtain in R1. (In using your god-like powers to intervene on R1, you must, *inter alia*, intervene on its first moment.) But remember that the counterfactual world we are constructing is one in which the *actual* fundamental dynamical laws obtain. So, that change to the initial state of R1 is going to have downstream *effects*. Some of those effects will be felt quite quickly–that is, *before* R1 is, as it were, finished, and what is more, will be felt *outside* of R1. So, on pain of violating the fundamental laws, we *cannot* just stipulate that conditions outside of R1 remain exactly as they actually are, for the duration of R1.

Examples make the point obvious. This one will do: what would have happened if it had rained heavily last night, from midnight until dawn? Lots of things, no doubt, and some of them would have been more or less immediate. Very soon after midnight, for instance, the ground would have been wet. It is quite clearly insane to think that, in asking this question, we mean to be considering a counterfactual situation in which it rains steadily, *but the ground remains dry* throughout the duration of the rainfall.

But this problem, such as it is, is fairly easy to address. Here is the most straightforward way to do so, applied to our example of the nighttime rainfall: consider the state of the world at midnight. In the region under consideration, conditions are, in actual fact, such that there *will not* be 6 hours of rainfall. So, a relevant counterfactual situation will be one in which, at that time, the state of the world *outside* the given region is, at midnight, just as it actually is, whereas the state of the world *inside* the region differs just enough to guarantee 6 hours of heavy rain. (Obviously, there will be many ways in which such a rain-guaranteeing state could be realized, so the issues discussed in Section 1.4.1 arise here as well.) So, would there, for example, have been flash-flooding, had it rained heavily for 6 hours last night? Well, if all (or most) of these counterfactual states would have evolved forward, under the laws, in such a way that flash-flooding occurred, then yes, there would have been. You can see at work, here, a sensible and generalizable strategy: do not try to hold the rest of the world fixed at its actual state throughout the *entire duration* of R1; rather, hold that state fixed only at the *beginning* of R1, while ringing changes within R1 just sufficient to make it the case that, in the counterfactual scenario, condition C1 obtains.

Unfortunately, the strategy does not generalize far enough. The basic problem is that we are assuming that we can get C1 to obtain *merely* by making localized changes to the initial state of R1; that will not always be the case. I will illustrate the problem with a pair of examples, one artificial and the other quite realistic. To start with the artificial example, suppose Suzy throws a rock at a window at 11:45, breaking it. Billy stops by half an hour later, intent on breaking the same window, only to leave disappointed. Consider a half-hour stretch of time from noon to 12:30. In fact, the window is in a broken state throughout that stretch of time. Suppose we wish to understand what else *depends* on that fact–on, that is, the window's being in a broken condition throughout that interval. Note what we are *not* asking: we are not asking what depends on the *breaking* of the window–the event brought about by Suzy's throw, at 11:45. (We *could* ask about that, of course. The relevant counterfactual situations are not difficult to construct: we simply ring localized changes on the state of the world at 11:45 so as to "undo" the breaking and evolve the resulting state forward. Thus, for example, had the window not broken at 11:45, it would have broken half an hour later.) No, we are focusing on a protracted condition of the window in its broken state, over the half hour from noon to 12:30, and investigating the relations of localized dependency *on that condition*. So, we need to consider counterfactual situations in which, throughout that stretch of time, the window is unbroken–but in which everything *else* is, in some reasonable sense, "the same."

Now, if Billy were not in the picture, no special problem would present itself, and we could treat this case exactly like the rainfall case. That is, we start with the state of the world at noon. We leave the state of everything *except* the window just as it actually is, while changing the state of the window so that it is unbroken. With Billy out of the picture, there are no threats to the window's integrity (at least, not for that half hour); so when we evolve this resulting state forward, condition C1 (that the window be unbroken) obtains in R1 (the little region of spacetime carved out by this half hour of the window's history). We can then see what other differences from actuality result and thereby get a partial fix on the localized dependence structure we are interested in.

But Billy *is* in the picture. That pretty much ruins this simple strategy. For we cannot secure condition C1 simply by counterfactually "altering" the state of the world at noon, in a manner localized to the window. For that leaves Billy's murderous intentions–which, remember, are realized, at noon, in parts of the world's state well *outside* the location of the window–unchanged. And so, the forward evolution of this counterfactual state will only leave the window unbroken until 12:15, not 12:30.

Now, it may seem obvious what to do in this artificial example: construct the needed counterfactual situation by modifying the state of the world, at noon, in *two* ways: first by returning the window to its unbroken state, second by altering Billy's intentions just enough to keep him from wanting to break the window. But this is much less principled than it looks. After all, why make the second alteration *here*? Why not, instead, add something to the environment that will *stop* Billy? Bear in mind what our strategy is: we want to make changes to the state of the world *localized to the time at which R1 begins*, sufficient to guarantee that C1 obtains in R1. We have already seen that we cannot *confine* these changes to R1 itself. (That is because Billy-at-noon, with his murderous intentions, is not part of R1.) So, in the abstract, our task is to find ways to change noontime conditions *outside* of R1, such that the resulting counterfactual state of the world will bring about C1 in R1. When we state the task at this level of generality, the problem that comes into view is that there are simply *too many* different ways–*very* different ways–to ring changes on the noon state of the world, compatible with this requirement. The "obvious" way–change Billy's intentions–is obvious only because the narrative structure of the example raises it to psychological salience. But we should not be interested in an account of localized dependence structure whose details turn on what strikes us as especially noteworthy about any given case. No, we want a more principled way of determining the character of the relevant counterfactual circumstances.

A second example deepens the problem, by showing that it in fact raises issues of "confounding." Remember our character Randolph from Section 1.3.4? The one who smoked heavily in his youth and now (in his 50s) has lung cancer? Suppose we wish to consider the hypothesis that his youthful smoking caused his cancer. We want to make this hypothesis precise, by means of suitable counterfactuals. Suppose his heavy smoking took place entirely in his 20s. Then we might try this, as a precisification: if Randolph had not smoked at all in his 20s, then he would not now have lung cancer. (Or, his probability of contracting lung cancer would have been much lower than in fact it was.) So, we need to consider possible worlds in which Randolph does not smoke in his 20s, but in which everything *else*, during that period, is "held fixed" as much as possible. A natural thought: we construct this counterfactual situation by taking the state of the world at a time *t* shortly before Randolph turns 20 and altering *just Randolph*, in psychological respects sufficient to prevent him from ever acquiring a taste for cigarettes. We then evolve the resulting state forward, via the laws, and see whether cancer results.[13]

But the problem is that any psychological changes significant enough to prevent Randolph from becoming a smoker are likely to have *other* effects on his behavior, and remember that our hypothesis was that his cancer is to be explained by his *smoking*, and not by his smoking-together-with-other-behavioral-tendencies. For example, suppose the alteration to his psychology that leads him to avoid cigarettes consists in a much increased concern for his health. As a result, he does not smoke. But he also starts going to the gym regularly, and avoiding excess sugar in his diet. He also moves out of the smog-laced city he had been living in. And so on. The result is that we are no longer considering, and making precise, the hypothesis that Randolph's *smoking* caused his cancer, but rather the hypothesis that his *overall health habits* caused his cancer. There is nothing wrong with that hypothesis; it might be a perfectly suitable subject for investigation, in its own right. It is just that it is not the one we had in mind.

I think the only solution is to introduce a departure from our official truth conditions for counterfactuals. We can think of the extended condition of *smoking in his 20s* as constituted by a large number of much more temporally localized conditions: roughly, one for each inhalation of cigarette smoke. Imagine, now, a massively multipronged intervention *on each of these conditions*: the first time Randolph puffs, you use your god-like powers to swoop in and change the state of his lungs, so that the toxic chemicals are removed.[14] Then you let the laws take over and evolve the resulting state forward until his next puff. Then you swoop in again. And so on. Once this 10 year history of interventions is over, you let the laws take over for good, and see whether a cancer-in-his-50s states results. Notice, by the way, that the same technique can be used to handle our case of Billy and the broken window.[15]

At this point, you might be forgiven for worrying that things have gone off the rails: do we really have to take seriously such an outlandish counterfactual situation? (How secure does that vaunted connection to good scientific practice remain, if we do?) But it is genuinely unclear to me[16] that, within the broadly interventionist framework we have adopted, we have a good alternative–at least, provided that we wish to treat our two hypotheses about what explains Randolph's cancer as legitimately scientifically distinct. One hypothesis, loosely stated, says that the explanation rests just with his history of smoking. The other says that it rests with his youthful health habits more generally. At this point, it is helpful to remember our remarks about good explanatory hygiene, back in Section 1.3.4: if we want to treat these hypotheses as genuine rivals, we ought to be able to say what, precisely, the distinction between them comes to. Relying just on talk of "explanation" will not give us the needed precision. But

[13]As before, there will in fact be a large range of t-states that meet this condition, and what will matter is the proportion of them that leads to Randolph-with-cancer states.

[14]You might also have to make subtle alterations to Randolph's subsequent mental state; after all, we do not mean to be considering a situation in which he is regularly frustrated at not experiencing the pleasant psychological effects of smoking!

[15]It can also be used to handle at least some cases where the laws themselves forbid R1 from manifesting condition C1, though I will not pursue that matter here.

[16]Which is to say, this is not just a cheap rhetorical use of "unclear to me."

relying on talk of dependence, cashed out in terms of counterfactuals, promised to: for example, we might have hoped to clarify the distinction by noting that the first hypothesis, but not the second, implies that if Randolph had not smoked, but his other health habits had remained unchanged, then he would not have contracted (or, would have had much less chance of contracting) cancer. But then we owe an account of the relevant counterfactuals–one, remember, that preserves the distinction between the hypotheses. And so, it seems we are led back to truth conditions that require us to consider worlds whose evolving states are being sequentially modified by localized interventions.

But it may be that even this (possibly extreme) measure will not suffice. For consider cases in which the condition about whose causal consequences we wish to hypothesize is *continuous*–not, that is, constituted by a discrete series of punctuated events. For example, the case of Alice the astronaut, who has come home from a year-long stay on the international space station and is suffering from a weakening of her immune system. Now, various distinctive conditions characterized her year in space. One of them–but not the only one–was prolonged weightlessness. So, we might entertain two distinct hypotheses: her health problem is solely due to her prolonged weightlessness; or, it is in fact due to the entire constellation of factors she experienced while in space. Of course, neither hypothesis may be correct; but the present issue is what, precisely, *distinguishes* them. Here we encounter a more virulent form of the problem we saw in the case of Randolph. For suppose we wish to clarify the first hypothesis by means of a counterfactual: "If Alice had not experienced a year of weightlessness, then she would not presently be suffering from a weakening of her immune system." That counterfactual is meant to direct us to a possible situation in which Alice experiences normal weight for the year, but in which other factors *remain as they are*. But what sort of possible situation could that be? Not one in which she remains on Earth; that introduces, as potentially confounding variables, all the other conditions that distinguish life on Earth from life in the space station. But–or at least so it seems–it cannot be a situation in which she inhabits the space station, either; for unlike the case of Randolph and his puffs, no sequence of localized interventions will simply restore normal terrestrial weight to her, in that situation, while leaving other conditions unchanged.

Just to be clear, it is not that I doubt that these hypotheses *are* legitimately scientifically distinct. For one thing, we seem to have a pretty good idea what sorts of evidence would count in favor of one over the other. Rather, what we lack–or more carefully, what our broadly interventionist approach is so far failing to adequately provide–is a clear account of what precisely *makes* them distinct. Myself, I think the overly cavalier attitude toward structural equations models promoted by the quietists I complained about back in Section 1.3.4 is partly to blame. For it is easy enough to write down some variables to characterize different aspects of Alice's overall condition on the space station, write down some other variables to characterize different aspects of her health, and then represent different hypotheses about their explanatory connections by means of different systems of structural equations. If, with the quietists, you refuse to ask for a foundational account of the correctness conditions of these equations, then you might easily miss that you do not really know what you are talking about. But we should all do better. So, I offer up this problem as one of the more important bits of unfinished business in carrying out the interventionist program.

1.4.3 Higher Level Taxonomies and "Normal" Conditions

Our next problem is broadly similar, in that it too involves counterfactuals which, if interpreted too naively, introduce unwanted confounds. But it arises not because the condition whose causal influence we wish to assess is protracted, but rather because it is the kind of condition that arises within a certain sort of high-level description. Again, I will illustrate by example.

Billy and Suzy are having lunch. Over the course of the lunch, their conversation takes various directions. Suppose we wish to entertain various hypotheses about the explanatory contribution of Billy's *attire* to the conversation. He is, in fact, wearing an outrageously bright pink shirt. Could that be making a difference to the course of the conversation–say, because it makes a difference to Suzy's mood to see him dressed so outlandishly? Our focus on causal structure as being constituted by localized dependence structure instructs us to clarify this question by means of counterfactuals such as the following:

If the color of Billy's shirt had been different, would the conversation have gone significantly differently?

So far, this is all fine–we should consider just such counterfactuals if we wish to clarify the content of our explanatory hypothesis. But now we run into trouble when we ask how we should evaluate the counterfactual itself. Suppose we proceed along very naïve interventionist lines. That is, we take it that we are to be considering a counterfactual situation in which, at the time of the conversation, there is a localized change to the color of Billy's shirt, but everything else about the situation remains as it actually is. Then things go haywire, for "everything else" will include the fact that Suzy *vividly remembers just seeing Billy in a pink shirt*. So, what will be true–at least, on *this* way of evaluating the counterfactual–is that if the color of Billy's shirt had been different, then Suzy would have exclaimed something along the lines of "my God–how did your shirt just change color?!"

That counterfactual situation is obviously not what we had in mind. Rather, we had in mind something like this: a situation in which the color of Billy's shirt is different, and suitable changes are made in the environment–particularly, the *psychological* environment–to keep everything "normal." The question, of course, is how to make this invocation of "normality" at all rigorous or precise.

When I bring this problem up in conversation, a certain move almost *always* gets suggested. What we should be considering is a counterfactual situation in which, say, Billy puts on a differently colored shirt in the morning, well before he meets up with Suzy. But this will not do. In general, when considering how later conditions depend on some time-t condition C, we cannot ring changes on the state of the world *earlier* than t so as to bring about C, since there is no guarantee that we will not thereby bring about *confounds*, in the form of changes to causally relevant time-t conditions other than C. Once you see the idea, it is child's play to augment our story accordingly.

Suppose Billy wore the pink shirt out of deference to his friend Sally, who had just given it to him as a present. Had he put on some other shirt, she would have been sufficiently put out that she would have slipped him a nausea-inducing drug shortly before his lunch. (She is oddly touchy about these sorts of things.) Then–reading the counterfactual *this* way–the course of the conversation will depend quite a lot on the color of Billy's shirt. But that is just a confusion, a conflation of a hypothesis about the causal-explanatory import of one condition–the color of Billy's shirt, during the lunchtime conversation–and a distinct hypothesis about the causal-explanatory import of an earlier condition that gave rise to it.

What we need, I suggest, is a general theory of *normal conditions*. That would allow for a simple and clean amendment to our original truth conditions for the canonical counterfactual:

If conditions C1 had obtained in region R1, then conditions C2 would have obtained in region R2.

The idea would be the following: to evaluate this counterfactual, we consider a state of the world at the time of R1, in which conditions within R1 have been altered so as to make C1 obtain, while conditions external to R1 "remain normal." In some cases–for example, when we are considering dependence relations at the level of physics or chemistry–the way to "remain normal" is simply to remain *exactly the same*. But in other cases–most obviously, ones in which we are capturing dependency relations at a psychological or sociological scale, remaining normal may in fact require widespread if subtle changes. Here, too, we have an important bit of unfinished business. (For an extremely important–and much broader–investigation of the problems facing interventionism in the psychological domain, see Prescott-Couch, 2015)

1.5 THE PERPLEXING NEWS

So much for the challenging news. The perplexing news concerns the striking way in which our focus on localized dependence structure as the ultimate basis for causal structure runs roughshod over a certain central feature of our ordinary thinking about causation. Should we worry? I am not sure, though I think so, for reasons I will sketch next. But first, let us get this central feature on the table.

1.5.1 The Centrality of "Causal Process"

Everyday thinking about cause and effect appears to lean heavily on the notion of *process*. When a cause has some downstream effect, we take for granted that this connection is mediated in some distinctive way–by, as we would naturally put it, a process connecting cause with effect. This focus on mediating causal processes seems crucial to how we understand some very basic causal distinctions. Consider our earlier example:

Suzy First: Suzy and Billy, two young vandals, throw rocks at a particularly choice window. Both throw with deadly accuracy, but Suzy is a bit quicker: her rock hits the window first, breaking it. Billy's rock flies through a now empty window pane.

Let us notice some features of this example. First, its causal structure is entirely unambiguous, and *asymmetric*: it was Suzy's throw, and not Billy's, that caused the window to break. (We may reasonably point out that, as did Suzy's, Billy's throw *guaranteed*–in a causal sense–that the window would break. But it was not in fact a *cause* of the breaking; rather, it was a causally idle backup.) Second, it is just as unambiguous what connects Suzy's throw to the breaking: namely, a process consisting of the flight of her rock through the air, followed by its contact with the glass. Third, imagine that a philosopher comes along to challenge our judgment that Suzy's throw caused the breaking, whereas Billy's was a mere backup: "You fail to notice the symmetry in the situation; for it's equally true of *each* throw that, had it not occurred, the window would still have broken." We would naturally reply that the symmetry is broken by the fact that Suzy's throw is connected to the breaking by a suitable causal process, whereas Billy's is not.[17] Fourth, it is precisely by reference to such intervening processes that we distinguish *asymmetric* cases such as **Suzy First** from *symmetric* ones, such as the following:

Same Time: Suzy and Billy, two young vandals, throw rocks at a particularly choice window. Both throw with deadly accuracy, and with remarkable synchronization: their rocks hit the window at precisely the same time, breaking it. (Note that each rock was thrown with enough force to break the window, all by itself.)

The distinction seems perfectly clear and readily generalizable: in the symmetric case (but not the asymmetric case), a suitable process connects *each* candidate cause to the given effect. Of course, there is room (as always, in philosophy!) for dispute. (e.g., maybe what we are really attending to, in **Suzy First**, is the fact that the timing of the breaking is much more sensitive to Suzy's throw than to Billy's[18]–a symmetry absent from **Same Time**.) Still, I suggest that we accept my diagnosis, if only for the sake of discussion: the idea that causal connections are (at least, typically) mediated by processes has a starring role in our ordinary thinking about causation.[19]

[17] We might also reply in a more teleological manner, noting that both throws initiate processes that "aim at," or have as their "natural end-point," the breaking of the window, but that only one of these processes "goes to completion." I think this observation, correct though it may be in the case at hand, is a red herring. What really matters to the structure of our causal judgments in cases such as this is that we can see clearly which candidate causes are connected to which candidate effects by which sorts of suitable processes. Consider a variant, in which the target is not a window but a bell: Suzy's rock strikes it first, ringing it once; Billy's strikes it a moment later, ringing it a second time. There is no question that Suzy's throw, and not Billy's, is a cause of the first ringing; similarly, that Billy's throw, and not Suzy's, is a cause of the second ringing. But in defending the first of these judgments, it will not do–because it is false–to say, "Well, Suzy's throw and Billy's throw each initiated processes aimed at bringing about a ringing, but only one of these processes went to completion." The correct defense is, rather, the same as that given in the main text: only Suzy's throw is connected to the first ringing by a suitable causal process (and only Billy's to the second).

[18] For observe that if she had not thrown, the window would have broken a split second later, whereas the same is not true of Billy's throw. For attempts to fashion, along these lines, a systematic account of the difference between genuine causes and idle backups, see Paul (1999) and especially Lewis (2004).

[19] Why the parenthetical qualification? Because of cases with the following generic structure: event A happens, initiating a process that results in event E. Meanwhile, event B occurs, initiating a process which, if not interrupted, will prevent E from occurring. But event C also happens, initiating a process that interrupts the B-process. It is thus the case that had C not occurred, E would not have occurred; and perhaps that is

But it plays no role–none–in our account of localized dependence structure. What mediate relations of localized dependence are, ultimately, two sorts of facts: the fundamental physical states that make up the history of our world and the fundamental laws that govern the evolution of these states. Nothing in our analysis has suggested that we can get from these ground-level facts to facts about localized dependence only *by way of* some intermediary analysis of "causal processes." So, while causal processes there may be, it does not seem as though our scientifically illuminating account of causal structure needs to mention them.

1.5.2 A Speculative Proposal

That may well be right. If so, some striking results follow–for example, that while **Suzy First** and **Same Time** may *strike* us as having starkly different causal structures, from a properly scientific standpoint, they *do not*. Learn to use causal concepts in a scientifically legitimate manner, and you will learn (*inter alia*) to treat these two cases as causally indistinguishable.

Still, I think that judgment may be too hasty, that it remains worthwhile to consider how an approach to causal structure broadly in the spirit of the one sketched here might distinguish these cases, in a principled manner. One of my reasons for optimism is that, perhaps immodestly, I think I have made some progress on this problem (see, e.g., Hall 2004). But setting my own efforts aside, there is another reason, which is that this problem appears deeply connected to a problem of obvious relevance to anyone who wants to command a clear view of scientific explanatory practice. Let me explain.

Outside of fundamental physics, the sciences routinely–and unavoidably–make use of causal generalizations that feature "no interference" clauses. Metal bars expand when heated in proportion to their change in temperature. That is, *provided nothing interferes*: if you heat a metal bar by blowing it up, it will *not* expand in proportion to its change in temperature. And notice that there can be interference, even in cases where the target effect is realized. Suppose some drug D1 reliably cures a given disease. Ahmed has the disease and is administered D1. His disease goes away. But in this case, someone nefarious *also* slipped him D2, which *counteracts* D1. Happily, someone beneficent gave him D3, which also cures the disease. The generalization that administering D1 cures the given disease (provided nothing interferes) has no instance here, any more than it would in a case where the interference prevented the cure.

When I mentioned generalizations about localized dependence structure in Section 1.3.1, I completely ignored this phenomenon. But it is ubiquitous and philosophically perplexing: since interference appears to be a *causal* notion, we should very much like our account of the content of causal generalizations in science not to have to presuppose it, unanalyzed.

enough to say that, in some sense, C is a cause of E. I explore the complications such cases give rise to in Hall (2004); but for the sake of simplicity, I am going to ignore them in this essay.

So, suppose we have some generalization that says that if certain conditions C1 obtain at one time, then those will cause a certain outcome C2 at such-and-such later time–provided nothing interferes. How do we tell whether something else is interfering? Well, we can at least advance a *sufficient* condition for noninterference, which is that *nothing else is happening* at the time that conditions C1 obtain.[20] (So, nothing is happening that could serve as an interference.) That is, we can think of our causal law as, in the first instance, making a claim about highly sanitized conditions in which *all* that is happening, at the given initial time, is that conditions C1 obtain. If the law is correct, those conditions should be followed by C2 (and the obtaining of C2 should in fact *depend* on the obtaining of C1).

It is precisely at this point that *processes* need to make an appearance. For in any ordinary situation where we might think our law applies, *lots* of things will be happening other than the obtaining of C1. How do we tell whether it applies? I think the right answer is this: it applies, just in case the processes that unfold from C1 not only terminate in C2, but *do so in the same way they would have*, if the prevailing circumstances had been of the highly sanitized, nothing-else-happening variety. Put another way, a generalization that correctly captures dependency relations in such sanitized circumstances thereby provides us with a *blueprint*–in the form of the processes that connect causes to effect–that can be deployed to map the structure of nonsanitized cases.

Now return to our puzzle about how to distinguish **Suzy First** from **Same Time**. What stands out is that *exactly the same approach* can be brought to bear: we focus on sanitized situations in which Suzy (respectively, Billy) is alone and throwing a rock at the window. We note the processes–sequences of events–that connect the throws to the breakings, in these situations. We observe, finally, that while **Same Time** contains "copies" of both the Billy-process and the Suzy-process, **Suzy First** contains a copy only of the latter. That is what it comes to, to say, as I did earlier, that the symmetry is broken by the fact that Suzy's throw is connected to the breaking by a suitable causal process in *both* cases, whereas Billy's is so connected only in **Same Time**.

Of course, this is all massively speculative (although I have attempted to make good on some of the key proposals in Hall, 2004, 2005). It may yet be that the process-centric intuitions evoked by cases such as **Suzy First** and **Same Time** ought to be dismissed, notwithstanding how widespread and strongly held they are. But on the other hand, maybe they are clues–important, nonmisleading ones–to the location of an account of causal structure richer and more satisfying than the one we have been investigating in this essay.

[20]Making sense of this notion of "nothing else happening" will, I think, involve introducing a distinction between a *default* state of the world and *deviations* from it. A situation in which the only things happening at a certain time consist in the obtaining of such-and-such conditions will just be a situation in which every part of the world save those involved in the instantiation of the conditions is in its default state. See Hitchcock (2007) as well as Hall (2006) for more discussion.

REFERENCES

Albert, D. (2000) *Time and Chance*, Harvard University Press, Cambridge.

Armstrong, D. (2004) Going through the open door again: counterfactual vs. singularist theories of causation, in *Causation and Counterfactuals* (eds J. Collins, N. Hall, and L.A. Paul), MIT Press, Cambridge, MA.

Frick, J. (2015) Probabilistic causation and the problem of aggregate effects, Working Paper.

Hájek, A. (2015) Most counterfactuals are false, MS.

Hall, N. (2004) Two concepts of causation, in *Causation and Counterfactuals* (eds J. Collins, N. Hall, and L.A. Paul), MIT Press, Cambridge, MA.

Hall, N. (2005) Causation and ceteris paribus laws. *Harvard Review of Philosophy*, **13**, 80–99.

Hall, N. (2006) Structural equations and causation, (extended version) MS.

Halpern, J. and Pearl, J. (2005) Causes and explanations: a structural-model approach-part I: causes. *British Journal for the Philosophy of Science*, **56**, 843–887.

Hitchcock, C. (2007) Prevention, preemption, and the principle of sufficient reason. *Philosophical Review*, **116**, 495–532.

Lewis, D. (1986) Postscripts to 'causation', in *Philosophical Papers*, Volume **II** (ed. D. Lewis), Oxford University Press, Oxford, pp. 172–213.

Lewis, D. (2004) Causation as influence, in *Causation and Counterfactuals* (eds J. Collins, N. Hall, and L.A. Paul), MIT Press, Cambridge, MA.

Maudlin, T. (2007) A modest proposal concerning laws, counterfactuals, and explanations, in *The Metaphysics Within Physics* (ed. T. Maudlin), Oxford University Press, Oxford, pp. 5–49.

Paul, L. (1999) Keeping track of the time: emending the counterfactual analysis of causation. *Analysis*, **58**, 191–198.

Paul, L.A. and Hall, N. (2013) *Causation: A User's Guide*, Oxford University Press, Oxford.

Pearl, J. (2009) *Causality: Models, Reasoning, and Inference*, Cambridge University Press, Cambridge.

Popper, K. (1959) *The Logic of Scientific Discovery*, Hutchinson & Co, London.

Prescott-Couch, A. (2015) Explanation and manipulation, MS.

Price, H. (1996) *Time's Arrow and Archimedes' Point: New Directions for the Physics of Time*, Oxford University Press, Oxford.

Robb, D. and Heil, J. (2014) Mental causation, in *The Stanford Encyclopedia of Philosophy* (ed. E.N. Zalta), Spring 2014 Edition. `http://plato.stanford.edu/archives/spr2014/entries/mental-causation/`.

Weslake, B. (2014) Statistical mechanical imperialism, in *Chance and Temporal Asymmetry* (ed. A. Wilson), Oxford University Press, Oxford, pp. 241–257.

Woodward, J. (2005) *Making Things Happen: A Theory of Causal Explanation*, 2nd edn, Oxford University Press, Oxford.

2

EVIDENCE AND EPISTEMIC CAUSALITY

MICHAEL WILDE AND JON WILLIAMSON

Department of Philosophy, School of European Culture and Languages, University of Kent, Kent, UK

The epistemic theory of causality maintains that causality is an epistemic relation, so that causality is taken to be a feature of the way a subject reasons about the world rather than a nonepistemological feature of the world. In this paper, we take the opportunity to briefly rehearse some arguments in favour of the epistemic theory of causality, and then present a version of the theory developed in Williamson (2005, 2006, 2009, 2011). Lastly, we provide some possible responses to an objection based upon recent work in epistemology.

This paper provides a broad overview of the issues, and in a number of places readers are directed towards work which provides the relevant details.

2.1 CAUSALITY AND EVIDENCE

Standardly, there are *mechanistic* and *difference-making* theories of causality. On the one hand, mechanistic theories maintain that variables in a domain are causally related if and only if they are connected by an appropriate sort of mechanism. On the other hand, difference-making theories maintain that variables are causally related if and only if one variable makes an appropriate sort of difference to the other. Stated in these terms, there might be a worry that both types of theory will be uninformative or

Statistics and Causality: Methods for Applied Empirical Research, First Edition.
Wolfgang Wiedermann and Alexander von Eye.

circular, since the appropriate sort of mechanism or the difference-making relationship might be less well understood than the causal relation or only understandable in terms of causality. Therefore, a good mechanistic theory should attempt to provide an account of the appropriate sort of mechanism in better-understood and noncausal terms, for example, the mechanism might be understood in terms of a process that possesses a conserved quantity (Dowe, 2000). Similarly, a good difference-making theory should attempt to provide an account of the appropriate sort of difference-making relationship in better-understood and noncausal terms, for example, in terms of probabilistic dependence conditional upon other causes (cf. Williamson, 2009).

The two different types of theory–mechanistic and difference-making–have conflicting implications regarding the epistemology of causality. On a mechanistic theory, one's body of evidence is sufficient to establish a causal claim if and only if one's evidence is sufficient to establish that there exists an appropriate mechanistic connection. On a difference-making theory, one's evidence is sufficient to establish a causal claim if and only if that evidence is sufficient to establish the existence of an appropriate sort of difference-making relationship.

However, there are well-known proposed counterexamples to mechanistic and difference-making theories of causality. There are cases involving absences, which seem to be cases of causality but without any appropriate sort of mechanism (Williamson, 2011). For example, it seems that missing my flight in London is a cause of my talk being canceled in Australia, even though they are not connected by any appropriate mechanism. The cases involving absences seem to demonstrate that the existence of a suitable mechanism is not a necessary condition for causality. But these cases also seem instructive regarding the epistemology of causality. Given these cases, it seems that establishing a causal claim does not require establishing the existence of an appropriate sort of mechanism, since here the causal claim is established and there exists no such mechanism. Rather, having established that there exists an appropriate difference-making relationship between missing my flight and my talk being canceled here seems sufficient to establish the causal claim.

There are also cases of overdetermination, which seem to be cases of causality but without any appropriate sort of difference-making relationship (Hall, 2004, pp. 232–241). For example, it seems that dropping the first bomb caused the end of the war, even though dropping this first bomb did not make the appropriate difference to the ending of the war, since a second bomb was dropped an instant later and would have ended the war regardless. The cases of overdetermination seem to demonstrate that the existence of an appropriate difference-making relationship is not a necessary condition for causality. Once again, these cases are instructive about the epistemology of causality. In overdetermination cases, it seems that establishing a causal claim cannot require establishing the existence of an appropriate sort of difference-making relationship, since here the causal claim is established and there exists no such a relationship. Instead, having established that there exists an appropriate sort of mechanism between the dropping of the first bomb and the ending of the war by itself seems sufficient to establish the causal claim.

How should one respond to these proposed counterexamples? There are two standard lines of response.

The first line of response is simply to dismiss the relevant suggested counterexamples. The proponent of a mechanistic theory could deny that cases involving absences are genuine cases of causality; see, for example, Dowe (2000, pp. 123–145). Similarly, the proponent of a difference-making theory could deny that cases of overdetermination are genuine cases of causality; see, for example, Coady (2004). This line of response, however, looks implausible, since cases involving absences look like straightforward cases of causality. Indeed, Schaffer (2004) argues that cases involving absences are treated as genuine cases of causality in both ordinary and theoretical contexts, and rightly so, since such cases have all the hallmarks of genuine cases of causality. Similarly, overdetermination cases look like paradigmatic cases of causality, and accordingly, proponents of a difference-making theory of causality typically accept such cases and attempt to accommodate them; see, for example, Paul and Hall (2013, pp. 70–172). This suggests another way of dismissing the suggested counterexamples. The proponent of a difference-making theory could suggest that overdetermination cases in fact do involve an appropriate sort of difference-making relationship, by suitably refining their account of the difference-making relationship. Similarly, the proponent of a mechanistic theory could suggest that cases involving absences do involve some appropriate sort of mechanism; see, for example, Thomson (2003, pp. 84–86). However, it is generally agreed that there is currently neither a difference-making nor a mechanistic theory of causality that can accommodate all the proposed counterexamples in this manner (Paul and Hall, 2013, p. 1).

The second line of response is to advocate pluralism, for example, by maintaining that there is both a mechanistic type and a difference-making type of causality; see, for example, Hall (2004). The idea here is that cases involving absences are cases of the difference-making type of causality without the mechanistic type of causality and vice versa for cases of overdetermination. Of course, this line of response has its own implications regarding the epistemology of causality. Presumably, establishing that there exists an appropriate sort of difference-making relationship is sufficient to establish a causal claim about the difference-making type of causality, and establishing that there exists an appropriate sort of mechanism suffices to establish a causal claim about the mechanistic type of causality. Reasons to doubt this line of response are presented in Williamson (2006). For instance, it is argued there that nonpluralist theories of causality should be preferred on the grounds of simplicity.

The main problem is that both these lines of response to the counterexamples–attempting to rebut them or moving to pluralism–have difficulty accounting for the practice of scientists when establishing causal claims (Williamson, 2006, pp. 73–74). In particular, when establishing a causal claim, health scientists typically require evidence both that there exists an appropriate difference-making relationship and that there exists an appropriate mechanism. (In this sense, cases involving absences and overdetermination are atypical.) Firstly, establishing only that there exists an appropriate sort of difference-making relationship is typically not sufficient for a health scientist to consider the corresponding causal claim established. For example, smoking was not established as a cause of heart disease, despite strong evidence that smoking makes an appropriate difference to the prevalence of heart disease, until there was also strong evidence that there is an appropriate mechanism linking smoking and disease (Gillies, 2011). Secondly, establishing only that there exists an appropriate mechanism is also typically not sufficient for a health scientist to consider the corresponding causal claim established. It was not established that the microorganism anthrax bacillus was the cause of anthrax, despite strong evidence of an appropriate mechanism, until there was also strong evidence that there exists an appropriate difference-making relationship between anthrax bacillus and anthrax (Clarke *et al.*, 2014a, p. 345).

The two lines of response to the counterexamples struggle to explain this need for both types of evidence. On the one hand, if a standard difference-making theory of causality is correct, it is difficult to explain the scientist's apparent need for evidence that there exists an appropriate mechanism, when there is already good evidence of an appropriate difference-making relationship available. On the other hand, if a standard mechanistic theory is correct, it is difficult to explain the scientist's apparent need for evidence that there exists an appropriate difference-making relationship when available evidence already establishes the existence of a suitable mechanism. The pluralist response is no better. The pluralist analyzes some causal claims as mechanistic, others as referring to a difference-making type of cause. In the former case, the pluralist cannot explain the need for evidence of an appropriate difference-making relationship; in the latter case, the pluralist cannot explain the need for evidence of the existence of a suitable mechanism.

Now, it might be objected that it is not a desideratum of a theory of causality that it accounts for the practice of scientists when establishing causal claims. For instance, it might be objected that scientists are getting the epistemology of causality wrong and thus that a theory of causality need not make sense of their practice. However, given the success of the sciences in establishing causal claims, it is likely that their practice is indicative of the correct epistemology of causality. But there is another reason to believe that scientists are doing things right. In particular, causal claims are used for prediction, explanation, and control. Russo and Williamson (2007) argue that the explanatory use of causal claims typically requires that there is an appropriate sort of mechanism linking the cause and the effect. This is because explanations are best given by appealing to mechanisms (cf. Williamson, 2013). Moreover, Russo and Williamson argue that the use of causal claims for prediction and control requires that a cause should typically make an appropriate sort of difference to its effects, for otherwise information about the presence of the cause would tell us nothing about the presence of its effects and vice versa, and also instigating a cause would not be a good strategy for achieving its effects (Russo and Williamson, 2007, p. 159). Given this, it is plausible that establishing a causal claim typically also requires establishing that there exists an appropriate difference-making relationship. Thus, it looks as if scientists are getting the epistemology right, so a theory of causality should account for their practice.

Howick (2011) objects that sometimes establishing that there exists an appropriate difference-making relationship is sufficient to establish the corresponding causal claim. He says that "[i]n many cases, tightly controlled comparative clinical studies suffice to establish causation" (2011, p. 933). This is intended to constitute an objection to the thesis that establishing a causal claim requires establishing an appropriate difference-making relationship *and* an appropriate sort of mechanism. However, he assumes that tightly controlled comparative clinical studies provide evidence only that there exists an appropriate difference-making relationship. In fact, instances in which the results of tightly controlled comparative clinical studies suffice to establish a causal claim are plausibly instances in which those results also establish that some mechanism exists to explain the difference-making relationship. In other instances, it might be unreasonable to consider a causal claim established on the basis of the results of clinical studies, since the established difference-making relationship might be due to confounding, if the relata are correlated effects of a common cause. In these instances, the additional evidence that there exists an appropriate mechanism would help to rule out confounding as an explanation of the difference-making relationship. Plausibly, instances in which a causal claim can be established on the basis of the results of clinical studies are instances in which the studies are of sufficient quality that they establish that confounding is not the likely explanation of the difference-making relationship, and thus also provide evidence that the existence of a suitable mechanism is the likely explanation.

One might also object that establishing the existence of an appropriate mechanism is sometimes sufficient for establishing the corresponding causal claim. In these instances, however, it is plausible that the evidence that establishes the existence of the mechanism is also sufficient to establish an appropriate difference-making relationship. The problem with establishing the causal claim only on the basis of the existence of an appropriate sort of mechanism is that there might be undiscovered mechanisms that counteract the action of the known mechanism, so that overall there is no appropriate difference-making relationship. This is called the problem of *masking*. In some instances, one can know enough about a mechanism that one can establish that counteracting mechanisms do not exist and thus that overall there exists an appropriate difference-making relationship. In other instances, additional evidence that there exists an appropriate difference-making relationship helps to overcome the problem of masking. Plausibly, instances in which a causal claim can be established on the basis of evidence of the existence of an appropriate mechanism are instances in which this evidence also suffices to overcome the problem of masking, and thus also provide evidence that there exists an appropriate difference-making relationship. These considerations provide more reason that both types of evidence are required in order to establish a causal claim, since each type of evidence compensates for the limitations of the other (Illari, 2011, pp. 144–148).

To conclude this section, it seems that standard theories of causality are susceptible to counterexamples and also struggle to explain good evidential practice in establishing causal claims. Similarly, a pluralist theory struggles to explain the epistemology of causality. How should one respond to this state of affairs?

2.2 THE EPISTEMIC THEORY OF CAUSALITY

One response is to plump for an epistemic theory of causality. In this section, we introduce the epistemic theory of causality developed in Williamson (2005, 2006,

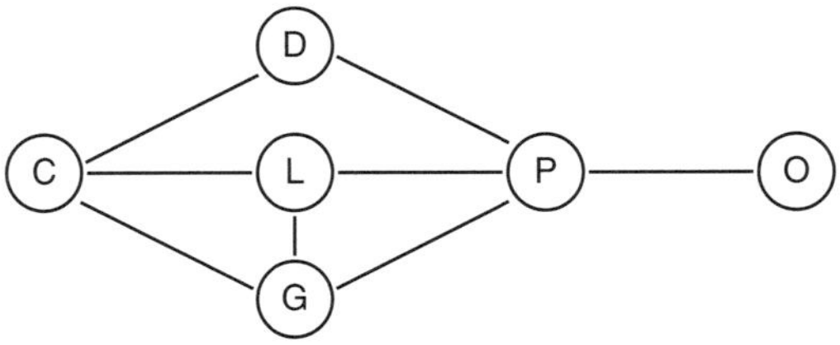

Figure 2.1 Trihoral relationships involving Canterbury (C), London (L), Gatwick (G), Dunkirk (D), Paris (P), and Orléans (O).

2009, 2011, 2013). According to this theory, causality is purely epistemic in the sense that our causal claims enable us to reason and interact with the world in certain ways; they are not claims about some causal relation that exists independently of us and our epistemic practices.

By way of analogy, consider the following relation, which we shall call the *trihoral* relation: two places stand in this relation if it is reasonable to expect to be able to travel between them within 3 hours. One can chart this relation, as in Fig. 2.1. It is not hard to see how such a chart might be useful–for parents of young children to plan breaks, for instance. Moreover, it seems clear that the edges between the nodes of such a graph do not correspond straightforwardly to a single sort of physical, or even nonepistemological, link between the places that correspond to the nodes. If the graph is correct, it is in virtue of a complex array of facts about the presence and absence of train, air, ferry, and road connections, as well as normal conditions relating to travel. In that sense, the trihoral relation is purely epistemic. This is not to say that there is no fact of the matter as to whether two places stand in the trihoral relation–there is, at least in nonborderline cases. It is to say that such propositions are not made true by the existence of some single, unified, worldly (nonepistemological) connection between places that we can call "trihorality."

Similarly with the causal relation, our causal claims are extremely useful–particularly for prediction, explanation, and control. It is this utility that accounts for our having the concept of cause: not the existence of some simple kind of worldly connection to which our causal claims refer that we can call causality. If a causal graph is correct, it is in virtue of a complex array of facts about the presence and absence of mechanisms, as well as the presence and absence of difference-making relationships and their magnitude.

Consider another analogy, to Bayesian probability. Bayesian probabilities are epistemic–rational degrees of belief, not directly physical entities–and they underwrite certain predictions and bets. Moreover, at least on the objective Bayesian view, there is typically a fact of the matter as to what the correct Bayesian probabilities are, given the extent and limitations of the evidence available. In the version of objective Bayesianism developed in Williamson (2010), for example, three norms constrain the strengths of one's beliefs:

Probability. One's degrees of belief should be representable by a probability function P_E.

Calibration. One's degrees of belief should fit evidence: $P_E \in \mathbb{E}$, the subset of probability functions that fit evidence. In particular, they should be calibrated to the corresponding empirical probabilities, insofar as one has evidence of them: if evidence determines just that the chance function $P^* \in \mathbb{P}^*$, then $P_E \in \langle \mathbb{P}^* \rangle$, the convex hull of the set of potential chance functions.

Equivocation. One's degrees of belief should otherwise equivocate as far as possible between the elementary outcomes. In particular, if there are finitely many elementary outcomes, then $P_E \in \text{maxent } \mathbb{E}$, the subset of those functions that fit evidence that have maximal entropy, as long as maxent $\mathbb{E} \neq \emptyset$.

These norms tend to be motivated by appealing to betting considerations, along the following lines. If one's degrees of belief do not meet the norms and one places bets in accordance with these beliefs, then one opens oneself up to potential losses: the possibility of sure loss in the case of the Probability norm, long-run loss in the case of the Calibration norm, and worst-case expected loss in the case of the Equivocation norm (Williamson, 2010, Chapter 3). On the other hand, one does not expose oneself to these losses if the norms are followed. Hence, one's degrees of belief must conform with the norms if one is to avoid avoidable losses. Arguably, it would be irrational not to avoid avoidable losses. So, the norms must hold for the strengths of one's beliefs to be apportioned in a rational way.

This view of probability is different to the epistemic view of causality in that it is pluralist, positing empirical, nonepistemic probabilities (chances), in addition to epistemic, Bayesian probabilities. Nevertheless, it is instructive in that it suggests a particular connection between evidence and epistemic probabilities. An epistemic theory of causality can posit similar norms that constrain one's causal claims:

Acyclicity. One's causal claims should be representable by an acyclic graph C.

Calibration. One's causal claims should fit evidence: $C \in \mathfrak{C}$, the subset of acyclic graphs that fit evidence.

Equivocation. C should otherwise be as noncommittal as possible about what causes what.

How might these norms be fleshed out? One simple recipe proceeds as follows. We can take C to be a graph whose nodes correspond to variables, which contains an arrow from variable A to variable B if it is claimed that A is a cause of B, a gap (i.e., no connection) between A and B if it is claimed that neither causes the other, and an undirected edge between A and B if neither of the aforementioned two claims is made involving A and B. Such a graph is acyclic if there is some way of orienting the undirected edges in the graph such that there is no chain of arrows in the graph that forms a cycle. It is plausible that causal claims can always be representable by an acyclic graph: in cases in which there are apparent causal cycles, one can eliminate these cycles by time-indexing the variables (see, e.g., Clarke *et al.*, 2014b). We may suppose that evidence imposes certain constraints on C. For example, if evidence establishes that A is a cause of B, represented by $A \rightsquigarrow B$, then there should be some chain of arrows from A to B in C; if evidence establishes that A is not a cause of B, $A \not\rightsquigarrow B$, then there should be no chain of edges and arrows from A to B. A causal graph C is maximally noncommittal, from all those in $\mathfrak{C}$, if there is no other causal graph D in $\mathfrak{C}$, which makes fewer causal claims (including both arrows and gaps) than C.

How does evidence impose a constraint of the form $A \rightsquigarrow B$? As discussed in Section 2.1, in order to establish that A is a cause of B, there would normally have to be evidence both that (i) there is an appropriate sort of difference-making relationship (or chain of difference-making relationships) between A and B–for example, that A and B are probabilistically dependent, conditional on B's other causes–, and that (ii) there is an appropriate mechanistic connection (or chain of mechanisms) between A and B–so that instances of B can be explained by a mechanism that involves A. (We saw earlier that there are some exceptions to this rule, which correspond to the counterexamples to standard theories of causality, discussed in Section 2.1.)

How does evidence impose a constraint of the form $A \not\rightsquigarrow B$? Typically, a causal relationship can be ruled out by either (i) evidence that there is no appropriate difference-making relationship (or chain of such relationships) between A and B or (ii) evidence that there is no mechanism (or chain of mechanisms) that can account for B in terms of A. Thus, a drug trial of sufficiently high quality that finds no association between treatment and cure would impose a $A \not\rightsquigarrow B$ constraint. On the other hand, we can rule out certain causal claims involving treatments (such as certain homeopathic treatments) where it is known that there is no possible mechanism by which the treatment can explain a cure.

In sum, then, the analogy with the trihoral relation suggests that causal claims are representational rather than real in the sense that they guide inference, explanation, and action but do not refer to a nonepistemic connection that we can call causality. Moreover, the analogy with Bayesian probability can shed some light on the link between evidence and epistemic causality, by suggesting three norms by which the extent and limitations of available evidence constrain one's causal claims. [1]

2.3 THE NATURE OF EVIDENCE

It appears that the aforementioned recipe for arriving at one's causal claims requires that an ideally rational subject has perfect access to her evidence. The recipe suggests that one's causal claims are rational if and only if they are appropriately constrained by one's body of evidence in accordance with the Acyclicity, Calibration, and Equivocation norms. Informally, if one's causal claims are thus constrained by what one takes to be one's body of evidence rather than what is in fact one's body of evidence, then one's causal claims are not appropriately constrained. Therefore, if an ideally rational subject's causal claims are appropriately constrained, it looks like an ideally rational subject must have perfect access to her body of evidence.

[1]Note that it is important to distinguish this task of determining, on the basis of current evidence, which set of causal claims is established by that evidence, from the task of formulating, on the basis of current evidence, a set of more tentative causal hypotheses that can be tested by collecting further evidence. A formal approach to developing a set of causal hypotheses is developed in (Williamson 2006, Appendix).

The problem is that certain recent work in epistemology purports to show that evidence is not as accessible as following the aforementioned recipe seems to require; see, for example, Williamson (2000, pp. 164–183). In particular, it can be claimed that even an ideally rational subject may not have perfect access to her evidence. Thus, the critic might object that the epistemic theory of causality is committed to a false theory of evidence. How should the proponent of an epistemic theory of causality respond to this objection?

One possibility is for the proponent of an epistemic theory to simply deny that evidence is not perfectly accessible in the relevant sense: evidence is such that an ideally rational subject has perfect access to her evidence. Indeed, it looks like there is room for this line of response. This is because many theorists assume something like the *ought implies can principle*, which says that one must be in a position to accomplish anything that one ought to accomplish. The principle might seem a reasonable assumption, since intuitively one is not failing to fulfill one's obligations if one could not have possibly fulfilled those putative obligations. Crucially, this intuitively plausible principle implies that evidence is perfectly accessible in the relevant sense. In particular, if an ideally rational subject's causal claims are constrained by her evidence in accordance with the aforementioned recipe, then it must be possible that her causal claims be thus constrained, given the ought implies can principle. In turn, in order for it to be possible that an ideally rational subject's causal claims be appropriately constrained, her evidence must be perfectly accessible. Thus, the proponent of the epistemic theory of causality might maintain that, *contra* recent work in epistemology, evidence is in fact accessible in the relevant sense. Of course, then there is the pressing matter of pointing out where the recent work in epistemology has gone wrong; see Williamson (2015) on this point.

Alternatively, the proponent of an epistemic theory might want to endorse the view that evidence is not perfectly accessible. Then, the question is whether a viable epistemic theory of causality can be proposed that dispenses with the requirement that evidence is perfectly accessible. Arguably, such an epistemic theory of causality can be proposed. Once again, the analogy with Bayesian probability is instructive.

Objective Bayesian probabilities are degrees of belief appropriately constrained by the evidence in accordance with the Probability, Calibration, and Equivocation norms. The problem is that the objective Bayesian theory also seems to require that evidence is such that an ideally rational subject has perfect access to her evidence. Thus, if evidence is not perfectly accessible in this sense, it looks like this objective Bayesian theory of probability cannot be correct.

In response to this state of affairs, Timothy Williamson proposes an alternative *evidential* theory of probability, a theory that dispenses with the requirement that evidence is perfectly accessible (Williamson, 2000, pp. 209–237). On this theory, there exists an objective degree to which a claim is entailed by a given body of evidence, and it is evidential probabilities that measure this partial entailment relation between one's body of evidence and specific claims. But the objective degree to which a claim is entailed by the evidence is not reducible to an ideally rational subject's degree of belief in that claim, where an ideally rational subject follows the Probability, Calibration, and Equivocation norms. This is because such ideally rational subjects might disagree with regard to their degrees of belief on the same body of evidence, if this body of evidence is not perfectly accessible to each of them. Rather, degrees of belief are rational given a body of evidence only insofar as they match the relevant evidential probabilities. Crucially, the theory of evidential probability remains an epistemic theory of probability, since evidential probabilities depend upon one's body of evidence rather than on purely nonepistemological features of the world.

In a similar manner, the proponent of the epistemic theory might propose an analogous evidential theory of causality, an epistemic theory that dispenses with the requirement that evidence is perfectly accessible in the relevant sense. This evidential theory of causality differs from the epistemic theory of causality presented in Section 2.2 in much the same way that the theory of evidential probability differs from the theory of objective Bayesian probability. In particular, this theory can hypothesize that one's body of evidence entails certain relationships between specific claims, relationships that license particular inferences concerning explanation, prediction, and control. These relationships can be charted by a unique causal graph given a body of evidence. This theory of causality remains epistemic, since the causal graph depends upon one's body of evidence rather than some nonepistemological feature of the world. But unlike the original epistemic theory, causality is not reducible to the causal claims arrived at by following a recipe that requires that evidence is perfectly accessible. Instead, one's causal claims are rational insofar as they match the unique causal graph. Arguably, then, a version of the epistemic theory of causality survives the objection based upon recent work in epistemology, namely, the evidential theory of causality.

2.4 CONCLUSION

In this paper, we have argued that standard theories of causality have a hard time making sense of the epistemology of causality or are susceptible to counterexamples (Section 2.1). One response to this state of affairs is to adopt an epistemic theory of causality, such as that developed in Williamson (2005, 2006, 2009, 2011) as outlined in Section 2.2. While some might object that such an epistemic theory of causality conflicts with some recent work in epistemology, in Section 2.3 we have suggested two lines of response to this objection: either to rebut worries about the inaccessibility of evidence or to adapt the epistemic theory to fit the view that evidence may be inaccessible.

REFERENCES

Clarke, B., Gillies, D., Illari, P., Russo, F., and Williamson, J. (2014a) Mechanisms and the evidence hierarchy. *Topoi*, **33**, 339–360.

Clarke, B., Leuridan, B., and Williamson, J. (2014b) Modelling mechanisms with causal cycles. *Synthese*, **191** (8), 1651–1681.

Coady, D. (2004) Preempting preemption, in *Causation and Counterfactuals* (eds J. Collins, N. Hall, and L. Paul), MIT Press, pp. 325–339.

Dowe, P. (2000) *Physical Causation*, Cambridge University Press, Cambridge.

Gillies, D. (2011) The Russo-Williamson thesis and the question of whether smoking causes heart disease, in *Causality in the Sciences* (eds P. Illari, F. Russo, and J. Williamson), Oxford University Press, Oxford, pp. 110–125.

Hall, N. (2004) Two concepts of causation, in *Causation and Counterfactuals* (eds N.H. John Collins and L. Paul), MIT Press, Cambridge, MA, pp. 225–276.

Howick, J. (2011) Exposing the vanities–and a qualified defense–of mechanistic reasoning in health care decision making. *Philosophy of Science*, **78**, 926–940.

Illari, P. (2011) Mechanistic evidence: disambiguating the Russo-Williamson thesis. *International Studies in the Philosophy of Science*, **25**, 139–157.

Paul, L.A. and Hall, N. (2013) *Causation: A User's Guide*, Oxford University Press, Oxford.

Russo, F. and Williamson, J. (2007) Interpreting causality in the health sciences. *International Studies in the Philosophy of Science*, **21**, 157–170.

Schaffer, J. (2004) Causes need not be physically connected to their effects: the case for negative causation, in *Contemporary Debates in Philosophy of Science* (ed. C. Hitchcock), Blackwell, Malden, MA, pp. 197–216.

Thomson, J.J. (2003) Causation: omissions. *Philosophy and Phenomenological Research*, **66** (1), 83–103.

Williamson, J. (2005) *Bayesian Nets and Causality*, Oxford University Press, Oxford.

Williamson, J. (2006) Causal pluralism versus epistemic causality. *Philosophica*, **77**, 69–96.

Williamson, J. (2009) Probabilistic theories of causality, in *The Oxford Handbook of Causation* (eds H. Beebee, C. Hitchcock, and P. Menzies), Oxford University Press, Oxford, pp. 185–212.

Williamson, J. (2010) *In Defence of Objective Bayesianism*, Oxford University Press, Oxford.

Williamson, J. (2011) Mechanistic theories of causality. *Philosophy Compass*, **6**, 433–444.

Williamson, J. (2013) How can causal explanations explain? *Erkenntnis*, **78**, 257–275.

Williamson, J. (2015) Deliberation, judgement and the nature of evidence. *Economics and Philosophy*, **31** (1), 27–65.

Williamson, T. (2000) *Knowledge and its Limits*, Oxford University Press, Oxford.

PART II

DIRECTIONALITY OF EFFECTS

3

STATISTICAL INFERENCE FOR DIRECTION OF DEPENDENCE IN LINEAR MODELS

Yadolah Dodge

Institute of Statistics, University of Neuchâtel, Neuchâtel, Switzerland

Valentin Rousson

Division of Biostatistics, Institute for Social and Preventive Medicine, University Hospital Lausanne, Lausanne, Switzerland

3.1 INTRODUCTION

In linear regression, it is not always clear which variable should be the predictor and which variable should be the response. Dodge and Rousson (2000, 2001) proposed criteria for choosing the direction of a regression line based on asymmetric properties of the correlation coefficient under a classical linear regression setting. Their main result among other findings was to introduce the cube of the correlation coefficient as the ratio of the skewness of the two variables involved. Their approach has attracted some attention in the literature, with potential for interesting applications, for example, in the domain of social sciences. See, for example, Muddapur (2003), Sungur (2005), von Eye and DeShon (2012a,b) Pornprasertmanit and Little (2012), or Wiedermann *et al.* (2014), among others. In this paper, we reconsider and further analyze the problematic of detecting the direction of dependence in a linear regression model. In particular, after a brief review of the key concepts that are given in Section 3.2, we examine, compare, simplify, and formalize some possible testing procedures in Section 3.3. In Sections 3.4, we discuss the problematic of causality and

Statistics and Causality: Methods for Applied Empirical Research, First Edition.
Wolfgang Wiedermann and Alexander von Eye.

the possible presence of lurking variables in the context considered. In Section 3.5, we revisit the brain and body data set, which had been used in Dodge and Rousson (2000). Some conclusions are given in Section 3.6.

3.2 CHOOSING THE DIRECTION OF A REGRESSION LINE

Consider two continuous variables X and Y with Pearson correlation ρ_{XY} and the two possible linear regression models:

$$Y = a + bX + \epsilon \tag{3.1}$$

and:

$$X = a' + b'Y + \epsilon' \tag{3.2}$$

where ϵ and ϵ' are residual terms, which are normally distributed, the former being independent of X and the latter being independent of Y. To avoid unnecessary complications, we also consider that X and Y are not perfectly correlated (i.e., $|\rho_{XY}| < 1$). In this paper, we shall assume that at least one of (3.1) and (3.2) holds. We have then the following possibilities:

- Both (3.1) and (3.2) hold, in which case both X and Y are normally distributed (since the normality of X and Y together with normal residuals in (3.1) and (3.2) is equivalent to the binormality of (X, Y)).
- Only (3.1) holds while (3.2) does not hold, in which case X is not normally distributed (since a normal X together with normal residuals in (3.1) would imply that X and Y are both normally distributed such that (3.2) would also hold). Note that Y is also nonnormal in that case unless $\rho_{XY} = 0$.
- Only (3.2) holds while (3.1) does not hold, in which case Y is not normally distributed. Note that X is also nonnormal in that case unless $\rho_{XY} = 0$.

In summary, X and Y are both normally distributed if both (3.1) and (3.2) hold, whereas X and Y are both nonnormal if only one of (3.1) and (3.2) holds, as soon as their correlation is nonzero. We are thus left with the following five possible and mutually exclusive situations (which we shall retrieve in the next section):

Situation 1 X and Y are normally distributed (both (3.1) and (3.2) hold).
Situation 2 X is nonnormal, Y is normal, and $\rho_{XY} = 0$ (only (3.1) holds).
Situation 3 X is normal, Y is nonnormal, and $\rho_{XY} = 0$ (only (3.2) holds).
Situation 4 X and Y are nonnormal, and only (3.1) holds.
Situation 5 X and Y are nonnormal, and only (3.2) holds.

In situations where at least one of X or Y is nonnormal, models (3.1) and (3.2) cannot hold simultaneously and one may hesitate between the two models. To help in this choice, Dodge and Rousson (2001) proposed to use the concept and properties of "cumulants" (which are functions of moments, see e.g., Kendall and Stuart, 1963). Denote by $\mathrm{Cum}_r(V)$ the rth cumulant of a random variable V. For any $r \geq 3$, one has the following properties:

$$\mathrm{Cum}_r(\omega + \upsilon V) = \upsilon^r \mathrm{Cum}_r(V) \text{ for any scalars } \omega \text{ and } \upsilon \tag{3.3}$$

$$\mathrm{Cum}_r(V + W) = \mathrm{Cum}_r(V) + \mathrm{Cum}_r(W) \text{ if } V \text{ and } W \text{ are independent random variables} \tag{3.4}$$

$$\mathrm{Cum}_r(V) = 0 \text{ if } V \text{ is normally distributed.} \tag{3.5}$$

In particular, property (3.5) is the reason why (standardized) cumulants can be used as measures of nonnormality in descriptive statistics. Assuming that model (3.1) holds, one gets from (3.3)–(3.5) that:

$$\mathrm{Cum}_r(Y) = \mathrm{Cum}_r(a + bX) + \mathrm{Cum}_r(\epsilon) = b^r \mathrm{Cum}_r(X) + 0 = b^r \mathrm{Cum}_r(X). \tag{3.6}$$

Denote by σ_X and σ_Y the standard deviations of X and Y. Dividing both terms of (3.6) by σ_Y^r yields:

$$\frac{\mathrm{Cum}_r(Y)}{\sigma_Y^r} = \frac{b^r \mathrm{Cum}_r(X)}{\sigma_Y^r} = b^r \cdot \frac{\sigma_X^r}{\sigma_Y^r} \cdot \frac{\mathrm{Cum}_r(X)}{\sigma_X^r}.$$

If we assume now that only (3.1) holds (i.e., (3.2) does not hold), X is nonnormal and thus $\mathrm{Cum}_r(X) \neq 0$, and one finally obtains (recalling that $\rho_{XY} = b\sigma_X/\sigma_Y$):

$$\rho_{XY}^r = \frac{\mathrm{Cum}_r(Y)/\sigma_Y^r}{\mathrm{Cum}_r(X)/\sigma_X^r}.$$

Thus, since $|\rho_{XY}^r| \leq 1$, assuming that only model (3.1) holds implies that the rth standardized cumulant of Y, $\mathrm{Cum}_r(Y)/\sigma_Y^r$, is smaller in magnitude than the rth standardized cumulant of X, $\mathrm{Cum}_r(X)/\sigma_X^r$. By symmetry, assuming that only model (3.2) holds implies that the rth standardized cumulant of X is smaller in magnitude than the rth standardized cumulant of Y. This means that the response variable should be closer to a normal distribution than the predictor variable. Therefore, if one hesitates between models (3.1) and (3.2), one may select (3.1) if one has some statistical evidence that $|\mathrm{Cum}_r(Y)/\sigma_Y^r| < |\mathrm{Cum}_r(X)/\sigma_X^r|$ and one may select (3.2) if one has some statistical evidence that $|\mathrm{Cum}_r(X)/\sigma_X^r| < |\mathrm{Cum}_r(Y)/\sigma_Y^r|$ for some $r \geq 3$.

Dodge and Rousson (2000, 2001) applied this strategy using $r = 3$, whereas Dodge and Yadegari (2010) considered $r = 4$. Recall that the third standardized cumulant is the skewness coefficient (the third central moment divided by the cube of the standard deviation), whereas the fourth standardized cumulant is known as the "excess kurtosis" (obtained by subtracting 3 from the fourth central moment divided by the fourth power of the standard deviation). In what follows, we shall denote by $\gamma_X = E((X - E(X))^3)/\sigma_X^3$ and $\gamma_Y = E((Y - E(Y))^3)/\sigma_Y^3$ the skewness coefficients of

X and Y and by $\kappa_X = E((X - E(X))^4)/\sigma_X^4 - 3$ and $\kappa_Y = E((Y - E(Y))^4)/\sigma_Y^4 - 3$ their excess kurtosis. Assuming that only (3.1) holds (i.e. (3.2) does not hold) yields:

$$\rho_{XY}^3 = \frac{\gamma_Y}{\gamma_X} \tag{3.7}$$

as well as:

$$\rho_{XY}^4 = \frac{\kappa_Y}{\kappa_X}. \tag{3.8}$$

If one hesitates between models (3.1) and (3.2), one may select (3.1) if one has some statistical evidence that $|\gamma_Y| < |\gamma_X|$ or/and that $|\kappa_Y| < |\kappa_X|$, and one may select (3.2) if one has some statistical evidence that $|\gamma_X| < |\gamma_Y|$ or/and that $|\kappa_X| < |\kappa_Y|$. In the literature, choosing between models (3.1) and (3.2) is sometimes referred to as "choosing the direction (dependence) of a regression line."

3.3 SIGNIFICANCE TESTING FOR THE DIRECTION OF A REGRESSION LINE

The results presented in the previous section are theoretical ones. In practice, one has a sample of n independent observations (X_i, Y_i) drawn from a bivariate random variable (X, Y) $(i = 1, \cdots, n)$, and the theoretical results (3.7) or (3.8) will not hold exactly in the sample even if (3.1) holds in the population. If one hesitates between models (3.1) and (3.2), one should take into account the uncertainty due to sampling variability. In other words, a significance test is needed and such attempts are currently discussed in the literature (e.g., von Eye and DeShon, 2012a,b). In particular, Pornprasertmanit and Little (2012) proposed to check the following four conditions to conclude either (3.1) or (3.2) based on skewness:

Condition 1 (skewness) The correlation ρ_{XY} should be significantly different from zero.

Condition 2 (skewness) At least one of the skewness γ_X and γ_Y should be significantly different from zero.

Condition 3 (skewness) If both γ_X and γ_Y are significant, they should not be significant in opposite directions if ρ_{XY} is significantly positive, and they should not be significant in the same direction if ρ_{XY} is significantly negative.

Condition 4 (skewness) The difference in absolute skewness $\Delta_\gamma = |\gamma_X| - |\gamma_Y|$ should be significantly different from zero.

If the four conditions are satisfied, one may conclude model (3.1) (X is the predictor and Y the response) if Δ_γ is significantly larger than zero, and one may conclude model (3.2) (Y is the predictor and X the response) if Δ_γ is significantly smaller than zero. If the first or third condition is not satisfied, neither X nor Y is the predictor. If the second or fourth condition is not satisfied, the direction of the regression line is said to be "undetermined." They proposed a similar testing procedure based on kurtosis:

Condition 1 (kurtosis) The correlation ρ_{XY} should be significantly different from zero.

Condition 2 (kurtosis) At least one excess kurtosis κ_X or κ_Y should be significantly different from zero.

Condition 3 (kurtosis) If both κ_X and κ_Y are significant, they should not be significant in opposite directions.

Condition 4 (kurtosis) The difference in absolute excess kurtosis $\Delta_\kappa = |\kappa_X| - |\kappa_Y|$ should be significantly different from zero.

Here also, if the four conditions are satisfied, one may conclude model (3.1) if Δ_κ is significantly larger than zero, and one may conclude model (3.2) if Δ_κ is significantly smaller than zero, while neither X nor Y is the predictor if the first or third condition is not satisfied, and the direction of the regression line is undetermined if the second or fourth condition is not satisfied. The authors further mentioned that all significance tests performed in this procedure can be done using a simple bootstrap estimation method and they defined the Type I error in this context as "the probability that the directional dependency test suggests that X, Y or neither is the explanatory variable when a decision of undetermined directional dependency should be selected based on the information at the population level" (Pornprasertmanit and Little, 2012, p. 317).

In what follows, we further explore the validity of such a global testing procedure. In particular, we discuss whether each of the aforementioned four conditions is really needed or whether the testing procedure could be simplified without affecting its performance and validity. For this, we simulated data under model (3.1) with sample sizes of $n = 50, 100, 500$ and with square correlations of $R^2 = \rho^2_{XY} = 0, 0.25, 0.5, 0.75$. We considered various skewness and kurtosis values for X, which was generated either according to a chi-square distribution, for which the skewness coefficient is $\gamma_X = \sqrt{8/df}$, or according to a Student's t-distribution, for which the excess kurtosis is $\kappa_X = 6/(df - 4)$, where df represents the degrees of freedom. The number of degrees of freedom was chosen such that the skewness, respectively, the excess kurtosis, was equal to 0, 0.5, 1, 2, 4, and 6, where the cases $\gamma_X = 0$ and $\kappa_X = 0$ correspond to a normal distribution. To assess Conditions 2, 3, and 4 in the aforementioned testing procedure, we used a percentile bootstrap with 500 replications. To assess Condition 1, we used a Spearman correlation test, which was found to better control the Type I error than a percentile bootstrap. A nominal significance level of 5% has been used for each of the tests performed. In each setting considered, we estimated the probability to conclude model (3.1) and the probability to conclude model (3.2) based on 1000 simulated data sets.

Besides the testing procedure proposed by Pornprasertmanit and Little (2012) described earlier (in what follows Procedure A), we also considered two simplified procedures (in what follows Procedures B and C). In Procedure B, we removed Conditions 2 and 3 (keeping only Conditions 1 and 4) from Procedure A, and in Procedure C, we removed Conditions 1–3 (keeping only Condition 4). We considered such simplifications both for the testing procedure based on skewness and for the testing procedure based on kurtosis. In summary, we conclude either model (3.1) or model (3.2) if:

Procedure A Conditions 1–4 are satisfied.

Procedure B Only Conditions 1 and 4 are satisfied (whatever the status of Conditions 2 and 3).

Procedure C Only Condition 4 is satisfied (whatever the status of Conditions 1–3).

Figures 3.1–3.3 show the estimated probabilities to conclude model (3.1), the model under which the data have been generated, when using Procedures A–C in various settings. Here is a summary of our results:

- There was almost no difference between Procedures A and B. The highest difference was observed for $R^2 = 0.25$ and $\kappa_X \geq 4$, where the probability to (correctly) conclude model (3.1) was 1% higher for Procedure B, compared to Procedure A. This suggests that Conditions 2 and 3 used in Procedure A but not

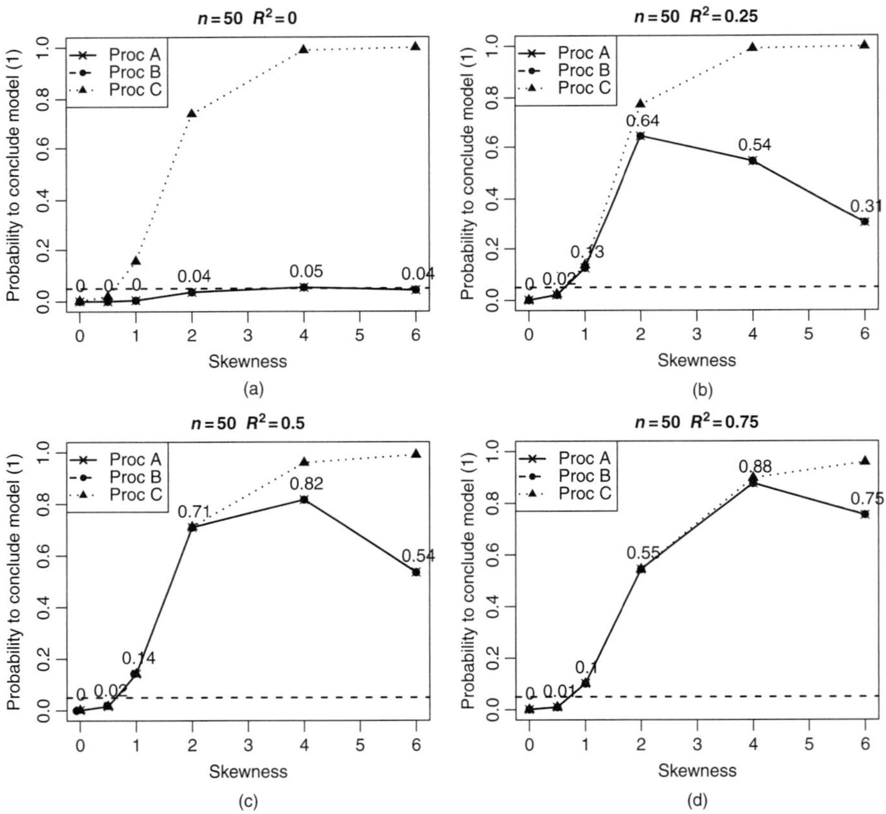

Figure 3.1 Estimated probability to conclude model (3.1) using tests based on skewness for $n = 50$, $R^2 = 0, 0.25, 0.5, 0.75$ (shown respectively in panels (a), (b), (c), and (d)) and various values of γ_X, with Procedure A (Conditions 1–4), Procedure B (Conditions 1 and 4), and Procedure C (only Condition 4). A horizontal line at 5% is plotted as reference line. The numbers plotted refer to the probabilities reached with Procedure A.

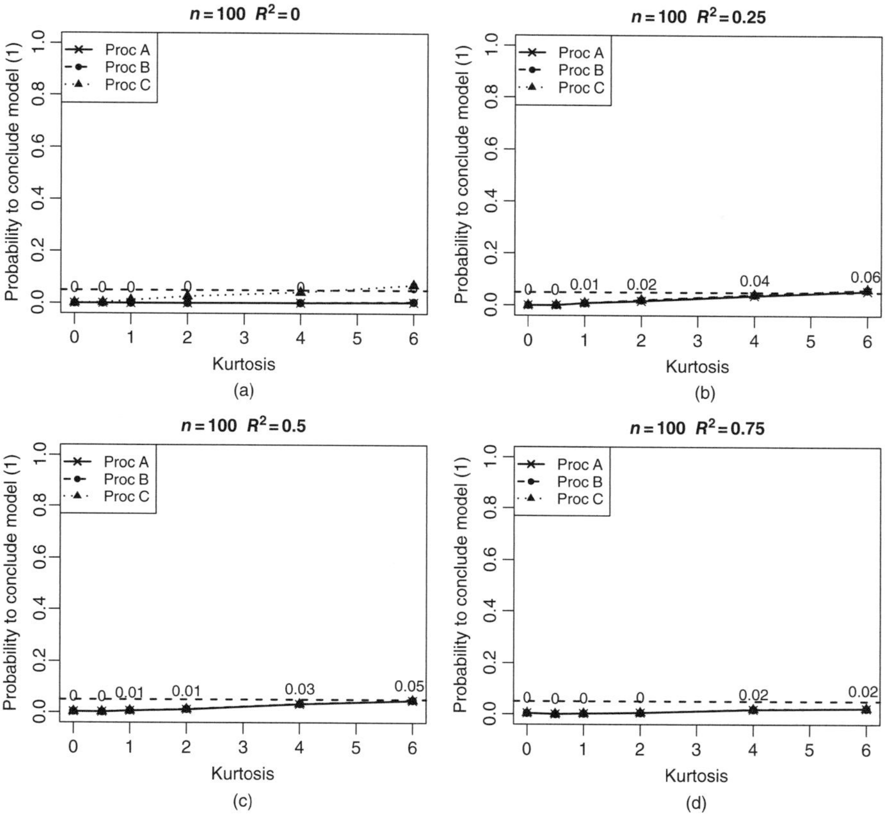

Figure 3.2 Estimated probability to conclude model (3.1) using tests based on kurtosis for $n = 100$, $R^2 = 0, 0.25, 0.5, 0.75$ (shown respectively in panels (a), (b), (c), and (d)) and various values of κ_X, with Procedure A (Conditions 1–4), Procedure B (Conditions 1 and 4), and Procedure C (only Condition 4). A horizontal line at 5% is plotted as reference line. The numbers plotted refer to the probabilities reached with Procedure A.

in Procedure B might not be needed. This is actually not surprising since Condition 4 implies Condition 2 in the population, $|\gamma_X| > |\gamma_Y|$ implying $|\gamma_X| > 0$ and $|\kappa_X| > |\kappa_Y|$ implying $|\kappa_X| > 0$, such that Condition 2 appears redundant, and since Condition 3 is always met in the population (if we assume a linear model) and was met in more than 99% of our simulations.

- The probability to conclude model (3.1) using Procedure A or B did not (or just barely) exceed the nominal significance level of 5% when $R^2 = 0$ or when X was normal, whereas it was in these cases much higher than 5% using Procedure C. This means that Condition 1 used in Procedures A and B but not in Procedure C is needed to control the Type I error, if the Type I error includes the probability to conclude model (3.1) when either $R^2 = 0$ or X is normal (corresponding to Situations 1–3 described in Section 3.2). By symmetry, the Type I error will also include the probability to conclude model (3.2) when either $R^2 = 0$ or Y is

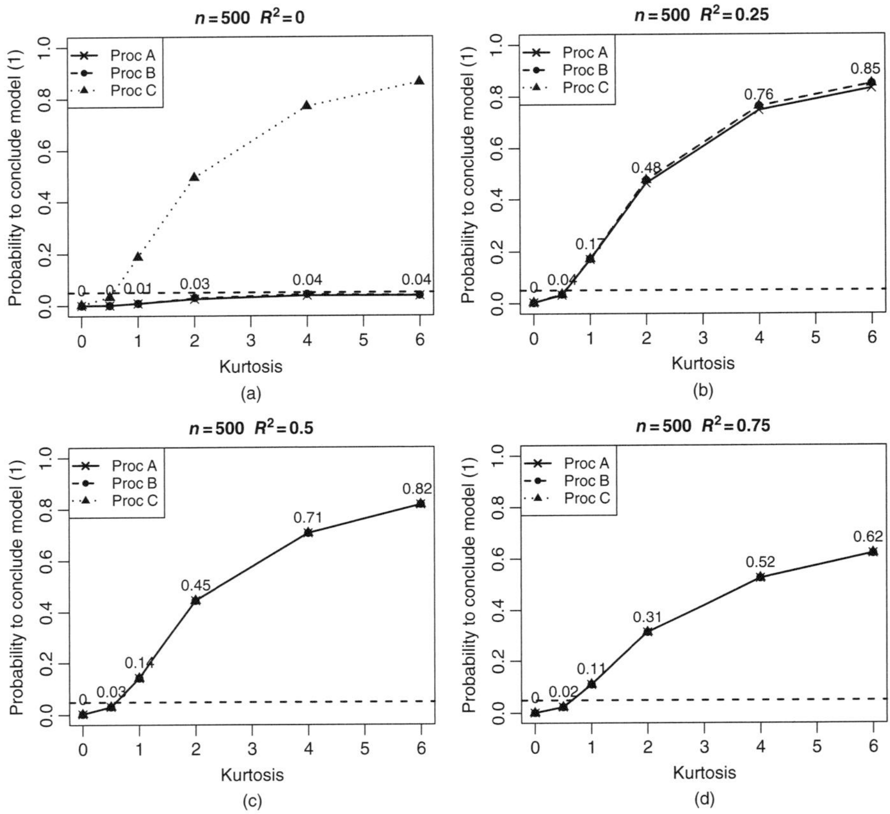

Figure 3.3 Estimated probability to conclude model (3.1) using tests based on kurtosis for $n = 500$, $R^2 = 0, 0.25, 0.5, 0.75$ (shown respectively in panels (a), (b), (c), and (d)) and various values of κ_X, with Procedure A (Conditions 1–4), Procedure B (Conditions 1 and 4), and Procedure C (only Condition 4). A horizontal line at 5% is plotted as reference line. The numbers plotted refer to the probabilities reached with Procedure A.

normal (corresponding to the same three situations). Taken together, the Type I error can be defined as the probability to conclude either (3.1) or (3.2) when we are in one of the first three situations described in Section 3.2.

- The probability to select model (3.2), which was a wrong model when X was nonnormal, did not exceed 0.2% in each setting (and for each procedure) considered, illustrating that the probability to select the wrong model, sometimes referred to as the Type III error (e.g. Kimball, 1957, Leventhal and Huynh, 1996), is well controlled.
- When X was normal, in which case Y was also normal such that we were in Situation 1 from Section 3.2 where both (3.1) and (3.2) hold, the probability to conclude model (3.1) and the probability to conclude model (3.2) were both almost zero, although they should be in principle both close to 2.5% (such that their sum would reach the nominal significance level of 5%). This suggests that

the percentile bootstrap that we have used to assess Condition 4 was overly conservative and that we would improve the statistical power of the testing procedure by using another method, for example, the (computationally intensive) BCa bootstrap. Using a more powerful test to assess Condition 4 in our simulations would, however, not alter our present conclusions.

- The probability to (correctly) select model (3.1) for a positive R^2 and a nonnormal X was much higher when using tests based on skewness than when using tests based on kurtosis. Whereas such probabilities were already fairly large with the former for $\gamma_X > 0$ and $n = 50$ (see Fig. 3.1), they were still close to zero with the latter for $\kappa_X > 0$ and $n = 100$ (See Fig. 3.2). It is only with $n = 500$ that these probabilities became large (see Fig. 3.3). Thus, detecting the direction of a regression line is much easier when the predictor is skewed (such as a chi-square distribution) than when the predictor is symmetric with an excess of kurtosis (such as a Student's t-distribution).
- Less importantly, we noted that the probability to conclude model (3.1) when using a testing procedure based on skewness was not a monotone function of neither R^2 nor γ_X, since this probability was increasing at small values of $R^2 > 0$ and $\gamma_X > 0$ while decreasing at higher values (see Fig. 3.1). This latter counterintuitive result was due to the fact that a Spearman correlation tends to get lower with increasing skewness (no such decreasing trend at high values of γ_X would be observed using a Pearson correlation to assess Condition 1, which would, however, not be robust against outliers). We did not observe such decreasing trends at high values of κ_X when using a testing procedure based on kurtosis (see Fig. 3.3).

Our main conclusion from these simulations is thus the following. If one is ready to assume a linear model between X and Y, either (3.1) or (3.2), and if the goal is to conclude either (3.1) or (3.2), Procedure B can be used instead (as a simplification) of Procedure A. One can write the null hypothesis as:

$$H_0: \text{ either } \rho_{XY} = 0 \text{ or both (3.1) and (3.2) hold}$$

which is the same as:

$$H_0: \text{ either } \rho_{XY} = 0 \text{ or both } X \text{ and } Y \text{ are normal}$$

or even as:

$$H_0: \text{ at least one of } X \text{ or } Y \text{ is normal.}$$

Referring to the five situations described in Section 3.2, one can also equivalently write:

$$H_0\text{: Situation 1, 2, or 3 applies.}$$

One can then proceed as follows:

(1) Reject H_0 and conclude Situation 4 (model (3.1) with $\rho_{XY} \neq 0$) if the correlation ρ_{XY} is significantly different from zero (either positive or negative) and if the difference in absolute skewness/kurtosis Δ is significantly larger than zero.

(2) Reject H_0 and conclude Situation 5 (model (3.2) with $\rho_{XY} \neq 0$) if the correlation ρ_{XY} is significantly different from zero (either positive or negative) and if the difference in absolute skewness/kurtosis Δ is significantly smaller than zero.

(3) Do not reject H_0 (and do not conclude anything) if either ρ_{XY} or Δ is not significantly different from zero.

In this procedure, one can use $\Delta = \Delta_\gamma = |\gamma_X| - |\gamma_Y|$ if one suspects a skewed predictor and one can use $\Delta = \Delta_\kappa = |\kappa_X| - |\kappa_Y|$ if one suspects a symmetric predictor with an excess of kurtosis. In either case, the Type I error is (as usual) the probability to reject H_0 when H_0 is true. As seen in our simulations, it is controlled at α if the tests on the nullity of ρ_{XY} and on the nullity of Δ are both conducted at the significance level α, while the probability to conclude the wrong model, that is, to conclude (3.1) when only (3.2) holds or to conclude (3.2) when only (3.1) holds, is also controlled. Note also that in the case where H_0 is not rejected, we do not have a statistical proof that there is no relationship ($\rho_{XY} = 0$) or that the direction of the regression line is undetermined (both (3.1) and (3.2) hold) since, as is well known, an absence of evidence does not imply an evidence of absence (e.g. Altman and Bland, 1995). This is also the reason why we have included these two cases (the case of no relationship, i.e., when neither X nor Y is the predictor, and the case of an undetermined direction of the regression line) in the null hypothesis, whereas Pornprasertmanit and Little (2012) only considered the latter when defining their Type I error. Note, finally, that since both X and Y are nonnormal if H_0 is wrong, we have to assume that both X and Y are nonnormal to have a chance to consistently detect the direction of a regression line.

When considering a statistical model, it may be useful to distinguish between the "assumptions," which are made (defining the model), and the "hypotheses," which are tested for (within the model). In our case, the assumptions are that there exists a linear model between X and Y, either (3.1) or (3.2). In this setting, one can formally test the null hypothesis H_0 as described earlier. Of course, a testing procedure is valid only (or mainly) under the assumptions under which it has been developed, and it may be useful or necessary to check these assumptions. For example, it is certainly useful to check in our case that the residuals are close to be normally distributed. In addition, one can also check Condition 3 of Pornprasertmanit and Little (2012), but this is only one among many possible checks. However, such checking is usually part of a goodness of fit procedure, not of an hypothesis testing procedure. This is also in this spirit that we have simplified the testing procedure of Pornprasertmanit and Little (2012).

3.4 LURKING VARIABLES AND CAUSALITY

When using the testing procedure described in the previous section, it is tempting to conclude that X is a cause of Y if we can conclude model (3.1), and it is tempting to conclude that Y is a cause of X if we can conclude model (3.2). Of course, there might exist a lurking (confounding) variable Z, which is a cause of both X and Y, such that we may have no association anymore between X and Y once conditioned on Z. In what follows, we argue that under the assumption of a linear model between X and Y, it is actually not very likely that such a lurking variable exists if we can conclude (3.1) or (3.2), at least when we consider the whole population (and not an artificial subpopulation, as discussed in Subsection 3.4.3 below).

3.4.1 Two Independent Predictors

Let us consider two continuous variables X and Y. A special case of lurking variable is a variable Z, which is independent of X and which is correlated with Y. Without loss of generality, we assume that X, Y, and Z have been standardized to have zero expectation and unit variance. Assume that we have:

$$Y = cX + c'Z + \epsilon^*$$

where $c \neq 0$ and $c' \neq 0$, where X and Z are two independent predictors (or even causes) of Y, and where ϵ^* is normally distributed and independent of both X and Z. This situation was considered by Pornprasertmanit and Little (2012). They argued that if X is normally distributed and Z is skewed, Y will be skewed as well, although less than Z but more than X, such that in a model involving only X and Y, one would select model (3.2) rather than model (3.1) using the testing procedure above, and one would thus wrongly conclude that Y is a cause of X (instead of the other way around). Recall, however, that the null hypothesis includes the case of a normal X, such that the probability to wrongly select model (3.2) would then be controlled. Even if X were slightly skewed, note that the residual term in (3.1) would be $\epsilon = c'Z + \epsilon^*$, which is skewed if Z is skewed and $c' \neq 0$, such that (3.1) would not hold. On the other hand, one has:

$$X = cY + (1 - c^2)X - cc'Z - c\epsilon^*$$

such that the residual term in (3.2) would be $\epsilon' = (1 - c^2)X - cc'Z - c\epsilon^*$. We argue that it is unlikely that such a combination of skewed X and Z yields a residual term ϵ', which is normally distributed. This means that neither (3.1) nor (3.2) would then hold, such that the testing procedure should in principle not be applied and there would be no such wrong conclusion that Y is a cause of X.

3.4.2 Confounding Variable

Alternatively, assume that we can write:

$$X = dZ + \delta$$

and:

$$Y = d'Z + \delta'$$

where $d \neq 0$, $d' \neq 0$, and where δ and δ' are not correlated with Z while being independent of each other, such that X and Y are conditionally independent given Z, the observed correlation between X and Y being due to the confounding variable Z. Without loss of generality, we further consider that the three variables have been standardized to have zero expectation and unit variance, such that d and d' are correlations between X and Z, respectively between Y and Z. One has in that case $Z = (X - \delta)/d$ and one can write:

$$Y = \left(\frac{d'}{d}\right)X + \delta' - \left(\frac{d'}{d}\right)\delta.$$

Note, however, that X is correlated with $\delta^* = \delta' - (d'/d)\delta$ since:

$$\text{Cov}(X, \delta^*) = \text{Cov}\left(dZ + \delta, \delta' - \left(\frac{d'}{d}\right)\delta\right) = -\left(\frac{d'}{d}\right)Var(\delta).$$

Thus, if model (3.1) is fitted using least squares, the slope will not be an estimate of d'/d but of:

$$b = \text{Cov}(X, Y) = \text{Cov}(dZ + \delta, d'Z + \delta') = dd'.$$

This means that the residual term ϵ in model (3.1) is in fact given by:

$$\epsilon = X\left(\frac{d'}{d} - dd'\right) + \delta' - \left(\frac{d'}{d}\right)\delta.$$

Since we exclude the case of a perfect correlation between X and Y, we have $(d'/d - dd') \neq 0$, and it appears unlikely to obtain a normal distribution for ϵ if X is nonnormal. By symmetry, the distribution of the residual term ϵ' in model (3.2) is given by:

$$\epsilon' = Y\left(\frac{d}{d'} - d'd\right) + \delta - \left(\frac{d}{d'}\right)\delta'$$

and it appears unlikely to obtain a normal distribution for ϵ' if Y is nonnormal. Putting this together, this means that neither (3.1) nor (3.2) would hold if both X and Y are nonnormal in the presence of a confounder Z. Thus, the testing procedure should in principle not be performed in those cases (whereas the case where either X or Y is normal is included in our null hypothesis).

3.4.3 Selection of a Subpopulation

It is, however, possible to construct an example with a lurking variable Z where either model (3.1) or model (3.2) holds and where X and Y are both nonnormal, by considering a nonnormal subpopulation from a normal population. Consider three continuous variables $\tilde{X}$, $\tilde{Y}$, and $\tilde{Z}$ with zero expectation, which follow a trivariate normal distribution on some population, where $\tilde{Z}$ is a cause of both $\tilde{X}$ and $\tilde{Y}$, where $\tilde{X}$ and $\tilde{Y}$ are conditionally independent given $\tilde{Z}$, where $\rho_{\tilde{X}\tilde{Z}} \neq 0$, $\rho_{\tilde{Y}\tilde{Z}} \neq 0$, and $\rho_{\tilde{X}\tilde{Y}} =$

$\rho_{\tilde{X}\tilde{Z}} \cdot \rho_{\tilde{Y}\tilde{Z}}$. This is a case where both (3.1) and (3.2) hold, which is included in our null hypothesis H_0 and which is not yet problematic. In particular, one has:

$$\tilde{Y} = \tilde{b}\tilde{X} + \tilde{\epsilon}$$

with $\tilde{b} = \rho_{\tilde{X}\tilde{Y}} = \rho_{\tilde{X}\tilde{Z}} \cdot \rho_{\tilde{Y}\tilde{Z}} \neq 0$, and where $\tilde{\epsilon}$ is normally distributed and independent of $\tilde{X}$. Let us consider now a subpopulation including only those individuals with some particular values of $\tilde{X}$ (e.g., the subpopulation of those individuals with $\tilde{X} < 0$), and let us redefine $\tilde{X}$, $\tilde{Y}$, and $\tilde{Z}$ on this subpopulation, obtaining X, Y, and Z. This is a case where X has a nonnormal distribution (e.g., a folded normal distribution, which is a skewed one), also implying the nonnormality of Y and Z, while the correlations among the three variables may still be nonzero. This is also a case where (3.1) holds (while (3.2) does not hold). This follows from the fact that if we have a linear relationship between $\tilde{X}$ and $\tilde{Y}$, we shall still have a linear relationship between the two variables if we restrict our attention to a particular domain of the values of $\tilde{X}$ (e.g., the domain where $\tilde{X} < 0$). Thus, we shall be able to conclude model (3.1) using our testing procedure, at least if the sample size is large enough, although X is not a cause of Y (as long as $\tilde{X}$ is not a cause of $\tilde{Y}$)! Using the notation from the previous subsection, this example shows that it is actually possible to get a normal $\epsilon = X(d'/d - dd') + \delta' - (d'/d)\delta$, while $(d'/d - dd') \neq 0$ and X is nonnormal. Although this is admittedly an artificial example (with an artificial selection of a subpopulation rather than a natural one), it shows that one should remain cautious when trying to make causality statements. From an applied perspective, this also underlines the importance of representatively sampling from the entire (a priori defined) population.

3.5 BRAIN AND BODY DATA REVISITED

Dodge and Rousson (2000) applied the strategy described in Section 3.2 with $r = 3$ (i.e., using the skewness coefficient) on a data set published by Crile and Quiring (1940), who recorded the weight of the brain and the weight of the body of 499 species, among which 91 birds, 56 carnivores, 56 fishes, 56 primates, 40 anthropoids, 28 reptiles, 55 rodents, 60 ungulates, and 57 odd-toed ungulates. The data are plotted in Figure 3.4 on the log scale to achieve a near linear relationship between the log brain X and the log body Y. The underlying scientific question was whether it is the brain that drives the body or whether it is the body that drives the brain in the evolution process of these species. Dodge and Rousson (2000) found that the absolute skewness of the log body was higher than the absolute skewness of the log brain for fishes, reptiles, and rodents, suggesting that the log body is the predictor of the log brain for these species (the body is driving the brain). For the other six groups of species (birds, carnivores, primates, anthropoids, ungulates, and odd-toed ungulates), the absolute skewness of the log brain was higher than the absolute skewness of the log body, suggesting that the log brain is the predictor of the log body (the brain is driving the body). However, no significance test had been applied.

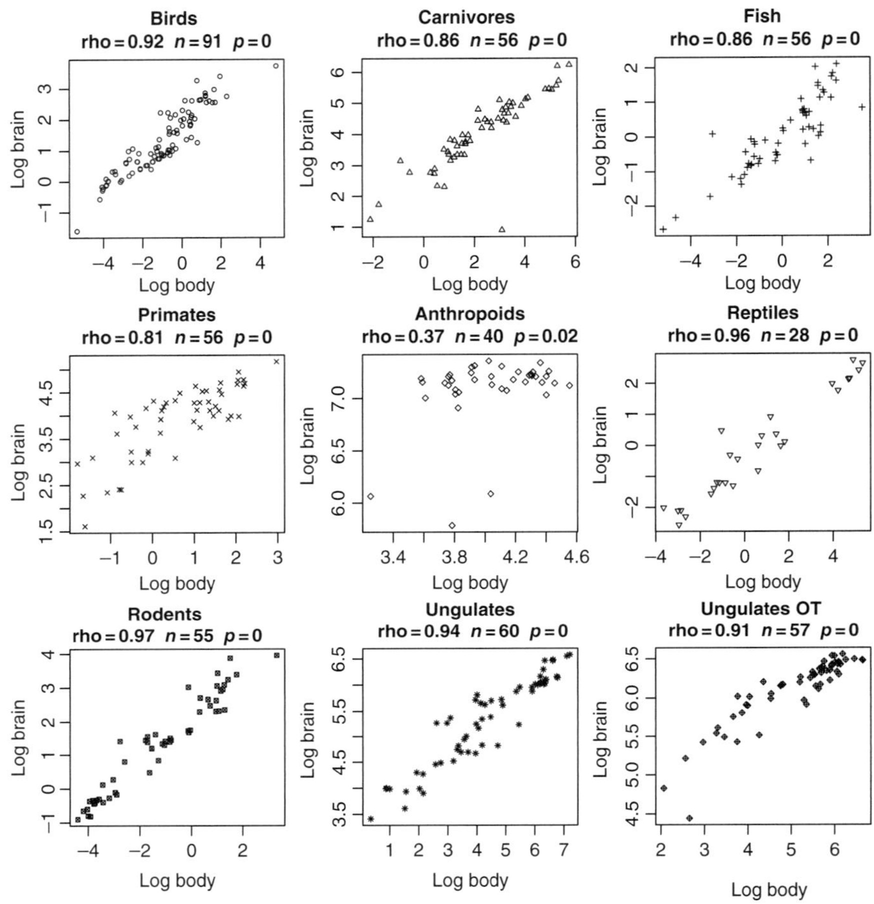

Figure 3.4 Log body versus log brain for nine groups of species collected by Crile and Quiring (1940), together with Spearman's rho correlation, sample size *n*, and *p*-value associated to Spearman's test.

In this section, we revisit this analysis using the testing procedure described in Section 3.3. We used a 5% significance level throughout. From Figure 3.4, we see that the log brain and the log body were pretty highly correlated, with Spearman correlations ranging from 0.81 to 0.97, except for anthropoids, where it was 0.37. The correlation was significant for each group of species, such that the first condition in the testing procedure was satisfied. On the other hand, the difference in absolute skewness was significant for carnivores, fishes, primates, anthropoids, and odd-toed ungulates, as seen in the top left panel of Figure 3.5. Note that to gain statistical power, we calculated here confidence intervals for the difference in absolute skewness using a BCa bootstrap, as implemented in the online Appendix from Pornprasertmanit and Little (2012). This suggests that the brain is significantly driving the body for carnivores,

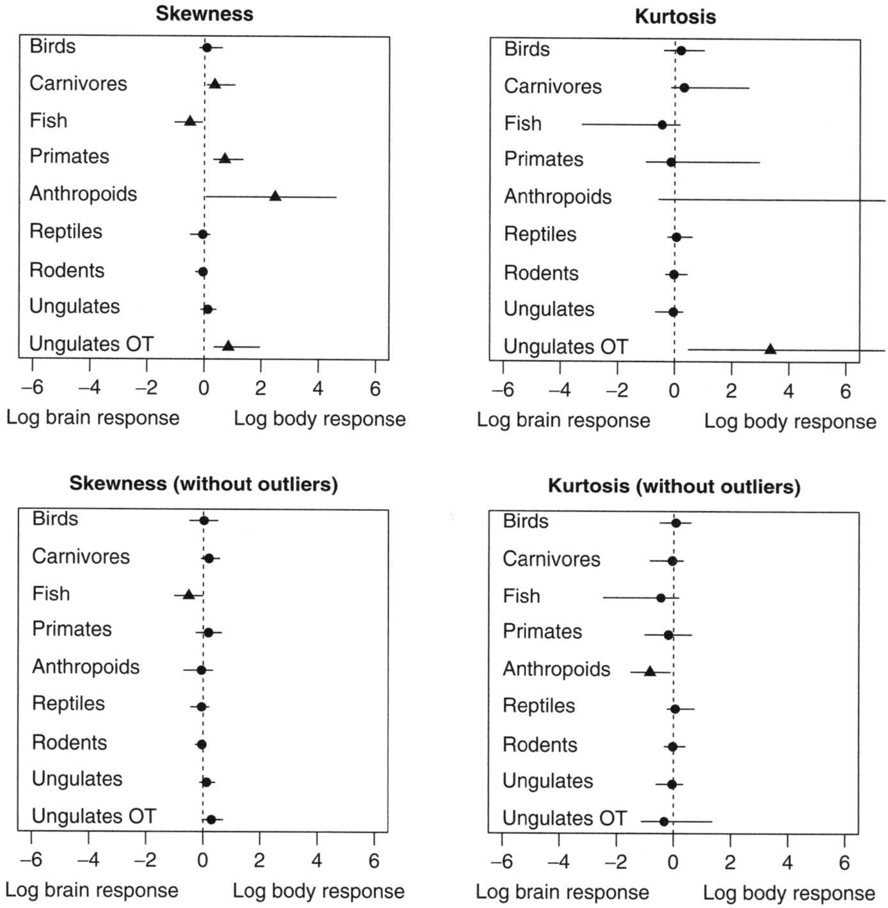

Figure 3.5 BCa bootstrap confidence intervals at the 95% level for the difference in absolute skewness (left panels), respectively, in absolute excess kurtosis (right panels), for nine groups of species, when calculated using all observations (top panels) and after removing 15 outliers (bottom panels). Estimated differences are plotted with a triangle when significant, with a circle when not significant.

primates, anthropoids, and odd-toed ungulates, whereas the body is significantly driving the brain for fishes. Note that these results nicely agree with some conclusions presented by Jerison (1973) using another approach. We have also repeated the testing procedure using kurtosis instead of skewness, where we could find only one single significant result for odd-toed ungulates (see the top right panel of Fig. 3.5).

Looking at Figure 3.4, however, we see that there are some potential outliers, especially for anthropoids where three species had a very light brain. In any statistical analysis, we do not like when our conclusions depend only on a few outlying observations. We have thus repeated the analysis without some outliers. To detect formally outliers, we used the "boxplot rule," removing those observations that were beyond the third quartile plus 1.5 time the interquartile range, or below the first quartile minus 1.5 times the interquartile range, repeating the process until no more outliers could be found, separately for log brain and for log body. We detected 15 such outliers (2 birds, 1 carnivore, 5 primates, 4 anthropoids, and 3 odd-toed ungulates). Without the outliers, Spearman correlations still ranged from 0.76 to 0.97 for eight groups of species while the correlation dropped to 0.2 for anthropoids, which was no longer significant. Confidence intervals for the difference in absolute skewness and excess kurtosis are shown in the bottom left and bottom right panels of Figure 3.5. We see that most of our significant results were lost after removing the outliers. The absolute excess kurtosis for anthropoids was significantly higher for log brain than for log body, suggesting that the body is driving the brain, but we were not allowed to draw here such a conclusion since the correlation was not significant. Thus, the only remaining significant result was for fishes, for which we had not detected any outliers, while even this single significant result would be lost if we would remove, for example, the two fishes with the lightest brain.

This example illustrates that it is not an easy task to detect a significant direction of a regression line if we do not want this result to be affected by a couple of influential observations. Moreover, removing outliers is in a sense in contradiction with our quest of nonnormality, which is necessary for the procedure to work. If the distributions are too close to a normal distribution, the procedure will lack statistical power. If there are far enough from normality, this means that we may have some influential (outlying) observations, which once removed take us back toward normality. There is indeed not much room left to get an approach that is both robust and powerful to detect the direction of a regression line. Of course, this issue is still amplified in the case of a small sample size, as we had in our example, where significance tests exhibit a low power and where the outliers are particularly influential.

3.6 CONCLUSIONS

We have reconsidered in this paper the problematic of the detection of the direction of a regression line as originally proposed by Dodge and Rousson (2000, 2001). In particular, we have suggested a simplification of the testing procedure of Pornprasertmanit and Little (2012) and a reformulation of the Type I error and the null hypothesis. We have seen that the testing procedure based on skewness is more powerful than the testing procedure based on kurtosis. We have also argued that, assuming that a linear model exists with nonnormal variables and normal residuals, the existence of a lurking variable is not very likely (unless one selects an artificial nonnormal subpopulation from a normal population). Thus, while testing for the direction of a regression line may indicate the direction of the relationship, assuming a linear model (with nonnormal variables and with normal residuals) protects us to some extent against the existence of a lurking variable. Taken together, we might be pretty close to a causality statement. This might be, however, more a theoretical result than a practical one since checking the assumption of a linear model, in particular the normality of residuals, is a noticeably delicate task. In the case of a binary confounder, for example, the distribution of the residuals might be a mixture of two normal distributions, which is known to be difficult to distinguish statistically from a normal distribution (see, e.g., Everitt, 1981). Finally, we have illustrated the fact that a few outlying observations may have a high influence on the testing procedure and that removing these observations may bring us back toward the normality, which we wish to avoid. At the end, proving statistically the direction of a regression line, let alone causality, certainly remains a major challenge.

REFERENCES

Altman, D.G. and Bland, J.M. (1995) Absence of evidence is not evidence of absence. *BMJ*, **311** (7003), 485.

Crile, G. and Quiring, D.P. (1940) A record of the body weight and certain organ and gland weights of 3690 animals. *Ohio Journal of Science*, **40**, 219–259.

Dodge, Y. and Rousson, V. (2000) Direction dependence in a regression line. *Communications in Statistics: Theory and Methods*, **29** (9–10), 1957–1972.

Dodge, Y. and Rousson, V. (2001) On asymmetric properties of the correlation coefficient in the regression setting. *The American Statistician*, **55** (1), 51–54.

Dodge, Y. and Yadegari, I. (2010) On direction of dependence. *Metrika*, **72** (1), 139–150.

Everitt, B. (1981) A Monte Carlo investigation of the likelihood ratio test for the number of components in a mixture of normal distributions. *Multivariate Behavioral Research*, **16** (2), 171–180.

von Eye, A. and DeShon, R.P. (2012a) Decisions concerning directional dependence. *International Journal of Behavioral Development*, **36** (4), 323–326.

von Eye, A. and DeShon, R.P. (2012b) Directional dependence in developmental research. *International Journal of Behavioral Development*, **36** (4), 303–312.

Jerison, H. (1973) *Evolution of the Brain and Intelligence*, Academic Press, New York.

Kendall, M. and Stuart, A. (1963) *The Advanced Theory of Statistics*, 2nd edn, Hafner, New York.

Kimball, A. (1957) Errors of the third kind in statistical consulting. *Journal of the American Statistical Association*, **52** (278), 133–142.

Leventhal, L. and Huynh, C.L. (1996) Directional decisions for two-tailed tests: power, error rates, and sample size. *Psychological Methods*, **1** (3), 278–292.

Muddapur, M. (2003) On directional dependence in a regression line. *Communications in Statistics: Theory and Methods*, **32** (10), 2053–2057.

Pornprasertmanit, S. and Little, T.D. (2012) Determining directional dependency in causal associations. *International Journal of Behavioral Development*, **36** (4), 313–322.

Sungur, E.A. (2005) A note on directional dependence in regression setting. *Communications in Statistics: Theory and Methods*, **34** (9–10), 1957–1965.

Wiedermann, W., Hagmann, M., and Eye, A. (2014) Significance tests to determine the direction of effects in linear regression models. *British Journal of Mathematical and Statistical Psychology*, **68** (1), 116–141.

4

DIRECTIONALITY OF EFFECTS IN CAUSAL MEDIATION ANALYSIS

WOLFGANG WIEDERMANN

Department of Educational, School & Counseling Psychology, College of Education, University of Missouri, Columbia, MO, USA

ALEXANDER VON EYE

Department of Psychology, Michigan State University, East Lansing, MI, United States

4.1 INTRODUCTION

Path modeling (Wright, 1921; Blalock, 1964; Duncan, 1975) constitutes one of the cornerstones of modern data analysis and enables researchers to estimate hypothesized structural relations of variables and decompose total effects of hypothesized causal relations into direct and indirect components. For the trivariate data setting, this effect decomposition is commonly known as mediation analysis (e.g. MacKinnon, 2008; Baron and Kenny, 1986). One variable is assumed to be the cause of variation (variable X), another variable is assumed to be the outcome (variable Y), and a third intervening variable (the mediator M) is assumed to mediate the predictor–outcome relation. An early discussion of mediation in the field of psychology was given by, for example, Lazarsfeld (1955); examples of mediation processes in various fields including, for example, psychology, sociology, communication research, agriculture, and epidemiology, are given by MacKinnon (2008).

A conceptual graph for a simple mediation model is given in the left panel of Figure 4.1. A predictor variable X is assumed to cause a mediator variable (M), which, in turn, causes an outcome (Y). The path from X to Y labeled with b'_{YX} is of particular

---

Statistics and Causality: Methods for Applied Empirical Research, First Edition.
Wolfgang Wiedermann and Alexander von Eye.

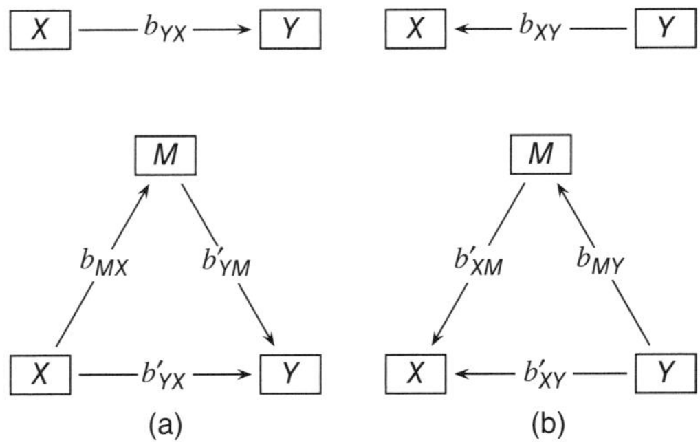

Figure 4.1 Path diagrams for two mediation models with reversed causal directions.

importance because it carries the necessary information to decide whether a full mediation or a partial mediation model is more likely to reflect the data-generating process. A full mediation process is empirically confirmed when the predictor–outcome path becomes zero while the indirect paths (from X to Y via M) statistically exist.

The majority of research on mediation analysis discusses mediation models from a purely statistical perspective. Several approaches have been proposed for the evaluation of mediational hypotheses. These approaches typically focus on statistical inference concerning the indirect, direct, and total effects of mediation models, and a number of studies have been performed to identify the optimal procedure for testing mediational hypotheses (Fritz and MacKinnon, 2007; Hayes and Scharkow, 2013; Judd and Kenny, 2010; Kenny and Judd, 2014; MacKinnon *et al.*, 2002; O'Rourke and MacKinnon, 2015). In empirical sciences, mediation analysis is dominated by linear regression models where direct and indirect effects are defined as linear regression coefficients or sums and products thereof. Because estimated regression coefficients and correlations do not allow causal statements per se, researchers in the empirical sciences are typically advised to avoid causal interpretations of empirically observed results (MacKinnon, 2008; Wang and Sobel, 2013). Overwhelming consensus exists that observed predictor–outcome relations can unambiguously be interpreted as causal if the predictor is under experimental control. Because failing to additionally randomize the mediator variable may seriously hamper causal interpretation (Judd and Kenny, 2010; Spencer *et al.*, 2005), blockage and enhancement designs have been discussed, which enable researchers to additionally manipulate the mediator (MacKinnon, 2008; Imai *et al.*, 2011). However, even when both, the predictor and the mediator variable, are experimentally manipulated, causal claims may still be questionable because randomization of a mediator variable requires certain assumptions (e.g., among the set of potential mediators, the experimental manipulation only affects the target mediator, see Bullock *et al.* 2010). Because of all these difficulties, tremendous effort has been invested by various researchers to develop *causal mediation analysis* (Holland, 1988; Mayer *et al.*, 2015; Pearl, 2001; Imai *et al.*, 2010; Imai *et al.*, 2011; Robins, 1986; Robins and Greenland, 1992; Valeri and VanderWeele, 2013). Causal mediation analysis is based on a redefinition of direct, indirect, and total effects using the language of structural counterfactuals, that is, a general solution to the mediation problem is based on embedding hypothesized structural equations in the context of counterfactual theory. However, causal mediation analysis does not make any statement concerning the correctness of a hypothesized causal mechanism. The term "causal" solely refers to inference on mediation effects using the counterfactual (potential outcome) framework and assumes that the hypothesized model structure (i.e., the causal ordering of the three variables) is correct. This is of particular importance in applications of causal mediation analysis in *nonexperimental* (observational) data settings. Here, correctness of path directions must be established using a priori theoretical arguments, for example, directional substantial theories or temporality (Mayer *et al.*, 2015; Pearl, 2012; Valeri and VanderWeele, 2013).

However, in practical applications, alternative theories for explaining underlying mediation mechanisms may exist and identifying the correct mediation model among a set of theoretically plausible models is not always possible. This, of course, applies, in particular, in nonexperimental data settings but is, in fact, not restricted to observational studies. Consider, for example, a standard experimental design in which the predictor variable is randomized. In this case, alternative theories concerning the mediator–outcome relation may exist, which justifies the estimation of competing mediation models, for example, $X \rightarrow M \rightarrow Y$ versus $X \rightarrow Y \rightarrow M$. Unfortunately, estimated direct, indirect, and total effects of competing models do not allow one to identify the correct path specification because estimated parameters are based on functions of pairwise correlation coefficients. For example, in the linear standardized case, the effect for the predictor–mediator relation, which is part of the indirect effect, is simply the Pearson correlation coefficient (ρ_{XM}), which does not depend on variable order, that is, $\rho_{XM} = \rho_{MX}$. Similarly, the total effect of competing models (b_{YX} and b_{XY} in Fig. 4.1a and b) both equal the correlation, $\rho_{YX} = \rho_{XY}$, in the standardized case.

Many previous studies on causal mediation analysis discuss conditions for identifying causal effects with a focus on experimental data scenarios (of course, principles of causal mediation are not restricted to experimental studies). In addition, methods have been proposed for the evaluation of underlying assumptions. The more fundamental assumption of causal ordering (i.e., assuming correctness of path specifications in a mediation model), which is particularly important in observational data scenarios, received considerably less attention. This chapter aims at filling this gap by discussing methods for the evaluation of underlying directionality assumptions of a hypothesized mediation model with a focus on cross-sectional nonexperimental settings. The remainder of the chapter is structured as follows: First, we introduce the key elements of causal mediation analysis and sensitivity analyses, developed to quantify the robustness of causal mediation effects under potential assumption violations. Second, we review common approaches to address the issue of causal ordering of variables. Third, we propose an empirical approach to test hypothesized directions of effects, which, under certain conditions, enables researchers to decide which one of two competing mediation models (i.e., a target model and an alternative model in which the causal flow is reversed) is more likely to reflect the true data-generating process. Fourth, results of a Monte Carlo simulation experiment are presented, which demonstrate the adequacy of the proposed directionality tests. Fifth, an empirical example on the development of numerical cognition in children is given to illustrate the application of the proposed directionality tests.

4.2 ELEMENTS OF CAUSAL MEDIATION ANALYSIS

In the past decades, various authors have discussed potential outcomes in the context of structural models (see, e.g., Holland, 1988, Robins, 1986, Robins and Greenland, 1992). Similarly, Pearl (2001, 2009, 2012) discussed general definitions of direct, indirect, and total effects in mediation settings through redefining causal mediation effects using the language of structural counterfactuals. Suppose that the model given in Figure 4.1a represents the true underlying data-generating process. The predictor variable X is assumed to cause a mediator variable M, which, in turn, is assumed to cause the outcome variable Y. The (natural) *direct effect* (DE) of the predictor–outcome relation can then be described as the expected change in the outcome (Y), which is induced by changing the predictor (X) from value x to x' while keeping the mediator (M) constant at whatever value it would have assumed under $X = x$. Assuming that certain ignorability assumptions hold (which will be discussed next), this can be expressed as

$$DE(Y) = \sum_m [E(Y|x', m) - E(Y|x, m)]P(m|x) \tag{4.1}$$

where E is the expectation operator.[1] Further, the *indirect effect* (IE) is defined as the expected change in the outcome (Y) when holding the predictor (X) constant at the value $X = x$ while changing the mediator (M) to whatever value it would have assumed had the predictor been set to $X = x'$, that is,

$$IE(Y) = \sum_m E(Y|x, m)[P(m|x') - P(m|x)] \tag{4.2}$$

Equation (4.2) represents a nonparametric definition of the well-known product-of-coefficients term commonly used to define the indirect effect, where $E(Y|x, m)$ describes the mediator–outcome relation for any fixed value $X = x$ and $P(m|x') - P(m|x)$ captures the predictor–mediator relation by describing the impact of changing x to x' on the probability of M. Finally, the *total effect* naturally arises from effect decomposition. Although (4.1) and (4.2) provide general formulas for the direct and indirect effects, which are applicable to linear and any nonlinear systems, we focus on linear relations among variables. In this case, the *total effect* (TE) can be written as the sum of the direct and the indirect effects, that is,

$$TE(Y) = DE(Y) + IE(Y). \tag{4.3}$$

[1]Following Pearl's (2012) notation, we use the summation sign while implicitly assuming that integrals should be used when variables are continuous.

Assuming linearity of effects, the following three equations can be used to describe the relevant elements of a simple mediation model (throughout the chapter, we assume, without loss of generality, that model intercepts have been set to zero):

$$Y = b_{YX}X + \epsilon_{Y(X)} \tag{4.4}$$

$$M = b_{MX}X + \epsilon_{M(X)} \tag{4.5}$$

$$Y = b'_{YX}X + b'_{YM}M + \epsilon_{Y(XM)} \tag{4.6}$$

The b parameters describe the slope coefficients, and ϵ denotes the error terms. In each term, the first subscript denotes the response variable and the remaining subscripts denote predictor variables of the corresponding linear model. Following Baron and Kenny's (1986) stepwise approach, regression coefficients can be estimated via ordinary least squares while error terms are assumed to be normally distributed, homoscedastic, serially independent, independent of the corresponding predictors, and independent of each other. In the linear system, the conditional expectations, $E(Y|x, m)$ and $E(Y|x', m)$, used in (4.1) are given as

$$E(Y|x, m) = b'_{YX}x + b'_{YM}m \tag{4.7}$$

and

$$E(Y|x', m) = b'_{YX}x' + b'_{YM}m. \tag{4.8}$$

Thus, one obtains the following well-known expressions for the direct, indirect, and total effects:

$$\begin{aligned} DE(Y) &= \sum_m [(b'_{YX}x' + b'_{YM}m) - (b'_{YX}x + b'_{YM}m)]P(m|x) \\ &= \sum_m [b'_{YX}x' - b'_{YX}x]P(m|x) \\ &= b'_{YX}x' - b'_{YX}x \\ &= b'_{YX}(x' - x) \end{aligned} \tag{4.9}$$

$$\begin{aligned} IE(Y) &= \sum_m (b'_{YX}x + b'_{YM}m)[P(m|x') - P(m|x)] \\ &= b'_{YM}[E(M|x') - E(M|x)] \\ &= b'_{YM}(b_{MX}x' - b_{MX}x) \\ &= b'_{YM}b_{MX}(x' - x) \end{aligned} \tag{4.10}$$

$$\begin{aligned} TE(Y) &= b'_{YX}(x' - x) + b'_{YM}b_{MX}(x' - x) \\ &= (b'_{YX} + b'_{YM}b_{MX})(x' - x). \end{aligned} \tag{4.11}$$

As mentioned earlier, a set of ignorability assumptions must be fulfilled to identify the (natural) direct and indirect effects (given in Eqs (4.1) and (4.2)), that is, to endow the estimated effects in (4.9)–(4.11) with valid causal interpretations. Imai *et al.* (2010) argued that causal effects can be estimated consistently when the assumption of *sequential ignorability* holds. The sequential ignorability assumption consists of two parts: first, given a set of pretreatment covariates, that is, covariates that are not affected by the treatment, the treatment assignment is assumed to be ignorable (i.e., statistically independent of potential mediators and potential outcomes). In other words, the set of covariates is assumed to deconfound the treatment-{mediator, outcome} relation. Second, given the actual treatment status and pretreatment confounders, the mediator status is ignorable, that is, both treatment status and confounders are assumed to deconfound the mediator–outcome relation. Pearl (2014a) discusses somewhat weaker assumptions of identification. Here, it is assumed that (i) pretreatment covariates deconfound the mediator–outcome relation (while holding X constant), (ii) the covariate-specific effect of the treatment on the mediator is identifiable, and (iii) the joint covariate-specific effect of treatment and mediator on the outcome is identifiable—for further discussion, see Imai *et al.* (2014) and Pearl (2014b).

Because the ignorability assumptions—necessary to identify causal mediation processes—cannot be statistically tested using observed data, sensitivity analyses have been proposed to evaluate the robustness of the results against potential violations of the ignorability assumptions (Imai *et al.*, 2010, 2011). In essence, this sensitivity analysis identifies the degree of violation, which reverses one's initial conclusion by systematically varying the correlation between the error terms of the mediator and the outcome model ($cor(\epsilon_{M(X)}, \epsilon_{Y(XM)})$) and examining how corresponding effect estimates change.

A fundamental assumption that received less attention in the mainstream literature on causal mediation analysis is the assumption of correct causal ordering. This may largely be attributed to the fact that causal mediation analysis is predominantly discussed in contexts of randomized studies where questions of directionality are less obvious (or do not occur at all), whenever the independent variable is under experimental control. Of course, correctness of the path directions of a hypothesized mediation model always has to be established theoretically (e.g., Pearl, 2012, Valeri and VanderWeele, 2013), which can particularly challenging in observational cross-sectional data settings. In the following sections, we discuss the assumption of causal ordering and introduce statistical techniques to empirically test the direction of effect in mediation models, which may help establish directional statements even in scenarios where data are gathered nonexperimentally.

4.3 DIRECTIONALITY OF EFFECTS IN MEDIATION MODELS

Suppose that the model in Figure 4.1a constitutes the true data-generating process and further suppose that there exists a substantive theory that justifies that X is the true cause of variation, M is the true mediator, and Y is the true outcome. In practice, however, the true data-generating mechanism (i.e., the exact way how nature assigns values to the three constructs of interest) is not always known. From a purely statistical perspective, six alternative mediation models can be estimated given the variable triple (X, M, Y). Of course, the majority of possible mediation models may be excluded because of, for example, lack of substantive theory, logical inconsistency, or arguments of face validity. However, at least a second alternative theory may exist, which justifies the estimation of a competing mediation model.[2] Suppose that this second theory involves exchanging the roles of predictor and outcome while M is still treated as the mediator variable. In this case, the mediation model conforms to the path diagram given in Figure 4.1b, which is in line with the recommendations of, for example, Iacobucci *et al.* (2007), who state that of various competing mediation models at least $Y \rightarrow M \rightarrow X$ should be tested. The linear equations for this (in the current setup) misspecified mediation model can be written as

$$X = b_{XY}Y + \epsilon_{X(Y)} \tag{4.12}$$

$$M = b_{MY}Y + \epsilon_{M(Y)} \tag{4.13}$$

$$X = b'_{XY}Y + b'_{XM}M + \epsilon_{X(YM)} \tag{4.14}$$

In this misspecified model, the direct, indirect, and total effects are b'_{XY}, $b_{MY}b'_{XM}$, and $b_{XY} = b'_{XY} + b_{MY}b'_{XM}$. Unfortunately, the estimated effects do not contain any information that could be used to answer the question whether path directions are correctly specified. In other words, through estimating the relevant effects of the two competing mediation models given in Figure 4.1a and b, no additional information is gained that could be used to distinguish between the two models (see also Fiedler *et al.*, 2011, Wiedermann and von Eye, 2015b). In the linear setting, for example, when all variables are standardized prior to mediation analysis (i.e., all variables have zero means and unit variances), knowledge of the three pairwise Pearson correlation coefficients (ρ_{XY}, ρ_{XM}, and ρ_{YM}) is sufficient to estimate the difference in mediation effects of competing models without even estimating the models, because indirect effects of both models can be written as functions of pairwise correlations, and thus, the differences in indirect effect estimates can also be written as a function of ρ_{XY}, ρ_{XM}, and ρ_{YM} (Wiedermann and von Eye, 2015b). Still, the rather exploratory approach of testing a series of alternative mediation models to infer on the underlying data-generating mechanism is recommended by various authors (e.g. Iacobucci *et al.*, 2007; Hayes and Scharkow 2013; Gelfand *et al.*, 2009), and applied researchers tend to use the presence or absence of statistically significant effects for model selection (e.g., Bizer *et al.*, 2012; Coyle *et al.*, 2011; Greitemeyer and McLatchie, 2011; Guendelman *et al.*, 2011; Huang *et al.*, 2011; Langer *et al.*, 2014; Oishi *et al.*, 2011; Shrum *et al.*, 2011; Usborne and Taylor, 2010). Wiedermann and von Eye (2015b) showed that this strategy can be highly

[2]Note that this situation may also occur even when the predictor (X) is randomized, because correctness of directionality of the mediator–outcome path cannot unambiguously established.

problematic because, depending on the observed correlational pattern of ρ_{XY}, ρ_{XM}, and ρ_{YM}, researchers run the high risk to erroneously retain a misspecified model.

Alternatively, instrumental variable techniques (Smith, 1982) have been proposed to empirically evaluate the hypothesis of a reversed causal flow (e.g., $Y \rightarrow X$ instead of $X \rightarrow Y$). For example, to test reverse causation for the mediator–outcome relation, two additional (instrumental) variables (I_1 and I_2) are incorporated into the hypothesized mediation model. Here, I_1 must be known to be related to the mediator but unrelated to the outcome and I_2 must be known to be unrelated to the mediator but related to the outcome. These requested features of the instrumental variables can hamper the applicability of the approach in practice, because such variables can sometimes be hard to come by.

Longitudinal or sequential approaches (e.g. Maxwell *et al.*, 2011, Mitchell and Maxwell, 2013)—for example, measuring the predictor at time point t_1, the mediator at t_2, and the outcome at t_3—are also commonly applied to counteract ambiguities concerning the causal ordering. The argument of temporality is, from a philosophical perspective, deeply rooted in a Humean tradition of causality (Hume, 1777/1975) where the cause must precede the effect. Temporal order is, for example, also a key element of mediation as defined by the MacArthur Foundation Network Group (Kraemer *et al.*, 2008). However, from a philosophical perspective, mechanistic approaches of causation exist, which emphasize that two variables can only be causally related if an appropriate underlying physical mechanism exists (Ney, 2009; Williamson, 2011). Focusing on an underlying physical mechanism renders the element of temporality unimportant, because even if the time between the cause and effect is split into infinitesimally small segments, one has to answer the question of what physically happens between the occurrence of the cause and the onset of the effect. Hicks (1979) proposed considering time intervals (instead of splitting time into smaller units) as a potential solution of this problem. Because time intervals are allowed to vary in breadth, cause and effect are also allowed to be contemporaneous.

From a purely statistical perspective, temporal ordering does not guarantee a correct causal ordering for at least two reasons: first, spuriousness may cause certain correlational patterns over time (e.g. Link and Shrout, 1992) and second, a particular time point of measurement is not sufficient to ensure correctness of causal ordering. For example, even if X is measured at t_1, M is measured at t_2, and Y is measured at t_3, one cannot rule out that, for example, earlier occurrences of Y have in fact caused X (MacKinnon, 2008).

As mentioned earlier, the issue of directionality results from the symmetry property of the Pearson correlation (i.e., $\rho_{XY} = \rho_{YX}$). Thus, recently, some authors have invested effort into analyzing asymmetry properties of the Pearson correlation (Dodge and Rousson, 2000, 2001). These asymmetry properties refer to data situations in which the two variables, X and Y, are no longer exchangeable in their status as predictor and outcome (see also Dodge and Rousson, 2016, Muddapur, 2003, Sungur, 2005). Dodge and Rousson (2000, 2001) showed, for example, that the cube of the Pearson correlation coefficient can be expressed as the ratio of the skewness of the outcome (e.g., γ_Y) and the skewness of the predictor (γ_X), $\rho^3_{XY} = \gamma_Y / \gamma_X$, when (i) the underlying data-generating mechanism can be described by a linear model, (ii) the error term of the linear model is normally distributed and independent of the predictor, and (iii) the true predictor is asymmetrically distributed. This result naturally arises from the fact that, in this particular setup, the outcome variable results from the convolution of a normal (the error term) and a nonnormal variate (the predictor) and will, thus, always be closer to the normal distribution than the predictor. Methods of testing direction of dependence based on this asymmetric facet of the Pearson correlation are discussed in von Eye and DeShon (2012) and in the related chapter of Dodge and Rousson (2016). Shimizu *et al.* (2006a) further generalized the result of Dodge and Rousson (2000, 2001) to the multivariate case through assuming nonnormal error terms (see also the related chapter of Shimizu, 2016). Determining the direction of effects (i.e., identifying which variable is more likely to be on the outcome side) becomes possible through considering the higher order moment structure of observed data (as long as at most one variable follows a normal distribution).

Wiedermann and von Eye (2015c) extended Dodge and Rousson's (2000, 2001) direction dependence approach to multiple variable cases under the assumption of normally distributed true error terms. In essence, the authors make use of the fact that the error term of the misspecified model will always show larger deviations from the normal distribution, compared to the true error term when the true predictor variable is skewed. Thus, evaluating the third moment structure of error terms of competing models can inform researchers who aim at making statements about the correctly specified model (provided that it exists). This *residual-based direction dependence* approach has also been extended to mediation models (Wiedermann and von Eye, 2015b). While the approach of Wiedermann and von Eye (2015b) makes use of distributional properties of error terms, this chapter discusses another approach to infer on the directionality of a hypothesized mediation model. The current approach focuses on independence properties of the error terms and the corresponding predictor variables and makes use of the fact that the independence assumption of predictor and error term will systematically be violated in the misspecified mediation model when the true predictor variable is skewed.

4.4 TESTING DIRECTIONALITY USING INDEPENDENCE PROPERTIES OF COMPETING MEDIATION MODELS

In this section, we first introduce the key component of the proposed method, specifically, the *Darmois–Skitovich theorem*, which is subsequently used to infer on directional hypotheses of a priori defined mediation models. Then, we discuss independence properties of predictors and error terms of both the bivariate and the multiple linear regression models, which are part of the estimation process according to Baron and Kenny (1986). Finally, we discuss significance tests that can be used to empirically test the directionality of a mediation model, and, we present decision rules that can be applied to select between competing mediation models (i.e., a target model and an alternative model).

Various authors have studied consequences of stochastic independence of linear functions of random variables (e.g., Basu, 1951, Darmois, 1953, Kac, 1939,

Gnedenko, 1948, Laha, 1957, Skitovich, 1953). One of the most prominent related results was independently proposed by Darmois (1953) and Skitovich (1953) (known as the Darmois–Skitovich theorem). These authors showed that if there exist two linear functions,

$$U = \sum_{j=1}^{k} \alpha_j Z_j \text{ and } V = \sum_{j=1}^{k} \beta_j Z_j \tag{4.15}$$

where $\alpha_j \beta_j \neq 0$ ($j = 1, 2, \ldots, k$) such that U and V are stochastically independent, then each Z_j follows a normal distribution. From this theorem, it straightforwardly follows that when a nonnormal variable Z_j exists for which $\alpha_j \beta_j \neq 0$, then U and V are stochastically dependent. This fundamental (in)dependence property constitutes the key element to be used when comparing competing regression models and has also successfully been used to identify the causal structure of sets of variables in the context of causal learning algorithms (see, e.g., Entner and Hoyer, 2011, Shimizu *et al.*, 2006b, 2011, Shimizu, 2016, Zhang and Hyvärinen, 2010). In the following subsections, we show that consequences of the Darmois–Skitovich theorem can be used to make decisions upon competing mediation models.

4.4.1 Independence Properties of Bivariate Relations

Let model (4.4) be the true data-generating process for the bivariate *predictor–outcome relation* (i.e., the total effect) of a mediation model, and let Equation (4.12) be the total effect of the misspecified mediation model. Further, we suppose that the true predictor X is nonnormal (more precisely, we suppose that X is asymmetrically distributed). For the misspecified model, it is then easy to show that there exists a joint nonnormal random variable Z_j with $\alpha_j \beta_j \neq 0$. The error term of the misspecified model can be written as

$$\begin{aligned} \epsilon_{X(Y)} &= X - b_{XY} Y \\ &= X - b_{XY} \left(b_{YX} X + \epsilon_{Y(X)} \right) \\ &= \left(1 - \rho_{XY}^2 \right) X - b_{YX} \frac{\sigma_X^2}{\sigma_Y^2} \epsilon_{Y(X)} \end{aligned} \tag{4.16}$$

where $b_{YX} b_{XY} = \rho_{XY}^2$ and $b_{XY} = b_{YX}(\sigma_X^2/\sigma_Y^2)$. In other words, the error term of the misspecified model is a linear combination of the true (nonnormally distributed) predictor variable (X) and the (normally distributed) true error term ($\epsilon_{Y(X)}$). Note that both b_{YX} and ρ_{XY} are assumed to be nonzero. Thus, it follows that there exists a nonnormal variable, X, that satisfies $b_{YX}(1 - \rho_{XY}^2) \neq 0$ (we exclude the case of $|\rho_{XY}| = 1$ because of practical irrelevance). From this, it follows that Y and $\epsilon_{X(Y)}$ are stochastically dependent. When estimating the misspecified regression model, Y and $\epsilon_{X(Y)}$ will be uncorrelated because estimation uses ordinary least squares. However, uncorrelatedness implies independence only in the normal case (which is excluded here due to the assumption of a nonnormal true predictor variable). In general, independence holds when

$$E[f_1(\epsilon_{X(Y)})f_2(Y)] - E[f_1(\epsilon_{X(Y)})]E[f_2(Y)] = 0 \quad (4.17)$$

for *any* functions f_1 and f_2. Uncorrelatedness refers to the special case of zero-covariance of $\epsilon_{X(Y)}$ and Y, that is,

$$E[\epsilon_{X(Y)}Y] - E[\epsilon_{X(Y)}]E[Y] = 0 \quad (4.18)$$

Thus, independence implies uncorrelatedness, but uncorrelatedness does not necessarily imply stochastic independence. We now use the fact that the independence assumption will systematically be violated in the misspecified model whenever observed variables deviate from normality. This enables researchers to meaningfully distinguish between the two competing models $X \rightarrow Y$ and $Y \rightarrow X$, provided that one of the two models reflects the true data-generating process and that the independence assumption holds for the true model. For example, when the independence assumption holds for the model $X \rightarrow Y$ (i.e., assuming the absence of latent confounders, the error term and the predictor X are stochastically independent) and, at the same time, the independence assumption is violated in the alternative model $Y \rightarrow X$ (i.e., the error term and Y are stochastically nonindependent), $X \rightarrow Y$ is more likely to reflect the true underlying data-generating process.

Further, let (4.5) be the correctly specified bivariate model ($X \rightarrow M$), which describes the *predictor–mediator relation*, and let $X = b_{XM}M + \epsilon_{X(M)}$ constitute the competing bivariate model. In this case, the error term of the misspecified model can be written as

$$\epsilon_{X(M)} = \left(1 - \rho^2_{XM}\right) X - b_{MX}\frac{\sigma^2_X}{\sigma^2_M}\epsilon_{M(X)} \quad (4.19)$$

which implies that again the true predictor, X, possesses particular features (nonnormality and $b_{MX}(1 - \rho^2_{XM}) \neq 0$), which lead to stochastic nonindependence of M and $\epsilon_{X(M)}$. Assuming that the independence assumption holds for the true mediation model (i.e., assuming the absence of latent confounders), observed independence properties of both models can inform researchers about which of the two models is more likely to be correct.

Finally, suppose that the true *mediator–outcome relation* is characterized through $Y = b_{YM}M + \epsilon_{Y(M)}$, and $M = b_{MY}Y + \epsilon_{M(Y)}$ constitutes the misspecified bivariate model. Then, the error term of the latter model can be rewritten as

$$\epsilon_{M(Y)} = \left(1 - \rho^2_{YM}\right) M - b_{YM}\frac{\sigma^2_M}{\sigma^2_Y}\epsilon_{Y(M)} \quad (4.20)$$

which implies that the true outcome Y and $\epsilon_{M(Y)}$ will be dependent as long as M is nonnormal and $(1 - \rho^2_{YM})b_{YM} \neq 0$. Note that potential dependencies in this bivariate model (which is usually not part of Baron and Kenny's, 1986, stepwise approach) are expected to be weaker than in the other bivariate models, because the mediator is expected to be closer to the normal distribution than the predictor due to the normally distributed true error $\epsilon_{M(X)}$. This directly follows from Dodge and Rousson's (2000, 2001) result. In the present context, Dodge and Rousson's (2000, 2001) result can be written as $\rho^3_{XM} = \gamma_M/\gamma_X$ (provided that $\epsilon_{M(X)}$ is normal and independent of X) from which follows that the skewness of M, $\gamma_M = \rho^3_{XM}\gamma_X$, will always be smaller than the skewness of X (again excluding $|\rho_{XM}| = 1$). The magnitude of dependence of Y and $\epsilon_{M(Y)}$ depends on the degree of nonnormality of M. However, theoretically, decisions concerning the two models $M \rightarrow Y$ and $Y \rightarrow M$ are possible as long as $\gamma_M \neq 0$, and Y is more likely to be the outcome when the independence assumption holds for $M \rightarrow Y$ and, at the same time, is violated for $Y \rightarrow M$.

4.4.2 Independence Properties of the Multiple Variable Model

So far, we have focused on the three possible bivariate regression models, that is, each single path in the specified mediation model was treated as an isolated bivariate regression problem. In the following sections, we discuss (in)dependence properties of the multiple regression parts of competing mediation models. Again, let (4.5) and (4.6) constitute the true underlying mediation mechanism and let Equations (4.13) and (4.14) describe the misspecified mediation model. Then, the error term for the misspecified multiple regression model ($\epsilon_{X(YM)}$) can be expressed through

$$\begin{aligned}\epsilon_{X(YM)} &= X - (b'_{XY} + b'_{XM}b_{MY})Y - b'_{XM}\epsilon_{M(Y)} \\ &= X - (b'_{XY} + b'_{XM}b_{MY})[(b'_{YX} + b'_{YM}b_{MX})X + b'_{YM}\epsilon_{M(X)} + \epsilon_{Y(XM)}] \\ &\quad - b'_{XM}\epsilon_{M(Y)} \\ &= (1 - b_{XY}b_{YX})X - b'_{YM}b_{XY}\epsilon_{M(X)} - b_{XY}\epsilon_{Y(XM)} - b'_{XM}\epsilon_{M(Y)} \qquad (4.21)\end{aligned}$$

with $b_{YX} = b'_{YX} + b'_{YM}b_{MX}$ and $b_{XY} = b'_{XY} + b'_{XM}b_{MY}$. Thus, provided that the true predictor, X, is nonnormal, Y and $\epsilon_{X(YM)}$ will be stochastically dependent whenever $(1 - b_{XY}b_{YX})b_{YX} \neq 0$. In a fashion analogous to the bivariate cases discussed earlier, decisions concerning directionality can be made if the independence assumption is fulfilled in the true model. Given that X and $\epsilon_{Y(XM)}$ are stochastically independent and, at the same time, evidence is found that Y and $\epsilon_{X(YM)}$ are stochastically nonindependent, Y is more likely to reflect the true outcome and X is more likely to be the cause of variation.

4.4.3 Measuring and Testing Independence

The theoretical implications of the Darmois–Skitovich theorem for nonnormal variables, together with the fact that the error term of a misspecified regression model can be expressed as a function of the true predictor, lead to the conclusion that the independence assumption of the error term and the explanatory variable will systematically be violated whenever a nonnormal true predictor variable is erroneously used as the outcome. Thus, measures of independence and methods for statistical inference are required to test the independence assumption of regression models, which can subsequently be used to evaluate hypotheses of directionality of effects. A straightforward approach to evaluate the independence of two variables, say Z_1 and Z_2, relies on the fundamental relation $E[f_1(Z_1)f_2(Z_2)] - E[f_1(Z_1)]E[f_2(Z_2)] = 0$ for any functions f_1 and f_2, given independence. For example, correlation coefficients applied to single nonlinear functions for f_1 and using the identity function for f_2 (i.e., $f_2(Z_2) = Z_2$) can be informative for the decision as to whether the independence assumption is likely to be violated (Entner and Hoyer, 2011; Hyvärinen, 1998; Shimizu *et al.*, 2011; Shimizu, 2014). Because stochastic independence is only confirmed if $E[f_1(Z_1)f_2(Z_2)] - E[f_1(Z_1)]E[f_2(Z_2)] = 0$ for *any* functions f_1 and f_2, a simple correlation test based on a single nonlinear function is, of course, not an ultimate proof of stochastic independence of two random variates. Failing to reject the null hypothesis of independence for one nonlinear function does not imply that no other nonlinear function exists for which independence can be rejected. In other words, this computationally simple approach introduces a Type II error when making decisions concerning directionality (in addition to Type II errors due to small sample sizes and low power). Alternatively, computationally more complex approaches are discussed by Gretton *et al.* (2005, 2008), as well as Gretton and Györfi (2010). These approaches are based on the concept of the Hilbert–Schmidt Independence Criterion (HSIC). The HSIC approach overcomes these forms of Type II errors because *any* statistical dependence can be detected in the large sample limit. For an application of the HSIC to test consistency of causal effects when adjusting for covariates see Entner *et al.* (2012).

In this chapter, we, however, focus on the computationally simple nonlinear correlation approach which, in essence, relies on computing the well-known Pearson correlation coefficient and is, thus, readily available in virtually every general-purpose statistical software package. We will show that this simple approach is well suited to identify serious model violations. Wiedermann and von Eye (2015a) discuss the use of the square function $f_1(Z_1) = Z_1^2$ as one possible nonlinear function for the evaluation of independence of regression residuals and the corresponding model predictors. The authors suggested squaring the estimated regression residuals and using the identity function for the values of the corresponding predictor, that is, $cor(X, \epsilon_{YX}^2)$ and $cor(Y, \epsilon_{XY}^2)$, which essentially corresponds to performing the well-known Breusch–Pagan test (Breusch and Pagan, 1979) of homoscedasticity for the two competing regression models. For a skewed true predictor, X, and a normally distributed true error term, $\epsilon_{Y(X)}$ (assumed to be independent from each other), the covariance of the true outcome and the squared error term of the misspecified model, $cov(Y, \epsilon_{X(Y)}^2)$, can be expressed as a function of the skewness of the true predictor. For standardized variables, this function can be written as $cov(Y, \epsilon_{X(Y)}^2) = \rho_{XY}(1 - \rho_{XY}^2)^2\gamma_X$. In this chapter, we focus on the reverse case of using the square function and the identity function and evaluate the covariance of squared predictors and untransformed regression residuals, $cov(X^2, \epsilon_{Y(X)})$ and $cov(Y^2, \epsilon_{X(Y)})$. We present derivations for (i) the covariance terms for the bivariate model to estimate the total effect ($Y \rightarrow X$) and the model to estimate the predictor–mediator relation and (ii) the covariance term for the multiple regression model to estimate the direct and indirect effects ($(Y, M) \rightarrow X$). Here, we do not focus on the third bivariate regression ($M \rightarrow Y$ versus $Y \rightarrow M$) because this model is not part of Baron and Kenny's (1986) stepwise approach and, for reasons explained earlier, the power for this bivariate model is expected to be low.

Bivariate Predictor–Outcome Relation. Again, let Equation (4.4) be the true model while (4.12) describes the misspecified causal flow. To simplify derivations, further assume that all variables have been standardized. Note that prior standardization does not affect decisions concerning strength of relation and significance of the considered association. The covariance of the squared predictor and the error term of the misspecified model, $cov(Y^2, \epsilon_{X(Y)})$, can be expressed by

$$cov\left(Y^2, \epsilon_{X(Y)}\right) = \rho_{XY}^2 \left(1 - \rho_{XY}^2\right) \gamma_X, \tag{4.22}$$

with γ_X being the skewness of X (a proof is given in Appendix A). From Equation (4.22), we derive the following conclusions: (i) the covariance of Y^2 and $\epsilon_{X(Y)}$ can be interpreted as the weighted skewness of the true predictor, (ii) the covariance of Y^2 and $\epsilon_{X(Y)}$ and the skewness of X will always have the same sign (note that the weighting term in (4.22), $\rho_{XY}^2(1 - \rho_{XY}^2)$, will always be positive), (iii) the covariance of Y^2 and $\epsilon_{X(Y)}$ is an inversely U-shaped function of the correlation between X and Y, and most important, (iv) excluding zero and perfect linear correlations because of practical irrelevance, the covariance of Y^2 and $\epsilon_{X(Y)}$ increases with the skewness of X. In contrast, the covariance of Y^2 and $\epsilon_{X(Y)}$ will always be zero for symmetrically distributed variables (i.e., $\gamma_X = 0$). For the true model, independence of the predictor and the error term is assumed, which implies $cov(X^2, \epsilon_{Y(X)}) = 0$. Thus, the following decision rules can be used to decide between competing models:

(1) If the null hypothesis H_0: $cov(X^2, \epsilon_{Y(X)}) = 0$ is retained and H_0: $cov(Y^2, \epsilon_{X(Y)}) = 0$ is rejected, then X is more likely to be the cause and Y is more likely to be the outcome.
(2) If the null hypothesis H_0: $cov(X^2, \epsilon_{Y(X)}) = 0$ is rejected and H_0: $cov(Y^2, \epsilon_{X(Y)}) = 0$ is retained, then Y is more likely to be the cause and X is more likely to be on the outcome side.
(3) If both null hypotheses, H_0: $cov(X^2, \epsilon_{Y(X)}) = 0$ and H_0: $cov(Y^2, \epsilon_{X(Y)}) = 0$, are retained/rejected, no decision can be made.

Bivariate Predictor–Mediator Relation. In analogy to the decision rules for testing the predictor–outcome relation, similar statements can be derived for the bivariate relation between the predictor (X) and the mediator (M). Let Equation (4.5), that is, $X \rightarrow M$ be the true model. Focusing on the simple bivariate association of X and M, $M \rightarrow X$, that is, $X = b_{XM}M + \epsilon_{X(M)}$, constitutes the corresponding misspecified model. Again, assume that X and M have been standardized prior to analysis. Then the covariance of M^2 and $\epsilon_{X(M)}$ can be written as

$$cov\left(M^2, \epsilon_{X(M)}\right) = \rho_{XM}^2 \left(1 - \rho_{XM}^2\right) \gamma_X. \tag{4.23}$$

The corresponding proof works in a fashion analogous to the proof for the bivariate predictor–outcome relation presented in Appendix A. Again, for the true model, independence of predictor and error term is assumed, which implies $cov(X^2, \epsilon_{M(X)}) = 0$, and which leads to the following set of decision rules:

(1) If the null hypothesis H_0: $cov(X^2, \epsilon_{M(X)}) = 0$ is retained and H_0: $cov(M^2, \epsilon_{X(M)}) = 0$ is rejected, then X is more likely to be the cause and M is more likely to be on the outcome side.

(2) If the null hypothesis H_0: $cov(X^2, \epsilon_{M(X)}) = 0$ is rejected and H_0: $cov(M^2, \epsilon_{X(M)}) = 0$ is retained, then M is more likely to be the cause and X is more likely to be on the outcome side.

(3) If both null hypotheses, H_0: $cov(X^2, \epsilon_{M(X)}) = 0$ and H_0: $cov(M^2, \epsilon_{X(M)}) = 0$, are retained/rejected, no decision can be made.

Multiple Variable Relation. In the multiple variable case, we use Equations (4.5) and (4.6) to define the true data-generating process, while Equations (4.13) and (4.14) describe the corresponding misspecified model. Again, for the true model, we assume normally distributed error terms, which are independent from each other and independent of the corresponding predictor(s). Then, for standardized variables, the covariance of the squared values of Y and the error term of the misspecified model ($\epsilon_{X(YM)}$) can also be expressed as a weighted version of the skewness of X, that is,

$$cov(Y^2, \epsilon_{X(YM)}) = \rho^2_{XY} \left[(1 - \rho^2_{XY}) - \frac{(\rho_{XM} - \rho_{XY}\rho_{YM})^2}{(1 - \rho^2_{YM})} \right] \gamma_X, \tag{4.24}$$

which implies that the covariance increases systematically with the skewness of the true predictor (a proof is given in Appendix B). In the multiple variable setting, the weighting term becomes more complex. Thus, we performed a small simulation study to empirically demonstrate properties of Equation (4.24) (details of the simulation algorithm are given in the section "Simulating the performance of directionality tests"). One thousand observations were generated according to the mediation model given in Figure 4.1a. The slope parameters of the true mediation model varied from −0.6 to 0.6 in increments of 0.2, the intercepts were fixed at zero, and the skewness of the true predictor was $\gamma_X = 0.5$, 1.5, and 2.5. The three effects were fully crossed resulting in 7 (effect sizes for b_{MX}) × 7 (effect sizes for b'_{YX}) × 7 (effect sizes of b'_{YM}) × 3 (magnitude of the skewness γ_X) = 1029 experimental conditions. For each condition, the average covariance of Y^2 and $\epsilon_{X(YM)}$ was computed.

Figures 4.2–4.4 summarize the results of the simulation experiment for $\gamma_X = 0.5$, 1.5, and 2.5. The depicted data points refer to the average covariance estimates based on the 1000 samples, and the solid black lines correspond to the theoretical values based on Equation (4.24). As expected, the covariance increases with the skewness of X, which is in line with the theoretical findings given in Equation (4.24). The magnitude of covariation of Y^2 and $\epsilon_{X(YM)}$ depends on the magnitude and the sign of all three b-coefficients. In general, the covariance increases with positive direct

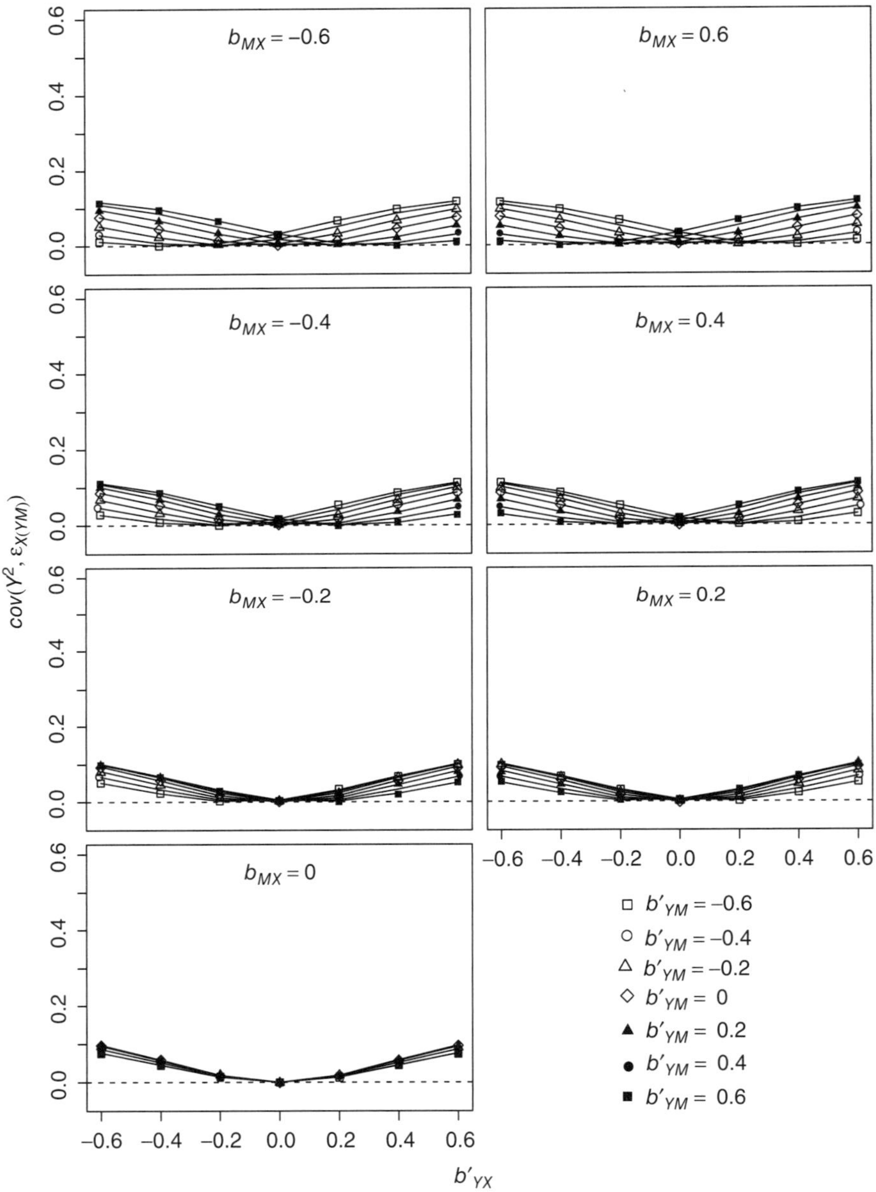

Figure 4.2 Theoretical (solid lines) and empirical values (data points) of $cov(Y^2, \epsilon_{X(YM)})$ for $\gamma_X = 0.5$ as a function of b_{MX}, b'_{YX}, and b'_{YM}.

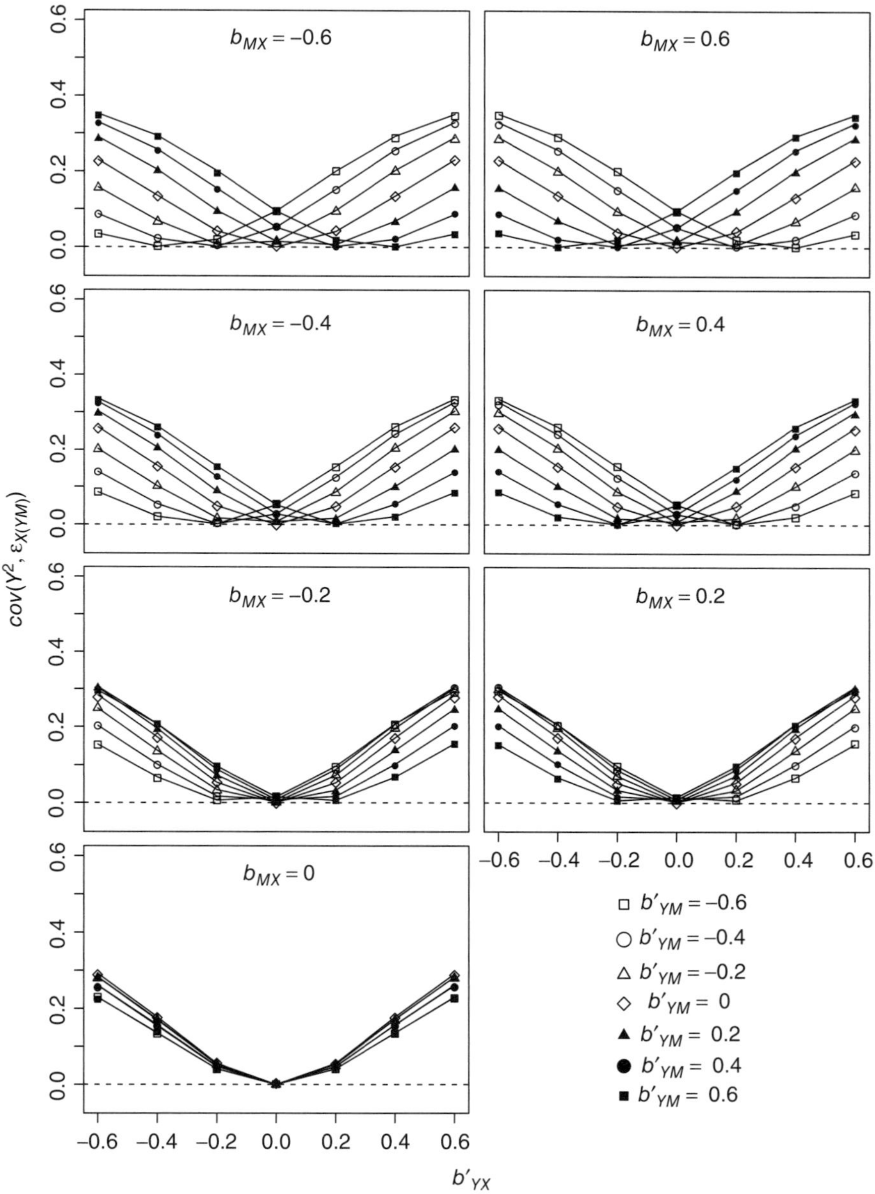

Figure 4.3 Theoretical (solid lines) and empirical values (data points) of $cov(Y^2, \epsilon_{X(YM)})$ for $\gamma_X = 1.5$ as a function of b_{MX}, b'_{YX}, and b'_{YM}.

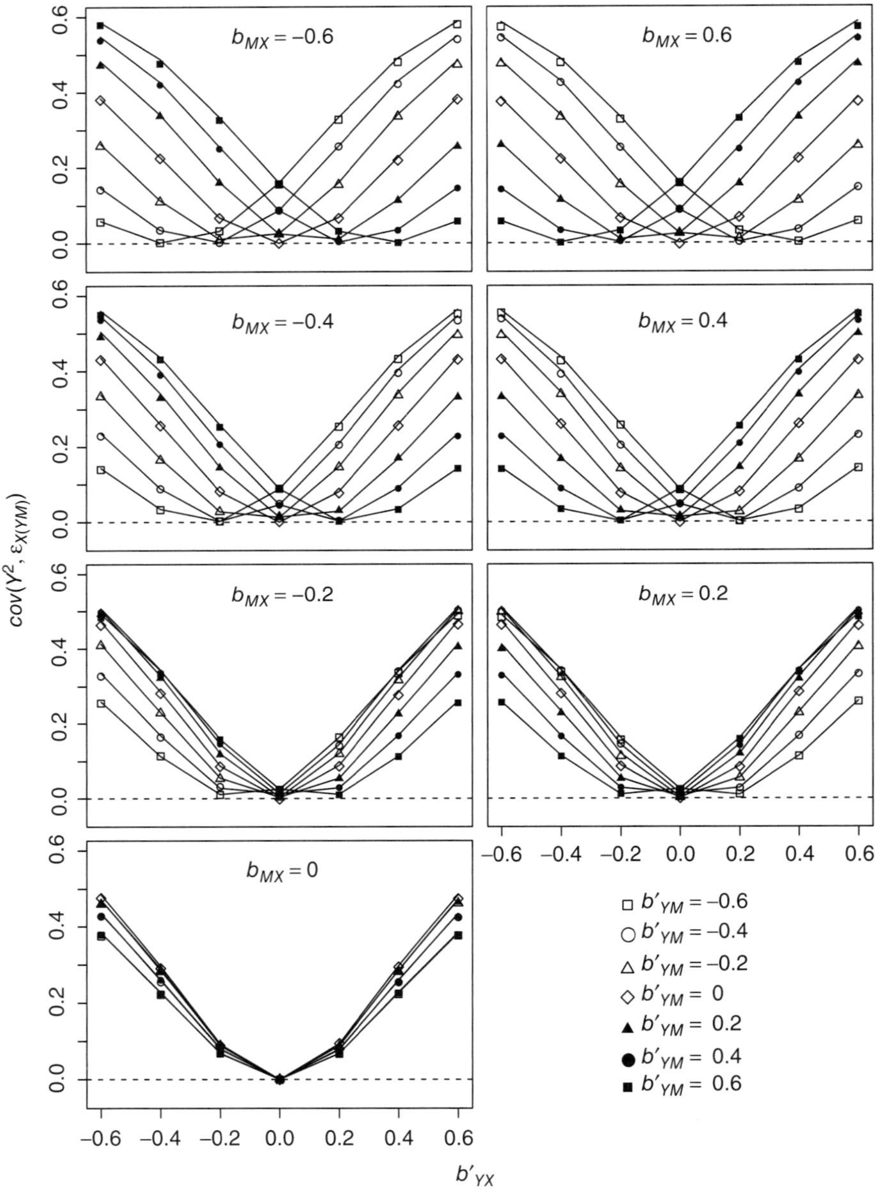

Figure 4.4 Theoretical (solid lines) and empirical values (data points) of $cov(Y^2, \epsilon_{X(YM)})$ for $\gamma_X = 2.5$ as a function of b_{MX}, b'_{YX}, and b'_{YM}.

effects ($b'_{YX} > 0$) together with positive indirect effects ($b_{MX}b'_{YM} > 0$) and decreases with positive direct effects together with negative indirect effects ($b_{MX}b'_{YM} < 0$). Conversely, the covariance increases with negative direct effects and negative indirect effects and decreases with negative direct effects together with positive indirect effects.

Based on the theoretical results and the empirical evidence, we formulate the following simple decision rules for the multiple variable case:

(1) If the null hypothesis H_0: $cov(X^2, \epsilon_{Y(XM)}) = 0$ is retained and H_0: $cov(Y^2, \epsilon_{X(YM)}) = 0$ is rejected, then X is more likely to be the cause and Y is more likely to be the outcome while adjusting for M.

(2) If the null hypothesis H_0: $cov(X^2, \epsilon_{Y(XM)}) = 0$ is rejected and H_0: $cov(Y^2, \epsilon_{X(YM)}) = 0$ is retained, then Y is more likely to be the cause and X is more likely to be on the outcome side while adjusting for M.

(3) If both null hypotheses, H_0: $cov(X^2, \epsilon_{Y(XM)}) = 0$ and H_0: $cov(Y^2, \epsilon_{X(YM)}) = 0$, are retained/rejected, no decision can be made.

Strictly following the stepwise approach proposed by Baron and Kenny (1986), evaluating the direction of effects for the predictor–outcome, the predictor–mediator, and the multiple variable relation would be necessary to empirically confirm the correctness of path specifications. However, assuming that M is known to be the mediator (e.g., based on logical arguments), testing the directionality of the multiple regression models would be sufficient to determine which mediation model is more likely to be correct. Because the selection of the necessary steps to test a tentative mediation model certainly has to be guided by prior knowledge, we will later evaluate the performance of all three steps (i.e., testing the multiple variable relation as well as the two bivariate relations) using Monte Carlo simulations to give a complete rendering of the accuracy of the proposed procedure.

Statistical Inference. In practical applications, decisions whether observed covariations of regression residuals and predictors are statistically different from zero are necessary for conclusions concerning the underlying mechanism, which may have generated the data. Here, various correlation tests can be used to evaluate the null hypothesis of zero covariation. In this chapter, we focus on three significance tests: (i) the ordinary Pearson correlation test, (ii) the Spearman rank correlation test, and (iii) a correlation test based on normal scores. However, instead of treating the three procedures as separate, unrelated methods for inference on the association of two random variates, we discuss the significance tests from a rank transformation perspective (Conover and Iman, 1981).

The most straightforward approach to evaluate the magnitude of correlation of squared predictors and the corresponding regression residuals is to perform the classic Pearson correlation significance test based on Student's t-distribution,

$$t = \rho\sqrt{\frac{N-2}{1-\rho^2}}, \tag{4.25}$$

with $df = N - 2$ degrees of freedom. However, the test statistic approximates a t-distribution in case of bivariate normality, which will be violated in the present context of testing, for example, the correlation between Y^2 and $\epsilon_{X(Y)}$. Although it is well known that the test is rather robust in the face of distributional violations, nonparametric alternatives may be superior (in terms of Type I error protection and power), in particular when sample sizes are small (see, for example, Fowler, 1987, Stuart, 1954).

For example, the Pearson correlation (ρ) in (4.25) might be replaced by the Spearman correlation coefficient (Zar, 1972). In the absence of ties, replacing Pearson's coefficient by Spearman's correlation is equivalent to performing a rank transformation in which continuous scores are replaced with the corresponding ranks (Conover and Iman, 1981). In other words, considering, for example, the two competing regression models for estimating the total effect: The variable Y^2 and the estimated regression residuals $\epsilon_{X(Y)}$ are converted to ranks, $R(Y^2)$ and $R(\epsilon_{X(Y)})$, and one asks whether $cor(R(Y^2), R(\epsilon_{X(Y)})) = 0$ can be rejected. An analogous procedure is used to evaluate the independence of X^2 and $\epsilon_{Y(X)}$. Transforming initial scores to ranks has the advantage of down-weighting potential outliers.

Previous studies have shown that tests based on the so-called normal scores outperform nonparametric tests based on ranks (e.g., Hoeffding, 1951, van der Waerden, 1952). Normal scores transformations have successfully been used to develop robust and powerful alternatives to classic nonparametric significance tests (e.g., Wiedermann and Alexandrowicz, 2011). Recently, Bishara and Hittner (2012) compared various approaches of testing the significance of correlations in cases of nonnormal data and concluded that, among all considered transformations, a correlation test based on normal scores was most adequate. The normal scores transformation of a random variate X can be defined as

$$NS(X) = \Phi^{-1}\left(\frac{R(X) - .5}{N}\right) \tag{4.26}$$

with Φ^{-1} being the inverse normal cumulative distribution, $R(X)$ being the ranks of X, and N being the sample size. The ranks of the variable are first used to compute probabilities, which are then converted into z-values of the standard normal distribution. Again, considering the case of deciding between the two competing models for the total effect, the variables Y^2 and $\epsilon_{X(Y)}$ are converted into normal scores, $NS(Y^2)$ and $NS(\epsilon_{X(Y)})$, using (4.26) and the correlation test based on the t-distribution is used to evaluate the null hypothesis $H_0 : cor(NS(Y^2), NS(\epsilon_{X(Y)})) = 0$. Analogous steps are used for X^2 and $\epsilon_{Y(X)}$.

4.5 SIMULATING THE PERFORMANCE OF DIRECTIONALITY TESTS

To analyze the accuracy of the three independence tests to identify the correct mediation model, we conducted a Monte Carlo simulation experiment using the R statistical environment (R Core Team, 2016). Data were generated according to the mediation model given in Figure 4.1a. Regression coefficients were fixed at 0, 0.14, 0.39, and 0.59, which is commonly used to mimic zero, small (2% of the variance), medium (13% of the variance), and large (26% of the variance) effects (MacKinnon *et al.*, 2002; MacKinnon *et al.*, 2004; Preacher and Selig, 2012; Tofighi *et al.*, 2009). In general, $b'_{YX} = 0$ refers to a full mediation model and $b'_{YX} > 0$ constitutes cases of a partial mediation process. The regression intercepts were fixed at zero throughout the simulation study, and all error terms were randomly sampled from the standard normal distribution. First, to evaluate Type I error performance, the true predictor X was randomly drawn from a standard normally distributed population. Because of the fact that uncorrelatedness implies independence when variables are normally distributed, the proposed correlation tests are expected to reject the null hypothesis of zero correlation of residuals and squared predictors only by chance for both competing mediation models. It is important to note that these Type I error scenarios do not imply that directionality of effects does not exist. Rather, they refer to scenarios where the causal mechanism is not reflected in correlational patterns of error terms and corresponding predictors.

In the next step, the true predictor X was randomly drawn from various χ^2-distributions with predefined degrees of freedom (*df*) to ensure $\gamma_X = 0.5$, 1, 1.5, 2, and 2.5. Marszalek *et al.* (2011) analyzed sample sizes in psychological research over the past 30 years and reported average sample sizes ranges of 180.49–211.03. Thus, $N = 200$ observations were generated, which may be considered typical for psychological studies. All simulation factors were fully crossed leading to 4 (effect size for b'_{YX}) × 4 (effect size for b'_{YM}) × 4 (effect size for b_{MX}) × 6 (level of skewness; γ_X) = 384 experimental conditions. For each condition, 1000 samples were generated. Following Baron and Kenny's (1986) stepwise approach, the two competing mediation models depicted in Figure 4.1 were estimated using a series of ordinary least squares regression models for each generated triple X, M, and Y. Correlation tests based on the Pearson, Spearman, and normal scores correlation were applied to evaluate the independence of estimated regression residuals and squared values of the corresponding predictor. All tests were performed two-sided under a nominal significance level of 5%. Empirical Type I error and power rates were computed for selecting the correct mediation model using the decision rules discussed earlier, that is, cases in which the null hypothesis of zero correlation was retained for the correct model and simultaneously rejected for the misspecified model were of particular interest. Type I error robustness of the correlation tests was assessed using the liberal robustness criterion of Bradley (1978). That is, a test is considered robust if empirical Type I error rates fall within the interval 2.5–7.5%.

4.5.1 Results

Figures 4.5–4.7 give the Type I error rates for the Pearson, the Spearman, and the normal scores correlation tests for the two bivariate regressions and the multiple linear regression models as a function of effect sizes of b_{MX}, b'_{YX}, and b'_{YM}. In general, empirical Type I error rates of all tests are within the robustness range. In other words, rejection rates for the null hypothesis of independence for both models are well in

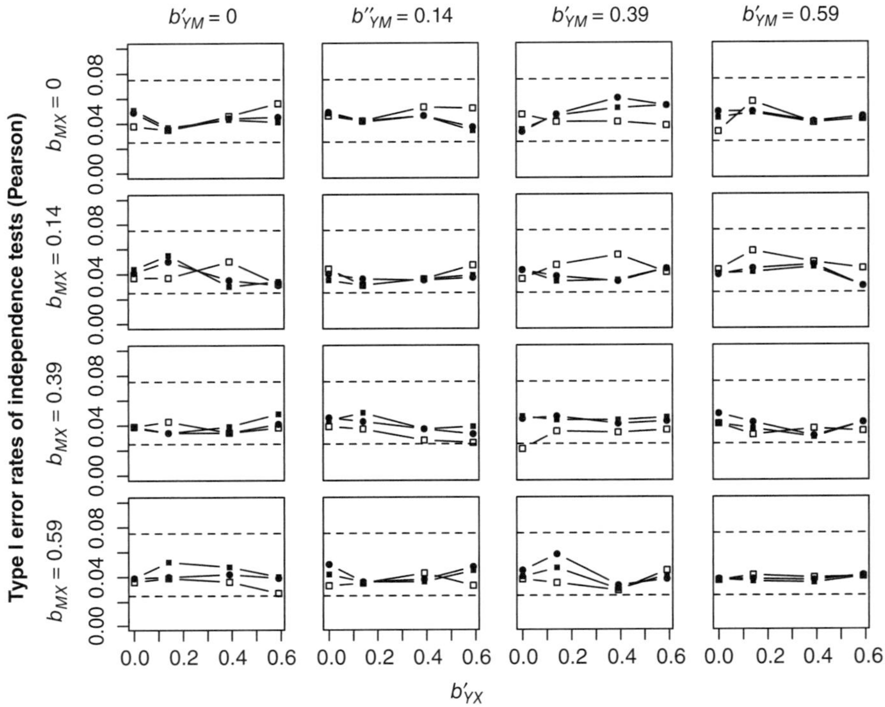

Figure 4.5 Type I error rates of the Pearson correlation test as a function of b'_{YM}, b'_{YX}, and b_{MX}. The dashed lines give Bradley's (1978) liberal robustness interval (2.5–7.5%); ■ = $X \rightarrow Y$, □ = $X \rightarrow M$, and • = $(XM) \rightarrow Y$.

accordance with the nominal significance level of 5% and no distinct decision concerning directionality can be made for normally distributed true predictors.

In the next step, we analyze the power of the tests to select the correct mediation model. In general, the power to identify the correct models increased with the skewness of the true predictor and the effect sizes of all regression coefficients (b'_{YM}, b'_{YX}, and b_{MX}). The effect size of b'_{YM} had the smallest impact on the power of the tests. Thus, to save space, we restrict the presentation of the power results to the simulation factors b_{MX}, b'_{YX}, and γ_X. Figures 4.8–4.10 give the empirically observed power curves for the three correlation tests for zero, small, medium, and large effects of b_{MX} and b'_{YX} across all considered levels of predictor skewness (γ_X). Again, the curves represent the relative frequencies of selecting the correct model based on the combined decision of separate tests of independence for the steps of Baron and Kenny's approach. In the case of $b_{MX} = b'_{YX} = 0$, which refers to the absence of an underlying mediational process, no decision can be made. The power of all correlation tests depends on the skewness of the true predictor and the effect sizes of regression coefficients. For $b_{MX} = 0$ and $b'_{YX} > 0$, that is, cases in which the model reduces to a simple multiple linear regression scenario, the power of the tests for evaluating directionality

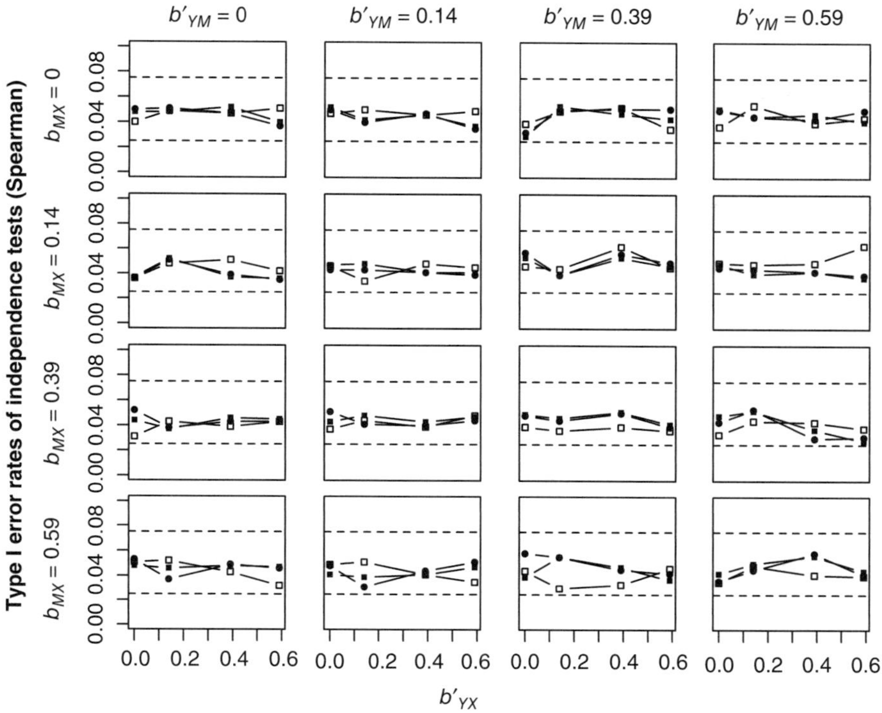

Figure 4.6 Type I error rates of the Spearman correlation test as a function of b'_{YM}, b'_{YX}, and b_{MX}. The dashed lines give Bradley's (1978) liberal robustness interval (2.5–7.5%); ■ = $X \rightarrow Y$, □ = $X \rightarrow M$, and • = $(XM) \rightarrow Y$.

of the sub-models $X \rightarrow Y$ versus $Y \rightarrow X$ and $(XM) \rightarrow Y$ versus $(YM) \rightarrow X$ increases with b'_{YX} and the skewness of X, while power curves for the submodel $X \rightarrow M$ remain unaffected. In contrast, for $b'_{YX} = 0$ and $b_{MX} > 0$, that is, full mediation processes, the power of the test that evaluates the submodel $X \rightarrow M$ systematically increases with b_{MX} and the skewness of X. Note that the power values for the models $X \rightarrow Y$ versus $Y \rightarrow X$ and $(XM) \rightarrow Y$ versus $(YM) \rightarrow X$ appear to be low due to aggregating statistical decisions across all effect sizes of b'_{YM}. For partial mediation processes (i.e., when both b_{MX} and b'_{YX} are larger than zero), the test power for all three submodels increases with the skewness of X. In general, the Pearson correlation test outperforms both nonparametric correlation tests, and the Spearman correlation test is more powerful than the normal scores test.

4.6 EMPIRICAL DATA EXAMPLE: DEVELOPMENT OF NUMERICAL COGNITION

In this section, we illustrate the application of the proposed methodology using data from Koller and Alexandrowicz (2010) and Gareiß (2010) on the development of

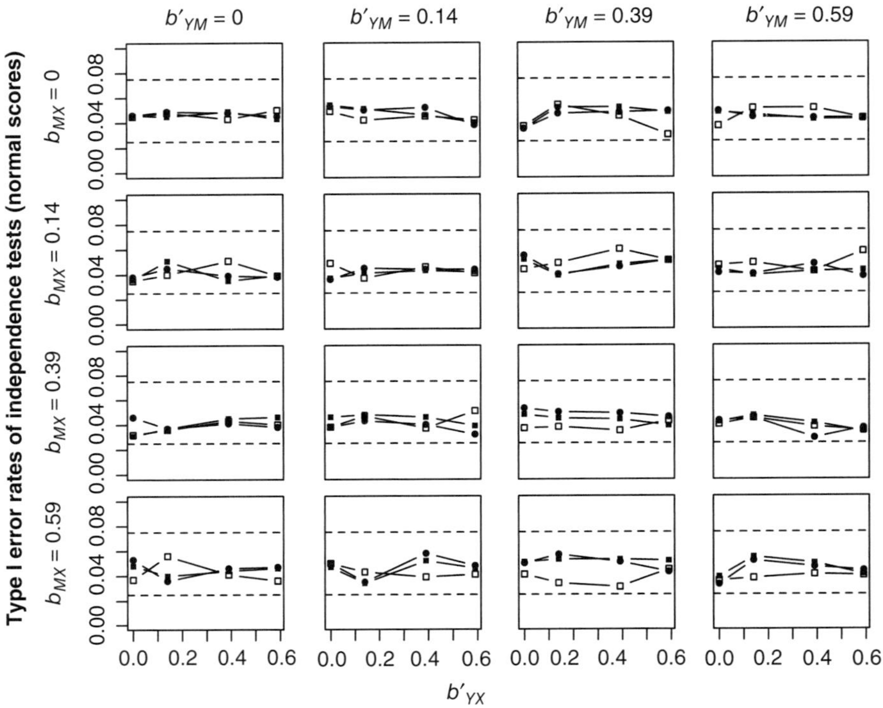

Figure 4.7 Type I error rates of the normal scores correlation test as a function of b'_{YM}, b'_{YX}, and b_{MX}. The dashed lines give Bradley's (1978) liberal robustness interval (2.5–7.5%); ■ = $X \rightarrow Y$, □ = $X \rightarrow M$, and • = $(XM) \rightarrow Y$.

numerical cognition and number processing in children. Dehaene and Cohen (1998) proposed the so-called triple code model, which posits that three different codes of number representation exist, which are of particular importance for the development of numerical cognition: (i) the *Analog Magnitude Code* (AMC), which describes a child's ability of number processing using abstract representations (e.g., used to make judgments of which of two numbers is smaller/larger); (ii) the *Auditory Verbal Code* (AVC), which consists of syntactic and lexical elements necessary for counting and retrieving memorized arithmetic facts; and (iii) the *Visual Arabic Code* (VAC) necessary for multidigit operations and parity judgments. Based on the triple code model, von Aster and Shalev (2007) proposed a hierarchical developmental model of numerical cognition where the AMC constitutes an inherited core system, which is necessary to further develop the AVC and, in turn, the VAC. In addition, a direct effect of the AMC on the development of the VAC can be assumed. Thus, in this mediation process, the AMC is assumed to be the cause, the AVC is assumed to be the mediator, and the VAC is assumed to be the outcome. In an alternative mediation model, we reverse the AMC–VAC path, that is, we assume that the VAC constitutes the predictor and the AMC is the outcome (while the AVC is still used as the mediator). In the

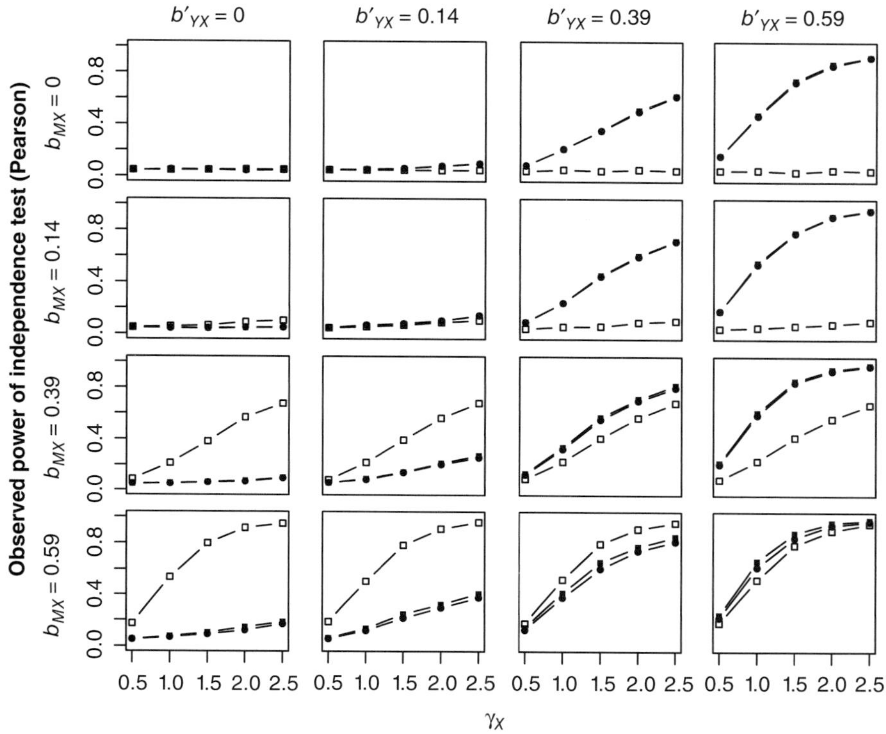

Figure 4.8 Empirical power rates of the Pearson correlation test as a function of b'_{YM}, b_{MX}, and γ_X; ■ = $X \rightarrow Y$, □ = $X \rightarrow M$, and • = $(XM) \rightarrow Y$.

present reanalysis, we use data from 101 girls and 71 boys between 7 and 11 years of age ($M = 8.66$, SD = 1.07) with preexisting difficulties with numbers to empirically test the directionality of the mediation model. The ability to deal with numeric information in terms of the AMC, the AVC, and the VAC was measured using the Neuropsychological Test Battery for Number Processing and Calculation in Children (ZAREKI-R; von Aster and Shalev, 2007). All variables were standardized prior to the mediation analysis.

In the first step, we evaluated distributional properties of the variables AMC, AVC, and VAC. The Shapiro–Wilk normality test confirmed that all three variables significantly deviated from normality (all p-values < 0.05). Table 4.1 shows the results of all regression models compatible with the two competing mediation processes. Following Baron and Kenny's (1986) approach, we first regressed the mediator (AVC) on the predictor (AMC; see Model I in Table 4.1), which suggests that a significant linear association exists between the AMC and the AVC. In addition, we observed nonsignificant independence tests for the extracted residuals of the model and the squared scores of AMC (Pearson: $\rho = 0.016$, $t(170) = 0.213$, $p = .832$; Spearman: $\rho = 0.063$, $t(170) = 0.827$, $p = .410$; Normal Scores: $\rho = 0.060$,

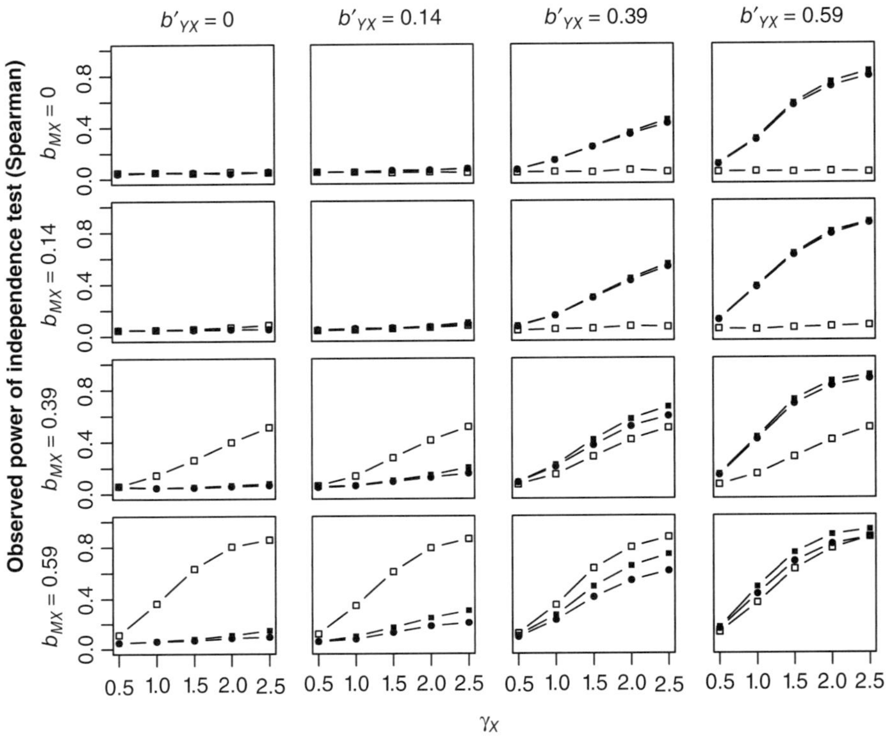

Figure 4.9 Empirical power rates of the Spearman correlation test as a function of b'_{YM}, b_{MX}, and γ_X; ■ = $X \rightarrow Y$, □ = $X \rightarrow M$, and • = $(XM) \rightarrow Y$.

$t(170) = 0.778$, $p = .438$), which suggests that the independence assumption is fulfilled. Next, we regressed the putative outcome variable (VAC) on the predictor AMC, which resulted in a significant total effect (Model II in Table 4.1). Again, all correlation tests suggested independence between corresponding regression residuals and squared values of AMC (Pearson: $\rho = -0.012$, $t(170) = -0.155$, $p = 0.8772$; Spearman: $\rho = -0.066$, $t(170) = -0.868$, $p = 0.387$; Normal Scores: $\rho = -0.027$, $t(170) = -0.349$, $p = 0.728$). Third, we regressed VAC on AVC (Model III in Table 4.1), which is usually not part of Baron and Kenny's (1986) stepwise approach. The results of this model will solely be used for testing the direction of effects. Again, all correlation tests suggested independence between the estimated regression residuals and the squared values of AVC (Pearson: $\rho = 0.089$, $t(170) = 1.165$, $p = 0.246$; Spearman: $\rho = 0.030$, $t(170) = 0.386$, $p = 0.700$; Normal Scores: $\rho = 0.044$, $t(170) = 0.580$, $p = 0.563$). Finally, we regressed VAC on AVC and AMC. Both independent variables significantly predict VAC scores. In addition, we observed a significant indirect effect of $IE = 0.220$ (95% percentile bootstrap CI = 0.131–0.314; Sobel z-test: $z = 5.00$, $p < 0.001$). The direct effect in Model IV remains significant after including the mediator, which suggests a

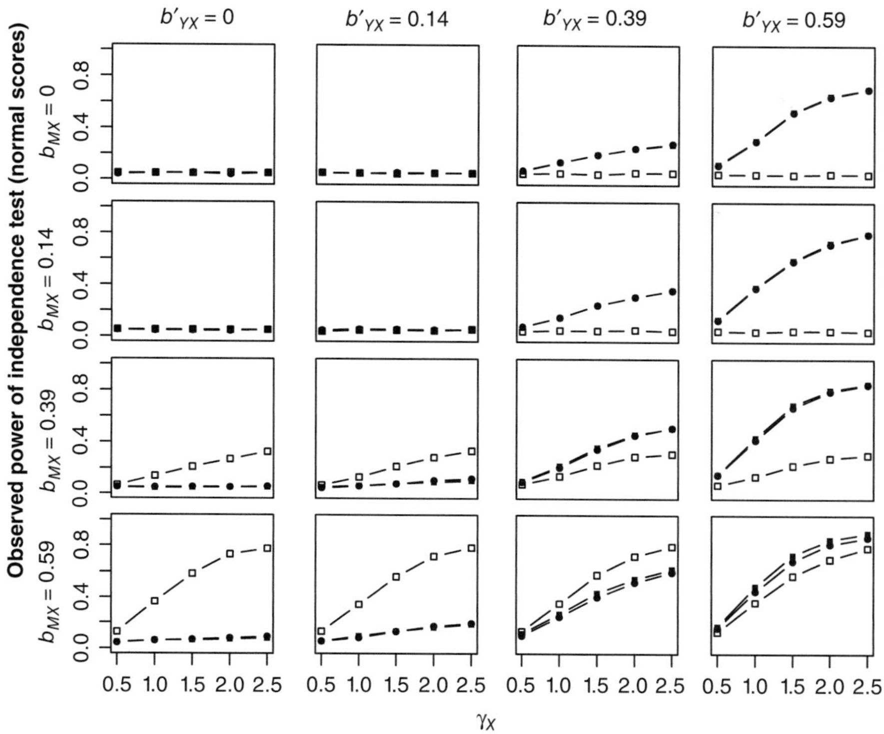

Figure 4.10 Empirical power rates of the normal scores correlation test as a function of b'_{YM}, b_{MX}, and γ_X; ■ = $X \rightarrow Y$, □ = $X \rightarrow M$, and • = $(XM) \rightarrow Y$.

partial mediation process. Although the common practice of using the significance of the direct effect to distinguish partial from full mediation has been criticized (Rucker *et al.*, 2011), the partial mediation model seems plausible in light of the proposed triple code model. Finally, we evaluated the directionality assumption of the multiple linear regression model using correlation tests for the estimated regression residuals and squared values of the putative predictor (AMC). Again, all correlation tests suggest independence (Pearson: $\rho = -0.020$, $t(170) = -0.259$, $p = 0.796$; Spearman: $\rho = -0.084$, $t(170) = -1.100$, $p = 0.273$; Normal Scores: $\rho = -0.044$, $t(170) = -0.575$, $p = 0.566$).

Models I–IV in Table 4.1 are the regression models that reflect our a priori hypothesized direction of the mediation model. To test the directionality assumption of von Aster and Shalev's (2007) hierarchical model, we next estimated the competing mediation model depicted in Figure 4.1b (see Table 4.1, Models V–VIII). Here, VAC serves as the predictor, AVC is again treated as the mediator, and AMC serves as the outcome variable. For the competing mediation model, we again observed a significant indirect effect of $IE = 0.167$ (95% percentile bootstrap CI = 0.084–0.258; Sobel's z-test: $z = 3.500$, $p < 0.001$). However, it is less surprising that the indirect

TABLE 4.1 Linear Regression Results for Two Competing Mediation Models.

Model	DV	IV	*b*	SE	*t*-Value	*p*-Value
I	AVC	AMC	0.662	0.057	11.514	<0.001
II	VAC	AMC	0.772	0.049	15.853	<0.001
III	VAC	AVC	0.698	0.055	12.721	<0.001
IV	VAC	AMC	0.552	0.060	9.207	<0.001
		AVC	0.333	0.060	5.556	<.001
V	AMC	AVC	0.662	0.057	11.514	<0.001
VI	AMC	VAC	0.772	0.049	15.853	<0.001
VII	AVC	VAC	0.698	0.055	12.721	<0.001
VIII	AMC	VAC	0.605	0.066	9.207	<0.001
		AVC	0.239	0.066	3.639	<0.001

Models I–IV are tentatively assumed to describe the correct mediation process; that is, AMC = predictor, AVC = mediator, and VAC = outcome (DV and IV denote the dependent and independent variables in the corresponding regression model; SE denotes the standard error).

effect for the competing mediation models is smaller than the indirect effect estimated from the target mediation model. Wiedermann and von Eye (2015b) showed that the difference in estimated indirect effects only depends on the correlational patterns between predictor, mediator, and outcome and cannot be used to infer on the directionality of mediation models.

Next, we evaluated the directionality assumption of the competing mediation model. We extracted the regression residuals of all estimated models and tested the correlation between estimated regression residuals and the squared values of the corresponding predictor. In addition, Figure 4.11 gives the estimated correlation coefficients using the Pearson, the Spearman, and the normal scores correlations together with the 95% nonparametric bootstrap confidence intervals. In Model V (i.e., AVC → AMC), the reversed directional flow of Model I (AMC → AVC) is considered. For Model V, all correlation tests suggest slightly larger correlations between the estimated regression residuals and the squared values of AVC, which, however, remain nonsignificant (based on the 5% significance criterion; Pearson: $\rho = 0.119$, $t(170) = 1.547$, $p = 0.124$; Spearman: $\rho = 0.143$, $t(170) = 1.886$, $p = 0.0612$; Normal Scores: $\rho = 0.144$, $t(170) = 1.892$, $p = 0.060$). Next, Model VI (VAC → AMC) estimates the reversed total effect (which, by necessity, is identical to the total effect of Model II due to prior standardization). Again, the Pearson correlation test suggests independence of the corresponding error term and the squared values of VAC ($\rho = 0.128$, $t(170) = 1.652$, $p = 0.100$). However, the Spearman correlation test and the test based on normal scores reject the null hypothesis of zero correlation (Spearman: $\rho = 0.181$, $t(170) = 2.400$, $p = 0.018$; Normal Scores: $\rho = 0.173$, $t(170) = 2.285$, $p = 0.024$). Thus, for the total effect, we have found empirical evidence that (compared to Model VI) Model II is more likely to reflect the correct data-generating process.

Next, we evaluated the directionality assumption of Model VII (VAC → AVC), which constitutes the competing bivariate model of the initial bivariate mediator–outcome relation. Here, no decision upon the correct directional flow can be made based on the result of the three correlation tests applied to the estimated residuals and the squared values of VAC (Pearson: $\rho = 0.053$, $t(170) = 0.697$, $p = 0.487$, Spearman: $\rho = 0.127$, $t(170) = 1.676$, $p = 0.096$; Normal Scores: $\rho = 0.091$, $t(170) = 1.190$, $p = 0.236$).

In the last step, we considered the multiple linear regression of the competing mediation model, which treats AMC as the outcome and AVC and VAC as the explanatory variables. Although the Pearson correlation test for the estimated residuals and the squared values of VAC suggests retaining the null hypothesis ($\rho = 0.116$, $t(170) = 1.518$, $p = 0.131$), both the Spearman correlation test and the normal scores test suggest rejecting the null hypothesis of zero-correlation (Spearman: $\rho = 0.161$, $t(170) = 2.134$, $p = 0.034$; Normal Scores: $\rho = 0.159$, $t(170) = 2.030$, $p = 0.044$). In sum, (i) in the majority of cases, the correlations between error terms and squared values of the corresponding explanatory variable were larger for the competing mediation model than for the target model (see also Fig. 4.11), (ii) no

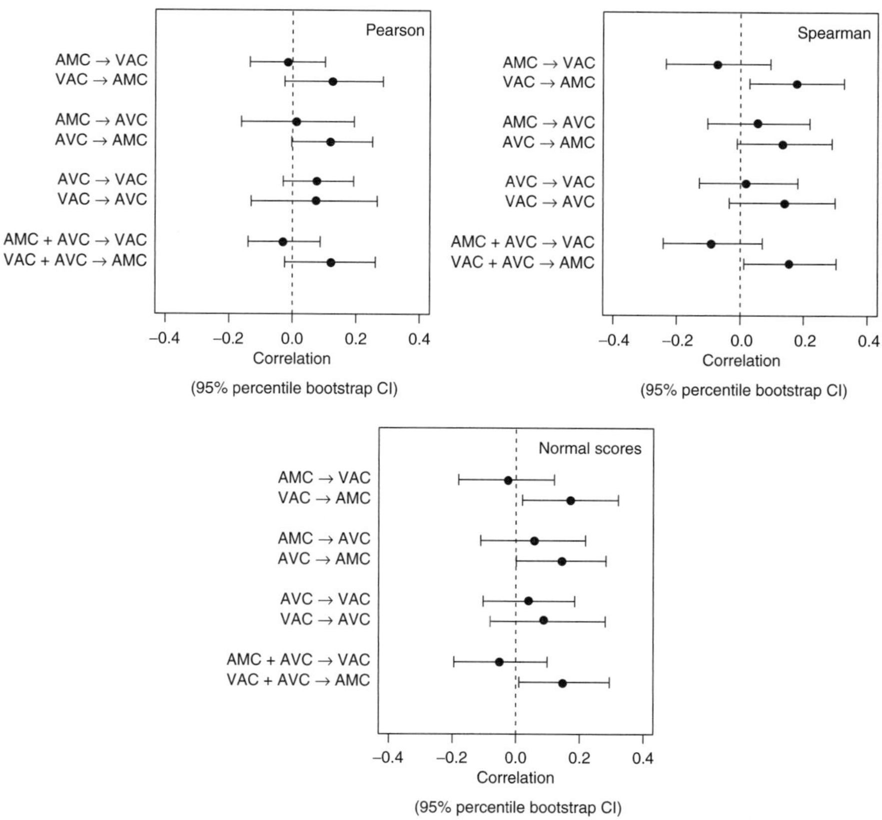

Figure 4.11 Pearson, Spearman, and normal scores correlations of estimated regression residuals and squared values of the corresponding predictor (Error bars give the 95% non-parametric bootstrap confidence interval).

cases exist in which a reversed model is considered more likely to reflect the data generating process, and, most importantly, (iii) the models that estimate the total and the direct effect suggest no violations for the tentative direction according to the theoretical developmental model and simultaneously reject the independence assumption for the competing models. Thus, empirical evidence exists that the tentative mediation model, in which AMC is the predictor and VAC is the outcome, is more likely to reflect the underlying data-generating process, which is line with the theoretical developmental model of von Aster and Shalev (2007).

4.7 DISCUSSION

Path modeling, or, more generally, structural equation modeling (SEM; Jöreskog, 1970) has been surrounded by various misunderstandings; for a critical examination of various myths about SEM, see Bollen and Pearl (2013). One of those misunderstandings is that path models are capable of establishing causal statements from associations alone, that is, only observed covariances are necessary to decide whether causal relations between variables exist. Consequently, various scholars cautioned against premature causal interpretations of path modeling results (e.g., Freedman, 1987, Sobel, 1996, Sobel, 2008, De Leeuw, 1985). In an attempt to dissolve this myth, Pearl (2009) and Bollen and Pearl (2013) follow the founding fathers of path modeling such as Wright (1921), Haavelmo (1943), and Duncan (1975) and explicitly declare that prior knowledge concerning the causal mechanism is *assumed* (in addition to statistical assumptions). Technically speaking, the equal sign in structural equations must be understood as assignment symbol (:=), and, thus, structural equations rely on and reflect causal assumptions of researchers. These causal assumptions must be derived from substantive sources such as theories that explain causal mechanisms, logical arguments, temporality, or prior studies. The redefinition of total, direct, and indirect effects using the counterfactual framework makes these causal assumptions more transparent, clearly describes necessary ignorability conditions to justify causal claims, and has subsequently led to the development of sensitivity analyses to evaluate the robustness of obtained results against potential violations of ignorability conditions (Imai *et al.*, 2010). From a purely statistical perspective, these ignorability conditions are additional assumptions that allow endowing estimated associations with causal meaning. Thus, other authors suggested to sharply distinguish between *regular SEMs* and *causal SEMs*, the latter describing structural models for which additional ignorability assumptions hold (Wang and Sobel, 2013). Following this dichotomy, causal SEMs are a subset of regular SEMs.

This chapter addresses causal assumptions that are used to substantially derive the particular structure of a hypothesized linear mediation model and presents a methodology designed to empirically validate the correctness of specified path directions. In particular, in observational data settings, competing substantive theories about the underlying causal flow may exist and researchers commonly estimate a series of competing mediation models that reflect alternative explanations of observed relations. Wiedermann and von Eye (2015b) caution against the use of this exploratory approach because researchers who use the significance of indirect effects as a selection criterion may run the risk to select a misspecified mediation model. As an alternative, we proposed three significance tests to evaluate independence assumptions in linear mediation models. While Imai's *et al.* (2010) sensitivity analysis focuses on the independence of error terms in the mediation model, the current approach evaluates the independence assumption of error terms and model predictors and, thus, enriches the statistical repertoire of causal mediation analysis. Simulation studies confirmed the adequacy of the proposed tests in terms of both Type I error protection and statistical power.

The majority of studies on causal mediation analysis is embedded into the context of randomized studies where, at least, the predictor, X, is under experimental control. In this case, questions concerning an unambiguous (directional) interpretation of the direct effect ($X \rightarrow Y$) and the predictor–mediator path ($X \rightarrow M$) do not occur. However, even in randomized studies where X is experimentally controlled, directionality issues between the mediator (M) and the outcome (Y) may arise. The presented approach can straightforwardly be applied to empirically evaluate competing directional theories about the mediator–outcome path (see also Wiedermann and von Eye, 2015b).

The key element for testing the directionality of effects is that observed data are assumed to deviate from normality. Thus, nonnormality (as a common phenomenon in empirical data; see, e.g., Micceri, 1989), or, more specifically, asymmetry of variable distributions, is not prematurely dismissed as a source of bias that threatens valid statistical inference. Instead, nonnormality is viewed as an informative source that can be used to gain deeper insight into the underlying data-generating mechanism. Advantages of using data information based on skewness and kurtosis beyond its capability of indicating deviations from normality have also been recognized by previous authors. For example, Bentler (1983) outlined that higher than second moments may be used to resolve issues of equivalence of structural models, which subsequently led to the development of nonnormal SEMs (Shimizu and Kano, 2008). In nonnormal SEMs, higher order moment structures systematically differ across competing models when variables are nonnormal, which can be used to resolve equivalence issues. Extensions of nonnormal SEMs to behavior genetic models (the ACE and the ACDE model for twin designs) were proposed by Ozaki and Ando (2009) and Ozaki *et al.* (2011). Based on Bentler's (1983) proposition, Mooijaart (1985) developed nonnormal factor analysis, which, under certain conditions, can be used to uniquely identify factor loadings. Mooijaart and Bentler (2010) proposed considering higher than second moments in estimating nonlinear latent variable models. Of course, previous research on mediation models also addressed the issue of nonnormality of observed variables. However, these studies typically focus on potential biases of statistical inference. For example, Yuan and MacKinnon (2014) proposed robust mediation analysis based on medians, Kisbu-Sakarya *et al.* (2014) used higher order moments to explain confidence interval coverage and imbalance of indirect effects, and Zhang (2014) proposed Monte Carlo–based statistical power analysis for accurate power estimates under nonnormality.

Within the line of research on causal discovery techniques, nonnormality and its statistical consequences have been used to develop, for example, linear non-Gaussian acyclic models (LiNGAM; Shimizu and Kano, 2006a, Shimizu *et al.*, 2011). For an overview of the LiNGAM approach, see also the related chapter of Shimizu (2016). It is important to note that, compared to LiNGAM, the presented approach differs both conceptually and methodologically. On a conceptual level, causal discovery algorithms (see, e.g., also model selection procedures based on conditional independence tests such as the PC-algorithm implemented in the TETRAD project; Scheines *et al.*, 1998)) exhibit a major exploratory element, in the sense, that these methods are ideally suited to deduce causal structures from multivariate data, which may further lead to new hypotheses about the relation of constructs. In contrast, the method proposed here is intended to be used as a confirmatory approach to empirically testing a theory-based target mediation model against a plausible alternative model. From a statistical perspective, the LiNGAM approach differs from the presented method in terms of distributional assumptions. In the currently considered linear mediation model, the error terms of the target model ($\epsilon_{M(X)}$ and $\epsilon_{Y(XM)}$) are assumed to be normally distributed, which constitutes a common distributional assumption within the framework of ordinary least square estimation. Distributional asymmetry is only assumed for the true predictor variable. In contrast, LiNGAMs assume that all error terms (i.e., $\epsilon_X(= X)$, $\epsilon_{M(X)}$, and $\epsilon_{Y(XM)}$) are nonnormal, and at most, one variable is allowed to be normal (Shimizu, 2016).

The current work focuses on the simple linear mediation model for manifest variables. In other words, we implicitly assume that measurement errors associated with the predictor, the mediator, and the outcome are negligible. It is well known that low reliability of variables affects significance tests (e.g. Zimmerman *et al.*, 1993). Similarly, it may be safe to assume that low reliability of variables also affects decisions concerning the direction of effects based on the proposed tests. Similarly, careful data cleaning and data modeling is necessary to guarantee best-practice applications of directionality tests. This implies that data analysts have to check for outliers in order to rule out spuriously inflated skewness values. When the linear mediation model is applied, researchers implicitly assume that the linear model, as a mathematical relation to explain observed associations, is justified and valid. Zhang and Hyvärinen (2010) discuss methods for causal discovery based on nonlinear models. Integrating these nonlinear models into the framework of analyzing mediation processes is up to future studies.

The proposed method relies on the assumption that the independence of predictors and corresponding error terms holds in the true mediation model. In other words, it is assumed that latent confounders do not exist. This assumption definitely warrants future research, and simulation studies are needed to quantify the robustness of the proposed tests under assumption violations. Currently, results of directionality tests have to be interpreted with great caution whenever theoretical arguments exists that suggest the existence of latent confounders. Furthermore, throughout the chapter, we assume that the error terms of the true model are normally distributed. Although normally distributed error terms constitute a standard assumption in ordinary least square estimation, this distributional assumption can be relaxed. All theoretical results are based on the assumption of *zero skewness*, which implies that no restrictions concerning the kurtosis of the error terms are imposed. The proposed methodology is expected to perform well even when the true error terms do not follow a normal distribution as long as the symmetry assumption is fulfilled. Similar robustness properties have been shown for the residual-based direction dependence approach (Wiedermann and von Eye, 2015b). Future simulation studies are planned to quantify the robustness of the current approach against violation of the normality assumption.

Throughout the manuscript, we used the terminology of "true" and "misspecified" models because in our theoretical derivations and Monte Carlo simulations, we actually know the true model. This data–reality consistency cannot be assumed in practical applications because the ultimately true model will never be known (Cudeck and Henly, 2003). Thus, in practice, the questions answered by the proposed directionality tests as well as the subsequent directional statements must be rephrased. Instead of asking which of two competing mediation models constitutes the true model, researchers have to ask whether the target model better approximates the underlying data-generating mechanism than an alternative model. However, this does not solely apply to the presented methodology. This issue applies to data modeling as a tool for scientific discovery in general. Overall, provided that certain data requirements are fulfilled, the presented approach can be considered a valuable tool for putting directional mediation theories to the test.

APPENDIX A: DEPENDENCE PROPERTY OF THE PREDICTOR–OUTCOME RELATION

Assuming standardized predictor and outcome variables (i.e., $E[X] = E[Y] = 0$ and $E[X^2] = E[Y^2] = 1$), the error term of the misspecified model $Y \rightarrow X$ is given through

$$\epsilon_{X(Y)} = \left(1 - \rho_{XY}^2\right) X - \rho_{XY}\epsilon_{Y(X)} \tag{A1}$$

with $E[\epsilon_{X(Y)}] = (1 - \rho_{XY}^2)E[X] - \rho_{XY}E[\epsilon_{Y(X)}] = 0$. The expected value of the squared true outcome variable is then defined as

$$\begin{aligned} E(Y^2) &= E[(b_{YX}X + \epsilon_{Y(X)})^2] \\ &= b_{YX}^2 E[X^2] + E[\epsilon_{Y(X)}^2] \\ &= \rho_{XY}^2 + \sigma_{\epsilon_{Y(X)}}^2 \end{aligned} \tag{A2}$$

Using the aforementioned expressions to derive the covariance of Y^2 and $\epsilon_{X(Y)}$ results in

$$\begin{aligned} cov(Y^2, \epsilon_{X(Y)}) &= E[(Y^2 - E[Y^2])(\epsilon_{X(Y)} - E[\epsilon_{X(Y)}])] \\ &= E[(b_{YX}^2 X^2 + 2b_{YX}X\epsilon_{Y(X)} + \epsilon_{Y(X)}^2 - (\rho_{XY}^2 + \sigma_{\epsilon_{Y(X)}}^2))\epsilon_{X(Y)}] \\ &= b_{YX}^2 E[X^2\epsilon_{X(Y)}] + 2b_{YX}E[X\epsilon_{Y(X)}\epsilon_{X(Y)}] + E[\epsilon_{Y(X)}^2\epsilon_{X(Y)}] \\ &\quad - \rho_{YX}^2 E[\epsilon_{X(Y)}] - \sigma_{\epsilon_{X(Y)}}^2 E[\epsilon_{X(Y)}] \end{aligned} \tag{A3}$$

Finally, making use of the fact that

$$\begin{aligned} E[X^2\epsilon_{X(Y)}] &= E[X^2[(1-\rho_{XY}^2)X - \rho_{XY}\epsilon_{Y(X)}]] \\ &= (1-\rho_{XY}^2)E[X^3] - \rho_{XY}E[X^2\epsilon_{Y(X)}] \\ &= (1-\rho_{XY}^2)E[X^3] \end{aligned} \tag{A4}$$

$$\begin{aligned} E[X\epsilon_{Y(X)}\epsilon_{X(Y)}] &= E[X\epsilon_{Y(X)}[(1-\rho_{XY}^2)X - \rho_{XY}\epsilon_{Y(X)}]] \\ &= (1-\rho_{XY}^2)E[X^2\epsilon_{Y(X)}] - \rho_{XY}E[X\epsilon_{Y(X)}^2] \\ &= 0 \end{aligned} \tag{A5}$$

and

$$\begin{aligned} E[\epsilon_{Y(X)}^2\epsilon_{X(Y)}] &= E[\epsilon_{Y(X)}^2[(1-\rho_{XY}^2)X - \rho_{XY}\epsilon_{Y(X)}]] \\ &= (1-\rho_{XY}^2)E[X\epsilon_{Y(X)}^2] - \rho_{XY}E[\epsilon_{Y(X)}^3] \\ &= -\rho_{XY}E[\epsilon_{Y(X)}^3] \end{aligned} \tag{A6}$$

while $E[X] = E[\epsilon_{Y(X)}] = E[\epsilon_{X(Y)}] = 0$, Equation (A3) simplifies to

$$\begin{aligned} cov(Y^2, \epsilon_{X(Y)}) &= b_{YX}^2(1-\rho_{XY}^2)E[X^3] - \rho_{XY}E[\epsilon_{Y(X)}^3] \\ &= \rho_{XY}^2(1-\rho_{XY}^2)\gamma_X - \rho_{XY}\sigma_{\epsilon_{Y(X)}}^3\gamma_{\epsilon_{Y(X)}} \end{aligned} \tag{A7}$$

with γ_X and $\gamma_{\epsilon_{Y(X)}}$ being the skewness of the true predictor and the true error term. Assuming a normally distributed true error term (i.e., $\gamma_{\epsilon_{Y(X)}} = 0$), Equation (A7) reduces to $cov(Y^2, \epsilon_{X(Y)}) = \rho_{XY}^2(1-\rho_{XY}^2)\gamma_X$.

APPENDIX B: DEPENDENCE PROPERTIES IN THE MULTIPLE VARIABLE MODEL

In this appendix, we show that the covariance of Y^2 and $\epsilon_{X(YM)}$ can be written as a weighted function of the skewness of the true predictor X. The error term of the misspecified mediation model (see Figure 4.1b) can be written as (intercepts fixed at zero)

$$\epsilon_{X(YM)} = (1 - b_{XY}b_{YX})X - b'_{YM}b_{XY}\epsilon_{M(X)} - b_{XY}\epsilon_{Y(XM)} - b'_{XM}\epsilon_{M(Y)} \tag{B1}$$

with b_{XY} and b_{YX} being the total effects of the competing mediation models. For convenience, we assume standardized variables X, M, and Y (i.e., $E[X] = E[M] = E[Y] = 0$ and $E[X^2] = E[M^2] = E[Y^2] = 1$), which implies $E[\epsilon_{X(YM)}] = 0$. Further, the squared values of the true outcome (Y) are given by

$$Y^2 = (b_{YX}X + b'_{YM}\epsilon_{M(X)} + \epsilon_{Y(XM)})^2 \tag{B2}$$

Because X, $\epsilon_{M(X)}$, and $\epsilon_{Y(XM)}$ are assumed to be independent of each other, the expected value of Y^2 can be written as

$$E[Y^2] = b^2_{YX}E[X^2] + b'^2_{YM}E[\epsilon^2_{M(X)}] + E[\epsilon^2_{Y(XM)}] \tag{B3}$$

The covariance of Y^2 and $\epsilon_{X(YM)}$ can be defined through

$$\begin{aligned} cov(Y^2, \epsilon_{X(YM)}) &= E[(Y^2 - E[Y^2])(\epsilon_{X(YM)} - E[\epsilon_{X(YM)}])] \\ &= E[(Y^2 - E[Y^2])\epsilon_{X(YM)}] \end{aligned} \tag{B4}$$

Inserting (B1), (B2), and (B3) into (B4) leads to

$$\begin{aligned} cov(Y^2, \epsilon_{X(YM)}) &= E[((b_{YX}X + b'_{YM}\epsilon_{M(X)} + \epsilon_{Y(XM)})^2 \\ &\quad - (b^2_{YX}E[X^2] + b'^2_{YM}E[\epsilon^2_{M(X)}] + E[\epsilon^2_{Y(XM)}]))\epsilon_{X(YM)}] \\ &= b^2_{YX}E[X^2\epsilon_{X(YM)}] + 2b_{YX}b'_{YM}E[X\epsilon_{M(X)}\epsilon_{X(YM)}] \\ &\quad + b'^2_{YM}E[\epsilon^2_{M(X)}\epsilon_{X(YM)}] + 2b_{YX}E[X\epsilon_{Y(XM)}\epsilon_{X(YM)}] \\ &\quad + 2b'_{YM}E[\epsilon_{M(X)}\epsilon_{Y(XM)}\epsilon_{X(YM)}] + E[\epsilon^2_{Y(XM)}\epsilon_{X(YM)}] \end{aligned} \tag{B5}$$

Next, we separately evaluate the expected value terms in (B5). Assuming independence of X, $\epsilon_{M(X)}$, and $\epsilon_{Y(XM)}$, the term $E[X^2\epsilon_{X(YM)}]$ can be rewritten as

$$\begin{aligned} E[X^2\epsilon_{X(YM)}] &= E[X^2[(1 - b_{XY}b_{YX})X - b'_{YM}b_{XY}\epsilon_{M(X)} - b_{XY}\epsilon_{Y(XM)} \\ &\quad - b'_{XM}\epsilon_{M(Y)}]] \\ &= (1 - b_{XY}b_{YX})E[X^3] - b'_{XM}E[X^2\epsilon_{M(Y)}] \\ &= (1 - b_{XY}b_{YX})E[X^3] - b'_{XM}(b_{MX} - b_{MY}b_{YX})E[X^3] \\ &= [(1 - b_{XY}b_{YX}) - b'_{XM}(b_{MX} - b_{MY}b_{YX})]E[X^3] \end{aligned} \tag{B6}$$

with $E[X^2\epsilon_{M(Y)}] = E[X^2(M - b_{MY}Y)] = (b_{MX} - b_{MY}b_{YX})E[X^3]$. Further, assuming normality of true error terms (i.e., $E[\epsilon^3_{M(X)}] = E[\epsilon^3_{Y(XM)}] = 0$), we obtain

$$\begin{aligned} E[X\epsilon_{M(X)}\epsilon_{X(YM)}] &= E[X\epsilon_{M(X)}[(1 - b_{XY}b_{YX})X - b'_{YM}b_{XY}\epsilon_{M(X)} - b_{XY}\epsilon_{Y(XM)} \\ &\quad - b'_{XM}\epsilon_{M(Y)}]] \\ &= (1 - b_{XY}b_{YX})E[X^2\epsilon_{M(X)}] - b'_{YM}b_{XY}E[X\epsilon^2_{M(X)}] \\ &\quad - b_{XY}E[X\epsilon_{M(X)}\epsilon_{Y(XM)}] - b'_{XM}E[X\epsilon_{M(X)}\epsilon_{M(Y)}] \\ &= 0 \end{aligned} \tag{B7}$$

$$\begin{aligned}
E[\epsilon^2_{M(X)}\epsilon_{X(YM)}] &= E[\epsilon^2_{M(X)}[(1 - b_{XY}b_{YX})X - b'_{YM}b_{XY}\epsilon_{M(X)} - b_{XY}\epsilon_{Y(XM)} \\
&\quad - b'_{XM}\epsilon_{M(Y)}]] \\
&= (1 - b_{XY}b_{YX})E[X\epsilon^2_{M(X)}] - b'_{YM}b_{XY}E[\epsilon^3_{M(X)}] \\
&\quad - b_{XY}E[\epsilon^2_{M(X)}\epsilon_{Y(XM)}] - b'_{XM}E[\epsilon^2_{M(X)}\epsilon_{M(Y)}] \\
&= -b'_{YM}b_{XY}E[\epsilon^3_{M(X)}] - b'_{XM}E[\epsilon^2_{M(X)}\epsilon_{M(Y)}] \\
&= 0
\end{aligned} \tag{B8}$$

(with $E[\epsilon^2_{M(X)}\epsilon_{M(Y)}] = E[\epsilon^3_{M(X)}] - b_{MY}b'_{YM}E[\epsilon^3_{M(X)}] = 0$ due to the normality assumption),

$$\begin{aligned}
E[X\epsilon_{Y(XM)}\epsilon_{X(YM)}] &= E[X\epsilon_{Y(XM)}[(1 - b_{XY}b_{YX})X - b'_{YM}b_{XY}\epsilon_{M(X)} \\
&\quad - b_{XY}\epsilon_{Y(XM)} - b'_{XM}\epsilon_{M(Y)}]] \\
&= (1 - b_{XY}b_{YX})E[X^2\epsilon_{Y(XM)}] - b'_{YM}b_{XY}E[X\epsilon_{Y(XM)}\epsilon_{M(X)}] \\
&\quad - b_{XY}E[X\epsilon^2_{Y(XM)}] - b'_{XM}E[X\epsilon_{Y(XM)}\epsilon_{M(Y)}] \\
&= 0
\end{aligned} \tag{B9}$$

$$\begin{aligned}
E[\epsilon_{M(X)}\epsilon_{Y(XM)}\epsilon_{X(YM)}] &= E[\epsilon_{M(X)}\epsilon_{Y(XM)}[(1 - b_{XY}b_{YX})X - b'_{YM}b_{XY}\epsilon_{M(X)} \\
&\quad - b_{XY}\epsilon_{Y(XM)} - b'_{XM}\epsilon_{M(Y)}]] \\
&= (1 - b_{XY}b_{YX})E[X\epsilon_{M(X)}\epsilon_{Y(XM)}] \\
&\quad - b'_{YM}b_{XY}E[\epsilon^2_{M(X)}\epsilon_{Y(XM)}] \\
&\quad - b_{XY}E[\epsilon_{M(X)}\epsilon^2_{Y(XM)}] - b'_{XM}E[\epsilon_{M(X)}\epsilon_{Y(XM)}\epsilon_{M(Y)}] \\
&= 0
\end{aligned} \tag{B10}$$

(with $E[\epsilon_{M(X)}\epsilon_{Y(XM)}\epsilon_{M(Y)}] = E[\epsilon_{M(X)}\epsilon_{Y(XM)}(M - b_{MY}Y)] = 0$), and

$$\begin{aligned}
E[\epsilon^2_{Y(XM)}\epsilon_{X(YM)}] &= E[\epsilon^2_{Y(XM)}[(1 - b_{XY}b_{YX})X - b'_{YM}b_{XY}\epsilon_{M(X)} - b_{XY}\epsilon_{Y(XM)} \\
&\quad - b'_{XM}\epsilon_{M(Y)}]] \\
&= (1 - b_{XY}b_{YX})E[X\epsilon^2_{Y(XM)}] - b'_{YM}b_{XY}E[\epsilon_{M(X)}\epsilon^2_{Y(XM)}] \\
&\quad - b_{XY}E[\epsilon^3_{Y(XM)}] - b'_{XM}E[\epsilon^2_{Y(XM)}\epsilon_{M(Y)}] \\
&= -b_{XY}E[\epsilon^3_{Y(XM)}] - b'_{XM}E[\epsilon^2_{Y(XM)}\epsilon_{M(Y)}] \\
&= 0
\end{aligned} \tag{B11}$$

with $E[\epsilon^2_{Y(XM)}\epsilon_{M(Y)}] = -b_{MY}E[\epsilon^3_{Y(XM)}]$ and $E[\epsilon^3_{Y(XM)}] = 0$ due to the normality assumption.

Finally, inserting the results of (B6–B11) into (B5), the covariance of Y^2 and $\epsilon_{X(YM)}$ reduces to

$$cov(Y^2, \epsilon_{X(YM)}) = b_{YX}^2[(1 - b_{XY}b_{YX}) - b'_{XM}(b_{MX} - b_{MY}b_{YX})]E[X^3]. \tag{B12}$$

For standardized variables, one obtains $b_{YX} = \rho_{XY}, b_{XY}b_{YX} = \rho_{XY}^2, b_{MX} = \rho_{XM}, b_{MY} = \rho_{YM}$, and $b'_{XM} = (\rho_{XM} - \rho_{XY}\rho_{YM})/(1 - \rho_{YM}^2)$, and Equation (B12) can be expressed through

$$cov(Y^2, \epsilon_{X(YM)}) = \rho_{XY}^2 \left[(1 - \rho_{XY}^2) - \frac{(\rho_{XM} - \rho_{XY}\rho_{YM})^2}{(1 - \rho_{YM}^2)}\right] \gamma_X \tag{B13}$$

with γ_X being the skewness of X.

REFERENCES

von Aster, M.G. and Shalev, R.S. (2007) Number development and developmental dyscalculia. *Developmental Medicine & Child Neurology*, **49** (11), 868–873, doi: 10.1111/j.1469-8749.2007.00868.x.

Baron, R.M. and Kenny, D.A. (1986) The moderator–mediator variable distinction in social psychological research: conceptual, strategic, and statistical considerations. *Journal of Personality and Social Psychology*, **51** (6), 1173–1182, doi: 10.1037/0022-3514.51.6.1173.

Basu, D. (1951) On the independence of linear functions of independent chance variables. *Bulletin of the International Statistics Institute*, **33**, 83–96.

Bentler, P.M. (1983) Some contributions to efficient statistics in structural models: specification and estimation of moment structures. *Psychometrika*, **48** (4), 493–517, doi: 10.1007/BF02293875.

Bishara, A.J. and Hittner, J.B. (2012) Testing the significance of a correlation with nonnormal data: comparison of Pearson, Spearman, transformation, and resampling approaches. *Psychological Methods*, **17** (3), 399–417, doi: 10.1037/a0028087.

Bizer, G.Y., Hart, J., and Jekogian, A.M. (2012) Belief in a just world and social dominance orientation: evidence for a mediational pathway predicting negative attitudes and discrimination against individuals with mental illness. *Personality and Individual Differences*, **52** (3), 428–432, doi: 10.1016/j.paid.2011.11.002.

Blalock, H.M. (1964) *Causal Inferences in Nonexperimental Research*, University of North Carolina Press, Chapel Hill, NC.

Bollen, K.A., Pearl, J. (2013) Eight myths about causality and structural equation models, in *Handbook of Causal Analysis for Social Research.* (eds. Morgan, S.L) Springer-Verlag, Dordrecht, pp. 301–328.

Bradley, J.V. (1978) Robustness? *British Journal of Mathematical and Statistical Psychology*, **31** (2), 144–152, doi: 10.1111/j.2044-8317.1978.tb00581.x.

Breusch, T.S. and Pagan, A.R. (1979) A simple test for heteroscedasticity and random coefficient variation. *Econometrica: Journal of the Econometric Society*, **47** (5), 1287–1294, doi: 10.2307/1911963.

Bullock, J.G., Green, D.P., and Ha, S.E. (2010) Yes, but what's the mechanism? (Don't expect an easy answer). *Journal of Personality and Social Psychology*, **98** (4), 550–558, doi: 10.1037/a0018933.

Conover, W.J. and Iman, R.L. (1981) Rank transformations as a bridge between parametric and nonparametric statistics. *American Statistician*, **35** (3), 124–129, doi: 10.1080/00031305.1981.10479327.

Coyle, T.R., Pillow, D.R., Snyder, A.C., and Kochunov, P. (2011) Processing speed mediates the development of general intelligence (g) in adolescence. *Psychological Science*, **22** (10), 1265–1269, doi: 10.1177/0956797611418243.

Cudeck, R. and Henly, S.J. (2003) A realistic perspective on pattern representation in growth data: comment on Bauer and Curran (2003). *Psychological Methods*, **8** (3), 378–383, doi: 10.1037/1082-989X.8.3.378.

Darmois, G. (1953) Analyse générale des liaisons stochastiques: etude particulière de l'analyse factorielle linéaire. *Revue de l'Institut International de Statistique*, **21** (1/2), 2–8, doi: 10.2307/1401511.

Dehaene, S. and Cohen, L. (1998) Levels of representation in number processing, in *The Handbook of Neurolinguistics* (eds B. Stemmer and H.A. Whitaker), Academic Press, New York, pp. 331–341.

De Leeuw, J. (1985) Review of four books on causal analysis. *Psychometrika*, **50**, 371–373.

Dodge, Y. and Rousson, V. (2000) Direction dependence in a regression line. *Communications in Statistics: Theory and Methods*, **29** (9-10), 1957–1972, doi: 10.1080/03610920008832589.

Dodge, Y. and Rousson, V. (2001) On asymmetric properties of the correlation coeffcient in the regression setting. *American Statistician*, **55** (1), 51–54, doi: 10.1198/000313001300339932.

Dodge, Y. and Rousson, V. (2016) Statistical inference for direction of dependence in linear models, in *Statistics and Causality: Methods for Applied Empirical Research* (eds W. Wiedermann and A. von Eye), John Wiley & Sons, Inc.

Duncan, O.D. (1975) *Introduction to Structural Equation Models*, Academic Press, New York.

Entner, D. and Hoyer, P.O. (2011) Discovering unconfounded causal relationships using linear non-Gaussian models, in New Frontiers in Artificial Intelligence: JSAI-isAI 2010 Workshops, LENLS, JURISIN, AMBN, ISS, Tokyo, Japan, November 18-19, 2010, Revised Selected Papers, Lecture Notes on Computer Science, vol. 6797, pp. 181–195, doi: 10.1007/978-3-642-25655-4-17.

Entner, D., Hoyer, P.O., and Spirtes, P. (2012) Statistical test for consistent estimation of causal effects in linear non-Gaussian models, in Journal of Machine Learning Research: Workshop and Conference Proceedings, vol. 22, pp. 364–372.

von Eye, A. and DeShon, R.P. (2012) Directional dependence in developmental research. *International Journal of Behavioral Development*, **36** (4), 303–312, doi: 10.1177/0165025412439968.

Fiedler, K., Schott, M., and Meiser, T. (2011) What mediation analysis can (not) do. *Journal of Experimental Social Psychology*, **47** (6), 1231–1236, doi: 10.1016/j.jesp.2011.05.007.

Fowler, R.L. (1987) Power and robustness in product-moment correlation. *Applied Psychological Measurement*, **11** (4), 419–428, doi: 10.1177/014662168701100407.

Freedman, D.A. (1987) As others see us: a case study in path analysis. *Journal of Educational and Behavioral Statistics*, **12** (2), 101–128, doi: 10.3102/10769986012002101.

Fritz, M.S. and MacKinnon, D.P. (2007) Required sample size to detect the mediated effect. *Psychological Science*, **18** (3), 233–239, doi: 10.3758/BRM.40.1.55.

Gareiß, M. (2010) Testing Dehaene's triple code model using linear structural equation models, Master's thesis, University of Klagenfurt.

Gelfand, L.A., Mensinger, J.L., and Tenhave, T. (2009) Mediation analysis: a retrospective snapshot of practice and more recent directions. *Journal of General Psychology*, **136** (2), 153–178, doi: 10.3200/GENP.136.2.153-178.

Gnedenko, B.V. (1948) On a theorem of S.N. Bernstein. *Izvestiya Akademii Nauk SSR*, **12**, 97–100.

Greitemeyer, T. and McLatchie, N. (2011) Denying humanness to others: a newly discovered mechanism by which violent video games increase aggressive behavior. *Psychological Science*, **22** (5), 659–665, doi: 10.1177/0956797611403320.

Gretton, A., Bousquet, O., Smola, A., and Schölkopf, B. (2005) Measuring statistical dependence with Hilbert-Schmidt norms, in *Algorithmic Learning Theory*, Lecture Notes in Computer Science, vol. **3734** (eds S. Jain, H. Simon, and E. Tomita) Springer-Verlag, Berlin, pp. 63–77, doi: 10.1007/11564089-7.

Gretton, A., Fukumizu, K., Teo, C.H., Song, L., Schölkopf, B., and Smola, A.J. (2008) A kernel statistical test of independence. *Advances in Neural Information Processing Systems*, **20**, 585–592.

Gretton, A. and Györfi, L. (2010) Consistent nonparametric tests of independence. *Journal of Machine Learning Research*, **11**, 1391–1423.

Guendelman, M.D., Cheryan, S., and Monin, B. (2011) Fitting in but getting fat: identity threat and dietary choices among us immigrant groups. *Psychological Science*, **22** (7), 959–967, doi: 10.1177/0956797611411585.

Haavelmo, T. (1943) The statistical implications of a system of simultaneous equations. *Econometrica*, **11** (1), 1–12, doi: 10.2307/1905714.

Hayes, A.F. and Scharkow, M. (2013) The relative trustworthiness of inferential tests of the indirect effect in statistical mediation analysis: does method really matter? *Psychological Science*, **24** (10), 1918–1927, doi: 10.1177/0956797613480187.

Hicks, J.R. (1979) *Causality in Economics*, Basil Blackwell, Oxford.

Hoeffding, W. (1951) 'Optimum' nonparametric tests, in Proceedings of the 2nd Berkeley Symposion on Mathematical Statistics, University of California Press, pp. 83–92.

Holland, P.W. (1988) Causal inference, path analysis, and recursive structural equations models. *Sociological Methodology*, **18**, 449–484, doi: 10.2307/271055.

Huang, J.Y., Sedlovskaya, A., Ackerman, J.M., and Bargh, J.A. (2011) Immunizing against prejudice: effects of disease protection on attitudes toward out-groups. *Psychological Science*, **22** (12), 1550–1556, doi: 10.1177/0956797611417261.

Hume, D. (1777/1975) *Enquiries Concerning Human Understanding and Concerning the Principles of Morals*, Clarendon Press, Oxford.

Hyvärinen, A. (1998) New approximations of differential entropy for independent component analysis and projection pursuit. *Advances in Neural Information Processing Systems*, **10**, 273–279.

Iacobucci, D., Saldanha, N., and Deng, X. (2007) A meditation on mediation: evidence that structural equations models perform better than regressions. *Journal of Consumer Psychology*, **17** (2), 139–153, doi: 10.1016/S1057-7408(07)70020-7.

Imai, K., Keele, L., and Tingley, D. (2010) A general approach to causal mediation analysis. *Psychological Methods*, **15** (4), 309–334, doi: 10.1037/a0020761.

Imai, K., Keele, L., Tingley, D., and Yamamoto, T. (2011) Unpacking the black box of causality: learning about causal mechanisms from experimental and observational studies. *American Political Science Review*, **105** (4), 765–789, doi: 10.1017/S0003055411000414.

Imai, K., Keele, L., Tingley, D., and Yamamoto, T. (2014) Comment on Pearl: practical implications of theoretical results for causal mediation analysis. *Psychological Methods*, **19** (4), 482–487, doi: 10.1037/met0000021.

Jöreskog, K.G. (1970) A general method for analysis of covariance structures. *Biometrika*, **57** (2), 239–251, doi: 10.1093/biomet/57.2.239.

Judd, C.M. and Kenny, D.A. (2010) Data analysis in social psychology: recent and recurring issues, in *Handbook of Social Psychology* (eds S.T. Fiske, D.T. Gilbert, and G. Lindzey), John Wiley & Sons, Inc., New York, pp. 115–139.

Kac, M. (1939) On a characterization of the normal distribution. *American Journal of Mathematics*, **61** (3), 726–728.

Kenny, D.A. and Judd, C.M. (2014) Power anomalies in testing mediation. *Psychological Science*, **25** (2), 334–339, doi: 0956797613502676.

Kisbu-Sakarya, Y., MacKinnon, D.P., and Miočević, M. (2014) The distribution of the product explains normal theory mediation confidence interval estimation. *Multivariate Behavioral Research*, **49** (3), 261–268, doi: 10.1080/00273171.2014.903162.

Koller, I. and Alexandrowicz, R. (2010) A psychometric analysis of the ZAREKI-R using Rasch-models. *Diagnostica*, **56**, 57–67, doi: 10.1026/0012-1924/a000003.

Kraemer, H.C., Kiernan, M., Essex, M., and Kupfer, D.J. (2008) How and why criteria defining moderators and mediators differ between the Baron & Kenny and MacArthur approaches. *Health Psychology*, **27** (2S), S101–S108, doi: 10.1037/0278-6133.27.2(Suppl.).S101.

Laha, R. (1957) On a characterization of the normal distribution from properties of suitable linear statistics. *Annals of Mathematical Statistics*, **28** (1), 126–139, doi: 10.1214/aoms/1177707041.

Langer, S.L., Romano, J.M., Mancl, L., and Levy, R.L. (2014) Parental catastrophizing partially mediates the association between parent-reported child pain behavior and parental protective responses. *Pain Research and Treatment*, **2014**, 1–9, doi: 10.1155/2014/751097.

Lazarsfeld, P.F. (1955) Interpretation of statistical relations as a research operation, in *The Language of social research* (eds P.F. Lazarsfeld and M. Rosenberg), Free Press, New York, pp. 115–125.

Link, B.G. and Shrout, P.E. (1992) Spurious associations in longitudinal research. *Research in Community and Mental Health*, **7**, 301–321.

MacKinnon, D.P. (2008) *Introduction to Statistical Mediation Analysis*, Lawrence Erlbaum Associates, New York.

MacKinnon, D.P., Lockwood, C.M., Hoffman, J.M., West, S.G., and Sheets, V. (2002) A comparison of methods to test mediation and other intervening variable effects. *Psychological Methods*, **7** (1), 83–104, doi: 10.1037/1082-989X.7.1.83.

MacKinnon, D.P., Lockwood, C.M., and Williams, J. (2004) Confidence limits for the indirect effect: distribution of the product and resampling methods. *Multivariate Behavioral Research*, **39** (1), 99–128, doi: 10.1207/s15327906mbr3901-4.

Marszalek, J.M., Barber, C., Kohlhart, J., and Holmes, C.B. (2011) Sample size in psychological research over the past 30 years. *Perceptual and Motor Skills*, **112** (2), 331–348, doi: 10.2466/03.11.PMS.112.2.331-348.

Maxwell, S.E., Cole, D.A., and Mitchell, M.A. (2011) Bias in cross-sectional analyses of longitudinal mediation: partial and complete mediation under an autoregressive model. *Multivariate Behavioral Research*, **46** (5), 816–841, doi: 10.1080/00273171.2011.606716.

Mayer, A., Thoemmes, F., Rose, N., Steyer, R., and West, S.G. (2015) Theory and analysis of total, direct, and indirect causal effects. *Multivariate Behavioral Research*, **49** (5), 425–442, doi: 10.1080/00273171.2014.931797.

Micceri, T. (1989) The unicorn, the normal curve, and other improbable creatures. *Psychological Bulletin*, **105** (1), 156–166, doi: 10.1037/0033-2909.105.1.156.

Mitchell, M.A. and Maxwell, S.E. (2013) A comparison of the cross-sectional and sequential designs when assessing longitudinal mediation. *Multivariate Behavioral Research*, **48** (3), 301–339, doi: 10.1080/00273171.2013.784696.

Mooijaart, A. (1985) Factor analysis for non-normal variables. *Psychometrika*, **50** (3), 323–342, doi: 10.1007/BF02294108.

Mooijaart, A. and Bentler, P.M. (2010) An alternative approach for nonlinear latent variable models. *Structural Equation Modeling*, **17** (3), 357–373, doi: 10.1080/10705511.2010.488997.

Muddapur, M. (2003) On directional dependence in a regression line. *Communications in Statistics: Theory and Methods*, **32** (10), 2053–2057, doi: 10.1081/STA-120023266.

Ney, A. (2009) Physical causation and difference-making. *British Journal for the Philosophy of Science*, **60** (4), 737–764, doi: 10.1093/bjps/axp037.

Oishi, S., Seol, K.O., Koo, M., and Miao, F.F. (2011) Was he happy? Cultural difference in conceptions of Jesus. *Journal of Research in Personality*, **45** (1), 84–91, doi: 10.1016/j.jrp.2010.11.018.

O'Rourke, H.P. and MacKinnon, D.P. (2015) When the test of mediation is more powerful than the test of the total effect. *Behavior Research Methods*, **47** (2), 424–442.

Ozaki, K. and Ando, J. (2009) Direction of causation between shared and non-shared environmental factors. *Behavior Genetics*, **39** (3), 321–336.

Ozaki, K., Toyoda, H., Iwama, N., Kubo, S., and Ando, J. (2011) Using non-normal SEM to resolve the ACDE model in the classical twin design. *Behavior Genetics*, **41** (2), 329–339.

Pearl, J. (2001) Direct and indirect effects, in *Proceedings of the 17th Conference on Uncertainty in Artificial Intelligence* (eds J. Breese and D. Koller), Morgan Kaufmann Publishers Inc., San Francisco, CA, pp. 411–420.

Pearl, J. (2009) *Causality: Models, Reasoning, and Inference*, 2nd edn, Cambridge University Press, Cambridge.

Pearl, J. (2012) The causal mediation formula—a guide to the assessment of pathways and mechanisms. *Prevention Science*, **13** (4), 426–436, doi: 10.1007/s11121-011-0270-1.

Pearl, J. (2014a) Interpretation and identification of causal mediation. *Psychological Methods*, **19** (4), 459–481, doi: 10.1037/a0036434.

Pearl, J. (2014b) Reply to commentary by Imai, Keele, Tingley, and Yamamoto concerning causal mediation analysis. *Psychological Methods*, **19** (4), 488–492, doi: 10.1037/met0000022.

Preacher, K.J. and Selig, J.P. (2012) Advantages of Monte Carlo confidence intervals for indirect effects. *Communication Methods and Measures*, **6** (2), 77–98, doi: 10.1080/19312458.2012.679848.

R Core Team (2016) *R: A Language and Environment for Statistical Computing*, R Foundation for Statistical Computing, Vienna, Austria, `http://www.R-project.org/` (accessed 18 December 2015).

Robins, J. (1986) A new approach to causal inference in mortality studies with a sustained exposure period—application to control of the healthy worker survivor effect. *Mathematical Modelling*, **7** (9), 1393–1512, doi: 10.1016/0270-0255(86)90088.6.

Robins, J. and Greenland, S. (1992) Identifiability and exchangeability for direct and indirect effects. *Epidemiology*, **3** (2), 143–155, doi: 10.1097/00001648-199203000-00013.

Rucker, D.D., Preacher, K.J., Tormala, Z.L., and Petty, R.E. (2011) Mediation analysis in social psychology: current practices and new recommendations. *Social and Personality Psychology Compass*, **5** (6), 359–371, doi: 10.1111/j.1751-9004.2011.00355.x.

Scheines, R., Spirtes, P., Glymour, C., Meek, C., and Richardson, T. (1998) The TETRAD project: constraint based aids to causal model specification. *Multivariate Behavioral Research*, **33** (1), 65–117, doi: 10.1207/s15327906mbr3301-3.

Shimizu, S. (2014) LiNGAM: Non-Gaussian methods for estimating causal structures. *Behaviormetrika*, **41** (1), 65–98, doi: 10.2333/bhmk.41.65.

Shimizu, S. (2016) Non-Gaussian structural equation models for causal discovery, in *Statistics and Causality: Methods for Applied Empirical Research* (eds W. Wiedermann and A. von Eye), John Wiley & Sons, Inc.

Shimizu, S. and Kano, Y. (2008) Use of non-normality in structural equation modeling: application to direction of causation. *Journal of Statistical Planning and Inference*, **138** (11), 3483–3491, doi: 10.1016/j.jspi.2006.01.017.

Shimizu, S., Hoyer, P.O., Hyvärinen, A., and Kerminen, A. (2006a) A linear non-Gaussian acyclic model for causal discovery. *Journal of Machine Learning Research*, **7**, 2003–2030.

Shimizu, S., Hyvärinen, A., Hoyer, P.O., and Kano, Y. (2006b) Finding a causal ordering via independent component analysis. *Computational Statistics and Data Analysis*, **50** (11), 3278–3293, doi: 10.1016/j.csda.2005.05.004.

Shimizu, S., Inazumi, T., Sogawa, Y., Hyvärinen, A., Kawahara, Y., Washio, T., Hoyer, P.O., and Bollen, K. (2011) DirectLiNGAM: a direct method for learning a linear non-Gaussian structural equation model. *Journal of Machine Learning Research*, **12**, 1225–1248.

Shrum, L., Lee, J., Burroughs, J.E., and Rindfleisch, A. (2011) An online process model of second-order cultivation effects: how television cultivates materialism and its consequences for life satisfaction. *Human Communication Research*, **37** (1), 34–57, doi: 10.1111/j.1468-2958.2010.01392.x.

Skitovich, . (1953) On a property of the normal distribution. *Doklady Akademii Nauk SSSR*, **89**, 217–219.

Smith, E.R. (1982) Beliefs, attributions, and evaluations: nonhierarchical models of mediation in social cognition. *Journal of Personality and Social Psychology*, **43** (2), 248–259, doi: 10.1037/0022-3514.43.2.248.

Sobel, M.E. (1996) An introduction to causal inference. *Sociological Methods & Research*, **24** (3), 353–379, doi: 10.1177/0049124196024003004.

Sobel, M.E. (2008) Identification of causal parameters in randomized studies with mediating variables. *Journal of Educational and Behavioral Statistics*, **33** (2), 230–251, doi: 10.3102/1076998607307239.

Spencer, S.J., Zanna, M.P., and Fong, G.T. (2005) Establishing a causal chain: why experiments are often more effective than mediational analyses in examining psychological processes. *Journal of Personality and Social Psychology*, **89** (6), 845–851, doi: 10.1037/0022-3514.89.6.845.

Stuart, A. (1954) Asymptotic relative efficiencies of distribution-free tests of randomness against normal alternatives. *Journal of the American Statistical Association*, **49** (265), 147–157, doi: 10.2307/2281041.

Sungur, E.A. (2005) A note on directional dependence in regression setting. *Communications in Statistics: Theory and Methods*, **34** (9-10), 1957–1965, doi: 10.1080/03610920500201228.

Tofighi, D., MacKinnon, D.P., and Yoon, M. (2009) Covariances between regression coefficient estimates in a single mediator model. *British Journal of Mathematical and Statistical Psychology*, **62** (3), 457–484, doi: 10.1348/000711008x331024.

Usborne, E. and Taylor, D.M. (2010) The role of cultural identity clarity for self-concept clarity, self-esteem, and subjective well-being. *Personality and Social Psychology Bulletin*, **36** (7), 883–897, doi: 10.1177/0146167210372215.

Valeri, L. and VanderWeele, T.J. (2013) Mediation analysis allowing for exposure–mediator interactions and causal interpretation: theoretical assumptions and implementation with SAS and SPSS macros. *Psychological Methods*, **18** (2), 137–150, doi: 10.1037/a0031034.

van der Waerden, B.L. (1952) Order tests for the two-sample problem and their power. *Indagationes Mathematicae*, **14**, 453–458.

Wang, X., Sobel, M.E., and Morgan, S.L. (2013) New perspectives on causal mediation analysis, in *Handbook of Causal Analysis for Social Research* (ed. S.L. Morgan), Springer-Verlag, Dordrecht, pp. 215–242.

Wiedermann, W. and Alexandrowicz, R.W. (2011) A modified normal scores test for paired data. *Methodology: European Journal of Research Methods for the Behavioral and Social Sciences*, **7** (1), 25–38, doi: 10.1027/1614-2241/a000020.

Wiedermann, W. and von Eye, A. (2015a) Direction-dependence analysis: a confirmatory approach for testing directional theories. *International Journal of Behavioral Development*, doi: 10.1177/0165025415582056.

Wiedermann, W. and von Eye, A. (2015b) Direction of effects in mediation analysis. *Psychological Methods*, **20** (2), 221–244, doi: 10.1037/met0000027.

Wiedermann, W. and von Eye, A. (2015c) Direction of effects in multiple linear regression models. *Multivariate Behavioral Research*, **50** (1), 23–40, doi: 10.1080/00273171.2014.958429.

Williamson, J. (2011) Mechanistic theories of causality Part I. *Philosophy Compass*, **6** (6), 421–432, doi: 10.1111/j.1747-9991.2011.00400.x.

Wright, S. (1921) Correlation and causation. *Journal of Agricultural Research*, **20** (7), 557–585.

Yuan, Y. and MacKinnon, D.P. (2014) Robust mediation analysis based on median regression. *Psychological Methods*, **19** (1), 1–20, doi: 10.1037/a0033820.

Zar, J.H. (1972) Significance testing of the spearman rank correlation coefficient. *Journal of the American Statistical Association*, **67** (339), 578–580, doi: 10.2307/2284441.

Zhang, Z. (2014) Monte Carlo based statistical power analysis for mediation models: methods and software. *Behavior Research Methods*, **46** (4), 1184–1198, doi: 10.3758/s13428-013-0424-0.

Zhang, K. and Hyvärinen, A. (2010) Distinguishing causes from effects using nonlinear acyclic causal models, in Journal of Machine Learning Research Workshop and Conference Proceedings, Vol. 6, pp. 157–164.

Zimmerman, D.W., Williams, R.H., and Zumbo, B.D. (1993) Reliability of measurement and power of significance tests based on differences. *Applied Psychological Measurement*, **17** (1), 1–9, doi: 10.1177/014662169301700101.

5

DIRECTION OF EFFECTS IN CATEGORICAL VARIABLES: A STRUCTURAL PERSPECTIVE

ALEXANDER VON EYE

Department of Psychology, Michigan State University, East Lansing, MI, United States

WOLFGANG WIEDERMANN

Department of Educational, School and Counseling Psychology, College of Education, University of Missouri, Columbia, MO, USA

5.1 INTRODUCTION

The concept of *independence* has been and still is playing an important role in categorical data analysis. A classical application comes in the form of the Pearson χ^2 test, in which the expected cell frequencies are estimated based on a log-linear model of independence of the row and column variables. In log-linear modeling, the independence model is often used as the base model with which more complex models are compared in the process of model development (see Agresti, 2013, von Eye and Mun, 2013). Other models of independence, for example, models of conditional independence, are frequently discussed and applied, and will be reviewed in the next section, along with independence as it is an integral element of latent class analysis.

In this chapter, we pursue two goals. First, we place the concept of independence in a new context. We embed this concept into an approach to testing hypotheses that are compatible with *direction of effects*. Methods for the analysis of direction of effects, as originally introduced by Dodge and Rousson (2000, 2001), have focused on the univariate distribution of an (continuous) outcome variable, as it is explained by one

Statistics and Causality: Methods for Applied Empirical Research, First Edition.
Wolfgang Wiedermann and Alexander von Eye.

(continuous) independent variable (see also the related chapter of Dodge and Rousson, 2016). The methods are based on the idea that the distribution of the outcome variable can be considered a convolution of two elements, a nonnormally distributed explanatory variable and a normally distributed random term, the error term, which is assumed to be independent of the explanatory variable (i.e., it is assumed that no latent confounders exist). In the context of linear regression models, this idea can be illustrated using the simple regression model, $Y = \beta_0 + \beta_1 X + \epsilon$, where Y is the outcome variable, β_0 is the intercept, β_1 is the slope parameter, X is the explanatory variable, and ϵ represents the residual term. If the regression model explains the distribution of the outcome variable (Y) and ϵ is normally distributed and independent of X, the distribution of Y will be closer to normal than the distribution of the explanatory variable (X), provided that the distribution of X is nonnormal. The concept of explaining the distribution of an outcome variable was introduced into the domain of categorical variables by Wiedermann and von Eye (2015a).

Second, we propose a new perspective on direction dependence. Instead of looking at univariate distributions, we also look at multivariate distributions of categorical variables. We ask whether explanatory variables allow one to capture multivariate distributions of manifest categorical variables, also including structural characteristics of categorical outcome variables.

The remainder of this chapter is structured as follows. First, we review concepts of independence and, second, concepts of direction dependence. We then discuss the explanation of effects in cross-classifications. To this end, a new principle is proposed, the *generalized direction of effect principle*. Further, we show that the direction dependence principle originally proposed by Dodge and Rousson (2000, 2001) can be considered a special case of the generalized direction of effect principle. Results of a Monte Carlo simulation study are presented, which illustrate the relation between Dodge and Rousson's (2000, 2001) direction dependence principle and the generalized direction of effect principle. Application of the latter principle is illustrated in a real-world data example.

5.2 CONCEPTS OF INDEPENDENCE IN CATEGORICAL DATA ANALYSIS

To introduce and illustrate concepts of independence in categorical variables (see Agresti, 2013, von Eye and Mun, 2013), we use the three manifest categorical variables X, Y, and Z with categories i, j, and k. Log-linear models that take into account selections of the main effects of these three variables include

$$\log\ \hat{m}_{ijk} = \lambda$$

$$\log\ \hat{m}_{ijk} = \lambda + \lambda_i^X$$

$$\log\ \hat{m}_{ijk} = \lambda + \lambda_j^Y$$

$$\log\ \hat{m}_{ijk} = \lambda + \lambda_k^Z$$

$$\log\ \hat{m}_{ijk} = \lambda + \lambda_i^X + \lambda_j^Y$$
$$\log\ \hat{m}_{ijk} = \lambda + \lambda_i^X + \lambda_k^Z$$
$$\log\ \hat{m}_{ijk} = \lambda + \lambda_j^Y + \lambda_k^Z$$

and

$$\log\ \hat{m}_{ijk} = \lambda + \lambda_i^X + \lambda_j^Y + \lambda_k^Z$$

In each of these models,[1] the log of the estimated cell frequencies ($\hat{m}_{ijk}$) equals an additive function of model parameters (λ), where the superscripts denote the variables involved in an effect. The first of these models posits that the cells of the $X \times Y \times Z$ cross-classification are equiprobable. This implies that X, Y, and Z are independent. The second model posits that the cells of the $Y \times Z$ subtable are equiprobable but that X can have a main effect. This also implies that X, Y, and Z are independent. This applies accordingly to the third and fourth models. The fifth model posits that the categories of Z are equiprobable and that X and Y can have main effects. The last model is the main effect model of all three variables, also known as the *model of mutual independence*.

Starting from the model of mutual independence, we can formulate the *model of joint independence* of X as

$$\log\ \hat{m}_{ijk} = \lambda + \lambda_i^X + \lambda_j^Y + \lambda_k^Z + \lambda_{jk}^{YZ}$$

This model posits that X is jointly independent of Y and Z. If one defines a variable that has the YZ combinations (configurations; see von Eye and Gutiérrez Peña, 2004) as its categories, then this model posits that X is independent of this variable.

The *model of conditional independence* of X and Y, given Z is

$$\log\ \hat{m}_{ijk} = \lambda + \lambda_i^X + \lambda_j^Y + \lambda_k^Z + \lambda_{ik}^{XZ} + \lambda_{jk}^{YZ}$$

This model posits that X and Y are independent in each subtable when Z is fixed. In other words, this model posits that X and Y are independent in each category of Z. There are other models of independence, for example, the model of marginal independence. This model, however, is not a log-linear model and will not be discussed in the remainder of this chapter, where we remain in the context of log-linear models.

It is interesting to realize that the model of latent class analysis (LCA; Lazarsfeld, 1955, Vermunt, 1997) uses a model of independence similar to the ones discussed here. LCA is used to explain the covariation among manifest categorical variables from one or more latent variables. Let there be a model with one latent variable, L, and the three manifest variables, X, Y, and Z. Then, the log-linear representation of the latent class model of these four variables is

$$\log\ \hat{m}_{ijk} = \lambda + \lambda_i^X + \lambda_j^Y + \lambda_k^Z + \lambda_l^L + \lambda_{il}^{XL} + \lambda_{jl}^{YL} + \lambda_{kl}^{ZL}$$

[1]Note that here and in the rest of this chapter, we define λ parameters under the zero-sum constraint (see von Eye and Mun, 2013).

Evidently, this model corresponds to a model of conditional independence. That is, the manifest variables X, Y, and Z are mutually independent in each category of L. In the following sections, we discuss direction dependence for metric and categorical variables with reference to models of conditional independence and then apply concepts of independence to direction dependence.

5.3 DIRECTION DEPENDENCE IN BIVARIATE SETTINGS: METRIC AND CATEGORICAL VARIABLES

Instead of repeating the details of Dodge and Rousson's (2000, 2001) approach, we start by illustrating the distributional characteristics of three manifest metric variables, X, Y, and ϵ. Let X be a skewed explanatory variable, Y denotes the outcome variable, and ϵ is the normally distributed residual. Figure 5.1 displays the kernel density plots of the two artificial variables X and Y. The skewed X variable was generated as the square of an $N(0, 1)$ variable (squaring resulted in a skewness of 2.81), the residual term (ϵ) was a standard normally distributed variable (with a skewness of 0.01), and Y was the sum of the skewed X and the residual ϵ (i.e., $\beta_0 = 0$ and $\beta_1 = 1$), which results in a skewness of 1.55.

Figure 5.1 shows that the outcome variable, $Y(= X + \epsilon)$, is less skewed than the explanatory variable, X. Based on Dodge and Rousson's (2000) results, we know that, of two variables, only the one with more skew can be the explanatory variable, provided that assumptions of the linear regression model are fulfilled. This applies to cross-sectional studies as well as to longitudinal studies in which the change in skewness can be examined when the explanatory variable (the cause) becomes active or ceases to be active between two observations (von Eye and DeShon, 2012). This applies accordingly in the multivariate case and when regression residuals of competing linear models are analyzed instead of raw scores (Wiedermann and Hagmann, 2015, Wiedermann *et al.*, 2015, Wiedermann and von Eye, 2015c).

Standard textbooks on log-linear modeling, such as, for example, Knoke and Burke (1980) or Agresti (2013), usually introduce the log-linear model in its general form in which no explicit distinction is made between predictor and outcome variables. In *general log-linear models*, all variables are considered as "responses" whose mutual associations are analyzed. If researchers distinguish between predictor and outcome variables, logistic regression models or multinomial logistic regression models are typically applied. The relations between, for example, a logistic regression model and a log-linear model are well-known. Every logistic regression model can be reformulated as a log-linear model; however, the opposite, that each log-linear model has a logistic regression correspondence, does not generally hold (Agresti, 2013, von Eye and Mun, 2013). When nominal-level categorical variables[2] are analyzed using log-linear models, the goal of analysis can be similar to the metric

[2]In the following sections, we focus on binary variables. When variables are multicategorical, not all coding forms guarantee independence of individual effects, which would impede parameter interpretation and model selection.

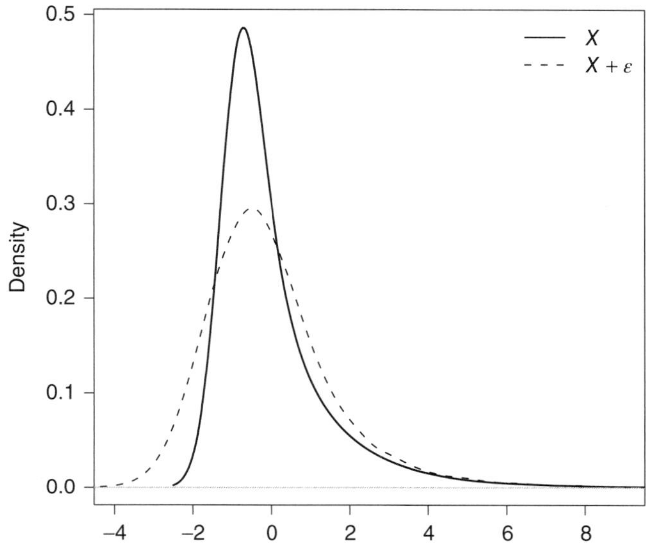

Figure 5.1 Kernel density plots of X (skewness = 2.81) and Y (skewness = 1.55).

case, that is, researchers can aim at explaining the univariate probability distribution of a putative outcome variable. In the following paragraphs, we discuss log-linear models, which can be used to answer these types of research questions.

In the first step, we need to distinguish between three types of information in a contingency table (for simplicity, we start with the bivariate case; X and Y with categories i and j): the observed frequencies (n_{ij}) of cells ij, the expected cell frequencies (m_{ij}), and the estimated values of expected cell frequencies ($\hat{m}_{ij}$) based on a particular log-linear model.

Consider the (saturated) log-linear model

$$\log\ \hat{m}_{ij} = \lambda + \lambda_i^X + \lambda_j^Y + \lambda_{ij}^{XY}$$

for the two categorical variables X and Y. In this special case, the number of parameters equals the number of cells and the model will, by necessity, perfectly describe any $m_{ij} > 0$. Thus, we have $n_{ij} = \hat{m}_{ij}$. Now, suppose X is hypothesized to be the explanatory and Y is hypothesized to be the outcome variable. Suppose, in addition, that the univariate probability distribution of X is not uniform (the necessity of this assumption is discussed next). If X explains the univariate probability distribution of Y, and if there is a relation between X and Y, the main effect term for Y is not needed any more. The model thus reduces to $\log\ \hat{m}_{ijk} = \lambda + \lambda_i^X + \lambda_{ij}^{XY}$, that is, a nonhierarchical model (for discussions of nonhierarchical log-linear models, see Mair and von Eye, 2007, Rindskopf, 1990, Vermunt, 1997, von Eye and Mun, 2013). Furthermore, provided that this nonhierarchical model with $\lambda_j^Y = 0$, $\lambda_i^X \neq 0$, and $\lambda_{ij}^{XY} \neq 0$ is indeed capable of validly describing the data, the competing log-linear model, that is, the model that posits that X is the outcome and Y is the predictor, is more likely to be rejected in terms of model fit.

Instead of expressing the "true" nonhierarchical model in terms of estimated values of expected frequencies ($\hat{m}_{ij}$), we can also write the model in terms of expected frequencies (m_{ij}), that is,

$$\log\ m_{ij} = \lambda + \lambda_i^X + \lambda_{ij}^{XY} + \epsilon_{ij}$$

with ϵ_{ij} being an error term that captures unexplained noise. Further, the misspecified log-linear model can be expressed as

$$\log\ m_{ij} = \lambda' + \lambda_j^Y + \lambda_{ij}^{XY} + \epsilon'_{ij}$$

which, due to $\lambda^Y = 0$, further reduces to

$$\log\ m_{ij} = \lambda' + \lambda_{ij}^{XY} + \epsilon'_{ij}$$

The error term of the misspecified model (ϵ'_{ij}) now conceptually captures the true error term, the main effect of X, and the difference in model intercepts, [3] that is, $\epsilon'_{ij} = \epsilon_{ij} + \lambda_i^X + (\lambda - \lambda')$, from which follows that the portion of unexplained variability of the misspecified model can expected to be larger than the portion of unexplained variability in the "true" model, provided that $\lambda_i^X \neq 0$.

The error term of a log-linear model basically reflects the difference between observed (n_{ij}) and predicted cell frequencies ($\hat{m}_{ij}$). Whether a particular difference value ($n_{ij} - \hat{m}_{ij}$) indicates a poor or a good prediction depends on the cell counts itself. Thus, the differences (i.e., the estimated residuals of the model) need to be rescaled or standardized. For example, dividing $n_{ij} - \hat{m}_{ij}$ by $\sqrt{\hat{m}_{ij}}$ leads to the standardized Pearson residuals (see, e.g., Christensen, 1990). The components of the error term of a log-linear model constitute the key element for evaluating the fit of a model. The commonly used likelihood ratio (LR) χ^2 statistic, for example, is defined as $\chi^2 = 2\sum n_{ij} \log\ (n_{ij}/\hat{m}_{ij})$ from which naturally follows that the χ^2 statistic increases with deviations of predicted and observed frequencies. Thus, for $\lambda_j^Y = 0$, $\lambda_i^X \neq 0$, and $\lambda_{ij}^{XY} \neq 0$, it follows that the LR-χ^2 statistic for the misspecified model will be larger than the LR-χ^2 statistic obtained from the correctly specified model. Model selection can be performed through separately inspecting the results of model fit tests. If the LR test retains the null hypothesis for the nonhierarchical log-linear model $X \rightarrow Y$ and, at the same time, rejects the null hypothesis of model fit for the competing model $Y \rightarrow X$, then Y is more likely to reflect the outcome variable and X is more likely to constitute the cause.

So far, we have discussed the ideal scenario in which one particular nonhierarchical model describes the true data-generating process. In the next step, we link Dodge and Rousson's (2000, 2001) direction dependence results for metric variables

[3]Formally, the intercepts represent the overall mean of the logarithms of the expected frequencies.

to the proposed directional (nonhierarchical) log-linear models and show that data requirements assumed in Dodge and Rousson (2000, 2001) lead to a special case of the generalized direction of effects principle. For this purpose, let X' and Y' be two latent continuous variables where X' constitutes the cause of Y'. Further, we assume that the functional relation between the two latent variables can validly be described by the (latent) linear regression model (for simplicity, we assume that variables have an expected value of $E[X'] = E[Y'] = 0$)

$$Y' = \beta_1 X' + \eta$$

The error term of this model (η) is assumed to be normally distributed and independent of X'. When X' is asymmetrically distributed (i.e., the skewness $\gamma_{X'} \neq 0$), the skewness of the outcome can be written as

$$\gamma_{Y'} = \rho^3 \gamma_{X'}$$

(with ρ being the Pearson correlation of X' and Y'), which directly follows from Dodge and Rousson's (2000, 2001) direction dependence results. From this relation, it follows that $|\gamma_{Y'}| < |\gamma_{X'}|$ given that $|\rho| < 1$ and $\gamma_{X'} \neq 0$. For a symmetrically distributed predictor ($\gamma_{X'} = 0$), one obtains $\gamma_{Y'} = 0$, independently of ρ. Similarly, in case of $\rho = 0$, one obtains $\gamma_{Y'} = 0$ independently of $\gamma_{X'}$.

Now, assume that the two latent continuous variables (X' and Y') are only partially observed as categorical variables, X and Y (e.g., we only know whether a respondent's response exceeds a critical threshold, τ). Assume that the predictor is observed according to the threshold rule:

$$X = \begin{cases} 0 & \text{if } X' \leq \tau \\ 1 & \text{if } X' > \tau, \end{cases}$$

with τ set equal to zero. Y is defined in a similar fashion.

Then, considering that $|\gamma_{Y'}|$ will always be smaller than $|\gamma_{X'}|$, it naturally follows that $\Pr(Y = 1)$ will be closer to 0.5 (i.e., the distribution of Y will be closer to uniformity) than $\Pr(X = 1)$. Conversely, the distribution of X will systematically deviate from uniformity given $\gamma_{X'} \neq 0$. In terms of parameters of a log-linear model, this implies that $\lambda_j^Y \to 0$ as $\gamma_{Y'} \to 0$, and $\lambda_i^X \neq 0$ given $\gamma_{X'} \neq 0$. Of course, given $\gamma_{Y'} = \rho^3 \gamma_{X'}$, one cannot expect that $\lambda_j^Y = 0$ will exactly hold (except for the practically irrelevant case of $\rho = 0$ or, equivalently, $\lambda_{ij}^{XY} = 0$); however, in terms of statistical inference, it is more likely that the main effect of the outcome variable (λ_j^Y) can be omitted without causing substantial harm to the model fit, which leads to the nonhierarchical model discussed earlier. To demonstrate the behavior of competing nonhierarchical log-linear models in this setup, we performed a Monte Carlo simulation experiment, which is discussed in detail in the next section.

5.3.1 Simulating the Performance of Nonhierarchical Log-Linear Models

To demonstrate that the direction dependence principle outlined by Dodge and Rousson (2000, 2001) generalizes to the case of categorical variables, we performed a Monte Carlo simulation using the R programming environment (R Core Team, 2015). Two latent continuous variables (X' and Y') were generated according to $Y' = \beta_1 X' + \epsilon$, with $E[X'] = E[Y'] = 0$ and $\epsilon \approx N(0, 1)$. The slope parameter was selected to obtain correlations of $\rho = 0, 0.2, 0.4, 0.6$, and 0.8. First, the true predictor, X', was sampled from a standard normally distributed population. Because in this case, one expects $\gamma_{X'} = 0$, no decision about directionality can be made using nonhierarchical log-linear models. Next, the predictor was sampled from various χ^2-distributions with $\gamma_{X'} = 0.5, \cdots, (0.5), \cdots, 3$. The simulation factors were fully crossed, leading to 5 (magnitude of ρ) $\times$ 7 (magnitude of $\gamma_{X'}$) = 35 experimental conditions. For each condition, 5000 samples were generated. Each generated sample was dichotomized using the decision rule described earlier (with the threshold set at $\tau = 0$), and the resulting binary variables were further analyzed using two competing nonhierarchical log-linear models. The model that posits $X \rightarrow Y$ consisted of the intercept (λ), the main effect of X (λ_i^X), and the interaction effect (λ_{ij}^{XY}). The second model, $Y \rightarrow X$, consisted of the intercept (λ'), the main effect of Y (λ_j^Y), and the interaction effect (λ_{ij}^{XY}). Type I error and power rates were determined in terms of model selection, that is, given that the LR test retains the null hypothesis for the model $X \rightarrow Y$, and, simultaneously, rejects the null hypothesis for the model $Y \rightarrow X$, the procedure correctly identified the underlying model. All tests were performed using a nominal significance level of 0.05.

Table 5.1 shows the empirically observed Type I error rates for the two LR tests and the model selection procedure based on the combined decisions of the LR tests for normally distributed (latent) predictors X' as a function of ρ. Type I error rates for all procedures fall within Bradley's liberal robustness interval of 0.025–0.075 given the nominal significance level of 0.05 Bradley (1978). Thus, as expected, for normally distributed true predictors, no decision about directionality can be made based on the two nonhierarchical log-linear models.

Next, we ask questions concerning the power of the nonhierarchical log-linear models to identify the correct direction of effects given nonnormal continuous predictors. Figure 5.2 shows the boxplots of the observed relative frequencies for $X = 1$

TABLE 5.1 Empirical Type I Error Rates of Model Goodness of Fit Tests and the Selection Procedure Based on the Combined Model Fit Decisions.

ρ	LR test ($X \rightarrow Y$)	LR test ($Y \rightarrow X$)	Model Selection
0.0	0.048	0.050	0.048
0.2	0.045	0.055	0.053
0.4	0.049	0.047	0.043
0.6	0.048	0.048	0.040
0.8	0.047	0.048	0.037

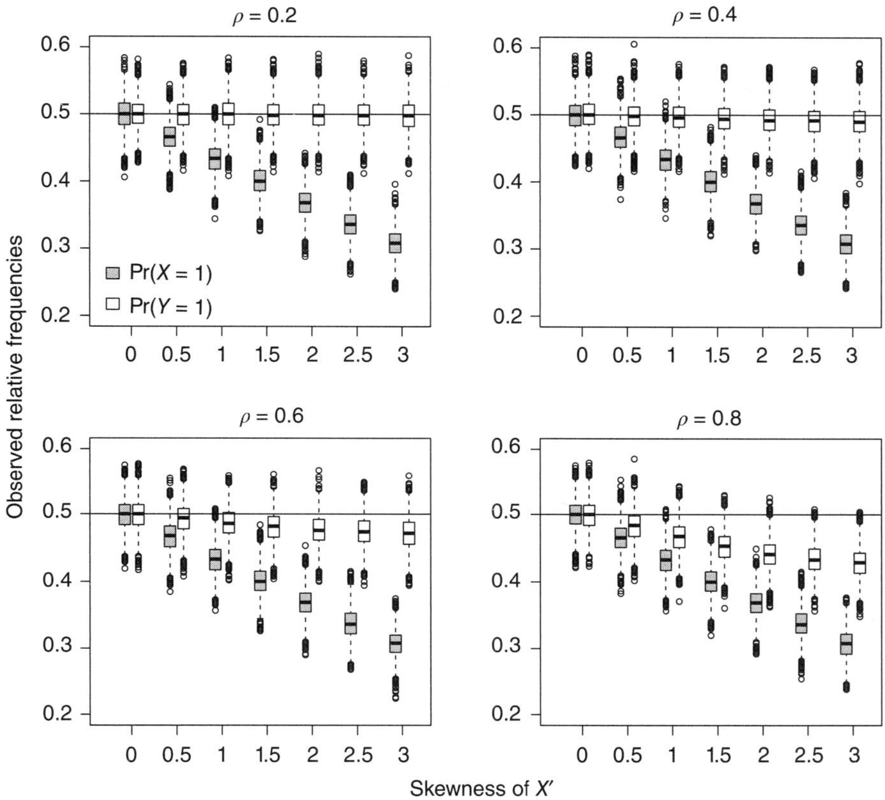

Figure 5.2 Observed relative frequencies for $X = 1$ and $Y = 1$ as a function of ρ and $\gamma_{X'}$.

and $Y = 1$, that is, measures for $\Pr(X = 1)$ and $\Pr(Y = 1)$, as a function of the true underlying correlation of X' and Y' and the true skewness of X'. As expected, the relative frequencies of $X = 1$ decrease with the skewness of X'. In contrast, for small to medium correlations, the relative frequencies of $Y = 1$ are close to 0.5 and systematically decrease for larger correlations. In terms of power of retaining the nonhierarchical log-linear model, which posits $X \rightarrow Y$ and, at the same time, rejecting the competing model ($Y \rightarrow X$), we expect an inverse U-shaped function, which depends on both the true underlying skewness of X' and the true underlying correlation ρ. Simulation results confirm this proposition. Figure 5.3 shows the empirically observed power curves for (i) the model selection procedure described earlier (top left panel), (ii) the LR test of the true model (top right panel), (iii) the LR test of the false model (middle left panel), (iv) the main effect λ_i^X of the true model (middle right panel), (v) the main effect λ_j^Y associated with the misspecified model (bottom left panel), and (vi) the interaction term λ_{ij}^{XY} of the true model (bottom right panel). In general, the power to identify the correct causal flow increases with the skewness of X'. However, the power of the model selection procedure is further moderated by the correlation

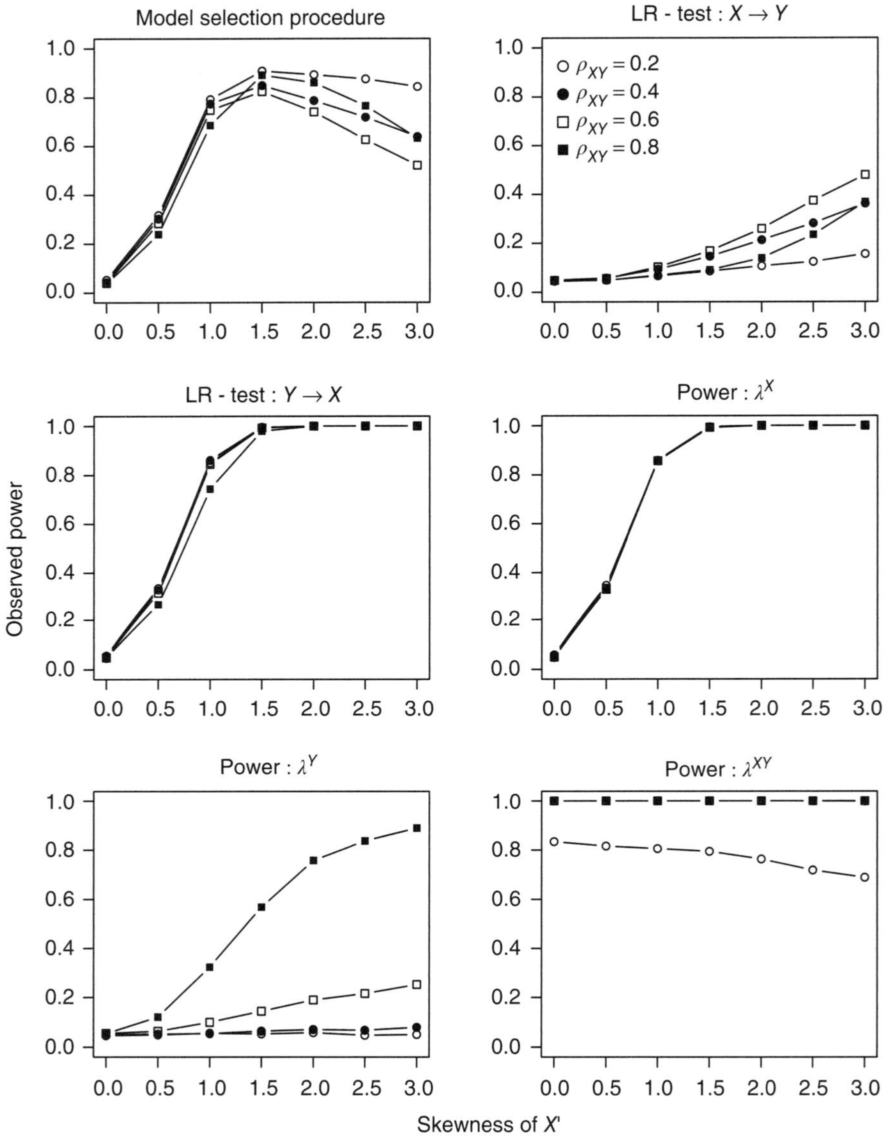

Figure 5.3 Observed power for the model selection procedure, the two LR tests, the main effects λ^X and λ^Y, and the interaction effect λ^{XY} of the true model (for simplicity, we omitted the subscripts of the λ terms).

between X' and Y' and decreases with ρ. Note that power loss is larger for medium correlations than for highly correlated variables. This effect can be explained by the fact that the observed power and the underlying correlation, ρ, are nonlinearly related. In general, the power loss can be explained by the behavior of the LR test of the correct model ($X \rightarrow Y$). Here, the probability of rejecting the null hypothesis increases for highly skewed continuous variables due to statistically significant main effects λ_j^Y (see bottom left panel). This effect is in good accordance with the theoretical relation of third moments of X' and Y', that is, $\gamma_{Y'} = \rho^3 \gamma_{X'}$. The skewness of Y' increases with the correlation ρ, which, in turn, leads to a nonuniformly distributed categorical variable Y. In other words, these are data settings where the main effect λ_j^Y exists and cannot be omitted from the log-linear model.

So far, we have focused on models that explain the univariate distribution of an outcome variable, that is, when $X \rightarrow Y$ reflects the data-generating process, the main effect of the outcome variable is not needed to explain observed frequency distributions. In the following sections, we extend the proposed methodology to answering structural questions of the data-generating process.

5.4 EXPLAINING THE STRUCTURE OF CROSS-CLASSIFICATIONS

In direction dependence analysis, the models for metric and nominal-level categorical variables share the characteristic that they are employed to explain the univariate distribution of an outcome variable. Models for multiple predictors have been proposed for metric variables (Wiedermann and von Eye, 2015c), and multivariate models have been proposed for categorical variables (Wiedermann and von Eye, 2015a). However, none of these models enables researchers to test hypotheses concerning structural elements of distributions. These elements would be reflected in interaction terms of first or higher order. We now discuss models for this purpose, in the context of categorical variables.

In more general terms, if the effects of a first set of variables explain the effects of a second set of variables, the terms needed to capture the effects of the second set of variables become redundant. They can be removed from the model without doing harm to model fit. More formally, consider the power set of the effects of variables, V, $\mathcal{P}(V) := \{U | U \subset V\}$, where U denotes the subsets of V. To exemplify, consider the three explanatory variables, X, Y, and Z. The power set of the effects of these three variables is depicted in Figure 5.4 (empty set depicted with no frame).

In Figure 5.4, one can see that the power set of the effects of the three variables, X, Y, and Z entirely consists of sets U that correspond with the effects that can be estimated in log-linear models. For example, the empty set $\emptyset$ corresponds with the null model. No effect is estimated. The set $\{X, Z\}$, corresponds to the effects estimated for the interaction between X and Z. In addition, the power set contains the relations between the subsets that correspond to hierarchical log-linear models. For example, the main effect of Y is part of the two-way interactions $[X, Y]$ and $[Y, Z]$, and the three-way interaction $[X, Y, Z]$ (we use brackets for log-linear terms and curly brackets (braces) for sets). The paths in the figure indicate how, in ascending order,

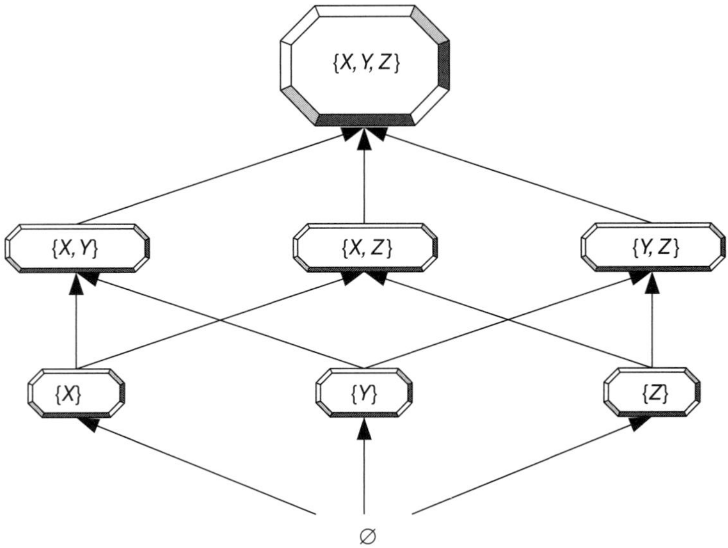

Figure 5.4 Power set of the three variables, X, Y, and, Z (presented in the form of a Hasse diagram).

higher order effects can be created for hierarchical models. Whenever a path from a lower order effect to a higher order effect does not exist, models that contain these two effects cannot be nested. For example, there is no path from $[X]$ to $[Y, Z]$. Therefore, models that contain these terms are not nested.

From a log-linear perspective, it should be noted that Figure 5.4 does contain all possible effects. However, it does not contain all possible models. Models can be created by including selections of effects, be they hierarchical or not (Mair and von Eye, 2007). In other words, the number of elements of $\mathcal{P}(V)$, denoted by $|\mathcal{P}(V)|$, is given by 2^t, where t is the number of variables. In the example with the three variables X, Y, and Z, $|\mathcal{P}(V)| = 2^3 = 8$. In contrast, the number of possible models is $t = 1 + \sum_g \binom{g}{1} + \sum_g \binom{g}{2} + \cdots$, where g is the number of variables in the analysis, and $|\mathcal{P}(V)| \leq t$.

In the following sections, we distinguish between two subsets of V, U_P, and U_C. U_P is a subset that contains effects of predictor variables, and U_C is a subset with effects of criterion or outcome variables. Together, U_P and U_C constitute V. The power set of V, $\mathcal{P}(V)$, which includes these two subsets, contains three groups of effects (null model implied):

(1) All possible effects of predictor variables; that is, all main effects and all interactions
(2) All possible effects of outcome variables; that is, all main effects and all interactions
(3) All interactions between predictor and outcome variables.

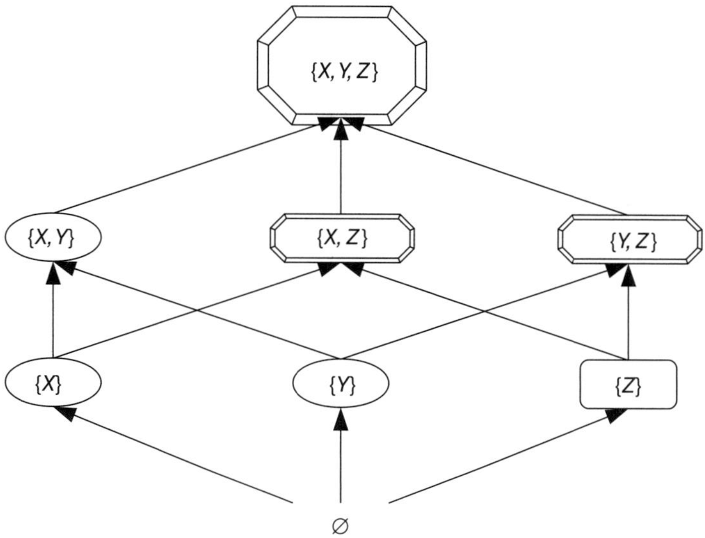

Figure 5.5 Hasse diagram of the power set of the two predictors, X, Y, and the outcome variable Z.

To illustrate, consider the situation in which X and Y are predictors, and Z is on the outcome side. Figure 5.5 displays the power set of these variables. The subsets that involve the predictors are indicated by oval frames, the subset that involves the outcome variable is indicated by a rectangular frame, and the subsets that involve predictors as well as the criterion are indicated by hexagonal frames. As in Figure 5.4, the empty set is depicted with no frame.

Now, to explain the univariate distribution of a variable, we propose omitting the main effect terms for the variable from a model. If the univariate probability distribution of an outcome variable is captured by the explanatory variables, only the random part is left, and the terms for the main effects are not needed for model fitting. In the present example, Z is the outcome variable. The corresponding log-linear model would include the terms that correspond to all subsets with oval, rectangular, and octagonal frames, but not the terms that correspond to the rectangular frame in Figure 5.5. The terms that correspond to the two shapes whose corresponding terms are part of a log-linear model are needed for different reasons. Specifically,

(1) the main effects of X and Y are needed because they help explain the marginal probabilities of X and Y;
(2) the $X \times Y$ interaction is needed because the model makes no assumption concerning the joint distribution of X and Y; therefore, any association between X and Y may prevail;
(3) the interactions between X, Y on one hand and Z on the other are needed because there must be a relation between the explanatory variables and the

outcome variable; without such an association, it would be pointless to test hypotheses concerning the direction of effect on Z that originates in X and Y.

The model, therefore, has three characteristics. First, it is saturated in the explanatory variables. Second, it contains all possible interactions between the explanatory and the outcome variables. Third, it does not include the main effects of the outcome variable. It is, therefore, nonhierarchical (Mair and von Eye, 2007).

Figure 5.5 can also be used to delineate the model for the opposite direction of effect, that is, the effect that originates in Z and goes to X and Y. In other words, with this model, we attempt to explain the univariate distributions of X and Y. The corresponding model would contain the following effects.

(1) All possible effects of the explanatory variable; in the present example, this includes no more than the main effects of Z
(2) All possible interaction effects within the subset of the outcome variables X and Y; that is, the $X \times Y$ interaction
(3) All interactions between predictor and outcome variables.

The comparison of the two models shows that they differ only in the main effects that are part of the model. The main effects that are to be explained are not part of the models.

Now, when it comes to making a decision concerning direction of effects, one of the two models must fit, and at least one of the interactions among explanatory and outcome variables must exist. In addition, the model for the reverse direction of effect either must fail to describe the data well or come with nonsignificant parameters for the interactions among the explanatory and the outcome variables. When these conditions are fulfilled, the explanatory variables of the fitting model can be considered explaining the univariate probability distribution of the outcome variable in the sense of Dodge and Rousson (2000, 2001).

In this chapter, we take one additional step. We propose generalizing the principle that underlies the analysis of direction dependence based on Dodge and Rousson (2000, 2001) to higher order effects. Specifically, we propose the *generalized direction of effect principle*, which posits that including an effect on the outcome variable side of a model is redundant when this effect is captured by (i) the main effects of explanatory variables, (ii) the interactions of explanatory variables with the outcome variables, or (iii) interactions among explanatory variables. An effect is considered captured when removing it from a model does not affect model fit. Among the main characteristics of the generalized direction of effect principle is that it applies to any effect on the outcome side as well as any effect on the explanatory side. Thus far, this principle has been applied to main effects only, that is, to univariate probability distributions in categorical variables (Wiedermann and von Eye, 2015a), and to univariate score distributions in metric variables (Dodge and Rousson, 2000, 2001; Wiedermann and von Eye, 2015b,c). In this chapter, we also consider higher order effects in categorical variables.

We just repeat the two additional prerequisites of interpretation of a result in the sense of direction of effect that originates in the explanatory variables and goes to the outcome variables. These prerequisites are the following: (i) the model describes the frequency distribution of all explanatory and outcome variables well, and (ii) interactions between putative explanatory and putative outcome variables exist.

To illustrate, consider the two predictors, X_1 and X_2, and the two outcome variables, Y_1 and Y_2. For these two sets of variables, we specify four models. The first tests the hypothesis that X_1 and X_2 allow one to capture the univariate distributions of Y_1 and Y_2 (Model 1). In other words, the first model tests the hypothesis that the direction of effect goes from the two X variables to the univariate distributions of the two Y variables. The second model tests the opposite direction of effect (Model 2). The third model tests the hypothesis that X_1 and X_2 capture the association between Y_1 and Y_2 (Model 3). Again, this model assumes that the direction of effect originates in the two X variables. However, instead of targeting the univariate distributions of the outcome variables, this model targets the association between the two outcome variables. We thus move from looking at univariate distributions to looking at structural elements of a cross-classification. The fourth model also focuses on structural elements, but it reverses direction of effect and tests the hypothesis that the direction of effect originates in the two Y variables (Model 4). It should be noted that these are not the only models that can be specified under the generalized direction of effect principle. Models that target both univariate probability distributions and structure are compatible with this principle as well.

Let M denote the set of four variables X_1, X_2, Y_1, and Y_2. The power set of these variables, $\mathcal{P}(M) := \{U_i | U_i \subset M\}$ can also be viewed as containing the three subsets U_1, U_2, and U_3. U_1 contains all terms that involve the predictors, X_1 and X_2. U_2 contains all terms that involve the outcome variables, Y_1 and Y_2, and U_3 contains all terms that connect X variables with Y variables.

When nonhierarchical log-linear models are specified that allow one to test the hypothesized direction of effect, terms from the three subsets are selected. Specifically, for Model 1, one selects from U_1 all possible terms that include X_1 and X_2. The model is thus saturated in the predictors. From U_2, one selects all terms that reflect interactions among the outcome variables. The model is, therefore, not saturated in the outcome variables. The main effects of the outcome variables are not part of the model, because the direction of effect hypothesis implies that these effects are redundant. From U_3, one selects all terms that reflect associations among X- and Y-variables. In other words, the log-linear model includes all possible $X \times Y$ interactions.

Model 2 is identical to Model 1 in the terms that are selected based on U_3. It differs, however, in the terms selected from U_1 and U_2. From U_1, only those terms are selected that represent interactions. Terms that represent main effects are omitted. From U_2, in contrast, all terms are used, that is, all main effects and all interactions. Model 2 is thus saturated in the Y-variables but nonhierarchical in the X-variables, reflecting that, for Model 2, variables have switched status as explanatory versus outcome.

Model 3, targeting structure, is identical to Model 1 in the selection of terms from U_1 as well as U_3. From U_2, however, only those terms are selected that represent main effects. In the present example, there are no terms that represent three-way or higher order interactions. Therefore, only the terms that represent main effects are used. The interaction between Y_1 and Y_2 is to be explained. This term is, therefore, omitted.

Model 4, specified for the hypothesis of reverse direction of effect, is identical to Models 1–3 in the selection of terms from U_3. It is saturated, however, in the terms selected from U_2, and it selects only those terms from U_1 that represent main effects. Here, the interaction between X_1 and X_2 is to be explained and, therefore, omitted from the model. The four log-linear models are (for the sake of simplicity, the subscripts have been omitted):

Model 1:

$$\begin{aligned}\log\ \hat{m} =& \lambda + \lambda^{X_1} + \lambda^{X_2} + \lambda^{X_1X_2} + \lambda^{Y_1Y_2}\\ &+ \lambda^{X_1Y_1} + \lambda^{X_1Y_2} + \lambda^{X_2Y_1} + \lambda^{X_2Y_2}\\ &+ \lambda^{X_1Y_1Y_2} + \lambda^{X_2Y_1Y_2} + \lambda^{X_1X_2Y_1} + \lambda^{X_1X_2Y_2}\\ &+ \lambda^{X_1X_2Y_1Y_2}\end{aligned}$$

Model 2:

$$\begin{aligned}\log\ \hat{m} =& \lambda + \lambda^{Y_1} + \lambda^{Y_2} + \lambda^{X_1X_2} + \lambda^{Y_1Y_2}\\ &+ \lambda^{X_1Y_1} + \lambda^{X_1Y_2} + \lambda^{X_2Y_1} + \lambda^{X_2Y_2}\\ &+ \lambda^{X_1Y_1Y_2} + \lambda^{X_2Y_1Y_2} + \lambda^{X_1X_2Y_1} + \lambda^{X_1X_2Y_2}\\ &+ \lambda^{X_1X_2Y_1Y_2}\end{aligned}$$

Model 3:

$$\begin{aligned}\log\ \hat{m} =& \lambda + \lambda^{X_1} + \lambda^{X_2} + \lambda^{Y_1} + \lambda^{Y_2} + \lambda^{X_1X_2}\\ &+ \lambda^{X_1Y_1} + \lambda^{X_1Y_2} + \lambda^{X_2Y_1} + \lambda^{X_2Y_2}\\ &+ \lambda^{X_1Y_1Y_2} + \lambda^{X_2Y_1Y_2} + \lambda^{X_1X_2Y_1} + \lambda^{X_1X_2Y_2}\\ &+ \lambda^{X_1X_2Y_1Y_2}\end{aligned}$$

Model 4:

$$\begin{aligned}\log\ \hat{m} =& \lambda + \lambda^{X_1} + \lambda^{X_2} + \lambda^{Y_1} + \lambda^{Y_2} + \lambda^{Y_1Y_2}\\ &+ \lambda^{X_1Y_1} + \lambda^{X_1Y_2} + \lambda^{X_2Y_1} + \lambda^{X_2Y_2}\\ &+ \lambda^{X_1Y_1Y_2} + \lambda^{X_2Y_1Y_2} + \lambda^{X_1X_2Y_1} + \lambda^{X_1X_2Y_2}\\ &+ \lambda^{X_1X_2Y_1Y_2}\end{aligned}$$

When researchers base model specification on the *Effect Sparsity Principle* (Box and Meyer, 1986, Wu and Hamada, 2000), higher order terms are rarely considered.

Therefore, and based on this principle, models can be considered that only contain the first two lines of these equations (for a discussion of this principle in the context of log-linear modeling, see von Eye, 2008, and von Eye and Mun, 2013). In the following sections, we give a data example and discuss parameter interpretation.

5.5 DATA EXAMPLE

In the following example, we use data from a longitudinal project on intimate partner violence (Bogat *et al.*, 2006). A sample of 204 women provided, in 1-year intervals, information on perpetration of severe violence by their intimate partners in the years before the third, fourth, and fifth birthday of a respondent's child. The observed variable, severe violence (S), was coded as 1 = did not occur, and 2 = did occur. The three observations will be labeled as S_1, S_2, and S_3.

The question we ask here is whether severe violence causes itself over time. A basis for this question can be found in theories that discuss self-reinforcing and, thus, self-perpetuating characteristics of violence (e.g., Bandura, 1973). Based on this type of theory, the prediction could be formulated that, over time, direction of effect of violence is such that earlier instances of violence cause later instances. This does not simply imply that there is a regression-type relation between earlier and later observations of violence. This can also imply that

(1) later (metric) observations of violence are closer to normally distributed than earlier observations; and

(2) the relations among later observations of violence disappear when earlier observations are causes of later observations.

Interestingly, the first implication is hard to access using direction dependence methodology. If, over time, distributions become more and more normal, statistical tests of this development will have less and less power. This has been demonstrated by Wiedermann and von Eye (2015b) in the context of mediation analysis. The second implication, more important in the present context, suggests that, in the case of three temporally ordered observations, the second and the third will look unrelated, given the first.

This prediction can be viewed as parallel to those that are made to examine spurious correlations. Observations 2 and 3 are both caused by Observation 1. Given Observation 1, the correlation between Observations 2 and 3 will vanish (see also the discussion of conditional independence in the Introduction). In the present example, we use observations of violence that an intimate partner perpetrated on women in the years before the third, fourth, and fifth birthday of her child. A model of the trajectory of violence will, if the aforementioned hypothesis prevails, not need the interaction between the reports on violence on the fourth- and fifth-year violence. In other words, the structure of later observations is caused by earlier observations.

From the power set of the effects of the three observations S_1, S_2, and S_3, we, therefore, need

(1) the main effects of all three observations, that is, $[S_1]$, $[S_2]$, and $[S_3]$;

(2) the interactions between the first and the second, and the first and the third observations, that is, $[S_1, S_2]$, and $[S_1, S_3]$.

However, we do not need the interaction between the second and the third observations, $[S_2, S_3]$. This term should be redundant if S_1 causes the structure of later observations. To analyze the cross-classification of S_1, S_2, and S_3, we, therefore, estimate the following two log-linear models:
Model 1:

$$\log\ \hat{m} = \lambda + \lambda^{S_1} + \lambda^{S_2} + \lambda^{S_3} + \lambda^{S_1S_2} + \lambda^{S_1S_3}$$

and Model 2:

$$\log\ \hat{m} = \lambda + \lambda^{S_1} + \lambda^{S_2} + \lambda^{S_3} + \lambda^{S_1S_2} + \lambda^{S_1S_3} + \lambda^{S_2S_3}$$

The second model does contain the structural element in question, $[S_2, S_3]$. If the hypothesis prevails, according to which $[S_1]$ captures $[S_2, S_3]$, this term will come with nonsignificant parameters, and the difference between the two models will be nonsignificant. The first model, that is, the more parsimonious one will, therefore, be retained (if it fits). Neither model includes the three-way interaction $[S_1, S_2, S_3]$. Table 5.2 contains the observed frequencies of the $S_1 \times S_2 \times S_3$ cross-classification and the frequencies that were estimated under the two competing models.

Both models fit the data well. For the first, the more parsimonious model, we obtain an LR-χ^2 of 3.401, which, for $df = 2$, suggests nonsignificant model-data discrepancies ($p = 0.182$). For the second model, we obtain an LR-χ^2 of 0.587, which, for $df = 1$, also suggests nonsignificant model-data discrepancies ($p = 0.443$). The difference in goodness-of-fit between these two nested models is Δ LR-$\chi^2 = 2.814$, which, for $\Delta df = 1$, suggests a nonsignificant difference ($p = 0.093$). We, therefore, retain the more parsimonious model, Model 1.

TABLE 5.2 Cross-Classification of S_1, S_2, and S_3: Observed Frequencies and Predicted Frequencies, Estimated under Two Log-Linear Models.

Cell Index			Frequencies		
S_1	S_2	S_3	Observed	Model 1	Model 2
1	1	1	157	156.16	156.56
1	1	2	2	2.84	2.44
1	2	1	8	8.84	8.44
1	2	2	1	0.16	0.56
2	1	1	20	18.94	20.44
2	1	2	2	3.06	1.56
2	2	1	11	12.06	10.56
2	2	2	3	1.94	3.44

This result is not sufficient for a decision concerning the direction of dependence of structural elements of a model. We need, in addition, to inspect the critical terms of the model. In the present example, the two-way interaction $[S_2, S_3]$ is the critical term. If this term comes with a nonsignificant parameter, the direction dependence hypothesis can be retained. The term is part of Model 2. We obtain a parameter estimate of $\lambda^{S_2,S_3} = 0.363$, a corresponding standard error of 0.214, and a parameter-standard error ratio of 1.692. This value suggests that this interaction is nonsignificant ($p = 0.091$). Considering that the two competing models do not differ significantly from each other, this result is as expected.

We, therefore, conclude that the hypothesis can be retained according to which earlier incidences of violence cause later incidences such that the relations among the later incidences can be explained from the earlier incidences. We interpret this result as providing the first example of direction of effect that targets the structure of variables instead of just univariate probability distributions.

Parameter interpretation. Parameters can be interpreted only if they represent the intended effects. Log-linear parameters can be interpreted based on the relation $\lambda = (W'W)^{-1}W' \log\ m$, where λ is the vector of parameters, W is the design matrix, and m is the vector of expected cell frequencies (von Eye and Mun, 2013). In Model 2, in which the interaction $S_2 \times S_3$ is predicted from S_1, the design matrix is

$$W = \begin{bmatrix} 1 & 1 & 1 & 1 & 1 & 1 \\ 1 & 1 & 1 & -1 & 1 & -1 \\ 1 & 1 & -1 & 1 & -1 & 1 \\ 1 & 1 & -1 & -1 & -1 & -1 \\ 1 & -1 & 1 & 1 & -1 & -1 \\ 1 & -1 & 1 & -1 & -1 & 1 \\ 1 & -1 & -1 & 1 & 1 & -1 \\ 1 & -1 & -1 & -1 & 1 & 1 \end{bmatrix}$$

where the first column represents the model constant, the next three columns represent the main effects of S_1, S_2, and S_3, the second to last column represents the $S_1 \times S_2$ interaction, and the last column represents the $S_1 \times S_3$ interaction. Inserting into $\lambda = (W'W)^{-1}W' \log\ m$ results in the interpretation of the parameters. We present two examples of how the parameters can be interpreted. The first example illustrates the interpretation of the first interaction parameter, which is specified in the second to last column in the design matrix:

$$\begin{aligned} \lambda^{S_1 S_2} = \frac{1}{8}(&\log\ m_{111} + \log\ m_{112} - \log\ m_{121} - \log\ m_{122} \\ &- \log\ m_{211} - \log\ m_{212} + \log\ m_{221} + \log\ m_{222}) \end{aligned}$$

The second example is for the second interaction parameter, specified in the last column of the aforementioned design matrix:

$$\begin{aligned} \lambda^{S_1 S_3} = \frac{1}{8}(&\log\ m_{111} - \log\ m_{112} + \log\ m_{121} - \log\ m_{122} \\ &- \log\ m_{211} + \log\ m_{212} - \log\ m_{221} + \log\ m_{222}) \end{aligned}$$

Evidently, the parameters in this model are interpretable as specified in the design matrix. This is a necessary precondition for an interpretation in the sense of direction dependence.

5.6 DISCUSSION

In developmental research, most statements about change concern means. Individuals show growth or decline. As was discussed repeatedly, however, (e.g., von Eye, 2010), change can manifest in just any parameter that can be estimated. For example, the correlation between variables can change with age, as was discussed in the context of the debate on intelligence divergence (Garrett, 1938). In the present work (applicable in both longitudinal and cross-sectional research), we propose considering that causal effects and, thus, direction dependence can manifest in just any parameter. Causes can affect more than the magnitude of scores. They can also affect cluster composition, the number of latent variables, manifest variable correlations, interactions, or just any structural parameter.

Existing methods for the analysis of direction dependence (Dodge and Rousson, 2000, 2001, von Eye and DeShon, 2012, von Eye and Wiedermann, 2014, Wiedermann *et al.*, 2013, 2015) have focused on univariate marginal probability distributions, at the expense of structural characteristics of multivariate distributions. The present work extends this line of work and discusses structural elements of variable relations.

Methods for the analysis of direction dependence have been developed for metric and categorical and for manifest and latent variables (Shimizu *et al.*, 2006, von Eye and Wiedermann, 2014). Models for single and for multiple predictors have been discussed (Wiedermann and von Eye, 2015c). This article extends this thinking into the domain of multiple dependent or outcome variables and into the domain of structural characteristics of distributions. The basic idea that propels the development of methods for the analysis of direction dependence is that the presence of causes allows one to explain effects. If this is the case, the elements of distributions that are explained disappear and are not needed in models.

In this chapter, we discuss models that explain interactions among categorical variables on the outcome side. This work can be extended in multiple ways, all compatible with the *generalized direction of effect principle*. We give three examples. First, this work can be extended into the domain of metric variables. Interactions among metric variables are routinely included in manifest variable or latent variable models. We propose developing models that allow researchers to test hypotheses according to which these interactions are caused by events on the explanatory side of a model.

Second, we propose extending this work so that multiple independent variables can be incorporated. Third, we propose extending this work so that causes can be not only variables or events but also structural elements. An interaction between independent variables could be considered the cause, and univariate probability distributions as well as the structure of relations among outcome variables can be the effects.

It is important to realize that the models presented here are more general than the well-known models of conditional independence (see Agresti, 2013, von Eye and Mun, 2013). Consider the three variables X, Y, and Z. The variables X and Y are *mutually independent*, if

$$\Pr(X \cap Y) = \Pr(X)\Pr(Y)$$

The corresponding log-linear model is the well-known main effects model $\log \hat{m}_{ij} = \lambda + \lambda_i^X + \lambda_j^Y$.

As discussed in the first section of this chapter, X and Y are conditionally independent given Z, if, for the joint probability of X and Y, it holds that

$$\Pr(X \cap Y|Z) = \Pr(X|Z)\Pr(Y|Z)$$

The corresponding log-linear model of X and Y, given Z, is

$$\log \hat{m}_{ijk} = \lambda + \lambda_i^X + \lambda_j^Y + \lambda_k^Z + \lambda_{ik}^{XZ} + \lambda_{jk}^{YZ}$$

This model is a special case of the models considered in this chapter in the sense that it proposes that the relation between X and Y does not exist when Z is known. In other words, the model proposes that the term λ_{ij}^{XY} can be removed from the model without affecting overall model fit. Models of conditional independence differ from the models of direction dependence discussed here, however, in three important aspects. First, when researchers attempt to make decisions about direction of effect at the level of univariate probability distributions, conditional independence models may be no longer applicable. The reason for this is that, in contrast to models of conditional independence, models of direction of effect do not contain the main effect of the putative dependent variable when a univariate frequency distribution is to be explained. Second, models of conditional independence cannot be applied when hypotheses are tested that imply that both univariate probability distributions are caused by another variable as well as interactions. Here again, whereas models of conditional independence are hierarchical, the lack of certain terms in models of direction of effect–in this case, main effect and interaction terms–results in nonhierarchical models of direction of effect. Conditional independence models have not been described for cases in which researchers attempt to predict univariate probability distributions and/or interactions from interactions of other variables.

Finally, to embed the proposed methodology into the broader framework of causal inference, it is important to realize that the questions posed by the presented nonhierarchical models and the models for causal inference differ fundamentally. Studies on causal inference typically deal with the identification of causally interpretable parameters. For example, Breen and Karlson (2013) discuss the identification of parameters of nonlinear probability models using the counterfactual framework. Similarly, Yamaguchi (2012) presents causal log-linear models for computing counterfactual odds ratios between treatment and outcome variables while eliminating influences of confounders (see also the related chapter of Yamaguchi, 2016). Of course, log-linear models for causally related categorical variables have been suggested earlier. For example, Goodman (1973) proposed a modified path analysis for nominal variables, which has later been integrated into the latent class and directed graphical modeling framework (known as *directed log-linear modeling*; see, for example, Hagenaars, 1998). In these models, it is assumed that joint probabilities of observed variables can be decomposed into the product of marginal and conditional probabilities (Vermunt, 1996). However, all these models assume that the causal ordering of variables (i.e., $X \rightarrow Y$ versus $Y \rightarrow X$) can validly be determined based on a priori theory. In contrast, the log-linear models proposed in this chapter can be used to empirically evaluate competing causal theories about the underlying data generating process. Such competing causal theories may concern both univariate probability distributions and structural characteristics of variables. Thus, whenever exchangeable elements in causal theories exist, the proposed method can–under certain conditions–be used to establish empirical evidence to support causal explanations.

REFERENCES

Agresti, A. (2013) *Categorical Data Analysis*, 3rd edn, John Wiley & Sons, Inc., New York.

Bandura, A. (1973) *Aggression: A Social Learning Analysis*, Prentice Hall, Englewood Cliffs, NJ.

Bogat, G.A., Levendosky, A.A., von Eye, A., and Davidson, W.S. (2006) The mental and physical health consequences of domestic violence for women and children receiving Medicaid. Medicaid Administrative Services Grant.

Box, G.E. and Meyer, R.D. (1986) An analysis for unreplicated fractional factorials. *Technometrics*, **28** (1), 11–18.

Bradley, J.V. (1978) Robustness? *British Journal of Mathematical and Statistical Psychology*, **31** (2), 144–152.

Breen, R. and Karlson, K.B. (2013) Counterfactual causal analysis and nonlinear probability models, in *Handbook of Causal Analysis for Social Research* (ed. S.L. Morgan), Springer-Verlag, Dordrecht, pp. 167–187.

Christensen, R. (1990) *Log-Linear Models*, Springer-Verlag, New York.

Dodge, Y. and Rousson, V. (2000) Direction dependence in a regression line. *Communications in Statistics - Theory and Methods*, **29** (9-10), 1957–1972.

Dodge, Y. and Rousson, V. (2001) On asymmetric properties of the correlation coefficient in the regression setting. *American Statistician*, **55** (1), 51–54.

Dodge, Y. and Rousson, V. (2016) Statistical inference for direction of dependence in linear models, in *Statistics and Causality: Methods for Applied Empirical Research* (eds W. Wiedermann and A. von Eye), John Wiley & Sons, Inc., Hoboken, NJ.

Garrett, H.E. (1938) Differentiable mental traits. *Psychological Records*, **2**, 259–298.

Goodman, L.A. (1973) The analysis of multidimensional contingency tables when some variables are posterior to others: a modified path analysis approach. *Biometrika*, **60** (1), 179–192.

Hagenaars, J.A. (1998) Categorical causal modeling latent class analysis and directed log-linear models with latent variables. *Sociological Methods & Research*, **26** (4), 436–486.

Knoke, D. and Burke, P.J. (1980) *Log-Linear Models*, Sage Publications, Thousand Oaks, CA.

Lazarsfeld, P.F. (1955) The logical and mathematical foundation of latent structure analysis, in *Measurement and Prediction* (ed. S.A. Stouffer), Princeton University Press, Princeton, NJ, pp. 362–472.

Mair, P. and von Eye, A. (2007) Application scenarios for nonstandard log-linear models. *Psychological Methods*, **12**, 139–156.

R Core Team (2015) *R: A Language and Environment for Statistical Computing*, R Foundation for Statistical Computing, Vienna, Austria, `http://www.R-project.org/` (accessed 21 December 2015).

Rindskopf, D. (1990) Nonstandard log-linear models. *Psychological Bulletin*, **108** (1), 150–162.

Shimizu, S., Hyvärinen, A., Hoyer, P.O., and Kano, Y. (2006) Finding a causal ordering via independent component analysis. *Computational Statistics and Data Analysis*, **50** (11), 3278–3293.

Vermunt, J.K. (1996) Causal log-linear modeling with latent variables and missing data, in *Analysis of Change: Advanced Techniques in Panel Data Analysis* (eds U. Engel and J. Reinecke), de Gruyter, Berlin, pp. 35–60.

Vermunt, J.K. (1997) *Log-Linear Models for Event Histories*, Sage Publications, Thousand Oakes, CA.

von Eye, A. (2008). Fractional factorial designs in the analysis of categorical data. InterStat, URL: `http://interstat.statjournals.net/YEAR/2008/articles/0804003.pdf`.

von Eye, A. (2010). The many parameters that can change. *ISSBD Bulletin*, **1** (57), 4–7.

von Eye, A., and DeShon, R.P. (2012). Directional dependence in developmental research. *International Journal of Behavioral Development*, **36** (4), 303–312.

von Eye, A., and Gutiérrez Peña, E. (2004). Configural frequency analysis: The search for extreme cells. *Journal of Applied Statistics*, **31** (8), 981–997.

von Eye, A., and Mun, E.-Y. (2013). Log-linear modeling: Concepts, interpretation and applications. New York: Wiley.

von Eye, A., Schuster, C., and Rogers, W.M. (1998). Modelling synergy using manifest categorical variables. *International Journal of Behavioral Development*, **22** (3), 537–557.

von Eye, A., and Wiedermann, W. (2014). On direction of dependence in latent variable contexts. Educational and Psychological Measurement, **74** (1), 5–30.

Wiedermann, W. and Hagmann, M. (2015) Asymmetric properties of the Pearson correlation coefficient: correlation as the negative association between linear regression residuals. *Communications in Statistics: Theory and Methods*, doi: 10.1080/03610926.2014.960582.

Wiedermann, W., Hagmann, M., Kossmeier, M., and von Eye, A. (2013) Resampling techniques to determine direction of effects in linear regression models, `http://interstat.statjournals.net/YEAR/2013/articles/1305002.pdf` (accessed 21 December 2015).

Wiedermann, W., Hagmann, M., and von Eye, A. (2015) Significance tests to determine the direction of effects in linear regression models. *British Journal of Mathematical and Statistical Psychology*, **68**, 116–141, doi: 10.1111/bmsp.12037.

Wiedermann, W. and von Eye, A. (2015a) Direction of effects in categorical variables. *under review*.

Wiedermann, W. and von Eye, A. Direction of effects in mediation analysis. (2015b) *Psychological Methods*, **20**, 221–244, doi: 10.1037/met0000027.

Wiedermann, W. and von Eye, A. (2015c) Direction of effects in multiple linear regression models. *Multivariate Behavioral Research*, **50**, 23–40, doi: 10.1080/00273171.2014.958429.

Wu, C.F.J. and Hamada, M. (2000) *Experiments: Planning, Analysis and Parameter Design Optimization*, John Wiley & Sons, Inc., New York.

Yamaguchi, K. (2012) Log-linear causal analysis of cross-classified categorical data. *Sociological Methodology*, **42** (1), 257–285.

Yamaguchi, K. (2016) Log-linear causal analysis of cross-classified categorical data, in *Statistics and Causality: Methods for Applied Empirical Research* (eds W. Wiedermann and A. von Eye), John Wiley & Sons, Inc., Hoboken, NJ.

6

DIRECTIONAL DEPENDENCE ANALYSIS USING SKEW–NORMAL COPULA-BASED REGRESSION

SEONGYONG KIM
Department of Applied Statistics, Hoseo University, Asan-si, Republic of Korea
DAEYOUNG KIM
Department of Mathematics and Statistics, University of Massachusetts, Amherst, MA, USA

6.1 INTRODUCTION

The analysis of directional dependence in nonexperimental study has been applied in various research contexts such as currency exchange rates (Dodge and Rousson, 2001), deficit hyperactivity disorder (Nigg *et al.*, 2008), gene networks from gene expression data (Kim *et al.*, 2008), and development of aggression in adolescence (von Eye and Wiedermann, 2014). By directional dependence, we mean asymmetric dependence/interaction between the variables, for example, the influence is from one variable to the other or two variables influence each other with different magnitude.

Approaches to the statistical study of directional dependence have been proposed from different perspectives. The first approach is to compare measures of skewness and kurtosis of the two variables of interest (Dodge and Rousson, 2000, 2001; Dodge and Yadegari, 2010; von Eye and DeShon, 2012). The basic idea of this method is as follows. In a valid simple linear regression where the error term is normally distributed, the dependent variable is a convolution of a nonnormal independent variable

Statistics and Causality: Methods for Applied Empirical Research, First Edition.
Wolfgang Wiedermann and Alexander von Eye.

with a normal error term, and thus, the distribution of the dependent variable is closer to a normal distribution. Wiedermann et al., (2013) further investigated properties of residuals of competing simple linear regression models and proposed using the skewness of residuals to statistically identify the direction of dependence.

The second approach proposed is the asymmetric copula-based regression model (Sungur, 2005a, b; Kim *et al.*, 2008; Kim and Kim, 2014). Copulas, useful for describing the dependence between the variables, arise from Sklar's theorem (Sklar, 1959): there exists a unique function, *copula*, that links the p univariate marginal distributions for the p continuous random variables to form the p-dimensional distribution function. Copulas allow us to model the dependence structure of the joint distribution and its marginal distributions separately, and they are invariant under strictly increasing transformations of the marginals. There are several important advantages of using the asymmetric copula-based regression approach to the analysis of directional dependence. The assumptions of normal error term and linearity between the variables of interest are not required in the models of copula regression. The asymmetric interaction between two variables is explicitly taken into account via asymmetric copula function. Furthermore, the copula regression approach enables us to study directional dependence stemming from not only marginal behavior of variables but also joint behavior of them.

In this chapter, we present a new methodology for analyzing directional dependence in the data, skew–normal copula-based regression. Recently, there has been a growing interest for more flexible parametric families of multivariate skew–elliptical distributions, such as multivariate skew–normal distribution and multivariate skew–t distribution (Azzalini and Capitanio, 2014; Genton, 2004) because there are many real applications where the data is nonelliptical and often skewed, heavy tailed. A skew–normal copula, derived from the skew–normal distribution with additional parameters allowing to regulate skewness, enables accounting for asymmetric dependence between the variables. For estimation of the skew–normal copula-based regression, two approaches are considered, fully parametric and semiparametric. The semiparametric method is useful in assessing directional dependence resulting from the joint behavior of the variables because it can remove the effect of the marginal distributions on the directional dependence.

The rest of the chapter is organized as follows: First, we briefly review the concept and properties of copula and copula-based regression. We then overview directional dependence using the copula regression setting in Sungur (2005a,b). Second, we introduce the skew–normal copula-based regression, consisting of two parts. In the first, we review the multivariate skew–normal distribution (Azzalini and Capitanio, 1999) and the multivariate skew–normal copula (Kollo *et al.*, 2013); in the second, we study their properties in terms of the conditional distribution and regression function. Third, we discuss the inference procedure of the directional dependence using skew–normal copula-based regression. To illustrate the proposed method, we provide a real data example.

6.2 COPULA-BASED REGRESSION

We briefly review the copula and copula-based regression. For a detailed overview, we refer to Joe (2004), McNeil *et al.* (2005) and Nelson (2006).

6.2.1 Copula

Let $\mathbf{X} = (X_1, X_2)'$ be a two-dimensional random vector with the joint continuous distribution function $\boldsymbol{H}(\mathbf{x}) := \boldsymbol{H}(x_1, x_2) = P(X_1 \leq x_1, X_2 \leq x_2)$ and the marginal distribution function $F_j(x_j) := P(X_j \leq x_j)$ where $j = 1, 2$ and $\mathbf{x} = (x_1, x_2)'$. Here, the marginal distribution $F_j(X_j)$ can be viewed as a transformed variable that is uniformly distributed on [0, 1].

From *Sklar's theorem* (Sklar, 1959), the joint distribution function $\boldsymbol{H}(\mathbf{x})$ of $\mathbf{X}$ can be expressed as the copula distribution of $\mathbf{X}$,

$$\boldsymbol{H}(x_1, x_2) = C(F_1(x_1), F_2(x_2)) \tag{6.1}$$

where C is a two-dimensional copula, the joint distribution function on $[0, 1]^2$ with standard uniform marginal distributions, defined by

$$C(\mathbf{u}) = C(u_1, u_2) = P(U_1 \leq u_1, U_2 \leq u_2) \tag{6.2}$$

where U_1 and U_2 are uniformly distributed over [0,1], and $\mathbf{u} = (u_1, u_2)'$. Let $U_j = F_j(X_j)$. When $F_j(X_j)$ is continuous and strictly increasing, it has an inverse, continuous, and strictly increasing quantile function, F_j^{-1}. If we evaluate Equations 6.1 and 6.2 at $x_j = F_j^{-1}(u_j)$, we obtain

$$\begin{aligned} C(F_1(x_1), F_2(x_2)) &= P(X_1 \leq F_1^{-1}(u_1), X_2 \leq F_2^{-1}(u_2)) \\ &= \boldsymbol{H}(F_1^{-1}(u_1), F_2^{-1}(u_2)) \end{aligned} \tag{6.3}$$

The copula distribution C of $\mathbf{X}$ in Equation 6.2 has the following basic properties:

(1) C is grounded and has univariate margins: $C(u_1, 0) = C(0, u_2) = 0$, $C(u_1, 1) = u_1$ and $C(1, u_2) = u_2$ for all $u_1, u_2 \in [0, 1]$
(2) C is two-increasing : $C(u_1^{**}, u_2^{**}) - C(u_1^{**}, u_2^{*}) - C(u_1^{*}, u_2^{**}) + C(u_1^{*}, u_2^{*}) \geq 0$ for all $u_1^{**}, u_1^{*}, u_2^{**}, u_2^{*} \in [0, 1]$ for which $u_1^{*} \leq u_1^{**}$ and $u_2^{*} \leq u_2^{**}$.
(3) C is invariant under strictly increasing transformations of the marginals: if T_1 and T_2 are two strictly increasing functions, the copula of (Y_1, Y_2) with $Y_1 = T_1(X_1)$ and $Y_2 = T_2(X_2)$ is the same as the copula of (X_1, X_2).

For an absolutely continuous $\boldsymbol{H}$ with strictly increasing and continuous marginal distribution functions, the joint density function of $\mathbf{X}$ is obtained from Equation 6.1:

$$h(x_1, x_2) = c(F_1(x_1), F_2(x_2))\, f_1(x_1)\, f_2(x_2), \tag{6.4}$$

where $c(u_1, u_2) := (\partial^2 C(u_1, u_2)/\partial u_1 \partial u_2)$ is some uniquely identified two-dimensional copula density and $f_j(x_j)$ is the density function of X_j. Thus, the corresponding copula density is given by

$$c(F_1(x_1), F_2(x_2)) = \frac{h(x_1, x_2)}{f_1(x_1)f_2(x_2)} = \frac{h(F_1^{-1}(u_1), F_2^{-1}(u_2))}{f_1(F_1^{-1}(u_1))f_2(F_2^{-1}(u_2))} \tag{6.5}$$

From Equations 6.1, 6.3, and 6.4, we can see that the joint distribution of the random vector $\mathbf{X}$ is created from two sources, the distribution of each random variable in $\mathbf{X}$, $U_j = F_j(X_j)$, and the copula C, which expresses the dependence structure between the random variables in $\mathbf{X}$ (their comovement) on a quantile scale.

Assume that the unknown copula C belongs to an absolutely continuous parametric copula family $C = \{C_\theta\}$ where θ represents a set of parameters measuring dependence between the variables. Then there are various types of copulas commonly used in many applied areas, including the normal copula, the t-copula, the Archimedean copula, Farlie–Gumbel–Morgenstern (FGM) copula, Plackett copula, and so on. For example, when $\mathbf{X} = (X_1, X_2)$ is a bivariate normal vector with zero mean vector and the correlation matrix Σ, then the bivariate normal copula and corresponding density are

$$C_\theta^n(u_1, u_2) = \Phi_\Sigma(\Phi^{-1}(u_1), \Phi^{-1}(u_2); \Sigma), \tag{6.6}$$

$$c_\theta^n(u_1, u_2) = |\ \Sigma\ |^{-1/2} \exp\left\{-\frac{1}{2}\begin{pmatrix}\Phi^{-1}(u_1)\\ \Phi^{-1}(u_2)\end{pmatrix}'(\Sigma^{-1} - \mathbf{I}_2)\begin{pmatrix}\Phi^{-1}(u_1)\\ \Phi^{-1}(u_2)\end{pmatrix}\right\} \tag{6.7}$$

where $\Phi_\Sigma(\cdot)$ denotes the joint distribution function of $\mathbf{X}$, $\Phi^{-1}(\cdot)$ denotes the quantile function of the univariate standard normal distribution function, and $\mathbf{I}_2$ is the 2×2 identity matrix. This shows that the bivariate normal density is expressed as the product of the bivariate normal copula, as given in Equation 6.6, and two standard normal marginal densities.

6.2.2 Copula-Based Regression

Using the definition and properties of copula function described earlier, we can express the conditional distribution and the corresponding conditional mean (i.e., regression function) in terms of copula and marginals. The conditional distribution function of X_1 given $X_2 = x_2$ and the corresponding conditional density are given by

$$\begin{aligned}F_{X_1|X_2}(x_1 \mid x_2) &\equiv P(X_1 \le x_1 \mid X_2 = x_2)\\ &= \lim_{\delta\to 0} P(X_1 \le x_1 \mid x_2 \le X_2 \le x_2 + \delta)\\ &= \lim_{\delta\to 0} \frac{H(x_1, x_2+\delta) - H(x_1, x_2)}{F_2(x_2+\delta) - F_2(x_2)}\end{aligned}$$

$$= \lim_{\delta \to 0} \frac{C(F_1(x_1), F_2(x_2 + \delta)) - C(F_1(x_1), F_2(x_2))}{F_2(x_2 + \delta) - F_2(x_2)}$$

$$= \frac{\partial C(u_1, u_2)}{\partial u_2} |_{u_1 = F_1(x_1), u_2 = F_2(x_2)} \tag{6.8}$$

$$f_{X_1|X_2}(x_1 \mid x_2) = c(F_1(x_1), F_2(x_2)) \quad f_1(x_1) \tag{6.9}$$

The conditional mean of X_1 or regression function of X_1 given $X_2 = x_2$ can be written as

$$\begin{aligned} m_{X_1|X_2}(x_2) &= E_C(X_1 \mid X_2 = x_2) \\ &= \int_R x_1 f_{X_1|X_2}(x_1 \mid x_2)\, dx_1 \\ &= \int_R x_1 c(F_1(x_1), F_2(x_2))\, f_1(x_1)\, dx_1 \\ &= \int_R x_1 c(F_1(x_1), F_2(x_2))\, dF_1(x_1) \end{aligned} \tag{6.10}$$

We can see from Equation 6.10 that the form of the regression function will be determined by the copula density of X_1 and X_2 (i.e., their dependence or comovement) and the marginal distribution of X_1. Hereafter, we call the regression function in Equation 6.10 as the copula-based regression.

When X_1 and X_2 are uniformly distributed over (0, 1), then Equations 6.8–6.10 become

$$F_{U_1|U_2}(u_1 \mid u_2) = \frac{\partial C(u_1, u_2)}{\partial u_2}, \qquad f_{U_1|U_2}(u_1 \mid u_2) = c(u_1, u_2)$$

$$m_{U_1|U_2}(u_2) = E_C(U_1 \mid U_2 = u_2) = 1 - \int_0^1 \frac{\partial C(u_1, u_2)}{\partial u_2} du_1$$

where $U_1 (= X_1)$ and $U_2 (= X_2) \sim U(0, 1)$. Note that the functional form of the copula regression $m_{U_1|U_2}(u_2)$ depends on only the choice of the copula function.

An interesting property of the copula-based regression is that it can capture not only linear dependence but also nonlinear dependence between the variables. By Taylor's Theorem, the copula-based regression in Equation 6.10 is represented as

$$\begin{aligned} m_{X_1|X_2}(x_2) &= E_C(X_1 \mid X_2 = x_2) \\ &= E_C(X_1 \mid x_2^o) + \sum_{s=1}^{S} \frac{E_C^{(s)}(X_1 \mid x_2^o)}{s!}(x_2 - x_2^o)^s + R_S \end{aligned}$$

where $E_C^{(s)}(X_1 \mid x_2^o) = (\partial^s E_C(X_1 \mid x_2)/\partial x_2^s)|_{x_2 = x_2^o}$, $R_S = (E_C^{(S+1)}(X_1 \mid x_2^\star)/(S+1)!)$ $(x_2 - x_2^\star)^{(S+1)}$ and $x_2^\star$ is an interior point between x_2 and x_2^o. Depending on the behavior of $E_C^{(s)}(X_1 \mid x_2^o)$ determined by the choice of the copula function and the marginal distribution of X_1, the copula regression can be either the linear or the sth order polynomial function of x_2. If $E_C^{(s)}(X_1 \mid x_2^o) = 0$ for all $s \geq 2$, $m_{X_1|X_2}(x_2)$ is linear in x_2. For example, suppose that $\mathbf{X} = (X_1, X_2)$ has the bivariate normal distribution with mean vector $\mu = (\mu_1, \mu_2)'$ and the correlation matrix, $\Sigma = \begin{pmatrix} 1 & \sigma_{12} \\ \sigma_{12} & 1 \end{pmatrix}$. Then the regression function in Equation 6.10 is linear in x_2,

$$m_{X_1|X_2}(x_2) = \mu_1 + \sigma_{12}(x_2 - \mu_2) \tag{6.11}$$

6.3 DIRECTIONAL DEPENDENCE IN THE COPULA-BASED REGRESSION

In this section, we briefly review directional dependence using the copula regression setting and the general measurements of the directional dependence in Sungur (2005a,b). Suppose that X_1 and X_2 are continuous random variables whose joint distribution function is given by $\mathbf{H}(X_1, X_2) = C(U_1, U_2)$ in Equation 6.1, where C is the copula function of X_1 and X_2, and $U_1 = F_1(X_1)$ and $U_2 = F_2(X_2)$ are the marginal distribution functions, respectively. Sungur (2005a) defined two types of directional dependence, one stemming from joint behavior of the variables through copula (i.e., dependence on a quantile scale) and the other from their marginal behaviors.

Definition 6.1 ***(Sungur, 2005a)***

(1) The random pair (U_1, U_2) is directionally dependent in joint behavior if $m_{U_1|U_2}(w) \neq m_{U_2|U_1}(w)$ where $m_{U_1|U_2}(w) = E_C(U_1 \mid U_2 = w)$ and $m_{U_2|U_1}(w) = E_C(U_2 \mid U_1 = w)$.

(2) The random pair (X_1, X_2) is directionally dependent in marginals if $m_{U_1|U_2}(w) = m_{U_2|U_1}(w)$ and $m_{X_1|X_2}(z) \neq m_{X_2|X_1}(z)$ where $m_{X_1|X_2}(z) = E_C(X_1 \mid X_2 = z)$ and $m_{X_2|X_1}(z) = E_C(X_2 \mid X_1 = z)$.

There are a few interesting results from Definition 6.1 (Sungur, 2005a, 2005b). Firstly, it is not possible to identify directional dependence in marginals if there is no directional dependence in joint behavior and the marginal distributions for X_1 and X_2 are the same. Secondly, if one is interested in directional dependence stemming from the joint behavior via copulas, one needs to consider a monotonic strictly increasing transformation of X_1 and X_2 (such as $U_1 = F_1(X_1)$, $U_2 = F_2(X_2)$) that leaves the joint dependence between the variables unchanged. In practice, as the marginal distributions of the variables are unknown, one can then use the rescaled empirical distributions of X_1 and X_2:

$$\hat{U}_1 = \hat{F}_1(x_1) = \frac{R_1}{n+1}, \quad \hat{U}_2 = \hat{F}_2(x_2) = \frac{R_2}{n+1} \tag{6.12}$$

where (X_{i1}, X_{i2}), $i = 1, \ldots, n$ be an independent and identically distributed sample of n observations from the distribution of (X_1, X_2), and R_1 and R_2 are the ranks of X_1 among $(X_{11}, \ldots, X_{n1})$ and X_2 among $(X_{12}, \ldots, X_{n2})$, respectively. Thirdly, if the copula is symmetric satisfying $C(u_1, u_2) = C(u_2, u_1)$ for every $(u_1, u_2) \in [0, 1]^2$, one can investigate only the directional dependence in marginals, not the directional dependence in joint behavior because $m_{U_1|U_2}(w) = m_{U_2|U_1}(w)$ for symmetric copula C. Lastly, if one is interested in directional dependence between the variables regardless of its source (marginal or joint behavior or both), one can check if $m_{X_1|X_2}(z)$ and $m_{X_2|X_1}(z)$ differ. Here, one should consider asymmetric copulas for better modeling the joint dependence structure. In next section, we introduce a flexible asymmetric copula, a skew–normal copula.

Sungur (2005a) also proposed the general measures for the directional dependence from marginal or joint behavior or both.

$$\rho^2(U_1 \to U_2) \equiv \frac{Var(m_{U_2|U_1}(U_1))}{Var(U_2)} = 12E[(m_{U_2|U_2}(U_1)^2] - 3$$

$$\rho^2(U_2 \to U_1) \equiv \frac{Var(m_{U_1|U_2}(U_2))}{Var(U_1)} = 12E[(m_{U_1|U_2}(U_2)^2] - 3$$

$$\rho^2(X_1 \to X_2) \equiv \frac{Var(m_{X_2|X_1}(X_1))}{Var(X_2)} = \frac{E[(m_{X_2|X_1}(X_1) - E(X_2))^2]}{Var(X_2)} \tag{6.13}$$

$$\rho^2(X_2 \to X_1) \equiv \frac{Var(m_{X_1|X_2}(X_2))}{Var(X_1)} = \frac{E[(m_{X_1|X_2}(X_2) - E(X_1))^2]}{Var(X_1)} \tag{6.14}$$

These measurements allow us to compare copula-based regressions in terms of predictive power, and each of them can be interpreted as the proportion of total variation of one variable that is explained by the copula regression based on the other variable. The difference is that the first two measurements, $\rho^2_{U_1 \to U_2}$ and $\rho^2_{U_2 \to U_1}$, take into account the dependence between the variables on a quantile scale, independent of the marginals of the variables, and they can measure the directional dependence in joint behavior.

Sungur (2005a) showed that (i) if the copula-based regressions on a quantile scale, $m_{U_1|U_2}(U_2)$ and $m_{U_2|U_1}(U_1)$, are both linear, (U_1, U_2) cannot be directionally dependent (i.e., $m_{U_1|U_2}(w) = m_{U_2|U_1}(w)$), and (ii) if regressions on the original scale, $m_{X_1|X_2}(X_2)$ and $m_{X_2|X_2}(X_1)$, are both linear, then $\rho^2(X_1 \to X_2) = \rho^2(X_2 \to X_1)$. This means that if the simple linear regression is valid for (X_1, X_2), one can use the approach based on skewness and kurtosis of the variables (Dodge and Rousson, 2000; Dodge and Rousson, 2001; Dodge and Yadegari, 2010; von Eye and DeShon, 2012) or regression residuals (Wiedermann *et al.*, 2013) to identify the directional dependence in marginals.

6.4 SKEW–NORMAL COPULA

There are various assumptions commonly made for the structures of copulas, such as symmetry, radial symmetry, joint symmetry, Archimedeanity, associativity, and max stability. For more details, refer to Joe (2004), Nelson (2006), and Li and Genton (2013). In this section, we are interested in asymmetric copulas because the copulas with symmetric structure lack the flexibility to examine the dependencies between the variables, particularly the directional (asymmetric) dependence and nonlinear forms of dependence.

The bivariate copula C is defined to be symmetric if $C(u_1, u_2) = C(u_2, u_1)$ for every $(u_1, u_2) \in [0, 1]^2$ and is asymmetric otherwise. We can easily see the effect of the symmetry by examining the conditional distributions of copulas. For the continuous variables X_1 and X_2 with the copula C in Equation 6.1, the conditional distribution of the copula (i.e., the conditional probability in Eq. 6.8) for $(U_1, U_2) = (F_1(X_1), F_2(X_2))$ is

$$P(U_1 \leq u_1 \mid U_2 = u_2) = \frac{\partial C(u_1, u_2)}{\partial u_2}$$

If C is symmetric, then

$$P(U_1 \leq u^* \mid U_2 = u^{**}) = P(U_2 \leq u^* \mid U_1 = u^{**})$$

This means that the probability that X_1 is smaller than or equal to the corresponding u^* quantile $(F_1^{-1}(u^*))$ given that X_2 is its u^{**} quantile $(F_2^{-1}(u^{**}))$ is the same as the probability that X_2 is smaller than or equal to the corresponding u^* quantile $(F_2^{-1}(u^*))$ given that X_1 is its u^{**} quantile $(F_1^{-1}(u^{**}))$ (McNeil *et al.*, 2005). Commonly used copulas including the normal copula in Equations 6.7, the t-copula, and the Archimedean copulas are symmetric.

This symmetric property is restrictive in reality because the associations between variables are not always identical between all variables involved. For example, as can be found in the insurance, operational risk, and finance problems, it is possible that the effect of the change in the marginal probability of X_1 is not the same as that of X_2, even assuming that the marginal distributions of X_1 and X_2 are the same. In this case, symmetric copulas cannot capture such asymmetric dependence. There are also several cases where the causal relations among the variables of interest occur in certain direction.

Compared to symmetric copulas, there are only few classes of asymmetric copulas, for example, skew–elliptical copulas (Genton, 2004; Kollo *et al.*, 2013), product of copulas proposed in Liebscher (2008), Marshall–Olkin family of copula (Marshall and Olkin, 1967), and Tawn copula (Tawn, 1988). Kim and Kim (2014) applied the product of Archimedean copulas in the areas of biological research and developmental research where the directional dependency between two variables is known to theoretically exist.

Recently, there has been a growing interest for more flexible parametric families of multivariate skew–elliptical distributions, including multivariate skew–normal distribution and multivariate skew–t distribution, with additional parameters allowing for regulation of skewness and tails. This is because, in real applications, there are many cases where the data is nonelliptical and often skewed, heavy tailed (Genton, 2004; Azzalini and Capitanio, 2014). In this section, we focus on the skew–normal copula belonging to the skew–elliptical copula family introduced in Kollo *et al.* (2013). The skew–normal copula is attractive in applications because it can deal with various types of skewness, while still maintaining mathematical tractability.

We first define the multivariate skew–normal distribution given in Azzalini and Capitanio (1999) because it is associated with the skew–normal copula.

Definition 6.2 *A p-dimensional random vector* $\boldsymbol{X} = (X_1, \ldots, X_p) \in \mathbb{R}^p$ *has skew–normal distribution with parameters* μ, Σ, *and* α, *denoted by* $SN_p(\mu, \Sigma, \alpha)$, *if* $\boldsymbol{X}$ *has the following density function*

$$h_{p,sn}(\boldsymbol{x};\mu,\Sigma,\alpha) = 2\phi_p(\boldsymbol{x};\mu,\Sigma)\Phi(\alpha'\Sigma^{-1/2}(\boldsymbol{x}-\mu)) \tag{6.15}$$

where $\mu = (\mu_1, \ldots, \mu_p)'$ *is a location parameter vector,* Σ *is a dispersion matrix whose jth diagonal element is* σ_{jj}, *and* (j,k)*th off-diagonal element is* σ_{jk}, $\alpha = (\alpha_1, \ldots, \alpha_p)'$ *is a shape parameter vector to account for skewness,* $\phi_p(\boldsymbol{x};\mu,\Sigma)$ *is the p-dimensional normal density with mean vector* μ *and dispersion matrix* Σ, *and* $\Phi(\cdot)$ *is the standard normal distribution function.*

Note that when $\alpha = \mathbf{0}$ in Equation 6.15, the multivariate skew–normal distribution reduces to the multivariate normal distribution.

By following the definitions in Kollo *et al.* (2013), the multivariate skew–normal copula and the corresponding density are given in Equations 6.16 and 6.17.

Definition 6.3 *A p-dimensional dimensional copula* C^{sn} *is called a skew–normal copula if*

$$C_\theta^{sn}(u_1, \ldots, u_p) = \boldsymbol{H}_{p,sn}(F_1^{-1}(u_1;0,1,\alpha_1), \ldots, F_p^{-1}(u_p;0,1,\alpha_p);\mathbf{0},\mathbf{R},\alpha) \tag{6.16}$$

where $F_j^{-1}(u_j;\mu_j,\sigma_{jj},\alpha_j)$ *denote the inverse of the univariate skew–normal distribution* $SN_1(\mu_j,\sigma_{jj},\alpha_j)$, $\mathbf{R}$ *is the correlation matrix whose diagonal elements are unity and whose* (j,k)*th off-diagonal element is* σ_{jk}, *and* $\boldsymbol{H}_{p,sn}(\cdot)$ *is the distribution function of the multivariate skew–normal distribution with the density given in Equation 6.15. The skew–normal copula density is*

$$c_\theta^{sn}(u_1, \ldots, u_p) = \frac{h_{p,sn}(F_1^{-1}(u_1;0,1,\alpha_1), \ldots, F_p^{-1}(u_p;0,1,\alpha_p);\mathbf{0},\mathbf{R},\alpha)}{\prod_{j=1}^{p} h_{1,sn}(F_j^{-1}(u_j;0,1,\alpha_j);0,1,\alpha_j)} \tag{6.17}$$

Note that the copula still remains invariant under a standardization of the marginal distributions as it is invariant under strictly increasing transformations of the marginals.

To better understand the features of the skew–normal copula defined above, we construct the contours of the bivariate skew–normal density for $(X_1, X_1) \sim SN_2(\mu, \Sigma, \alpha)$ in Equation 6.15, the corresponding copula function in Equation 6.16, the corresponding copula density in Equation 6.17, and the plots of two conditional probabilities, $P(U_1 \leq u^* \mid U_2 = u^{**})$ and $P(U_2 \leq u^* \mid U_1 = u^{**})$ where $U_1 = F_1(X_1; \mu_1, \sigma_{11}, \alpha_1)$ and $U_2 = F_2(X_2; \mu_2, \sigma_{22}, \alpha_2)$. The sets of the parameter values used are $(\mu_1 = \mu_2 = 0, \sigma_{11} = \sigma_{22} = 1, \sigma_{12} = 0.5, \alpha_1 = \alpha_2 = 0)$ in Figure 6.1 (i.e., the bivariate normal distribution), $(\mu_1 = \mu_2 = 0, \sigma_{11} = \sigma_{22} = 1, \sigma_{12} = 0.5, \alpha_1 = 2, \alpha_2 = -4)$ in Figure 6.2, $(\mu_1 = \mu_2 = 0, \sigma_{11} = \sigma_{22} = 1, \sigma_{12} = 0.5, \alpha_1 = -4, \alpha_2 = 4)$ in Figure 6.3.

Figure 6.1 contains the plots of the bivariate normal distribution (i.e., bivariate skew–normal distribution with α=(0,0)) whose copula is symmetric. The contour of the density in Figure 6.1a is elliptical in the (X_1, X_2) space, and the copula function in (b) and the copula density in (c) are symmetric along the diagonal line $U_2 = U_1$ in the (U_1, U_2) space. As the bivariate normal copula is symmetric, we can see that two conditional probabilities computed for the same values of u^{**} in (d) are the same over the value of u^*.

On the other hand, as shown in Figure 6.2 and 6.3, the behaviors of bivariate skew–normal distribution and corresponding copula are different. The contours of bivariate skew–normal density in (a) are both nonelliptical, and so the contours of the copula functions in (b) are not symmetric in the (U_1, U_2) space. Furthermore, the two conditional distributions in (d) are different over the value of u^* for a given value of u^{**}.

From the conditional mean regression analysis, we can also see the difference between the skew–normal distribution and the normal distribution. Consider the bivariate skew–normal random vector $\mathbf{X} = (X_1, X_2) \sim SN_2(\mu, \Sigma, \alpha)$ where $\mu = \begin{pmatrix} \mu_1 \\ \mu_2 \end{pmatrix}, \Sigma = \begin{pmatrix} 1 & \sigma_{12} \\ \sigma_{12} & 1 \end{pmatrix}, \alpha = \begin{pmatrix} \alpha_1 \\ \alpha_2 \end{pmatrix}$. This means that $\mathbf{X}$ has the two-dimensional skew–normal copula with the correlation matrix Σ, as given in Equation 6.17, and two skew–normal marginals (i.e, $X_j \sim SN_1(\mu_j, 1, \alpha_j)$). By Azzalini and Capitanio (1999), the mean regression of X_1 given $X_2 = x_2$ is

$$m_{X_1|X_2}(x_2) = \mu_1 + \sigma_{12}(x_2 - \mu_2) + \frac{\phi(x_2^*)}{\Phi(x_2^*)}\tau \tag{6.18}$$

where $x_2^* = ((\alpha_2 + \alpha_1\sigma_{12})(x_2 - \mu_2)/\sqrt{1 + \alpha_1^2(1 - \sigma_{12}^2)})$ and $\tau = (\alpha_1(1 - \sigma_{12}^2)/\sqrt{1 + \alpha_1^2(1 - \sigma_{12}^2)})$.

The mean regression of X_1 on X_2 for the bivariate skew–normal distribution shows the nonlinear dependence with respect to the conditioning variable, X_2 (see the third term in Eq. 6.18). Note that the first two terms are the same as the conditional mean for the bivariate normal distribution in Equation 6.11. There are two cases where $m_{X_1|X_2}(x_2)$ in Equation 6.18 is linear in x_2: (i) $\alpha_1 = 0$ (and so τ=0) and (ii) $\alpha_2 + \alpha_1\sigma_{12} = 0$ (and so x_2^*=0).

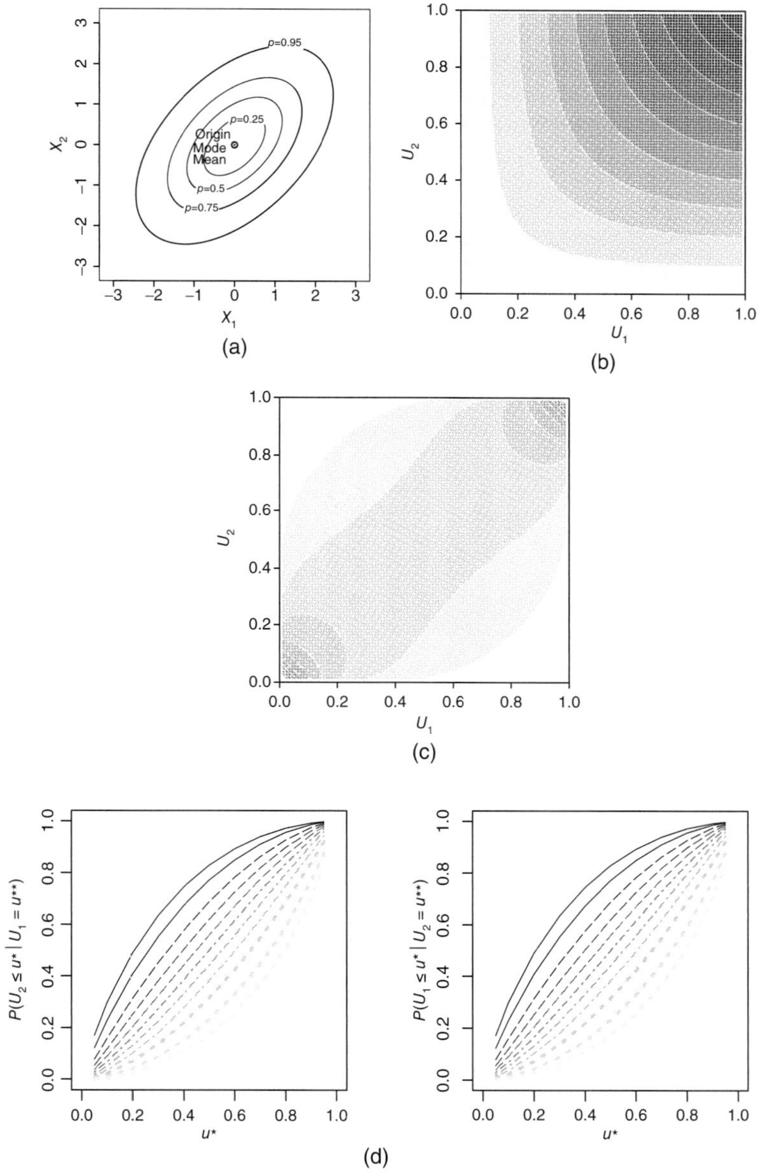

Figure 6.1 $(X_1, X_2) \sim SN_2(\mu{=}(0,0),\ \sigma_{11}{=}\sigma_{22}{=}1,\ \sigma_{12}{=}0.5,\ \alpha{=}(0,0))$: (a) contour plot for (X_1, X_2); (b) and (c) contour plots of the copula for (X_1, X_2) and its copula density; (d) conditional distributions of the copula for (X_1, X_2), $P(U_2 \le u^* \mid U_1 = u^{**})$ (left) and $P(U_1 \le u^* \mid U_2 = u^{**})$ (right) where $u^{**} = (0.05, 0.1, 0.2, 0.3, 0.4, 0.5, 0.6, 0.7, 0.8, 0.9, 0.95)$ with the line types and colors range from solid/black (0.05) to dotdash/dark gray (0.5) to dotted/dim gray (0.95).

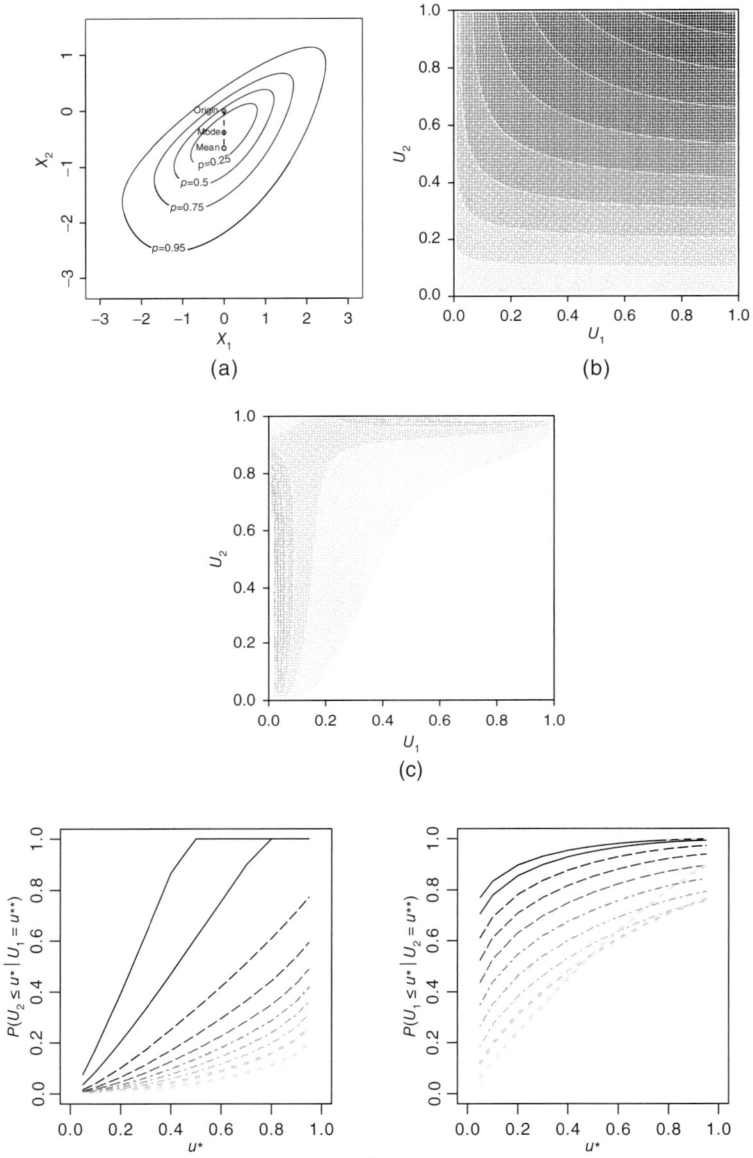

Figure 6.2 $(X_1, X_2) \sim SN_2(\mu{=}(0,0),\ \sigma_{11}{=}\sigma_{22}{=}1,\ \sigma_{12}{=}0.5,\ \alpha{=}(2,\text{-}4)\ )$ - (a) contour plot for (X_1, X_2); (b) and (c) contour plots of the copula for (X_1, X_2) and its copula density; (d) conditional distributions of the copula for (X_1, X_2), $P(U_2 \leq u^* \mid U_1 = u^{**})$ (left) and $P(U_1 \leq u^* \mid U_2 = u^{**})$ (right) where $u^{**} = (0.05, 0.1, 0.2, 0.3, 0.4, 0.5, 0.6, 0.7, 0.8, 0.9, 0.95)$ with the line types and colors range from solid/black (0.05) to dotdash/dark gray (0.5) to dotted/dim gray (0.95).

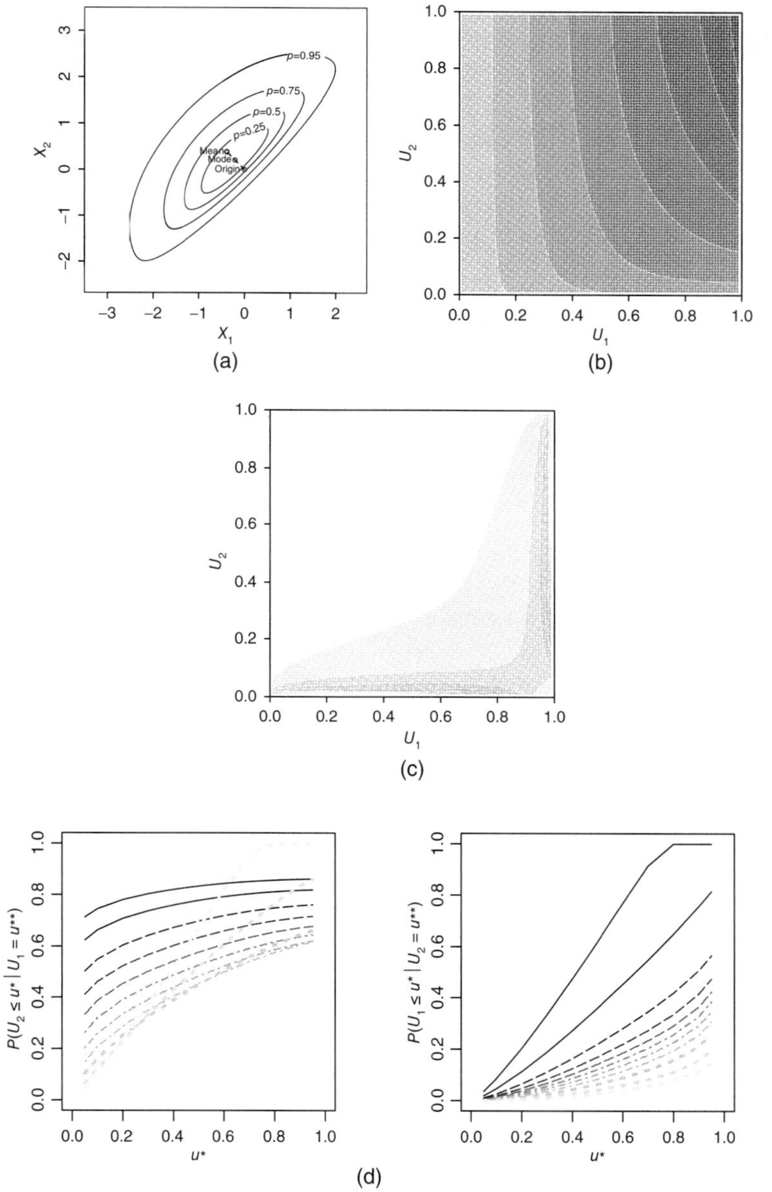

Figure 6.3 $(X_1, X_2) \sim SN_2(\mu{=}(0,0),\ \sigma_{11}{=}\sigma_{22}{=}1,\ \sigma_{12}{=}0.5,\ \alpha{=}(\text{-}4,4)\ )$ - (a) contour plot for (X_1, X_2); (b) and (c) contour plots of the copula for (X_1, X_2) and its copula density; (d) conditional distributions of the copula for (X_1, X_2), $P(U_2 \le u^* \mid U_1 = u^{**})$ (left) and $P(U_1 \le u^* \mid U_2 = u^{**})$ (right) where $u^{**} = (0.05, 0.1, 0.2, 0.3, 0.4, 0.5, 0.6, 0.7, 0.8, 0.9, 0.95)$ with the line types and colors range from solid/black (0.05) to dotdash/dark gray (0.5) to dotted/dim gray (0.95).

6.5 INFERENCE OF DIRECTIONAL DEPENDENCE USING SKEW–NORMAL COPULA-BASED REGRESSION

In this section, we discuss the inference of the directional dependence (in marginal or joint behavior or both) using skew–normal copula-based regression. The inference procedure consists of two stages: (i) estimation of skew–normal copula-based regression and (ii) empirical identification of directional dependence and computation of the directional dependence measures.

6.5.1 Estimation of Copula-Based Regression

The copula-based regressions for (X_1, X_2) can be estimated as long as the estimators for the marginal distributions and the copula density are given. The copula-based regression function of X_1 given $X_2 = x_2$ in Equation 6.10 can be estimated by

$$\hat{m}_{X_1|X_2}(x_2) = \int_{\mathcal{R}} x_1 \hat{c}(\hat{F}_1(x_1), \hat{F}_2(x_2)) d\hat{F}_1(x_1)$$

where $\hat{c}$, $\hat{F}_1$, and $\hat{F}_2$ are any given estimators for c, F_1 and F_2.

Assume that the unknown copula C belongs to an absolutely continuous parametric copula family $C = \{C_\theta\}$ where θ represents a set of parameters measuring dependence between the variables. Then, depending on the estimation method for F_1 and F_2, the estimator of the copula-based regression can be fully parametric or semiparametric. A fully parametric approach is that the marginals are chosen from a parametric family. A semiparametric approach is that the copula is still estimated parametrically but the marginal distributions are estimated nonparametrically, for example, by the rescaled empirical distribution or a kernel-smoothing estimator, in order to take into account the impact of misspecifications in the marginal distributions on the estimation of a parametric copula density.

Let $\mathbf{X}_i = (X_{i1}, X_{i2})'$ be an independent and identically distributed sample of n observations generated from the distribution $\mathbf{H}$ of $\mathbf{X} = (X_1, X_2)'$ where $i = 1, \ldots, n$. As the joint density function of $\mathbf{X}$ can be represented using the copula density and the marginal densities for the variables as shown in Equation 6.4, the log likelihood function of n random samples is

$$\begin{aligned} \ell &= \sum_{i=1}^{n} \log\ h(x_{i1}, x_{i2}) \\ &= \sum_{i=1}^{n} \log\ c_\theta(F_1(x_{i1}), F_2(x_{i2})) + \sum_{i=1}^{n} [\log\ f_1(x_{i1}) + \log\ f_2(x_{i2})] \end{aligned} \tag{6.19}$$

For the fully parametric estimation, we here use the bivariate skew–normal density in Equation 6.15 for $h(x_1, x_2)$ because it has the shape parameter vector $\alpha = (\alpha_1, \alpha_2)$ to take into account skewness in the bivariate data. Furthermore, when the distribution of (X_1, X_2) is the bivariate skew–normal distribution, the marginal distributions and the conditional distributions are still skew–normal distributions (Azzalini and Capitanio, 1999). The likelihood method maximizing the log likelihood of Equation 6.19 over the parameters in Equation 6.15 and the estimation of the regression functions (i.e., conditional means) can be done using the functions `selm`, `makeSECdistr`, `conditionalSECdistr`, and `sn.cumulants` in the **sn** package (Azzalini, 2014) of R (R Core Team, 2014). We denote the *fully parametric estimator of the skew–normal copula-based regression function* of X_1 given X_2 and of X_2 given X_1 as $\hat{m}^p_{X_1|X_2}(x_2)$ and $\hat{m}^p_{X_2|X_1}(x_1)$, respectively.

For the semiparametric estimation of the copula-based regression, we propose using a novel and easy-to-use approach proposed by Noh *et al.* (2013). The nonparametric estimator for the marginal distribution is the normalized rank, that is, a rescaled empirical distribution function in Equation 6.12. Given the nonparametric estimator $\hat{F}_j(x_j)$ of the univariate margin $F_j(x_j)$, we employ the bivariate skew–normal copula density in Equation 6.17 for the dependence structure between X_1 and X_2. We then obtain the maximum pseudo-likelihood (MPL) estimator (Genest *et al.*, 1995; Kim *et al.*, 2007), which maximizes the log-likelihood contribution from dependence structure in data over the parameters

$$\ell_p(\theta) = \sum_{i=1}^{n} \log\ c_\theta(\hat{F}_1(x_{i1}), \hat{F}_2(x_{i2}))$$

where $\theta = (\sigma_{12}, \alpha_1, \alpha_2)'$. Here, we obtain the MPL estimate of θ using a general-purpose optimizer in R, function `optim` from the built-in package **stats**.

Once we obtain the nonparametric estimator for F_j and the MPL estimator for θ in the copula density, we construct the *semiparametric estimator of the skew–normal copula-based regression function* of X_1 for a given $X_2 = x_2$, denoted by $\hat{m}^{sp}_{X_1|X_2}(x_2)$:

$$\hat{m}^{sp}_{X_1|X_2}(x_2) = \frac{1}{n} \sum_{i=1}^{n} x_{i1} c_{\hat{\theta}_{MPL}}(\hat{F}_1(x_{i1}), \tilde{F}_2(x_2)) \tag{6.20}$$

where $\hat{\theta}_{MPL}$ is the MPL estimator of θ and $\tilde{F}_2(x_2)$ is a kernel-smoothing estimator of $F_2(x_2)$ estimated using the function `kcde` in the R package **ks** (Duong, 2007). The *semiparametric estimator of the skew–normal copula-based regression function* of X_2 for a given $X_1 = x_1$, denoted as $\hat{m}^{sp}_{X_2|X_1}(x_1)$, can be estimated in a similar fashion. We use a kernel-smoothing estimator for the marginal distribution of the conditioning variable ($F_1(x_1)$ in $\hat{m}^{sp}_{X_2|X_1}(x_1)$ and $F_2(x_2)$ in $\hat{m}^{sp}_{X_1|X_2}(x_2)$) to obtain a smooth regression curve. For theoretical properties of the aforementioned semiparametric estimator, we refer to Noh *et al.* (2013).

6.5.2 Detection of Directional Dependence and Computation of the Directional Dependence Measures

Suppose that the two regression models for (X_1, X_2) are estimated using the method either fully parametrically ($\hat{m}^p_{X_1|X_2}(x_2)$ and $\hat{m}^p_{X_2|X_1}(x_1)$) or semiparametrically ($\hat{m}^{sp}_{X_1|X_2}(x_2)$ and $\hat{m}^{sp}_{X_2|X_1}(x_1)$), as described earlier. Then, from Definition 6.1, we can empirically identify directional dependence between X_1 and X_2 (marginal or joint behavior or both) by constructing a conditional regression plot, a plot of $\hat{m}_{X_1|X_2}(z)$ versus $\hat{m}_{X_2|X_1}(z)$ with a diagonal reference line, $X_2 = X_1$. Any deviation from the reference line, above or below the reference line, indicates that X_1 and X_2 are directionally dependent. An interesting property of a conditional regression plot based on semiparametric regression estimators (i.e., a plot of $\hat{m}^{sp}_{X_1|X_2}(z)$ vs $\hat{m}^{sp}_{X_2|X_1}(z)$) is that it is useful in evaluating if the directional dependence stemming from the joint behavior empirically exists. This is because the semiparametric regression estimator does not require specification of the marginal distributions and thus one can remove the influence of the marginal distribution on the directional dependence.

Once the directional dependence between X_1 and X_2 is empirically identified, we then compute directional dependence measures in Equations 6.13 and 6.14, depending on the regression estimators used:

$$\hat{\rho}^2_p(X_1 \to X_2) = n^{-1} \sum_{i=1}^{n} \frac{(\hat{m}^p_{X_2|X_1}(x_{i1}) - \hat{E}(X_2))^2}{\widehat{Var}(X_2)}, \tag{6.21}$$

$$\hat{\rho}^2_p(X_2 \to X_1) = n^{-1} \sum_{i=1}^{n} \frac{(\hat{m}^p_{X_1|X_2}(x_{i2}) - \hat{E}(X_1))^2}{\widehat{Var}(X_1)}, \tag{6.22}$$

$$\hat{\rho}^2_{sp}(X_1 \to X_2) = n^{-1} \sum_{i=1}^{n} \frac{(\hat{m}^{sp}_{X_2|X_1}(x_{i1}) - \widetilde{E}(X_2))^2}{\widetilde{Var}(X_2)}, \tag{6.23}$$

$$\hat{\rho}^2_{sp}(X_2 \to X_1) = n^{-1} \sum_{i=1}^{n} \frac{(\hat{m}^{sp}_{X_1|X_2}(x_{i2}) - \widetilde{E}(X_1))^2}{\widetilde{Var}(X_1)} \tag{6.24}$$

where $\hat{E}(X_j)$ and $\widehat{Var}(X_j)$ are the estimates of the mean and variance of X_j under the estimated bivariate skew–normal distribution for (X_1, X_2), respectively, and $\widetilde{E}(X_j)$ and $\widetilde{Var}(X_j)$ are the sample mean and the sample variance of a random sample $\{X_{1j}, \ldots, X_{nj}\}$, respectively. We further construct the 95% confidence interval for the difference in two measures, $\rho^2(X_1 \to X_2) - \rho^2(X_2 \to X_1)$, using the nonparametric bootstrap bias corrected and accelerated (BCa) method (Efron and Tibshirani, 1993; Davison and Hinkley, 1997).

Note that the R codes used in this chapter are available upon request.

6.6 APPLICATION

In this section, we present an example concerning a study on development of aggression in adolescence analyzed in Finkelstein *et al.* (1994) and von Eye and Wiedermann (2014) to illustrate the proposed skew–normal copula-based regression. In this study, there are 38 boys and 76 girls who responded to an aggression questionnaire in 1983, 1985, and 1987, and two aggression variables of interest are Verbal Aggression Against Adults (VAAA) and Physical Aggression Against Peers (PAAP). von Eye and Wiedermann (2014) analyzed this data to examine if physical aggression can predict verbal aggression or vice versa, using several approaches including comparison of skewness and kurtosis of the variables (Dodge and Rousson, 2000; Dodge and Rousson, 2001; Dodge and Yadegari, 2010; von Eye and DeShon, 2012) and evaluation of residuals of competing regression models (Wiedermann *et al.*, 2013) under the linear regression model with normally distributed errors.

We here employ the skew–normal copula-based regression to investigate directional dependence between two variables, VAAA and PAAP, without imposing the linear assumption. As there are the three measures of VAAA and PAAP from 1983, 1985, and 1987, respectively, as done in von Eye and Wiedermann (2014), we also carry out two principal component analyses, one for the three measures of VAAA and the other for the three measures of PAAP. We then obtain the component scores of the principal components of VAAA and PAAP. Figure 6.4 shows the scatter plot of component scores of VAAA and PAAP. Note that von Eye and Wiedermann (2014) reported that the component scores of PAAP are nonnormally distributed, while the component scores of VAAA are closer to a normal distribution.

In order to examine the directional dependence between the component scores of VAAA and PAAP, we first obtain a fully parametric regression function under the bivariate skew–normal distribution. Figure 6.4 displays (a) the fitted bivariate skew–normal density and its contour levels and (b) Healy-type graphical diagnostics in the form of Q–Q plot (Azzalini and Capitanio, 1999). We here consider the Q–Q plot diagnostic to compare the data distribution with the bivariate skew–normal distribution. Although the fitted bivariate skew–normal density appears to be satisfactory, the Q–Q plot indicates that there is some deviation at the tails (i.e., a few points deviating from the reference dotted line). As future work, it would be worthwhile to consider the multivariate skew–*t* distribution with long tail. Note that estimated bivariate skew–normal density is $SN_2(\hat{\mu}, \hat{\Sigma}, \hat{\alpha})$ where $\hat{\mu} = (-0.957, -1.685)'$, $\hat{\Sigma} = \begin{pmatrix} 2.755 & 2.866 \\ 2.866 & 4.912 \end{pmatrix}$ and $\hat{\alpha} = (-0.179, 3.492)'$.

Figure 6.5 shows the conditional regression plot (i.e., $\hat{m}^p_{PAAP|VAAA}(z)$ vs $\hat{m}^p_{VAAA|PAAP}(z)$) and the two fitted copula-based regressions using the fully parametric estimation, $\hat{m}^p_{PAAP|VAAA}$ and $\hat{m}^p_{VAAA|PAAP}$. From Figure 6.5a, we see that directional dependence exists because there is clear deviation from the reference line (dotted line). Figure 6.5b and c show that the forms of the regression functions are different, depending on the conditioning variable. Note that the solid line is a parametric copula-based regression and the dotted line is a simple linear regression. The

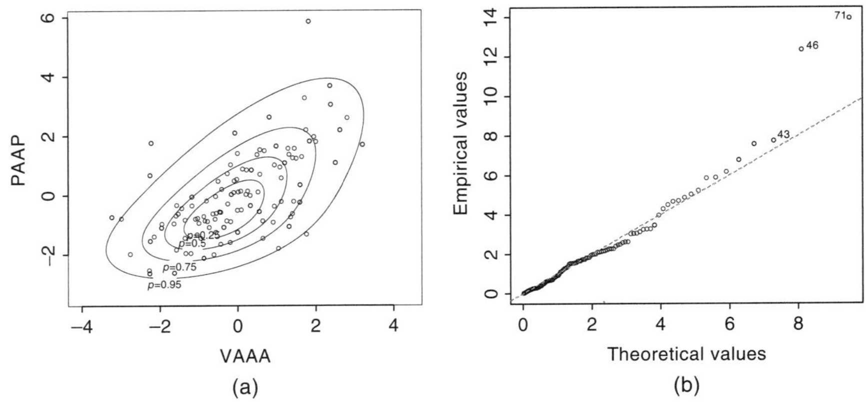

Figure 6.4 (a) Scatter plot of component scores of VAAA and PAAP and a fitted skew–normal density; (b) Q–Q plot.

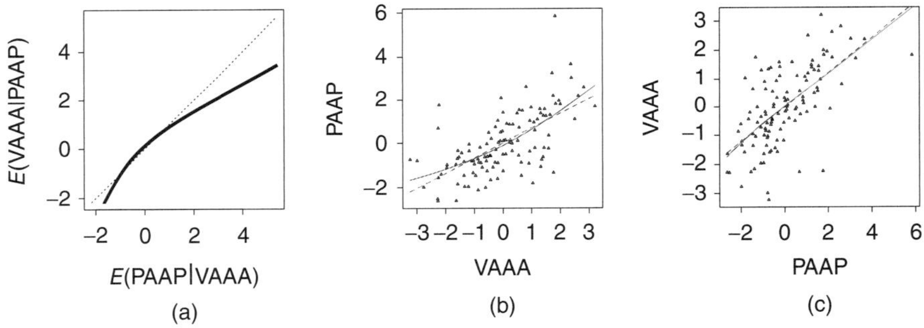

Figure 6.5 Parametric estimation: (a) conditional regression plot, $\hat{m}^{p}_{\text{PAAP|VAAA}}(z)$ versus $\hat{m}^{p}_{\text{VAAA|PAAP}}(z)$; (b) copula-based regression estimate for VAAA, $\hat{m}^{p}_{\text{PAAP|VAAA}}$ (solid line); (c) copula-based regression estimate for PAAP, $\hat{m}^{p}_{\text{VAAA|PAAP}}$ (solid line).

estimated regression of PAAP on VAAA seems to be nonlinear especially when VAAA is either small or large, unlike the estimated regression of VAAA on PAAP.

Figure 6.6 displays the plots of semiparametric copula regression analysis. From the conditional regression plot, $\hat{m}^{\text{sp}}_{\text{PAAP|VAAA}}(z)$ versus $\hat{m}^{\text{sp}}_{VAAA|\text{PAAP}}(z)$, in Figure 6.6a, there appears to be no directional dependence because of no deviation from the reference line (dotted line). This means that the directional dependence in the data stems from the marginal behavior of the variables, not the joint behavior. We observe from Figure 6.6b and c that both estimated copula-based regressions, $\hat{m}^{\text{sp}}_{\text{PAAP|VAAA}}$ and $\hat{m}^{\text{sp}}_{\text{VAAA|PAAP}}$, are not linear. Note that the solid line is a semiparametric copula-based regression and the dotted line is a simple linear regression.

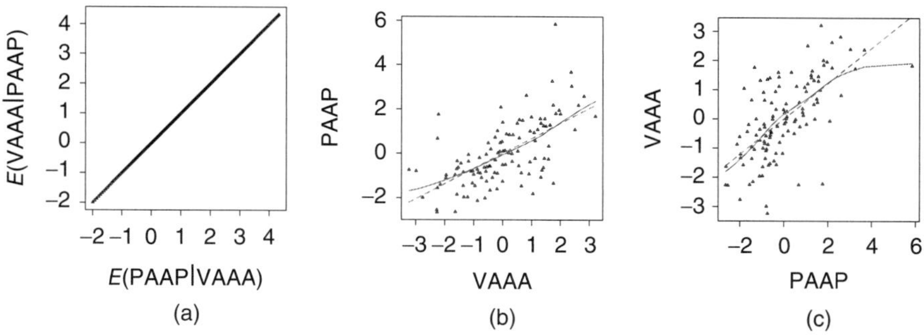

Figure 6.6 Semiparametric estimation: (a) conditional regression plot, $\hat{m}^{\text{sp}}_{\text{PAAP|VAAA}}(z)$ versus $\hat{m}^{\text{sp}}_{\text{VAAA|PAAP}}(z)$; (b) copula-based regression estimate for VAAA, $\hat{m}^{\text{sp}}_{\text{PAAP|VAAA}}$ (solid line); (c) copula-based regression estimate for PAAP, $\hat{m}^{\text{sp}}_{\text{VAAA|PAAP}}$ (solid line).

TABLE 6.1 Estimates for Directional Dependence Measures and the 95% Nonparametric Bootstrap BCa Interval for $\rho^2(PAAP \to VAAA) - \rho^2(VAAA \to PAAP)$ (Number of Nonparametric Bootstrap Samples = 999).

	Parametric	Semiparametric
$\rho^2(VAAA \to PAAP)$	0.404	0.364
$\rho^2(PAAP \to VAAA)$	0.415	0.389
$\rho^2(PAAP \to VAAA)$	0.010	0.025
$-\rho^2(VAAA \to PAAP)$	(−0.0283, 0.0496)	(0.0030, 0.0726)

Finally, Table 6.1 shows the estimates of the directional dependence measures in Equations 6.24–6.23, and the 95% adjusted bootstrap percentile (BCa) intervals for ρ^2(PAAP → VAAA) − ρ^2(VAAA → PAAP) under fully parametric and semiparametric estimation of skew–normal copula-based regressions, respectively. The proportion of total variation of VAAA that is explained by PAAP is 41.5% for the parametric copula-based regression and 38.9% for the semiparametric copula-based regression. The proportion of total variation of PAAP that is explained by VAAA is 40.4% for the parametric copula-based regression and 36.4% for the semiparametric copula-based regression estimate. The 95% nonparametric bootstrap BCa interval for ρ^2(PAAP → VAAA) − ρ^2(VAAA → PAAP) using the semiparametric copula-based regression does not include 0, unlike the confidence interval computed using the parametric copula-based regression, which requires restrictive assumptions on the marginals. Thus, it appears that it is more edged for VAAA to be explained by PAAP rather than vice versa (38.9% vs 36.4%, under the semiparametric estimation). Note that von Eye and Wiedermann (2014) in fact concluded that VAAA is the response variable and PAAP is the explanatory variable.

6.7 CONCLUSION

In this chapter, we presented the skew–normal copula-based regression for the inference of the directional dependence and demonstrated it using real data. Unlike the elliptical copulas such as the normal copula and the t-copula, the skew–normal copula-based regression can capture the asymmetric and nonlinear dependence between variables due to the shape parameters accounting for skewness in the data.

The example used in the application section was the bivariate case. Some of the tools presented in this chapter can be easily extended to the multivariate case. For instance, the skew–normal copula-based regression is available in a multiple regression setting because the multivariate skew–normal distribution/copula is available as shown in this chapter and the copula-based regression model can be estimated under multiple covariate models (Noh *et al.*, 2013). A valuable extension, which is currently under investigation, is to generalize the copula-based inference procedure for the analysis of directional dependence to the multivariate data.

In addition, another future work would be to use skew–t copulas in the regression model. The skew–t copula has additional parameters regulating skewness and kurtosis in the data, and it would be very useful for applications where the data has heavy tails.

REFERENCES

Azzalini, A. (2014) `sn` package: the skew-normal and skew-t distributions (version 1.1-0), `http://CRAN.R-project.org/package=sn` (accessed 21 December 2015).

Azzalini, A. and Capitanio, A. (1999) Statistical applications of the multivariate skew normal distribution. *Journal of the Royal Statistical Society Series B*, **61**, 579–602.

Azzalini, A. and Capitanio, A. (2014) *The Skew-Normal and Related Families*, IMS Monographs, Cambridge University Press, Cambridge.

Davison, A. and Hinkley, D. (1997) *Bootstrap Methods and their Application*, Cambridge University Press, Cambridge.

Dodge, Y. and Rousson, V. (2000) Direction dependence in a regression line. *Communications in Statistics: Theory and Methods*, **29**, 1957–1972.

Dodge, Y. and Rousson, V. (2001) On asymmetric properties of the correlation coefficient in the regression setting. *American Statistician*, **55**, 51–54.

Dodge, Y. and Yadegari, I. (2010) On direction of dependence. *Metrika*, **72**, 139–150.

Duong, T. (2007) `ks`: Kernel density estimation and kernel discriminant analysis for multivariate data in R. *Journal of Statistical Software*, **21**, 1–16.

Efron, B. and Tibshirani, R.J. (1993) *An Introduction to the Bootstrap*, Chapman and Hall, New York.

von Eye, A. and DeShon, R. (2012) Directional dependence in developmental research. *International Journal of Behavioral Development*, **36**, 303–312.

von Eye, A. and Wiedermann, W. (2014) On direction of dependence in latent variable contexts. *Educational and Psychological Measurement*, **74**, 5–30.

Finkelstein, J., von Eye, A., and Preece, M. (1994) The relationship between aggressive behavior and puberty in normal adolescents: a longitudinal study. *Journal of Adolescent Health*, **15**, 319–326.

Genest, C., Ghoudi, K., and Rivest, L.P. (1995) A semiparametric estimation procedure of dependence parameters in multivariate families of distribution. *Biometrika*, **82**, 543–552.

Genton, M. (2004) *Skew-Elliptical Distributions and their Applications: A Journey Beyond Normality*, Chapman and Hall/CRC, Boca Raton, FL.

Joe, H. (2004) *Multivariate Models and Dependence Concepts*, Chapman and Hall, London.

Kim, D. and Kim, J.M. (2014) Analysis of directional dependence using asymmetric copula-based regression models. *Journal of Statistical Computation and Simulation*, **84**, 1990–2010.

Kim, G., Silvapulle, M.J., and Silvapulle, P. (2007) Comparison of semiparametric and parametric methods for estimating copulas. *Computational Statistics and Data Analysis*, **51**, 2836–2850.

Kim, J.M., Jung, Y.S., Sungur, E.A., Han, K.H., Park, C., and Sohn, I. (2008) A copula method for modeling directional dependence of genes. *BMC Bioinformatics*, **9**, 225.

Kollo, T., Selart, A., and Visk, H. (2013) From multivariate skewed distributions to copulas, in *Combinatorial Matrix Theory and Generalized Inverses of Matrices* (eds R.B. Bapat, S.J. Kirkland, K.M. Prasad, and S. Puntanen), Springer-Verlag, New York, pp. 63–72.

Li, B. and Genton, M. (2013) Nonparametric identification of copula structures. *Journal of the American Statistical Association*, **108**, 666–675.

Liebscher, E. (2008) Construction of asymmetric multivariate copulas. *Journal of Multivariate Analysis*, **99**, 2234–2250.

Marshall, A. and Olkin, I. (1967) A multivariate exponential distribution. *Journal of the American Statistical Association*, **62**, 30–44.

McNeil, A., Frey, R., and Embrechts, P. (2005) *Quantitative Risk Management: Concepts, Techniques and Tools*, Princeton University Press, Princeton, NJ.

Nelson, R. (2006) *An Introduction to Copulas*, 2nd edn, Springer-Verlag, New York.

Nigg, J.T., Knottnerus, G.M., Martel, M.M., Nikolas, M., Cavanagh, K., Karmaus, W., and Rappley, M.D. (2008) Low blood lead levels associated with clinically diagnosed attention-deficit/hyperactivity disorder and mediated by weak cognitive control. *Biological Psychiatry*, **63**, 325–331.

Noh, H., El Ghouch, A., and Bouezmarni, T. (2013) Copula-based regression estimation and inference. *Journal of the American Statistical Association*, **108**, 676–688.

R Core Team (2014) *R: A Language and Environment for Statistical Computing*, R Foundation for Statistical Computing, Vienna, Austria, `http://www.R-project.org` (accessed 21 December 2015).

Sklar, A. (1959) Fonctions de répartition à n-dimensions et leurs marges. *Publications de l'Institut de Statistique de l'Université de Paris*, **8**, 229–231.

Sungur, E.A. (2005a) A note on directional dependence in regression setting. *Communications in Statistics: Theory and Methods*, **34**, 1957–1965.

Sungur, E.A. (2005b) Some observations on copula regression functions. *Communications in Statistics: Theory and Methods*, **34**, 1967–1978.

Tawn, J. (1988) Bivariate extreme value theory: models and estimation. *Biometrika*, **75**, 397–415.

Wiedermann, W., Hagmann, M., Kossmeier, M., and von Eye, A. (2013) Resampling techniques to determine direction of effects in linear regression models, *InterStat*. `http://interstat.statjournals.net/YEAR/2013/articles/1305002.pdf`.

7

NON-GAUSSIAN STRUCTURAL EQUATION MODELS FOR CAUSAL DISCOVERY

Shohei Shimizu

Institute of Scientific and Industrial Research, Osaka University, Osaka, Japan

7.1 INTRODUCTION

Structural equation models (SEMs) are mathematical systems that can be used to represent data-generation processes and causal relations of variables (Bollen, 1989, Pearl, 2000). An example of an SEM is

$$x_1 = e_1 \tag{7.1}$$

$$x_2 = b_{21}x_1 + e_2 \tag{7.2}$$

where x_1 and x_2 are observed continuous random variables, e_1 and e_2 are latent (unobserved) continuous random variables, or error variables, and b_{21} is a constant. The associated causal graph of the SEM in Equations (7.1) and (7.2) is shown on the left side of Figure 7.1. The structural equations (7.1) and (7.2) represent a data-generation process in which the values of e_1 and e_2 are first generated, x_1 is generated as the same value as e_1, and the value of x_2 is then given as a linear combination of x_1 and e_2. The error variables e_1 and e_2 are exogenous, as the other variables in the model do not contribute to their generation.

Statistics and Causality: Methods for Applied Empirical Research, First Edition.
Wolfgang Wiedermann and Alexander von Eye.

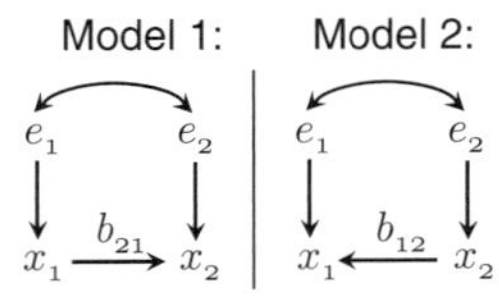

Figure 7.1 Which of these two models, with opposing directions of causation, is better? The errors e_1 and e_2 may be dependent.

A causal interpretation of this SEM is that, if the value of x_1 has been changed (i.e., is manipulated) from some constant c to another constant d, that of x_2 changes by an average of $b_{21}(d - c)$. Furthermore, the value of x_1 remains constant on average, even if x_2 is changed. In other words, x_1 is the cause and x_2 is its effect.

We want to estimate such a cause-and-effect relationship between x_1 and x_2 based only on their observed data, which are collected simultaneously, under minimal assumptions on the distributions of the exogenous latent error variables e_1 and e_2. More formally, the basic problem is defined as follows: let us denote the number of observations as n and collect them in a $2 \times n$ matrix $\mathbf{X}$ whose (i, m)-th element $x_i^{(m)}$ is the m-th observation of the variable x_i $(i = 1, 2; m = 1, \ldots, n)$. Suppose that the observed data $\mathbf{X}$ are randomly generated from either of the following two SEMs with opposing directions of causation:

$$\text{Model 1}: \begin{cases} x_1 = e_1 \\ x_2 = b_{21}x_1 + e_2, \end{cases} \tag{7.3}$$

$$\text{Model 2}: \begin{cases} x_1 = b_{12}x_2 + e_1 \\ x_2 = e_2, \end{cases} \tag{7.4}$$

where b_{21} and b_{12} are nonzero constants. In the former model, x_1 causes x_2, whereas in the latter, x_2 causes x_1. The associated causal graphs of the two models are shown in Figure 7.1. We wish to determine which of these two models generates the observed data $\mathbf{X}$. For example, x_1 and x_2 may represent verbal aggression against adults (VAAA) and physical aggression against peers (PAAP), respectively, and could be measured for many children at the same time (von Eye and Wiedermann, 2014).

In the example above, the classical Gaussian assumption on the error variables is not particularly helpful in determining causal direction (Bollen, 1989), because the two models' data have the same mean and covariance structures. Fortunately, however, many applications obtain non-Gaussian data (Micceri, 1989, Hyvärinen *et al.*, 2001, Smith *et al.*, 2011, Sogawa *et al.*, 2011, Moneta *et al.*, 2013), which means that the distribution contains information besides the mean and covariance structures.

Dodge and Rousson (2000) showed that it is possible to estimate the causal direction of two observed variables x_1 and x_2 based on the non-Gaussian structure of data (the third-order moment structure) when the following hold:

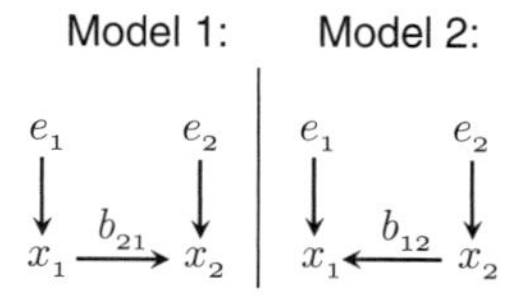

Figure 7.2 Which of the two models with opposing directions of causation is better? Here, the errors e_1 and e_2 are independent, which implies that Models 1 and 2 have no latent common causes.

- The error variables e_1 and e_2 are independent.
- The error variable corresponding to the cause is skewed, but the error variable corresponding to the effect is not.[1]

The independence assumption implies that there are no latent common causes (unobserved confounding variables) in the two observed variables x_1 and x_2. This changes the causal graphs in Figure 7.1 to those in Figure 7.2. The second assumption means that, when Model 1 generates the data, the true direction of causation can be determined if e_1 is skewed and e_2 is not skewed; if Model 2 generates the data, the true direction of causation can be determined if the skewness of e_2 is nonzero and that of e_1 is zero. (The skewness is zero if a variable is Gaussian and is nonzero for asymmetric non-Gaussian variables.)

Models 1 and 2 provide the same conditional independence structure of data, regardless of whether the error variables are Gaussian or non-Gaussian. Therefore, classical causal discovery methods based on their conditional independence (Pearl, 2000, Spirtes *et al.*, 1993) are not able to distinguish between the two models.[2]

Shimizu *et al.* (2006) generalized the results of Dodge and Rousson (2000) to any non-Gaussian distribution, and further extended the theory to multivariate cases. They also proved that the causal direction, and the existence of causal connections, among p observed variables $x_1, x_2, \ldots, x_p$ can be uniquely determined, that is, they are identifiable, under the following assumptions:

- The error variables e_i $(i = 1, \ldots, p)$ are independent.
- The error variables e_i $(i = 1, \ldots, p)$ have non-Gaussian probability density functions.[3]
- The causal relations are acyclic.

[1]Wiedermann and Hagmann (2014) recently considered another interesting case in which the error variable corresponding to the effect is skewed, but the error variable corresponding to the cause is not.

[2]It is possible to effectively combine these conditional independence-based methods with the non-Gaussian methods discussed in this chapter. See Hoyer *et al.* (2008a), Hyvärinen and Smith (2013), Ramsey *et al.* (2011) for such combinations.

[3]At most one error variable may be Gaussian.

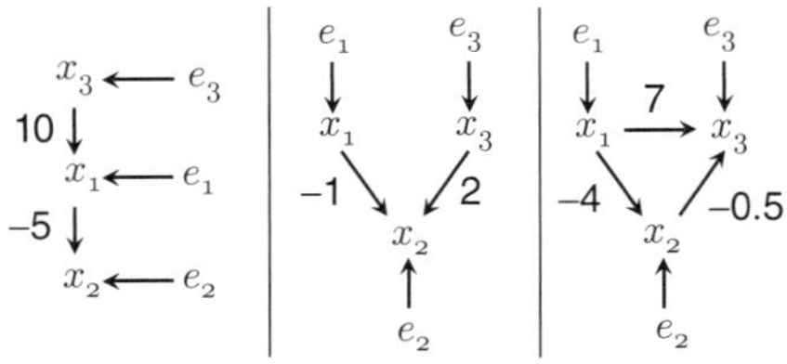

Figure 7.3 Causal graphs of basic causal discovery models with no latent common causes.

A model with these three properties is called a Linear Non-Gaussian Acyclic Model, or LiNGAM. Several examples of causal graphs belonging to this class of non-Gaussian model are shown in Figure 7.3.

These basic non-Gaussian models have led to the development of many extended models (Hoyer *et al.*, 2008b, 2009, Shimizu and Hyvärinen, 2008, Shimizu *et al.*, 2009, Lacerda *et al.*, 2008, Hyvärinen *et al.*, 2010, Zhang and Hyvärinen, 2009, Tillman *et al.*, 2010, Peters *et al.*, 2011, Shimizu, 2012). Among others, Hyvärinen *et al.* (2010) combined the basic non-Gaussian causal discovery model developed by Shimizu *et al.* (2006) with the classic autoregressive model (Granger, 1969) to analyze both contemporaneous and lagged causal relations based on time series data. Hoyer *et al.* (2008b) relaxed the independence assumption, thus allowing latent common causes, and showed that under assumptions of linearity and non-Gaussianity, it is possible to uniquely determine the causal direction of two observed variables, even in the presence of latent common causes.

In subsequent sections of this chapter, we discuss the following three fundamental LiNGAM models to elaborate the concepts and methods underlying this non-Gaussian causal discovery approach:

- Basic LiNGAM (Shimizu *et al.*, 2006) (Section 7.3);
- LiNGAM for time series (Hyvärinen *et al.*, 2010) (Section 7.4);
- LiNGAM with latent common causes (Hoyer *et al.*, 2008b) (Section 7.5).

These non-Gaussian methods have been applied to data studied in several fields, including economics (Ferkingsta *et al.*, 2011, Moneta *et al.*, 2013, Coad and Binder, 2014), environmental science (Niyogi *et al.*, 2010), epidemiology (Rosenström *et al.*, 2012, Helajärvi *et al.*, 2014), neuroscience (Smith *et al.*, 2011, Boukrina and Graves, 2013, Manelis and Reder, 2014, Mills-Finnerty *et al.*, 2014), and chemistry (Campomanes *et al.*, 2014).

7.2 INDEPENDENT COMPONENT ANALYSIS

The concepts and methods involved in the signal processing method known as independent component analysis (ICA) are closely related to those of LiNGAM. Thus, we first provide a brief overview of ICA (Jutten and Hérault, 1991, Comon, 1994, Hyvärinen *et al.*, 2001) before moving on to the detail of LiNGAM methods.

7.2.1 Model

The ICA model represents a data-generation process in which latent independent source signals are linearly mixed with one another to generate observed signals. This can be viewed as an SEM. An instance of such a model is given by

$$x_1 = a_{11}s_1 + a_{12}s_2 \tag{7.5}$$

$$x_2 = a_{21}s_1 + a_{22}s_2 \tag{7.6}$$

where x_1 and x_2 are observed continuous random variables, s_1 and s_2 are latent non-Gaussian continuous random variables, and a_{11}, a_{12}, a_{21}, and a_{22} are constants. This model first generates the values of the latent independent variables s_1 and s_2 and then generates values of the observed variables x_1 and x_2 as linear combinations of s_1 and s_2.

In general, an ICA model (Jutten and Hérault, 1991, Comon, 1994) for p observed variables x_i $(i = 1, \ldots p)$ can be defined as follows:

$$x_i = \sum_{j=1}^{q} a_{ij}s_j \tag{7.7}$$

where s_j $(j = 1, \ldots, q)$ are continuous latent random variables that are mutually independent. We refer to these as independent components. The latent independent variables s_j follow non-Gaussian distributions.

In matrix form, the ICA model in Equation (7.7) can be represented by

$$\boldsymbol{x} = \mathbf{A}\boldsymbol{s} \tag{7.8}$$

where the $p \times q$ mixing matrix $\mathbf{A}$ collects coefficients a_{ij}, and the vectors x and s collect the observed variables x_i and the independent components s_j, respectively. It is commonly assumed that there is more than one observed variable ($p \geq 2$) and that no column of the mixing matrix $\mathbf{A}$ is pairwise linearly dependent (Comon, 1994, Eriksson and Koivunen, 2004).

7.2.2 Identifiability

It has been shown that the mixing matrix $\mathbf{A}$ is identifiable up to some permutation, scaling, and the sign of the columns (Comon, 1994, Eriksson and Koivunen, 2004). Thus, the mixing matrix identified by ICA, which is denoted by $\mathbf{A}_{\text{ICA}}$, can be written as

$$\mathbf{A}_{\text{ICA}} = \mathbf{APD} \tag{7.9}$$

where $\mathbf{P}$ is an unknown $q \times q$ permutation matrix, and $\mathbf{D}$ is an unknown $q \times q$ diagonal matrix with no zeros on the diagonal.

7.2.3 Estimation

Most ICA estimation methods assume that the mixing matrix $\mathbf{A}$ is square, which means that the number of observed variables is equal to the number of independent components, that is, $p = q$. These methods estimate a matrix known as a separating matrix $\mathbf{W} = \mathbf{A}^{-1}$ (Hyvärinen *et al.*, 2001). Furthermore, most of these methods minimize the mutual information (or its approximation) of the estimated independent components in $\hat{s} = \mathbf{W}_{\text{ICA}}\boldsymbol{x}$, that is, $I(\hat{s}) = \{\sum_{j=1}^{d} H(\hat{s}_j)\} - H(\hat{s})$, where $H(\hat{s})$ is the differential entropy of $\hat{s}$ defined by $E\{-\log\ p(\hat{s})\}$. It can be shown that the mutual information of these estimated independent components is zero if and only if they are independent. Thus, ICA methods estimate a separating matrix that minimizes the independence among the estimated independent components. Following this, the separating matrix $\mathbf{W}$ is estimated up to the permutation $\mathbf{P}$, and the scaling and the sign $\mathbf{D}$ of the rows

$$\mathbf{W}_{\text{ICA}} = \mathbf{PDW}(= \mathbf{PDA}^{-1}) \tag{7.10}$$

Note that ICA estimation methods provide a random permutation of the rows. Consistent and computationally efficient estimation algorithms have been developed in which there is no need to specify the independent component distributions (Amari, 1998, Hyvärinen, 1999).

A relevant method is principal component analysis (PCA), which is commonly used as a preprocessing technique for ICA. PCA estimates a matrix $\mathbf{W}_{PCA}$ such that the principal components in $z = \hat{\mathbf{W}}_{\text{PCA}}\boldsymbol{x}$ are uncorrelated. This is because noncorrelation is a necessary condition for non-Gaussian variables to be independent, whereas for Gaussian variables, being uncorrelated is equivalent to independence. Thus, ICA can be seen as a method for determining a factor rotation that makes latent factors independent (Hyvärinen and Kano, 2003).

For further details on ICA, readers are referred to the textbook of Hyvärinen *et al.* (2001) and its recent update Hyvärinen (2013).

7.3 BASIC LINEAR NON-GAUSSIAN ACYCLIC MODEL

In this section, we first review the basic LiNGAM (Shimizu *et al.*, 2006), before going on to discuss its extensions to cases with temporal structures and latent common causes. Although the assumptions made in the basic LiNGAM may appear to be restrictive, they can be relaxed to develop more general methods based on the information obtained from the basic setup.

7.3.1 Model

Shimizu *et al.* (2006) proposed a linear non-Gaussian acyclic SEM, known as LiNGAM:

$$x_i = \sum_{k(j)<k(i)} b_{ij}x_j + e_i \tag{7.11}$$

where e_i are continuous latent variables that are exogenous and b_{ij} represent the strengths of the causal connections from x_j to x_i. With a causal ordering denoted by $k(i)$, the causal relations among the variables x_i are acyclic. The exogenous variables e_i follow non-Gaussian distributions and are independent of one another. Recall that the independence assumption among e_i implies that there are no latent common causes. The LiNGAM model in Equation (7.11) can be written in matrix form as

$$x = \mathbf{B}x + e \tag{7.12}$$

where the connection strength matrix **B** collects the causal connection strengths b_{ij}, and the vectors x and e collect the observed variables x_i and the exogenous variables e_i, respectively. The zero/nonzero pattern of b_{ij} corresponds to the absence/existence pattern of the directed edges. That is, if $b_{ij} \neq 0$, there is a directed edge from x_j to x_i in the causal graph; however, if this is not the case, there is no directed edge from x_j to x_i. Note that acyclicity implies that the connection strength matrix **B** can be permuted to become lower triangular with all zeros on the diagonal (i.e., strictly lower triangular) if simultaneous, equal row and column permutations are made according to the causal ordering $k(i)$.

We provide two examples to illustrate the notion of the causal ordering $k(i)$, as provided by Shimizu (2014). The SEM corresponding to the leftmost causal graph of Figure 7.3 is written as

$$\underbrace{\begin{bmatrix} x_1 \\ x_2 \\ x_3 \end{bmatrix}}_{x} = \underbrace{\begin{bmatrix} 0 & 0 & 10 \\ -5 & 0 & 0 \\ 0 & 0 & 0 \end{bmatrix}}_{\mathbf{B}} \underbrace{\begin{bmatrix} x_1 \\ x_2 \\ x_3 \end{bmatrix}}_{x} + \underbrace{\begin{bmatrix} e_1 \\ e_2 \\ e_3 \end{bmatrix}}_{e} \tag{7.13}$$

In this example, the causal ordering that makes **B** strictly lower triangular has x_3 in the first position, x_1 in the second position, and x_2 in the third position, that is, $k(3) = 1$, $k(1) = 2$, and $k(2) = 3$. If we permute variables x_1 to x_3 according to the causal ordering, we obtain

$$\begin{bmatrix} x_3 \\ x_1 \\ x_2 \end{bmatrix} = \begin{bmatrix} 0 & 0 & 0 \\ 10 & 0 & 0 \\ 0 & -5 & 0 \end{bmatrix} \begin{bmatrix} x_3 \\ x_1 \\ x_2 \end{bmatrix} + \begin{bmatrix} e_3 \\ e_1 \\ e_2 \end{bmatrix} \tag{7.14}$$

It can be seen that the resulting connection strength matrix is strictly lower triangular. In this example, there is no other causal ordering of variables that results in a strictly lower triangular structure. In contrast, there are two such causal orderings in the center causal graph of Figure 7.3: (i) $k(1) = 1$, $k(3) = 2$, and $k(2) = 3$; and (ii) $k(3) = 1$, $k(1) = 2$, and $k(2) = 3$, because there is no directed path between x_1 and x_3. A directed path from x_i to x_j is a sequence of directed edges such that x_j is reachable from x_i.

7.3.2 Identifiability

It has been shown that LiNGAM is identifiable (Shimizu *et al.*, 2006), that is, the connection strength matrix $\mathbf{B}$ can be uniquely identified based only on the data x. We now describe the method for identifying the connection strength matrix $\mathbf{B}$ of LiNGAM in Equation (7.12), as shown by Shimizu *et al.* (2006). Let us first solve Equation (7.12) for x. From this, we obtain

$$\boldsymbol{x} = \mathbf{A}\boldsymbol{e}, \tag{7.15}$$

where $\mathbf{A} = (\mathbf{I} - \mathbf{B})^{-1}$. Because the components of $\boldsymbol{e}$ are independent and non-Gaussian, Equation (7.15) defines the ICA model, which, as stated in Section 7.2, is known to be identifiable.

ICA is capable of estimating the mixing matrix $\mathbf{A}$ (and $\mathbf{W} = \mathbf{A}^{-1} = \mathbf{I} - \mathbf{B}$) up to some permutation, scaling, and sign indeterminacies of its columns. ICA gives $\mathbf{W}_{\text{ICA}} = \mathbf{PDW}$, where $\mathbf{P}$ is an unknown permutation matrix and $\mathbf{D}$ is an unknown diagonal matrix. However, in LiNGAM, the correct permutation matrix $\mathbf{P}$ can be found as follows (Shimizu *et al.*, 2006):

(1) First, the correct $\mathbf{P}$ is the only one that contains no zeros on the diagonal of $\mathbf{DW}$, because $\mathbf{B}$ should be a matrix that can be permuted to become lower triangular with all zeros on the diagonal (strictly lower triangular), and $\mathbf{W} = \mathbf{I} - \mathbf{B}$.

(2) Furthermore, the correct scaling and signs of the independent components can be determined using the unity values on the diagonal of $\mathbf{W} = \mathbf{I} - \mathbf{B}$. To obtain $\mathbf{W}$, it is only necessary to divide the rows of $\mathbf{DW}$ by their corresponding diagonal elements.

(3) Finally, the connection strength matrix $\mathbf{B} = \mathbf{I} - \mathbf{W}$ may be computed.

We now illustrate the concept of determining the correct permutation. Consider the following LiNGAM model:

$$x_1 = e_1 \tag{7.16}$$

$$x_2 = b_{21}x_1 + e_2 \tag{7.17}$$

where e_1 and e_2 are non-Gaussian and independent. In matrix form, the above model can be written as follows:

$$\underbrace{\begin{bmatrix} \boldsymbol{x}_1 \\ x_2 \end{bmatrix}}_{x} = \underbrace{\begin{bmatrix} 0 & 0 \\ b_{21} & 0 \end{bmatrix}}_{\mathbf{B}} \underbrace{\begin{bmatrix} x_1 \\ x_2 \end{bmatrix}}_{\boldsymbol{x}} + \underbrace{\begin{bmatrix} e_1 \\ e_2 \end{bmatrix}}_{\boldsymbol{e}} \tag{7.18}$$

Rewriting this in the form of ICA, we obtain

$$\underbrace{\begin{bmatrix} x_1 \\ x_2 \end{bmatrix}}_{x} = \underbrace{\left(\begin{bmatrix} 1 & 0 \\ 0 & 1 \end{bmatrix} - \begin{bmatrix} 0 & 0 \\ b_{21} & 0 \end{bmatrix} \right)^{-1}}_{(\mathbf{I}-\mathbf{B})^{-1}} \underbrace{\begin{bmatrix} e_1 \\ e_2 \end{bmatrix}}_{e} \tag{7.19}$$

$$= \underbrace{\begin{bmatrix} 1 & 0 \\ -b_{21} & 1 \end{bmatrix}^{-1}}_{\mathbf{W}^{-1}} \underbrace{\begin{bmatrix} e_1 \\ e_2 \end{bmatrix}}_{e} \tag{7.20}$$

In this case, the correct $\mathbf{W} = \mathbf{I} - \mathbf{B}$ is

$$\mathbf{W} = \begin{bmatrix} 1 & 0 \\ -b_{21} & 1 \end{bmatrix} \tag{7.21}$$

which is lower triangular and contains no zeros on the diagonal. Premultiplying $\mathbf{W}$ by a diagonal matrix $\mathbf{D}$ with no zeros on the diagonal does not have an effect on the zero/nonzero pattern of $\mathbf{W}$, as

$$\mathbf{DW} = \begin{bmatrix} d_{11} & 0 \\ -d_{22}b_{21} & d_{22} \end{bmatrix} \tag{7.22}$$

However, by exchanging the first and second rows, we obtain

$$\mathbf{P}^{12}\mathbf{DW} = \begin{bmatrix} -d_{22}b_{21} & d_{22} \\ d_{11} & 0 \end{bmatrix} \tag{7.23}$$

where $\mathbf{P}^{12}$ is the permutation matrix defined by

$$\mathbf{P}^{12} = \begin{bmatrix} 0 & 1 \\ 1 & 0 \end{bmatrix} \tag{7.24}$$

The permuted matrix $\mathbf{P}^{12}\mathbf{DW}$ contains a zero on the diagonal. This observation can be generalized to cases involving more than two variables (Shimizu *et al.*, 2006): any permutation of the rows of $\mathbf{DW}$ other than the correct one introduces a zero onto the diagonal. Therefore, we can determine the correct permutation matrix $\mathbf{P}$ by finding one that contains no zeros on the diagonal.

A Graphical Illustration Figure 7.4 provides an illustration of identifiability. The left scatterplot in the figure is generated by the following SEM, where x_1 causes x_2:

$$x_1 = e_1 \tag{7.25}$$

$$x_2 = 0.8x_1 + e_2 \tag{7.26}$$

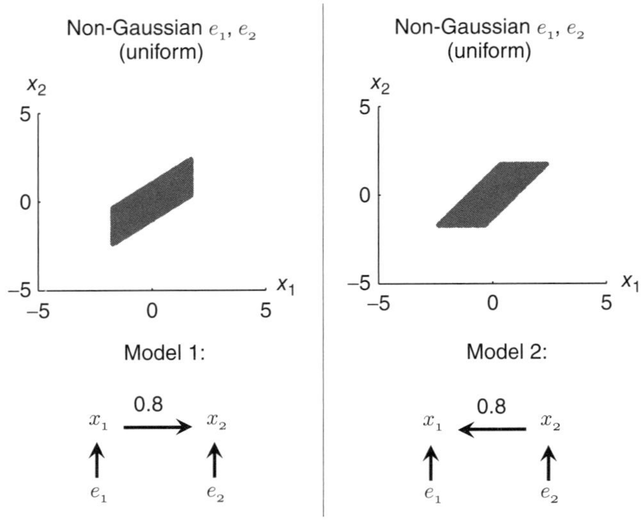

Figure 7.4 Different causal directions give different data distributions.

The right scatterplot is generated by the following model, where x_2 causes x_1:

$$x_1 = 0.8x_2 + e_1 \tag{7.27}$$

$$x_2 = e_2 \tag{7.28}$$

where the means and variances of e_1 and e_2 are chosen so that the observed variables x_1 and x_2 have a mean of 0 and a variance of 1. That is, the means $E(e_1)$ and $E(e_2)$ are zero, and the variances var(e_1) and var(e_2) are 1 and $\sqrt{1-0.8^2}$ in the left model, and $\sqrt{1-0.8^2}$ and 1 in the right model. e_1 and e_2 obey a uniform distribution, and the sample size is 5000. Although the two models have different causal directions that give the same mean and covariance structures, they exhibit different data distributions, as shown in the figure. This implies that the difference in data distributions can be used to identify the underlying causal direction. In the following section, we show how non-Gaussianity and independence are useful in determining the causal structure of observed variables.

7.3.3 Estimation

Analogous to ICA, we want to estimate the connection strength matrix **B** that minimizes the independence of the estimated error variables under the constraint that **B** can be permuted to be strictly lower triangular. Three estimation principles (Shimizu *et al.*, 2006, 2011, Hyvärinen and Smith, 2013) have been proposed in the literature, each of which focuses on determining a causal ordering $k(i)$ $(i = 1, \ldots, p)$ that makes the connection strength matrix **B** strictly lower triangular. The existence of such a

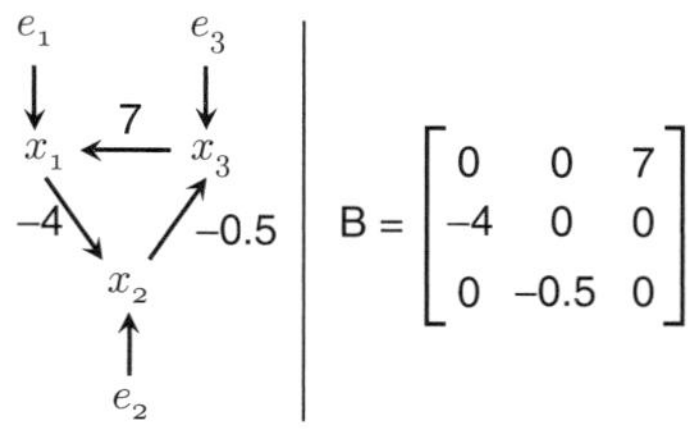

Figure 7.5 An example of a causal graph for a cyclic SEM and its corresponding connection strength matrix.

causal ordering of variables is ensured by the assumption of acyclicity (Bollen, 1989). Figure 7.5 provides an example of a cyclic structure, and its corresponding connection strength matrix, in which such a causal ordering does not exist. Once a causal ordering of variables is found in this way, we can prune redundant connection strengths, that is, find actual zero coefficients using ordinary well-developed sparse regularization methods (Tibshirani, 1996, Hastie *et al.*, 2001, Bach *et al.*, 2012), such as the adaptive lasso (Zou, 2006).

7.3.3.1 ***ICA-Based Estimation Approach*** The first estimation approach for LiNGAM, known as ICA-based LiNGAM (Shimizu *et al.*, 2006), involves the same process as that of demonstrating identifiability. ICA is first applied to estimate the separating matrix **W**, and then this estimated matrix is permuted so that the absolute values of the diagonal elements are as large as possible (in order to avoid zeros on the diagonal). This permutation problem can be efficiently solved using the classical linear assignment method (Burkard and Cela, 1999). Finally, a causal ordering of variables is estimated so as to make the permuted separating matrix as close to being strictly lower triangular as possible. This permutation search is computationally challenging, but an efficient approximation algorithm has been proposed by Hoyer *et al.* (2006). See Shimizu *et al.* (2006) for details of the entire algorithm.

The ICA-LiNGAM algorithm is computationally efficient because well-developed ICA techniques are available. One drawback is that most ICA algorithms, including FastICA (Hyvärinen, 1999) and gradient-based algorithms (Amari, 1998), may converge to local optima if the initial point is poorly chosen (Himberg *et al.*, 2004), or if the step size is unsuitable in the gradient-based methods. Another potential problem is that the approximate permutation algorithm (Hoyer *et al.*, 2006) may need improvements to ensure it is well posed.

7.3.3.2 ***Regression-Based Estimation Approach*** There is a second estimation approach known as DirectLiNGAM (Shimizu *et al.*, 2011). This does not use ICA estimation methods but instead estimates the causal order $k(i)$ $(i = 1, \ldots, p)$ of the variables by repeated least-squares linear regression and the assessment of independence between each observed variable and its residuals. In contrast to

ICA-based LiNGAM, DirectLiNGAM is guaranteed to converge to the desired solution in a fixed number of steps (which is equal to the number of variables), provided that all model assumptions are satisfied and the sample size is infinite.

Two-Variable Cases To explain the underlying concept of DirectLiNGAM, we consider the following two-variable cases in the framework of LiNGAM. We first consider the case where x_1 is the cause and x_2 is the effect:

$$x_1 = e_1 \tag{7.29}$$

$$x_2 = b_{21}x_1 + e_2 \tag{7.30}$$

where $b_{21} \neq 0$. Regressing x_2 on x_1,

$$r_2^{(1)} = x_2 - \frac{\text{cov}(x_2, x_1)}{\text{var}(x_1)}x_1 \tag{7.31}$$

$$= x_2 - b_{21}x_1 \tag{7.32}$$

$$= e_2 \tag{7.33}$$

Thus, if $x_1(= e_1)$ is the cause, the fact that e_1 and e_2 are independent implies x_1 and $r_2^{(1)}(= e_2)$ are also independent.

We now consider the case where x_1 is the effect and x_2 is the cause:

$$x_1 = b_{12}x_2 + e_1 \tag{7.34}$$

$$x_2 = e_2 \tag{7.35}$$

where $b_{12} \neq 0$. Regressing x_2 on x_1,

$$r_2^{(1)} = x_2 - \frac{\text{cov}(x_2, x_1)}{\text{var}(x_1)}x_1 \tag{7.36}$$

$$= x_2 - \frac{\text{cov}(x_2, x_1)}{\text{var}(x_1)}(b_{12}x_2 + e_1) \tag{7.37}$$

$$= \left\{1 - \frac{b_{12}\text{cov}(x_2, x_1)}{\text{var}(x_1)}\right\}x_2 - \frac{\text{cov}(x_2, x_1)}{\text{var}(x_1)}e_1 \tag{7.38}$$

$$= \left\{1 - \frac{b_{12}\text{cov}(x_2, x_1)}{\text{var}(x_1)}\right\}e_2 - \frac{b_{12}\text{var}(x_2)}{\text{var}(x_1)}e_1 \tag{7.39}$$

In contrast to the previous case, the effect x_1 and its residual $r_2^{(1)}$ can be shown to be interdependent. This is intuitively obvious, because e_1 contributes to the determination of both x_1 and $r_2^{(1)}$, as implied in Equations (7.34) and (7.39), and introduces some degree of dependency between them. This can be rigorously proved (Shimizu *et al.*, 2011). Therefore, the cause-and-effect relationship between x_1 and x_2 can be determined by examining the independence between observed variables and their residuals.

In practice, an exogenous variable may be identified by determining the most independent observed variable among its residuals. This independence can be evaluated by mutual information and nonparametric independence measures (Bach and Jordan, 2002, Gretton *et al.*, 2005, Kraskov *et al.*, 2004).

Cases Involving More Than Two Variables In the two-variable case, the observed variable representing the cause is exogenous. Based on this observation, we can extend the idea stated above to cases involving more than two variables using the following lemma. This lemma shows how an exogenous observed variable, which is not caused by any other observed variable in the model, can be found.

Lemma 7.1 ***(Lemma 1 of Shimizu et al. (2011))*** *Assume that all model assumptions of LiNGAM in Equation (7.2) are met and that the sample size is infinite. Denote by $r_i^{(j)}$ the residual when x_i is regressed on x_j: $r_i^{(j)} = x_i - cov(x_i, x_j)/var(x_j)x_j$ $(i \neq j)$. Then, a variable x_j is exogenous if and only if x_j is independent of its residuals $r_i^{(j)}$ for all $i \neq j$.* □

To explain the application of this method to cases involving more than two variables, we recall the following example considered in Shimizu (2014):

$$\begin{bmatrix} x_3 \\ x_1 \\ x_2 \end{bmatrix} = \begin{bmatrix} 0 & 0 & 0 \\ 10 & 0 & 0 \\ 0 & -5 & 0 \end{bmatrix} \begin{bmatrix} x_3 \\ x_1 \\ x_2 \end{bmatrix} + \begin{bmatrix} e_3 \\ e_1 \\ e_2 \end{bmatrix} \tag{7.40}$$

where e_1, e_2, and e_3 are non-Gaussian and independent. The associated causal graph is shown on the left side of Figure 7.3. DirectLiNGAM first seeks an exogenous observed variable. This is an observed variable that is not determined inside the model, that is, has no causal variable in the model, and hence the corresponding row of **B** contains only zeros. In the example considered in Equation (7.40), x_3 is an exogenous variable, and the corresponding (i.e., first) row of **B** consists entirely of zeros. Therefore, the exogenous variable x_3 $(= e_3)$ can be at the top of a causal ordering, meaning that **B** is lower triangular with zeros on the diagonal. Following this, the effect of the exogenous variable x_3 on the other variables is eliminated using least-squares regression. In other words, we compute the residuals $r_i^{(3)}$ when the other variables x_i $(i = 1, 2)$ are regressed on the exogenous x_3. It can be shown that the residuals $r_i^{(3)}$ $(i = 1, 2)$ follow a LiNGAM model if the relevant assumptions are met and the sample size is infinite (Shimizu *et al.*, 2011). Thus, we have

$$\begin{bmatrix} r_1^{(3)} \\ r_2^{(3)} \end{bmatrix} = \begin{bmatrix} 0 & 0 \\ -5 & 0 \end{bmatrix} \begin{bmatrix} r_1^{(3)} \\ r_2^{(3)} \end{bmatrix} + \begin{bmatrix} e_1 \\ e_2 \end{bmatrix} \tag{7.41}$$

The causal ordering of the residuals $r_1^{(3)}$ and $r_2^{(3)}$ is equivalent to that of the corresponding observed variables x_1 and x_2. Following this, DirectLiNGAM determines an exogenous *residual* (in this case, $r_1^{(3)}$). This implies that the corresponding original variable x_1 is second in the causal ordering, and the remaining variable, x_2, will then be in third position. Thus, we can estimate the causal ordering $k(i)$ $(i = 1, \dots, p)$ of the observed variables from the top down by repeatedly performing simple least-squares linear regression and evaluating the pairwise independence between each observed variable and its residuals.

7.3.3.3 Likelihood Ratio-Based Estimation Approach The third estimation approach is known as Pairwise-LiNGAM (Hyvärinen and Smith, 2013). Two models can be compared in a statistical manner by computing their likelihoods and considering the ratio of these values. Hyvärinen and Smith (2013) proposed such an estimation method for LiNGAM.

Consider the following two LiNGAM models with opposite causal directions. In the first model, x_1 causes x_2:

$$x_1 = e_1 \tag{7.42}$$

$$x_2 = b_{21}x_1 + e_2 \tag{7.43}$$

where $b_{21} \neq 0$. In the second model, x_2 causes x_1:

$$x_1 = b_{12}x_1 + e_1 \tag{7.44}$$

$$x_2 = e_2 \tag{7.45}$$

where $b_{12} \neq 0$. We denote the standardized versions of x_1 and x_2, which have zero means and unit variance, as $\overline{x}_1$ and $\overline{x}_2$, respectively. e_1 and e_2 are non-Gaussian and independent. Moreover, the log-likelihoods of the two models are denoted by $\log\ L(\overline{x}_1 \to \overline{x}_2)$ and $\log\ L(\overline{x}_2 \to \overline{x}_1)$, respectively. The log-likelihood ratio of the two models divided by the number of observations n is then given by

$$R = \frac{1}{n} \log\ L(\overline{x}_1 \to \overline{x}_2) - \frac{1}{n} \log\ L(\overline{x}_2 \to \overline{x}_1) \tag{7.46}$$

If $R > 0$, that is, $\log\ L(\overline{x}_1 \to \overline{x}_2) > \log\ L(\overline{x}_2 \to \overline{x}_1)$, the model in which x_1 causes x_2 is preferred; otherwise, the model where x_2 causes x_1 is preferred.

Hyvärinen and Smith (2013) showed that the asymptotic limit is

$$R \to -H(\overline{x}_1) - H\left\{\frac{(\overline{x}_2 - \rho\overline{x}_1)}{\overline{\sigma}_2}\right\} + H(\overline{x}_2) + H\left\{\frac{(\overline{x}_1 - \rho\overline{x}_2)}{\overline{\sigma}_1}\right\} \tag{7.47}$$

where $\rho = corr(\overline{x}_1, \overline{x}_2)$, and $\overline{\sigma}_1$, $\overline{\sigma}_2$ are the standard deviations of the regression residuals $\overline{x}_2 - \rho\overline{x}_1, \overline{x}_1 - \rho\overline{x}_2$, respectively. They further approximated the differential entropy $H(\cdot)$ in a computationally efficient way (developed by Hyvärinen (1998)) to obtain

$$H(u) \approx H(v) - k_1[E\{\log\cosh u\} - \gamma]^2 - k_2[E\{u\exp(-u^2/2)\}]^2 \tag{7.48}$$

where $H(\nu) = (1/2)(1 + \log\ 2\pi)$ is the differential entropy of the standardized Gaussian variable, and k_1, k_2, γ are constants. This third approach is computationally simpler than DirectLiNGAM because we need only to evaluate the one-dimensional differential entropies of observed variables and their residuals, and do not have to evaluate their independence.

This likelihood-ratio-based method can be used to determine an exogenous variable by comparing all pairs of observed variables and finding that which is most likely to have no causal variables. Thus, this approach is capable of estimating the causal orders of more than two variables.

7.4 LINGAM FOR TIME SERIES

We next discuss an extension of the basic LiNGAM to time series.

7.4.1 Model

Hyvärinen *et al.* (2010) analyzed both lagged and contemporaneous (instantaneous) causal effects in time series data. Such an approach is both necessary and useful if the measurements can have a lower time resolution than the causal influences. Basic LiNGAM is used to model contemporaneous causal effects, whereas a classic autoregressive model is used to model lagged causal effects. The combination of the two leads to the following model:

$$\boldsymbol{x}(t) = \sum_{\tau=0}^{h} \mathbf{B}_\tau \boldsymbol{x}(t-\tau) + \mathbf{e}(t) \tag{7.49}$$

where $\boldsymbol{x}(t)$ and $\boldsymbol{e}(t)$ are the observed variable vectors and the exogenous variable vectors at time point t, respectively. $\mathbf{B}_\tau = [b_{ij}^{(\tau)}]$ $(\tau = 0, \ldots, h; i, j = 1, \ldots, p)$ denotes the connection strength matrices with time lag τ. Note that the time lag τ starts from zero, and $\mathbf{B}_0$ can be permuted to become strictly lower triangular, that is, the contemporaneous causal relations are acyclic. An example of a causal graph is provided in Figure 7.6. The model described above is widely known in econometrics as a structural vector autoregressive model (Swanson and Granger, 1997).

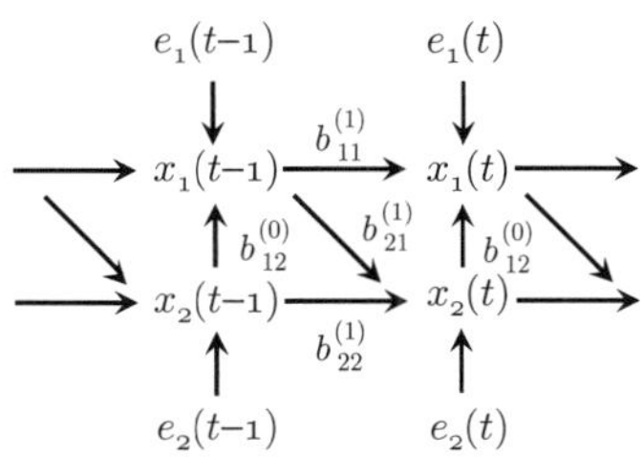

Figure 7.6 A causal graph of LiNGAM for time series.

7.4.2 Identifiability

Hyvärinen *et al.* (2010) showed that the model in Equation (7.49) is identifiable based on data $\boldsymbol{x}(t)$ if $e_i(t)$ are both non-Gaussian and mutually and temporally independent. We now explain a method for identifying the connection strength matrices $\mathbf{B}_\tau$ ($\tau = 0, \ldots, h$) of the time series LiNGAM in Equation (7.49), as provided by Hyvärinen *et al.* (2010). Let us first solve Equation (7.49) for $\boldsymbol{x}(t)$. From this, we obtain

$$\boldsymbol{x}(t) = \sum_{\tau=1}^{h} \underbrace{(\mathbf{I} - \mathbf{B}_0)^{-1}\mathbf{B}_\tau}_{\mathbf{M}_\tau} \boldsymbol{x}(t-\tau) + \underbrace{(\mathbf{I} - \mathbf{B}_0)^{-1}\boldsymbol{e}(t)}_{\boldsymbol{n}(t)} \tag{7.50}$$

$$= \sum_{\tau=1}^{h} \mathbf{M}_\tau \boldsymbol{x}(t-\tau) + \boldsymbol{n}(t) \tag{7.51}$$

where $\mathbf{M}_\tau = (\mathbf{I} - \mathbf{B}_0)^{-1}\mathbf{B}_\tau$ ($\tau = 1, \ldots, h$) represent autoregressive matrices and $\boldsymbol{n}(t) = (\mathbf{I} - \mathbf{B}_0)^{-1}\boldsymbol{e}(t)$ denotes the innovation vector. Because the time lag τ now starts from 1, Equation (7.51) defines the classic autoregressive model, which is known to be identifiable in general conditions. Therefore, we can compute the innovation vector $\boldsymbol{n}(t)$ as

$$\boldsymbol{n}(t) = \boldsymbol{x}(t) - \sum_{\tau=1}^{h} \mathbf{M}_\tau \boldsymbol{x}(t-\tau) \tag{7.52}$$

We are now ready to show that the time series LiNGAM is identifiable. From the relation $\boldsymbol{n}(t) = (\mathbf{I} - \mathbf{B}_0)^{-1}\boldsymbol{e}(t)$ in Equation (7.50), we obtain

$$(\mathbf{I} - \mathbf{B}_0)\boldsymbol{n}(t) = \boldsymbol{e}(t) \Longleftrightarrow \boldsymbol{n}(t) = \mathbf{B}_0\boldsymbol{n}(t) + \boldsymbol{e}(t) \tag{7.53}$$

As the components of $\boldsymbol{e}(t)$ are non-Gaussian, as well as mutually and temporally independent, this defines the basic LiNGAM model described in Section 7.3 for $\boldsymbol{n}(t)$. Thus, we can estimate $\mathbf{B}_0$ by applying a basic LiNGAM estimation method to the innovation vector $\boldsymbol{n}(t)$. Then, from the relation $\mathbf{M}_\tau = (\mathbf{I} - \mathbf{B}_0)^{-1}\mathbf{B}_\tau$ in Equation (7.50), we can estimate $\mathbf{B}_\tau$ ($\tau = 1, \ldots, h$) by

$$\mathbf{B}_\tau = (\mathbf{I} - \mathbf{B}_0)\mathbf{M}_\tau \tag{7.54}$$

7.4.3 Estimation

There is a simple and practical two-stage estimation method for this model that involves the same process of demonstrating identifiability as above. That is, a classic autoregressive model is applied to $\boldsymbol{x}(t)$, followed by the application of the basic LiNGAM estimation method discussed in Section 7.3 on the residuals or innovations (Hyvärinen *et al.*, 2010). Other extensions of the basic LiNGAM may be used to model contemporaneous causal relations, including the cyclic extension (Lacerda *et al.*, 2008, Hyvärinen and Smith, 2013). Moreover, other time series models can be employed to estimate the residuals, including an autoregressive moving average model and a vector error correction model, as shown in Kawahara et al. (2011); Ferkingsta et al. (2011); Moneta et al. (2013). An approach considered in Hyvärinen *et al.* (2010) uses a convolutive version of ICA, called multichannel blind deconvolution (Cichocki and Amari, 2002), to estimate the connection strength matrices directly, instead of using the two-stage method.

7.5 LINGAM WITH LATENT COMMON CAUSES

Many causal discovery methods, including LiNGAM, make the strong assumption that there are no latent common causes (Spirtes and Glymour, 1991, Chickering, 2002, Dodge and Rousson, 2000, Shimizu *et al.*, 2006, Hoyer *et al.*, 2009, Zhang and Hyvärinen, 2009). A latent common cause is an unobserved variable that contributes to determining the value of more than one observed variable (Hoyer *et al.*, 2008b). These methods are applied in various fields (Ferkingsta *et al.*, 2011, Moneta *et al.*, 2013, Coad and Binder, 2014, Niyogi *et al.*, 2010, Rosenström *et al.*, 2012, Helajärvi *et al.*, 2014, Boukrina and Graves, 2013, Manelis and Reder, 2014, Mills-Finnerty *et al.*, 2014, Campomanes *et al.*, 2014). However, in many empirical sciences, it can be difficult to accept the estimation results, because there often exist latent common causes that have been ignored. It is well known that neglecting such latent common causes seriously biases the estimation results (Pearl, 2000, Bollen, 1989). In this section, we discuss a non-Gaussian approach that considers latent common causes in the LiNGAM framework.

7.5.1 Model

Hoyer *et al.* (2008b) proposed a model for LiNGAM with latent common causes. This can be formulated as follows:

$$x_i = \sum_{k(j)<k(i)} b_{ij}x_j + \sum_{\ell=1}^{L} \lambda_{i\ell}f_\ell + e_i \tag{7.55}$$

where f_ℓ ($\ell = 1, \ldots, L$) are non-Gaussian latent common causes with zero mean and unit variance, and $\lambda_{i\ell}$ denote the causal connection strengths from f_ℓ to x_i. This model is expressed in matrix form as

$$\boldsymbol{x} = \mathbf{B}\boldsymbol{x} + \boldsymbol{\Lambda}\boldsymbol{f} + \boldsymbol{e} \tag{7.56}$$

This is different from the basic LiNGAM in Equation (7.56) because of the existence of a latent common cause vector f that incorporates the f_ℓ. The matrix $\boldsymbol{\Lambda}$ collects the $\lambda_{i\ell}$, and is assumed to be of full-column rank.

Moreover, it is assumed that x_i and f_ℓ are faithful to the generating graph. This assumption of faithfulness (Spirtes *et al.*, 1993) means that, when multiple causal

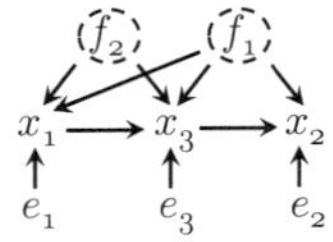

Figure 7.7 A causal LiNGAM graph with latent common causes f_1 and f_2.

paths exist from one variable to another, the combined effect is not exactly equal to zero (Hoyer *et al.*, 2008b). The faithfulness assumption is not considered to be very restrictive from a Bayesian perspective (Spirtes *et al.*, 1993), because the probability of obtaining the exact parameter values that do not satisfy faithfulness is zero (Meek, 1995). Without loss of generality, the latent common causes f_ℓ are assumed to be mutually independent, as shown below. An example of a causal LiNGAM graph with latent common causes is shown in Figure 7.7.

7.5.1.1 On the Independence Assumption for Latent Common Causes Without loss of generality, the latent common causes f_ℓ are assumed to be mutually independent. This is because any dependent latent common causes can be remodeled by linear combinations of independent latent variables if the underlying model is linear acyclic and the error variables are independent (Hoyer *et al.*, 2008b). To illustrate this, we recall the following example considered by Shimizu and Bollen (2014):

$$\bar{f}_1 = e_{\bar{f}_1} \tag{7.57}$$

$$\bar{f}_2 = \omega_{21}\bar{f}_1 + e_{\bar{f}_2} \tag{7.58}$$

$$x_1 = \lambda_{11}\bar{f}_1 + e_1 \tag{7.59}$$

$$x_2 = \lambda_{21}\bar{f}_1 + e_2 \tag{7.60}$$

$$x_3 = \lambda_{32}\bar{f}_2 + e_3 \tag{7.61}$$

$$x_4 = b_{43}x_3 + \lambda_{42}\bar{f}_2 + e_4 \tag{7.62}$$

where errors $e_{\bar{f}_1}$ $(=\bar{f}_1)$, $e_{\bar{f}_2}$, and e_1–e_4 are non-Gaussian and independent. The associated causal graph is shown in Figure 7.8. The relations of $\bar{f}_1$, $\bar{f}_2$, and x_1–x_4 are acyclic, and latent common causes $\bar{f}_1$ and $\bar{f}_2$ are dependent. In matrix form, this example model can be written as

$$\begin{bmatrix} x_1 \\ x_2 \\ x_3 \\ x_4 \end{bmatrix} = \begin{bmatrix} 0 & 0 & 0 & 0 \\ 0 & 0 & 0 & 0 \\ 0 & 0 & 0 & 0 \\ b_{43} & 0 & 0 & 0 \end{bmatrix} \begin{bmatrix} x_1 \\ x_2 \\ x_3 \\ x_4 \end{bmatrix} + \begin{bmatrix} \lambda_{11} & 0 \\ \lambda_{21} & 0 \\ 0 & \lambda_{32} \\ 0 & \lambda_{42} \end{bmatrix} \begin{bmatrix} \bar{f}_1 \\ \bar{f}_2 \end{bmatrix} + \begin{bmatrix} e_1 \\ e_2 \\ e_3 \\ e_4 \end{bmatrix} \tag{7.63}$$

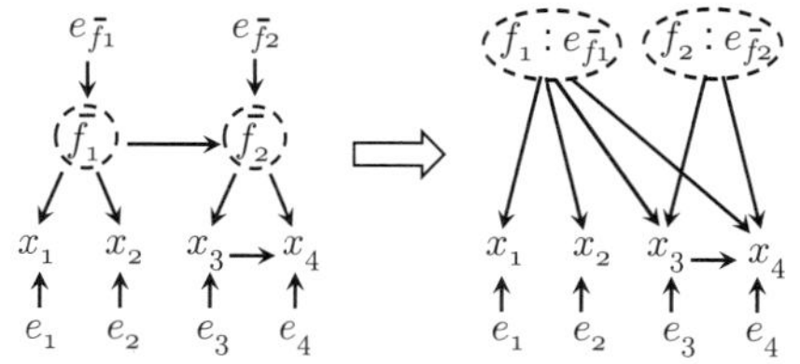

Figure 7.8 A causal graph to illustrate the idea of independent latent common causes.

The relations of $\bar{f}_1$ and $\bar{f}_2$ to $e_{\bar{f}_1}$ and $e_{\bar{f}_2}$, respectively, in Equations (7.57) and (7.58) can be written as

$$\begin{bmatrix} \bar{f}_1 \\ \bar{f}_2 \end{bmatrix} = \begin{bmatrix} 1 & 0 \\ \omega_{21} & 1 \end{bmatrix} \begin{bmatrix} e_{\bar{f}_1} \\ e_{\bar{f}_2} \end{bmatrix} \tag{7.64}$$

from which we obtain

$$\underbrace{\begin{bmatrix} x_1 \\ x_2 \\ x_3 \\ x_4 \end{bmatrix}}_{\boldsymbol{x}} = \underbrace{\begin{bmatrix} 0 & 0 & 0 & 0 \\ 0 & 0 & 0 & 0 \\ 0 & 0 & 0 & 0 \\ b_{43} & 0 & 0 & 0 \end{bmatrix}}_{\mathbf{B}} \underbrace{\begin{bmatrix} x_1 \\ x_2 \\ x_3 \\ x_4 \end{bmatrix}}_{\boldsymbol{x}} \tag{7.65}$$

$$+ \underbrace{\begin{bmatrix} \lambda_{11} & 0 \\ \lambda_{21} & 0 \\ \lambda_{32}\omega_{21} & \lambda_{32} \\ \lambda_{42}\omega_{21} & \lambda_{42} \end{bmatrix}}_{\boldsymbol{\Lambda}} \underbrace{\begin{bmatrix} e_{\bar{f}_1} \\ e_{\bar{f}_2} \end{bmatrix}}_{\boldsymbol{f}} + \underbrace{\begin{bmatrix} e_1 \\ e_2 \\ e_3 \\ e_4 \end{bmatrix}}_{\boldsymbol{e}} \tag{7.66}$$

Taking $f_1 = e_{\bar{f}_1}$ and $f_2 = e_{\bar{f}_2}$, this represents the LiNGAM with latent common causes from Equation (7.56), because $e_{\bar{f}_1}$ and $e_{\bar{f}_2}$ are non-Gaussian and independent.

7.5.2 Identifiability

Within the framework of LiNGAM with latent common causes given by Equation (7.56), it has been shown (Hoyer *et al.*, 2008b) that the following two models, which have opposite directions of causation in the presence of latent common causes, are distinguishable based on the observed data, that is, the two different causal directions induce different data distributions:

$$\text{Model } 1' : \begin{cases} x_1 = \sum_{\ell=1}^{L} \lambda_{1\ell} f_\ell + e_1 \\ x_2 = b_{21} x_1 + \sum_{\ell=1}^{L} \lambda_{2\ell} f_\ell + e_2, \end{cases} \tag{7.67}$$

$$\text{Model } 2' : \begin{cases} x_1 = b_{12} x_2 + \sum_{\ell=1}^{L} \lambda_{1\ell} f_\ell + e_1 \\ x_2 = \sum_{\ell=1}^{L} \lambda_{2\ell} f_\ell + e_2, \end{cases} \tag{7.68}$$

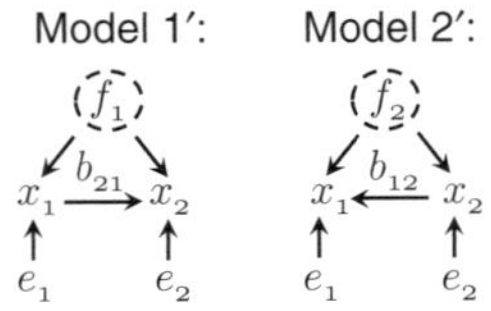

Figure 7.9 The associated causal graphs of Models 1′ and 2′. For simplicity, only one latent common cause is shown in the causal graph.

where $b_{21}b_{12} \neq 0$. In Model 1′, x_1 causes x_2, whereas in Model 2′, x_2 causes x_1. Moreover, from our definition of latent common causes (i.e., that they contribute to determining the values of more than one variable) and the faithfulness assumption, $\lambda_{1\ell}\lambda_{2\ell} \neq 0$. The associated causal graphs of Models 1′ and 2′ are shown in Figure 7.9.

To explain the concept of identifying the correct causal direction, as provided by Hoyer *et al.* (2008b), we first consider Model 1′. For simplicity, and without loss of generality, we take the number of latent common causes $L = 1$. We then have

$$x_1 = \lambda_{11}f_1 + e_1 \tag{7.69}$$

$$x_2 = b_{21}x_1 + \lambda_{21}f_1 + e_2 \tag{7.70}$$

where e_1 and e_2 are non-Gaussian and mutually independent. In matrix form, this can be written as

$$\underbrace{\begin{bmatrix} x_1 \\ x_2 \end{bmatrix}}_{\boldsymbol{x}} = \underbrace{\begin{bmatrix} 0 & 0 \\ b_{21} & 0 \end{bmatrix}}_{\mathbf{B}} \underbrace{\begin{bmatrix} x_1 \\ x_2 \end{bmatrix}}_{\boldsymbol{x}} + \underbrace{\begin{bmatrix} \lambda_{11} \\ \lambda_{21} \end{bmatrix}}_{\boldsymbol{\Lambda}} \underbrace{\begin{bmatrix} f_1 \end{bmatrix}}_{\boldsymbol{f}} + \underbrace{\begin{bmatrix} e_1 \\ e_2 \end{bmatrix}}_{\boldsymbol{e}} \tag{7.71}$$

Rewriting this in the ICA form discussed in Section 7.2, we obtain

$$\underbrace{\begin{bmatrix} x_1 \\ x_2 \end{bmatrix}}_{\boldsymbol{x}} = \underbrace{\begin{bmatrix} (\mathbf{I}-\mathbf{B})^{-1} & (\mathbf{I}-\mathbf{B})^{-1}\boldsymbol{\Lambda} \end{bmatrix}}_{\mathbf{A}} \underbrace{\begin{bmatrix} \boldsymbol{e} \\ \boldsymbol{f} \end{bmatrix}}_{\boldsymbol{s}} \tag{7.72}$$

$$= \underbrace{\begin{bmatrix} 1 & 0 & \lambda_{11} \\ b_{21} & 1 & b_{21}\lambda_{11} + \lambda_{21} \end{bmatrix}}_{\mathbf{A}} \underbrace{\begin{bmatrix} e_1 \\ e_2 \\ f_1 \end{bmatrix}}_{\boldsymbol{s}} \tag{7.73}$$

where $\mathbf{A}$ is the mixing matrix and $\boldsymbol{s}$ is the vector of independent components in ICA.

Similarly, we have the following for Model 2′ in the opposite direction of causation to that of Model 1′:

$$\underbrace{\begin{bmatrix} x_1 \\ x_2 \end{bmatrix}}_{\boldsymbol{x}} = \underbrace{\begin{bmatrix} 1 & b_{12} & \lambda_{11} + b_{12}\lambda_{21} \\ 0 & 1 & \lambda_{21} \end{bmatrix}}_{\mathbf{A}} \underbrace{\begin{bmatrix} e_1 \\ e_2 \\ f_1 \end{bmatrix}}_{\boldsymbol{s}} \tag{7.74}$$

Because the latent variables e_1, e_2, and f_1 are non-Gaussian and independent, these two models define the ICA models. Although the mixing matrices **A** are identified only up to the permutation, scaling, and sign of the columns (Comon, 1994, Eriksson and Koivunen, 2004), it can be seen that the zero/nonzero patterns of the mixing matrices in the two models are different in Equations (7.73) and (7.74). In Model 1′, only the (1, 2)-th element is zero, whereas only the (2, 1)-th element is zero in Model 2′. This difference cannot be eliminated by permuting the columns of the mixing matrices and changing their scale and sign. Note that the coefficients $b_{21}\lambda_{11} + \lambda_{21}$ and $\lambda_{11} + b_{12}\lambda_{21}$ cannot accidentally become zero because of the faithfulness assumption. Using this difference in the zero/nonzero patterns of the mixing matrices, we can identify the causal direction.

A Graphical Illustration Figure 7.10 illustrates identifiability. The left scatterplot is generated by the following SEM, where x_1 causes x_2 in the presence of latent common cause f_1:

$$x_1 = 0.3f_1 + e_1 \tag{7.75}$$

$$x_2 = 0.8x_1 + 0.3f_1 + e_2 \tag{7.76}$$

The figure on the right is generated by the following model, where x_2 causes x_1 in the presence of latent common cause f_1:

$$x_1 = 0.8x_2 + 0.3f_1 + e_1 \tag{7.77}$$

$$x_2 = 0.3f_1 + e_2 \tag{7.78}$$

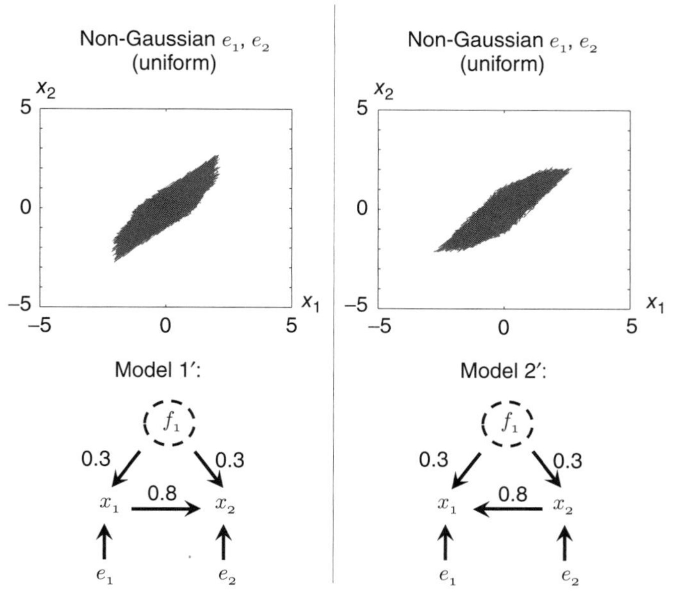

Figure 7.10 Different causal directions give different data distributions.

where the means and variances of errors e_1 and e_2 are chosen so that the observed variables x_1 and x_2 have zero means and variances of 1. That is, the means $E(e_1)$ and $E(e_2)$ are zero, and the variances $\mathrm{var}(e_1)$ and $\mathrm{var}(e_2)$ are $\sqrt{1-0.3^2}$ and $\sqrt{1-(0.8\times 0.3+0.3)^2-0.8^2\mathrm{var}(e_1)}$ in Model 1′, and $\sqrt{1-(0.8\times 0.3+0.3)^2-0.8^2\mathrm{var}(e_2)}$ and $\sqrt{1-0.3^2}$ in Model 2′. e_1, e_2, and f_1 follow a uniform distribution, and the sample size is 5000. The two models with different causal directions provide the same mean and covariance structures. However, they exhibit different data distributions, as shown in the figure. This shows that it is possible to identify the underlying causal direction using the differences in the data distributions.

7.5.3 Estimation

We provide a succinct exposition of estimation methods for LiNGAM with latent common causes in Equation (7.56).

7.5.3.1 ICA-Based Estimation Approach Hoyer *et al.* (2008b), Henao and Winther (2011) proposed estimation approaches that explicitly model latent common causes and compare two models with opposite directions of causation, that is, Models 1′ and 2′. Similar to the ICA-based approach for the basic LiNGAM, Hoyer *et al.* (2008b) proposed an estimation method that involves the same demonstration of identifiability as above, that is, an ICA model with more independent components than observed variables is first applied on $\boldsymbol{x}$ to estimate the $p \times q$ mixing matrix $\mathbf{A}$ ($p < q$). Such an extension of ICA is known as an overcomplete ICA (Lewicki and Sejnowski, 2000). The zero/nonzero pattern of the mixing matrix can then be investigated to determine which of the two models is better.

In the method proposed by Hoyer *et al.* (2008b), the distributions of independent components are modeled using a Gaussian mixture model. An expectation-maximization-type ICA algorithm (Lewicki and Sejnowski, 2000) is then used to estimate the mixing matrix, and a bootstrapping method (Efron and Tibshirani, 1993) is applied to discover its zero/nonzero pattern. However, current overcomplete ICA estimation algorithms often become stuck in local optima, and the estimates are not sufficiently reliable (Entner and Hoyer, 2011). Furthermore, such algorithms need to select the number of latent common causes, which can be quite high. This can lead to computational issues and statistically unreliable estimates. Henao and Winther (2011) proposed a more sophisticated method to investigate the zero/nonzero pattern of the mixing matrix using a Bayesian approach based on a sparse prior distribution of the elements in the mixing matrix. However, this method also requires the number of latent common causes to be selected.

7.5.3.2 Mixed Model-Based Estimation Approach Shimizu and Bollen (2014) proposed an alternative Bayesian approach that does not require the number of latent common causes to be specified.

LiNGAM with Observation-Specific Intercepts Shimizu and Bollen (2014) proposed a variant of LiNGAM that uses observation-specific intercepts. The model for observation m is formulated as follows:

$$x_i^{(m)} = \mu_i + \mu_i^{(m)} + \sum_{k(j)<k(i)} b_{ij} x_j^{(m)} + e_i^{(m)} \tag{7.79}$$

where μ_i are intercepts common to all observations $(i = 1, \ldots, p; m = 1, \ldots, n)$. The key difference from the basic LiNGAM of Equation (7.55) is the existence of observation-specific intercepts $\mu_i^{(m)}$. Error variables e_i $(i = 1, \ldots, p)$ are non-Gaussian and independent. This means that the observations are generated from the identifiable basic LiNGAM, possibly with different parameter values for the intercepts $\mu_i + \mu_i^{(m)}$ of different observations. Note that the causal ordering $k(i)$ $(i = 1, \ldots, p)$ of the variables is identical for all observations in the sample. Moreover, the distributions of $e_i^{(m)}$ $(m = 1, \ldots, n)$ are identical for every i. The model in Equation (7.79) can be viewed as a mixed model (Demidenko, 2004), as it has effects μ_i and b_{ij} that are common to all observations and observation-specific effects $\mu_i^{(m)}$.

Shimizu and Bollen (2014) then related the above LiNGAM model with observation-specific intercepts to that with latent common causes in Equation (7.56). For observation m, the LiNGAM with latent common causes in Equation (7.56) is written as follows:

$$x_i^{(m)} = \mu_i + \sum_{k(j)<k(i)} b_{ij} x_i^{(m)} + \underbrace{\sum_{\ell=1}^{L} \lambda_{i\ell} f_\ell^{(m)}}_{\mu_i^{(m)}} + e_i^{(m)} \tag{7.80}$$

where the common intercepts μ_i are explicitly modeled, unlike in Equation (7.56). Taking $\mu_i^{(m)} = \sum_{\ell=1}^{L} \lambda_{i\ell} f_\ell^{(m)}$, this is simply the LiNGAM with observation-specific intercepts of Equation (7.79). Graphical representations of the LiNGAM with latent common causes in Equation (7.56) and LiNGAM with observation-specific intercepts $\mu_i^{(m)}$ in Equation (7.79) are shown in Figures 7.11 and 7.12, respectively.

In contrast to the ICA-based approaches described above (Hoyer *et al.*, 2008b, Henao and Winther, 2011), the mixed model-based approach does not explicitly model the latent common causes f_ℓ. Instead, it simply includes the sums of the latent common causes $\mu_i^{(m)} = \sum_{\ell=1}^{L} \lambda_{i\ell} f_\ell^{(m)}$ as model parameters. Thus, it is not necessary to estimate the causal connection strengths $\lambda_{i\ell}$ or the number of latent common causes L. However, this leads to many additional parameters, that is,

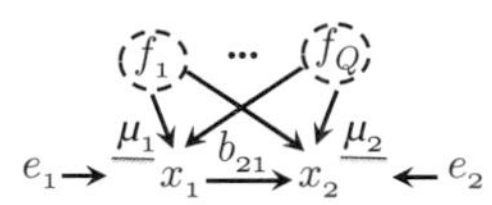

Figure 7.11 LiNGAM with latent common causes f_ℓ $(\ell = 1, \ldots, L)$.

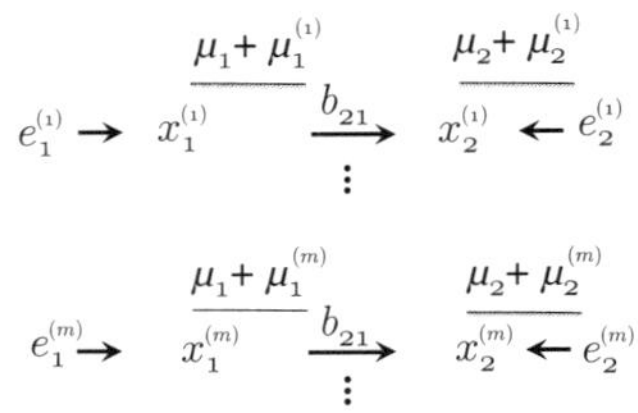

Figure 7.12 LiNGAM with observation-specific intercepts $\mu_1^{(m)}$ and $\mu_2^{(m)}$ $(m = 1, \ldots, n)$.

observation-specific intercepts $\mu_i^{(m)}$, the number of which is of the same order as the sample size.

Therefore, these observation-specific intercepts $\mu_i^{(m)}$ are assumed to follow an informative prior. This manner of modeling priors is often used in the field of mixed models (Demidenko, 2004) and multilevel models (Kreft and De Leeuw, 1998), although the estimation of causal direction is not a topic studied within these areas. Shimizu and Bollen (2014) proposed using a bell-shaped curve distribution (such as a t-distribution) for the $\mu_i^{(m)}$ priors. This was motivated by the central limit theorem (Billingsley, 1986), as the observation-specific intercepts $\mu_i^{(m)}$ are the sums of independent variables f_ℓ.

The error variables $e_1^{(m)}$ and $e_2^{(m)}$ are modeled by Laplace distributions, but they may be modeled by other non-Gaussian distributions, including the generalized Gaussian family and a finite mixture of Gaussians. The other parameters that are common to all observations are assumed to follow noninformative priors.

Following this observation, Shimizu and Bollen (2014) applied standard Bayesian model selection techniques (Kass and Raftery, 1995) to compare two models with opposite causal directions between two observed variables (i.e., Models 1′ and 2′ in Equations (7.67) and (7.68), respectively). Their approach was based on the log-marginal likelihoods from the framework of LiNGAM with observation-specific intercepts in Equation (7.79). Once the possible causal direction has been estimated, it can be seen whether the common connection strength b_{21} or b_{12} is likely to be zero by examining the posterior distribution.

Cases Involving More Than Two Variables For cases where more than two variables are involved, the mixed model-based method can be applied to every pair of variables, and the resulting estimations integrated to determine the causal ordering, as mentioned by Shimizu and Bollen (2014). Computationally, this approach is much simpler than attempting to compare all possible causal orderings of variables. Once a causal ordering has been estimated, the problem is to estimate causal connection strengths or their posterior distributions. We can then examine whether there are direct causal connections between these variables. Although this can still be computationally challenging for large numbers of variables, the problem is made significantly simpler by identifying the causal orders of variables.

An Empirical Example Using the mixed model-based approach, we reanalyzed data from Finkelstein *et al.* (1994) on the development of aggression in adolescence.

The data set was collected from 38 boys and 76 girls, that is, there are a total of 114 observations. von Eye and Wiedermann (2014) used the approaches proposed in Dodge and Rousson (2000), Wiedermann *et al.* (2015) to analyze the causal direction of two aggression variables: VAAA and PAAP. They analyzed the principal component scores of the VAAA and PAAP measurement variables and suggested that it is more likely that PAAP causes VAAA than vice versa. Some investigators might question the assumption made in their analysis that there are no latent common causes. Interestingly, the mixed model-based approach with t-distributed observation-specific intercepts (Shimizu and Bollen, 2014), which allows latent common causes, also indicated that PAAP causes VAAA using the same principal component scores.

7.6 CONCLUSION AND FUTURE DIRECTIONS

The utilization of non-Gaussianity in estimating SEMs is useful for causal discovery, because a wider variety of causal structures can be estimated in this manner than through classical methods. Non-Gaussian data is widely encountered, and the non-Gaussian approach based on SEMs discussed in this chapter can be useful in such applications. Download links to papers and codes on this topic are available online: `https://sites.google.com/site/sshimizu06/home/lingampapers`.

Research is underway to refine non-Gaussian causal discovery methods based on observational data. In particular, the following lines of research are expected to be important in future:

- Developing methods for efficiently estimating causal directions in the presence of latent common causes (Hoyer *et al.*, 2008b, Henao and Winther, 2011, Shimizu and Bollen, 2014) under minimal model assumptions is a fundamental problem that requires more research. Computational and statistical efficiencies need to be further improved.
- Methods for evaluating model assumptions and performing sensitivity analyses for the choice of model assumptions are necessary to apply causal discovery methods to real-world applications. Nevertheless, extensive research in the context of causal discovery has not been conducted. Several studies along these lines can be found in Shimizu and Kano (2008), Entner and Hoyer (2011). Sokol *et al.* (2014) considered quantifying the difficulty of identifying the ICA model when the distribution is close to Gaussian, and discussed the implications for non-Gaussian causal discovery methods.
- Evaluating the statistical reliability of estimation results is also necessary. Research in this direction can be found in Komatsu *et al.* (2010), Thamvitayakul *et al.* (2012), Wiedermann *et al.* (2015).
- Extending general-purpose methods to tailor-made techniques for specific applications is also useful and important in many application areas. Discussions specific to brain imaging data and neuroscience can be found in Ramsey *et al.* (2014), Xu *et al.* (2014).

- Methods for analyzing causal relations between discrete variables should be more extensively studied, as many variables in the social sciences are categorical. Peters *et al.* (2011), Inazumi *et al.* (2011) have considered extending the idea of LiNGAM to discrete variable cases by assuming additive error models. Nevertheless, a more promising direction might be to use generalized linear models, including the logit model, to determine the causal relations of discrete variables. Such logit model-based methods should benefit from classic structural equation modeling (Muthén, 2002), and would lead to a more natural extension of LiNGAM for analyzing causal relations between both continuous and categorical variables.
- Finally, developing freely available software that implements the above methods and is easy to use for empirical researchers will be key to the proliferation of applications for causal discovery methods.

REFERENCES

Amari, S. (1998) Natural gradient learning works efficiently in learning. *Neural Computation*, **10**, 251–276.

Bach, F., Jenatton, R., Mairal, J., and Obozinski, G. (2012) Optimization with sparsity-inducing penalties. *Foundations and Trends® in Machine Learning*, **4** (1), 1–106.

Bach, F.R. and Jordan, M.I. (2002) Kernel independent component analysis. *Journal of Machine Learning Research*, **3**, 1–48.

Billingsley, P. (1986) *Probability and Measure*, Wiley-Interscience.

Bollen, K. (1989) *Structural Equations with Latent Variables*, John Wiley & Sons, Inc., New York.

Boukrina, O. and Graves, W.W. (2013) Neural networks underlying contributions from semantics in reading aloud. *Frontiers in Human Neuroscience*, **7**, 518.

Burkard, R.E. and Cela, E. (1999) Linear assignment problems and extensions, in *Handbook of Combinatorial Optimization: Supplement Volume A* (eds P.M. Pardalos and D.Z. Du), Springer, US, pp. 75–149.

Campomanes, P., Neri, M., Horta, B.A., Roehrig, U.F., Vanni, S., Tavernelli, I., and Rothlisberger, U. (2014) Origin of the spectral shifts among the early intermediates of the rhodopsin photocycle. *Journal of the American Chemical Society*, **136** (10), 3842–3851.

Chickering, D. (2002) Optimal structure identification with greedy search. *Journal of Machine Learning Research*, **3**, 507–554.

Cichocki, A. and Amari, S.I. (2002) *Adaptive Blind Signal and Image Processing: Learning Algorithms and Applications*, John Wiley & Sons, Inc., New York.

Coad, A. and Binder, M. (2014) Causal linkages between work and life satisfaction and their determinants in a structural VAR approach. *Economics Letters*, **124** (2), 263–268.

Comon, P. (1994) Independent component analysis, a new concept? *Signal Processing*, **36**, 62–83.

Demidenko, E. (2004) *Mixed Models: Theory and Applications*, Wiley-Interscience.

Dodge, Y. and Rousson, V. (2000) Direction dependence in a regression line. *Communications in Statistics—Theory and Methods*, **29** (9-10), 1957–1972.

Efron, B. and Tibshirani, R. (1993) *An Introduction to the Bootstrap*, Chapman and Hall, New York.

Entner, D. and Hoyer, P.O. (2011) Discovering unconfounded causal relationships using linear non-Gaussian models, in *New Frontiers in Artificial Intelligence*, Lecture Notes in Computer Science, vol. **6797**, Springer-Verlag, Berlin, pp. 181–195.

Eriksson, J. and Koivunen, V. (2004) Identifiability, separability, and uniqueness of linear ICA models. *IEEE Signal Processing Letters*, **11**, 601–604.

von Eye, A. and Wiedermann, W. (2014) On direction of dependence in latent variable contexts. *Educational and Psychological Measurement*, **74** (1), 5–30.

Faes, L., Erla, S., Porta, A., and Nollo, G. (2013) A framework for assessing frequency domain causality in physiological time series with instantaneous effects. *Philosophical Transactions of the Royal Society of London, Series A: Mathematical, Physical and Engineering Sciences*, **371**, 20110618.

Ferkingsta, E., Lølanda, A., and Wilhelmsen, M. (2011) Causal modeling and inference for electricity markets. *Energy Economics*, **33** (3), 404–412.

Finkelstein, J.W., von Eye, A., and Preece, M.A. (1994) The relationship between aggressive behavior and puberty in normal adolescents: a longitudinal study. *Journal of Adolescent Health*, **15** (4), 319–326.

Granger, C.W.J. (1969) Investigating causal relations by econometric models and cross-spectral methods. *Econometrica*, **37** (3), 424–438.

Gretton, A., Bousquet, O., Smola, A.J., and Schölkopf, B. (2005) Measuring statistical dependence with Hilbert-Schmidt norms, in Proceedings of the 16th International Conference on Algorithmic Learning Theory (ALT2005), pp. 63–77.

Hastie, T., Tibshirani, R., and Friedman, J.H. (2001) *The Elements of Statistical Learning*, Springer-Verlag, New York.

Helajärvi, H., Rosenström, T., Pahkala, K., Kähönen, M., Lehtimäki, T., Heinonen, O.J., Oikonen, M., Tammelin, T., Viikari, J.S., and Raitakari, O.T. (2014) Exploring causality between TV viewing and weight change in young and middle-aged adults. The cardiovascular risk in young Finns study. *PloS ONE*, **9** (7), e101 860.

Henao, R. and Winther, O. (2011) Sparse linear identifiable multivariate modeling. *Journal of Machine Learning Research*, **12**, 863–905.

Himberg, J., Hyvärinen, A., and Esposito, F. (2004) Validating the independent components of neuroimaging time-series via clustering and visualization. *NeuroImage*, **22**, 1214–1222.

Hoyer, P.O., Hyvärinen, A., Scheines, R., Spirtes, P., Ramsey, J., Lacerda, G., and Shimizu, S. (2008a) Causal discovery of linear acyclic models with arbitrary distributions, in Proceedings of the 24th Conference on Uncertainty in Artificial Intelligence (UAI2008), pp. 282–289.

Hoyer, P.O., Shimizu, S., Kerminen, A., and Palviainen, M. (2008b) Estimation of causal effects using linear non-Gaussian causal models with hidden variables. *International Journal of Approximate Reasoning*, **49** (2), 362–378.

Hoyer, P.O., Janzing, D., Mooij, J., Peters, J., and Schölkopf, B. (2009) Nonlinear causal discovery with additive noise models, in Advances in Neural Information Processing Systems 21, pp. 689–696.

Hoyer, P.O., Shimizu, S., Hyvärinen, A., Kano, Y., and Kerminen, A. (2006) New permutation algorithms for causal discovery using ICA, in Proceedings of International Conference on Independent Component Analysis and Blind Signal Separation, Charleston, SC, USA, pp. 115–122.

Hyvärinen, A. (1998) New approximations of differential entropy for independent component analysis and projection pursuit, in Advances in Neural Information Processing Systems 10, pp. 273–279.

Hyvärinen, A. (1999) Fast and robust fixed-point algorithms for independent component analysis. *IEEE Transactions on Neural Networks*, **10**, 626–634.

Hyvärinen, A. (2013) Independent component analysis: recent advances. *Philosophical Transactions of the Royal Society of London, Series A: Mathematical, Physical and Engineering Sciences*, **371**, 20110 534.

Hyvärinen, A. and Kano, Y. (2003) Independent component analysis for non-normal factor analysis, in *New Developments in Psychometrics* (eds H. Yanai, A. Okada, K. Shigemasu, Y. Kano, and J.J. Meulman), Springer-Verlag, Tokyo, pp. 649–656.

Hyvärinen, A., Karhunen, J., and Oja, E. (2001) *Independent Component Analysis*, John Wiley & Sons, Inc., New York.

Hyvärinen, A. and Smith, S.M. (2013) Pairwise likelihood ratios for estimation of non-Gaussian structural equation models. *Journal of Machine Learning Research*, **14**, 111–152.

Hyvärinen, A., Zhang, K., Shimizu, S., and Hoyer, P.O. (2010) Estimation of a structural vector autoregressive model using non-Gaussianity. *Journal of Machine Learning Research*, **11**, 1709–1731.

Inazumi, T., Washio, T., Shimizu, S., Suzuki, J., Yamamoto, A., and Kawahara, Y. (2011) Discovering causal structures in binary exclusive-or skew acyclic models, in Proceedings of the 27th Conference on Uncertainty in Artificial Intelligence, pp. 373–382.

Jutten, C. and Hérault, J. (1991) Blind separation of sources, part I: an adaptive algorithm based on neuromimetic architecture. *Signal Processing*, **24** (1), 1–10.

Kass, R.E. and Raftery, A.E. (1995) Bayes factors. *Journal of the American Statistical Association*, **90** (430), 773–795.

Kawahara, Y., Shimizu, S., and Washio, T. (2011) Analyzing relationships among ARMA processes based on non-Gaussianity of external influences. *Neurocomputing*, **4** (12–13), 2212–2221.

Komatsu, Y., Shimizu, S., and Shimodaira, H. (2010) Assessing statistical reliability of LiNGAM via multiscale bootstrap, in Proceedings of International Conference on Artificial Neural Networks (ICANN2010), pp. 309–314.

Kraskov, A., Stögbauer, H., and Grassberger, P. (2004) Estimating mutual information. *Physical Review E*, **69** (6), 066 138.

Kreft, I.G.G. and De Leeuw, J. (1998) *Introducing Multilevel Modeling*, Sage Publications.

Lacerda, G., Spirtes, P., Ramsey, J., and Hoyer, P.O. (2008) Discovering cyclic causal models by independent components analysis, in Proceedings of the 24th Conference on Uncertainty in Artificial Intelligence (UAI2008), pp. 366–374.

Lewicki, M. and Sejnowski, T.J. (2000) Learning overcomplete representations. *Neural Computation*, **12** (2), 337–365.

Manelis, A. and Reder, L.M. (2014) Effective connectivity among the working memory regions during preparation for and during performance of the n-back task. *Frontiers in Human Neuroscience*, **8**, 598.

Meek, C. (1995) Strong completeness and faithfulness in Bayesian networks, in Proceedings of the 11th Conference on Uncertainty in Artificial Intelligence, pp. 411–418.

Micceri, T. (1989) The unicorn, the normal curve, and other improbable creatures. *Psychological Bulletin*, **105** (1), 156–166.

Mills-Finnerty, C., Hanson, C., and Hanson, S.J. (2014) Brain network response underlying decisions about abstract reinforcers. *NeuroImage*, **103**, 48–54.

Moneta, A., Entner, D., Hoyer, P., and Coad, A. (2013) Causal inference by independent component analysis: theory and applications. *Oxford Bulletin of Economics and Statistics*, **75** (5), 705–730.

Muthén, B.O. (2002) Beyond SEM: general latent variables modeling. *Behaviormetrika*, **29**, 81–117.

Niyogi, D., Kishtawal, C., Tripathi, S. and Govindaraju, R.S. (2010) Observational evidence that agricultural intensification and land use change may be reducing the Indian summer monsoon rainfall. *Water Resources Research*, **46**, W03 533.

Pearl, J. (2000) *Causality: Models, Reasoning, and Inference*, 2nd edn, 2009, Cambridge University Press, USA.

Peters, J., Janzing, D., and Schölkopf, B. (2011) Causal inference on discrete data using additive noise models. *IEEE Transactions on Pattern Analysis and Machine Intelligence*, **33** (12), 2436–2450.

Ramsey, J., Hanson, S., and Glymour, C. (2011) Multi-subject search correctly identifies causal connections and most causal directions in the DCM models of the Smith et al. simulation study. *NeuroImage*, **58** (3), 838–848.

Ramsey, J.D., Sanchez-Romero, R., and Glymour, C. (2014) Non-Gaussian methods and high-pass filters in the estimation of effective connections. *NeuroImage*, **84** (1), 986–1006.

Rosenström, T., Jokela, M., Puttonen, S., Hintsanen, M., Pulkki-Råback, L., Viikari, J.S., Raitakari, O.T., and Keltikangas-Järvinen, L. (2012) Pairwise measures of causal direction in the epidemiology of sleep problems and depression. *PloS ONE*, **7** (11), e50 841.

Scheines, R., Spirtes, P., Glymour, C., Meek, C., and Richardson, T. (1998) The TETRAD project: constraint based aids to causal model specification. *Multivariate Behavioral Research*, **33** (1), 65–117.

Shimizu, S. (2012) Joint estimation of linear non-Gaussian acyclic models. *Neurocomputing*, **81**, 104–107.

Shimizu, S. (2014) LiNGAM: Non-Gaussian methods for estimating causal structures. *Behaviormetrika*, **41** (1), 65–98.

Shimizu, S. and Bollen, K. (2014) Bayesian estimation of causal direction in acyclic structural equation models with individual-specific confounder variables and non-Gaussian distributions. *Journal of Machine Learning Research*, **15**, 2629–2652.

Shimizu, S. and Kano, Y. (2008) Use of non-normality in structural equation modeling: application to direction of causation. *Journal of Statistical Planning and Inference*, **138**, 3483–3491.

Shimizu, S., Hoyer, P.O., and Hyvärinen, A. (2009) Estimation of linear non-Gaussian acyclic models for latent factors. *Neurocomputing*, **72**, 2024–2027.

Shimizu, S., Hoyer, P.O., Hyvärinen, A., and Kerminen, A. (2006) A linear non-Gaussian acyclic model for causal discovery. *Journal of Machine Learning Research*, **7**, 2003–2030.

Shimizu, S. and Hyvärinen, A. (2008) Discovery of linear non-Gaussian acyclic models in the presence of latent classes, in Proceedings of the 14th International Conference on Neural Information Processing (ICONIP2007), pp. 752–761.

Shimizu, S., Inazumi, T., Sogawa, Y., Hyvärinen, A., Kawahara, Y., Washio, T., Hoyer, P.O., and Bollen, K. (2011) DirectLiNGAM: a direct method for learning a linear non-Gaussian structural equation model. *Journal of Machine Learning Research*, **12**, 1225–1248.

Smith, S., Miller, K., Salimi-Khorshidi, G., Webster, M., Beckmann, C., Nichols, T., Ramsey, J., and Woolrich, M. (2011) Network modelling methods for FMRI. *NeuroImage*, **54** (2), 875–891.

Sogawa, Y., Shimizu, S., Shimamura, T., Hyvärinen, A., Washio, T., and Imoto, S. (2011) Estimating exogenous variables in data with more variables than observations. *Neural Networks*, **24** (8), 875–880.

Sokol, A., Maathuis, M.H., and Falkeborg, B. (2014) Quantifying identifiability in independent component analysis. *Electronic Journal of Statistics*, **8**, 1438–1459.

Spirtes, P. and Glymour, C. (1991) An algorithm for fast recovery of sparse causal graphs. *Social Science Computer Review*, **9**, 67–72.

Spirtes, P., Glymour, C., and Scheines, R. (1993) *Causation, Prediction, and Search*, 2nd edn, MIT Press 2000, Springer-Verlag, New York.

Swanson, N. and Granger, C. (1997) Impulse response functions based on a causal approach to residual orthogonalization in vector autoregressions. *Journal of the American Statistical Association*, **92** (437), 357–367.

Thamvitayakul, K., Shimizu, S., Ueno, T., Washio, T., and Tashiro, T. (2012) Bootstrap confidence intervals in DirectLiNGAM, in Proceedings of 2012 IEEE 12th International Conference on Data Mining Workshops (ICDMW2012), IEEE, pp. 659–668.

Tibshirani, R. (1996) Regression shrinkage and selection via the lasso. *Journal of Royal Statistical Society: Series B*, **58** (1), 267–288.

Tillman, R.E., Gretton, A., and Spirtes, P. (2010) Nonlinear directed acyclic structure learning with weakly additive noise models, in Advances in Neural Information Processing Systems 22, pp. 1847–1855.

Wiedermann, W. and Hagmann, M. (2014) Asymmetric properties of the Pearson correlation coefficient: correlation as the negative association between linear regression residuals. *Communications in Statistics—Theory and Methods*, In press.

Wiedermann, W., Hagmann, M., and von Eye, A. (2015) Significance tests to determine the direction of effects in linear regression models. *British Journal of Mathematical and Statistical Psychology*, **68** (1), 116–141.

Xu, L., Fan, T., Wu, X., Chen, K., Guo, X., Zhang, J., and Yao, L. (2014) A pooling-LiNGAM algorithm for effective connectivity analysis of fMRI data. *Frontiers in Computational Neuroscience*, **8**, 125.

Zhang, K. and Hyvärinen, A. (2009) On the identifiability of the post-nonlinear causal model, in Proceedings of the 25th Conference on Uncertainty in Artificial Intelligence (UAI2009), pp. 647–655.

Zou, H. (2006) The adaptive Lasso and its oracle properties. *Journal of the American Statistical Association*, **101**, 1418–1429.

CODE AVAILABILITY

Codes for the non-Gaussian causal discovery methods discussed in this chapter are available on the web:

- ICA-based LiNGAM:
  - Matlab code: `http://www.cs.helsinki.fi/group/neuroinf/lingam/`
  - R code: `https://sites.google.com/site/dorisentner/publications/VARLiNGAM`
  - The TETRAD project (Scheines *et al.*, 1998): `http://www.phil.cmu.edu/projects/tetrad/`
- DirectLiNGAM
  - Matlab code: `https://sites.google.com/site/sshimizu06/Dlingamcode`
- Pairwise LiNGAM
  - Matlab code: `http://www.cs.helsinki.fi/u/ahyvarin/code/pwcausal/`
- LiNGAM for time series
  - R code: `https://sites.google.com/site/dorisentner/publications/VARLiNGAM`
  - Matlab code (Faes *et al.*, 2013): `http://www.science.unitn.it/biophysicslab/research/sigpro/eMVAR.html`
- ICA-based LiNGAM for latent common cause cases
  - ICA-based approach
    - Hoyer's method (Matlab): `http://www.cs.helsinki.fi/u/phoyer/code/lvlingam.tar.gz`
    - Henao's method (Matlab): `http://cogsys.imm.dtu.dk/slim/`
  - Mixed model-based approach
    - Matlab code: `https://sites.google.com/site/sshimizu06/mixedlingamcode`

8

NONLINEAR FUNCTIONAL CAUSAL MODELS FOR DISTINGUISHING CAUSE FROM EFFECT

KUN ZHANG

Max Planck Institute for Intelligent Systems, Tübingen, Germany and Carnegie Mellon University, Pittsburgh, PA, USA

AAPO HYVÄRINEN

Department of Computer Science, University of Helsinki, Helsinki, Finland

8.1 INTRODUCTION

Finding causal directions is a fundamental problem in scientific data analysis and other fields. In general, finding causal directions is extremely complex, but we can make progress by assuming that the causal relationships can only take some special forms.

For simplicity, let us assume that we have only two observed random variables, x and y, where either x is causing y or y is causing x. In particular, we exclude the possibility that there is some kind of bidirectional influence (feedback) between the two; we also exclude the case where both are actually caused by some further, unobserved variable (confounder).

Let us start by considering the very simplest case, where the relationship is assumed *linear*. We thus need to choose between the following two models. The first model assumes that x causes y, and is given by

$$y = \rho x + n \tag{8.1}$$

Statistics and Causality: Methods for Applied Empirical Research, First Edition.
Wolfgang Wiedermann and Alexander von Eye.

while the second model assumes that y causes x and is given by

$$x = \rho y + \tilde{n} \tag{8.2}$$

In both models, the disturbances (also called external influences or noise), denoted by n or $\tilde{n}$, are assumed independent of the regressors x and y, respectively. Without restriction of generality, we can assume that x and y are standardized to zero mean and unit variance. The parameter ρ is then the same in the two models because it is equal to the correlation coefficient.

Choosing between these two models, that is, identifying the causal direction, is a well-known problem that is widely encountered in statistics and machine learning. The problem is usually considered very difficult and perhaps unsolvable, since most analysis assumes that the variables x and y are *Gaussian*, which also implies that the disturbances are Gaussian. Under the Gaussian assumption, the two models are completely symmetric in the sense that the variance explained is equal for the two models, and further, the likelihood is the same for both models (both quantities being simple functions of ρ).

The symmetry between the two models is illustrated in Figure 8.1a, where the points were generated by $y = 2x + n$ and x and n follow the standard Gaussian distribution. We see that the Gaussian data cloud generated by any of the two models looks just the same, which underlines the unidentifiability of the causal direction.

The inability to decide between these two models under the assumptions of linearity and Gaussianity is one of the motivations for the well-known saying that "correlations does not equal causality." However, it is possible to change the situation by modifying any of these two assumptions.

For example, we can make the causal direction identifiable by assuming at least one of the variables (the regressor or the disturbance) in the true model is non-Gaussian. This leads to the theory whose main model is the linear non-Gaussian acyclic model (Shimizu *et al.*, 2006), treated in another chapter of this volume. It is worth noting that there have been several pieces of work in statistics about the asymmetry between two variables in the linear non-Gaussian case. Dating back to 2000, Dodge and Rousson (2000, 2001) considered identifying the correct bivariate linear regression model under the assumption of a non-Gaussian true predictor, which is also addressed in a related chapter of this volume (Dodge and Rousson, 2016). The case of a normal predictor and a non-Gaussian disturbance has been discussed by Wiedermann and Hagmann (2015).

Here, we merely show how the symmetry between the two models is broken, as is illustrated in Figure 8.1b, where the data were generated by $y = 2x + n$ and x and n were obtained by taking the square of standard Gaussian random samples and keeping their original sign. We see that scatterplots for the two models are quite different from each other (note the thin "arms" along either the vertical or the horizontal axis), which gives hope that the causal direction could be identifiable.

Another modification that makes the causal direction identifiable is to assume a nonlinear relationship, which is the topic of this chapter. We start by defining the basic nonlinear model, show how it can be estimated, and then go to a more general theory with more complex nonlinear relationships.

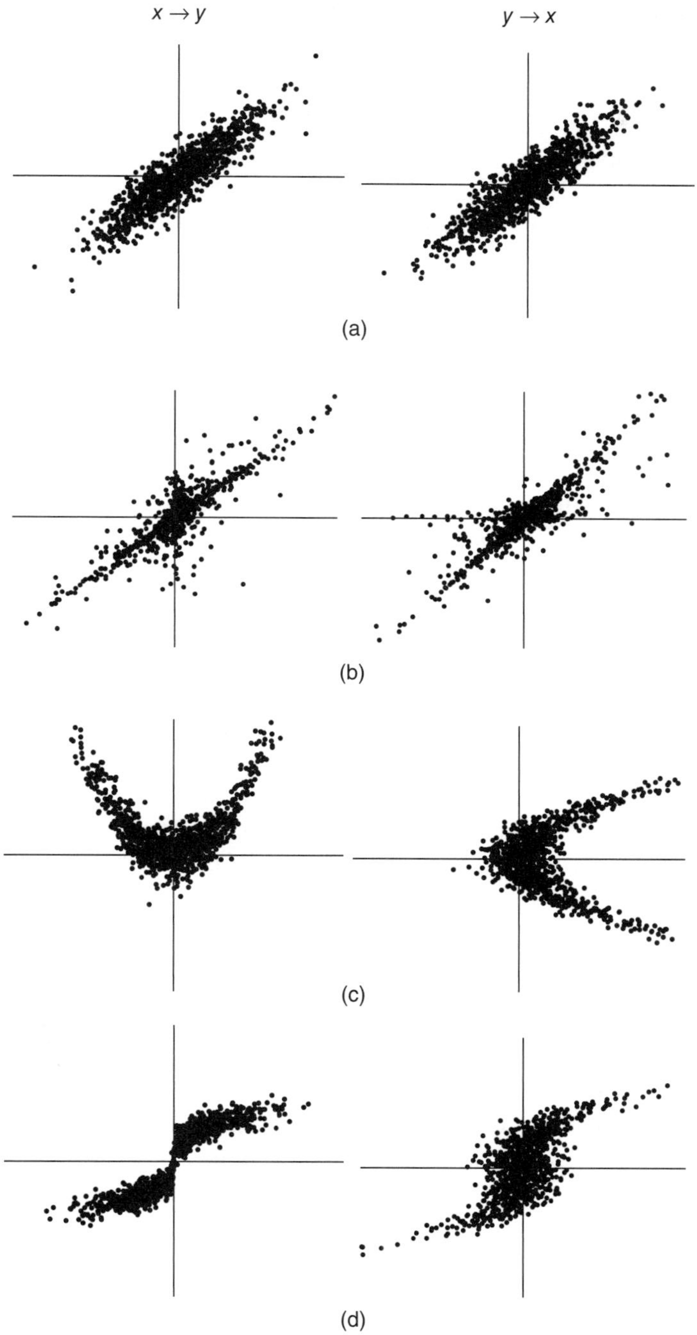

Figure 8.1 Illustration of the effect of the assumptions of linearity and Gaussianity on the identifiability. On the left, we have data generated for the causal direction $x \rightarrow y$, and on the right, data generated for the causal direction $y \rightarrow x$. The rows correspond to different models: (a) linear Gaussian model, (b) linear non-Gaussian model, (c) the nonlinear model, with two squaring nonlinearity in both directions, (d) the nonlinear model, with cubic root nonlinearity and cubic nonlinearity.

8.2 NONLINEAR ADDITIVE NOISE MODEL

8.2.1 Definition of Model

We start with a particularly simple form of nonlinear relationship, where a scalar-valued nonlinear function, f or g, is taken of the regressor, and the disturbance is added in an additive manner. Thus, we obtain the following two models to choose from. The first model, which we denote by $x \to y$, assumes that x causes y and is given by

$$y = f(x) + n \tag{8.3}$$

while the second model that assumes that y causes x, which we denote by $y \to x$, and is given by

$$x = g(y) + \tilde{n} \tag{8.4}$$

Again, in both models, the disturbances n and $\tilde{n}$, are independent of the regressors x and y, respectively. Here, unlike in the linear case, the functions f and g are likely to be very different from each other. Furthermore, no assumption is made on the distributions of the residuals.

To start with a graphical illustration, see Figure 8.1c. Here, on the left side, we have generated data from $x \to y$ with $f(x) = x^2$ and used Gaussian noise n. We use exactly the same nonlinearity and noise on the right in the direction $y \to x$. This is to illustrate the intuitively quite obvious idea that if the nonlinearity is not invertible (such as squaring), the model is intuitively easy to choose since in the wrong direction, it is not at all of the desired form: nonlinear transform plus independent noise. In fact, it is implausible that y on the right-hand side (generating direction $y \to x$) could have been obtained by adding independent noise on some function of x since any function of x cannot predict y well; in fact here the function of x that predicts y best is $y = 0$. Thus, only the true generating model is at all plausible.

In Figure 8.1d, we have a less drastic, and invertible, nonlinearity (third power) for $x \to y$. Now, since f and g need not be the same in general, a more realistic illustration would take $g = f^{-1}$, which we do here. In particular, we take $f(x) = x^3$ and $g(y) = |y|^{1/3}\text{sign}(y)$. The breaking of symmetry between x and y is seen in the fact that the data distributions are slightly different for the two models, even though they attempted to create the same kind of joint distribution. (We also tried to match the noise levels to make the distributions as similar as possible.) In particular, the noise is seen to "fatten" the regression curve either vertically or horizontally, depending on the direction of causal influence.

8.2.2 Likelihood Ratio for Nonlinear Additive Models

An attractive way of deciding between the two models in Equations (8.1) and (8.2) is to compute their likelihoods and compare them in terms of their ratio. Essentially, we choose the model that has the larger likelihood.

The log-likelihood of the model $x \to y$, for a single data point, can be obtained as the sum of the log-prior of the variable x and the log-likelihood of the residual:

$$\log p(x, y) = \log p_x(x) + \log p_n(y - f(x)) \tag{8.5}$$

Here, it is crucial that we have some kinds of estimates of the log-probability density functions (log-pdf's) of the regressor and the disturbance. Since it is usually more convenient to operate with standardized quantities, so let us denote the log-pdf of the standardized residual by G_n and the log-pdf of x by G_x. Then, the log-likelihood for a single data point can be written as

$$\log p(x, y) = G_x(x) + G_n\left(\frac{y - f(x)}{\sigma_n}\right) - \log \sigma_n \tag{8.6}$$

where we denote the variance of the disturbance by σ_n^2. Choosing the standardized log-pdf's G_x, G_n could be done by modeling the relevant log-pdf's by parametric (Karvanen and Koivunen, 2002) or nonparametric (Pham and Garat, 1997) methods. It is also possible that we have enough prior information on them, so we can fix the G in advance.

Consider a sample $(x_1, y_1), \cdots, (x_T, y_T)$ of data. Let us add together the log-likelihoods of the data points and take the difference, and we obtain the logarithm of the likelihood ratio for the sample as

$$R = \frac{1}{T}\sum_t \left[G_x(x_t) + G_n\left(\frac{y_t - f(x_t)}{\sigma_n}\right) - G_y(y_t) - G_{\tilde{n}}\left(\frac{x_t - g(y_t)}{\sigma_{\tilde{n}}}\right)\right]$$
$$- \log \sigma_n + \log \sigma_{\tilde{n}} \tag{8.7}$$

where we also need the standardized log-pdf's of y and the disturbance $\tilde{n}$.

The likelihood ratio further depends on the estimated nonlinearities f, g. The estimation of f and g can be done with classic least-squares estimation methods, fitting some nonlinear (nonparametric) regression model on the sample. Such regression methods are independent of any developments in this chapter. A large number of nonparametric methods have been developed in the literature; see Hoyer *et al.* (2009) for an example.

8.2.3 Information-Theoretic Interpretation

The likelihood ratio has a simple information-theoretic interpretation, which also means we can use well-known information-theoretic approximations for its practical computation in the case where we do not want to postulate functional forms for the G's.

In fact, if we take the asymptotic limit of the likelihood ratio ($T \to \infty$), we obtain asymptotically

$$R \to -H(x) - H\left(\frac{n}{\sigma_n}\right) + H(y) + H\left(\frac{\tilde{n}}{\sigma_{\tilde{n}}}\right) - \log \sigma_n + \log \sigma_{\tilde{n}} \tag{8.8}$$

where we denote differential entropy by H. The differential entropy, defined as $H(x) = -\int p(x) \log p(x) dx$, is the fundamental information-theoretic quantity for continuous-valued variables.

We can go back to the nonstandardized quantities (which can here be done by using the fundamental transformation formula $H(\alpha x) = H(x) + \log \alpha$, and we further obtain a very simple equivalent expression:

$$R \to -H(x) - H(n) + H(y) + H(\tilde{n}) \tag{8.9}$$

Here, we see that determining causal direction is related to finding the direction that corresponds to minimum entropy in the sense of the sum of the marginal entropies of the regressor and the disturbance; we have more on this connection next.

The practical utility of this connection is that we can approximate the likelihood ratio using any general, possibly nonparametric, approximations of differential entropy. Several such approximations have been developed; for example, we can use the maximum entropy approximations by Hyvärinen (1998), which are computationally simple. In fact, we only need to approximate one-dimensional differential entropies, which is much simpler than approximating two-dimensional entropies.

This information-theoretic formulation also leads to a simple intuitive interpretation of the likelihood ratio. It is well known that in the space of probability distributions of unit variance, differential entropy is maximized by Gaussian distribution. This is why (negative) differential entropy is often used as a measure of non-Gaussianity. In our case, we can thus interpret the asymptotic limit of the log-likelihood ratio in terms of non-Gaussianities and errors in regression:

$$\begin{aligned} R \to \\ &\text{nongaussianity}(x) + \text{nongaussianity}(\text{residual in } x \to y) - \log(\text{error in } x \to y) \\ &- [\text{nongaussianity}(y) + \text{nongaussianity}(\text{residual in } y \to x) - \log(\text{error in } y \to x)] \end{aligned}$$

Intuitively, this means that

(1) if the non-Gaussianities are negligible, we choose the direction in which the error in the regression is smaller;
(2) if the errors in the regression are almost equal, we choose the direction of causality in which the sum of non-Gaussianities of the regressor and residual is maximized;
(3) in the general case, we have a sum of these two criteria: error in regression and non-Gaussianity.

An interesting point here is that in the linear non-Gaussian case, the errors in the regression are always equal, and thus, choosing the direction is solely based on maximizing the non-Gaussianity. In contrast, in the nonlinear case, the errors in the regressions can be the decisive factor in the identification. This is intuitively clear in Figure 8.1c, where the right direction leads to a small regression error, while in the wrong direction, the regression is catastrophically bad.

8.2.4 Likelihood Ratio and Independence-Based Methods

An alternative approach to nonlinear additive models is provided by the independence-based method by Hoyer *et al.* (2009). In such methods, the idea is to use the independence of the regressor and the disturbance in each model as the selection criterion: the model in which the residual (i.e., estimate of disturbance) is more independent of the regressor is chosen (again, assuming that some nonparametric regression method is used to estimate f and g; see Section 8.3.3.)

The fundamental information-theoretic quantity for measuring the independence of two random variables is mutual information, defined as

$$I(u, v) = H(u) + H(v) - H(u, v) \tag{8.10}$$

which is always nonnegative and zero if and only if the two variables are independent. Here, $H(u, v)$ is the joint entropy, which is simply the entropy of the random vector consisting of (u, v).

In fact, the likelihood ratio can be interpreted from the viewpoint of such maximization of independence. Using basic information-theoretic properties, we have under $x \rightarrow y$

$$\begin{aligned} H(x, y) &= H(x) + H(y|x) = H(x) + H(y - f(x)|x) \\ &= H(x) + H(n|x) = H(x, n) \end{aligned} \tag{8.11}$$

and by symmetry, this is equal to $H(y, \tilde{n})$. Now if we can consider the difference between the mutual information of the regressors and residuals in the two directions and obtain

$$\begin{aligned} I(x, n) - I(y, e) &= H(x) + H(n) - H(x, n) - [H(y) + H(e) - H(y, e)] \\ &= H(x) + H(n) - H(y) - H(e) \\ &= H(x) + H\left(\frac{n}{\sigma_n}\right) - H(y) - H\left(\frac{\tilde{n}}{\sigma_{\tilde{n}}}\right) \\ &\quad + \log \sigma_n - \log \sigma_{\tilde{n}} \end{aligned} \tag{8.12}$$

where two terms equal to $H(x, y)$ cancel. Here, we see that asymptotically, the objective derived from the likelihood ratio is equal to the difference of the two mutual information (with sign reversed). Its sign tells which mutual information is larger and, in particular, in which direction the residual of the regression is more independent. Thus, using the likelihood ratio is equivalent to using mutual information as independence measure in the methods by Hoyer *et al.* (2009). We will elaborate more on this in Section 8.4.

The aforementioned developments thus show that when comparing independencies of the residuals such as Hoyer *et al.* (2009), it is not necessary to explicitly estimate mutual information; estimation of one-dimensional entropies leads to an equivalent result. This is very important from a practical viewpoint, since estimating one-dimensional entropies is much easier than estimating two-dimensional quantities such as mutual information.

8.3 POST-NONLINEAR CAUSAL MODEL

Obviously, it is important to use sufficiently general functional models in causal discovery: if the assumed functional causal model is too restrictive to properly approximate the true data-generating process, the causal discovery results may be misleading. Therefore, if specific knowledge about the data-generating mechanism is not available, we should attempt to fit a model that is as general as possible. Post-nonlinear (PNL) causal models offer an interesting generalization of the nonlinear additive model of the previous section.

8.3.1 The Model

The PNL causal model consists of a nonlinear influence from the cause, a noise or disturbance, and–in contrast to the model above–a possible sensor or measurement distortion in the observed variables (Zhang and Hyvärinen, 2009b, 2010). The effect y is generated by a post-nonlinear transformation of the nonlinear effect of the cause x with additive noise n:

$$y = f_2(f_1(x) + n) \tag{8.13}$$

where both f_1 and f_2 are nonlinear functions and f_2 is assumed to be invertible. The post-nonlinear transformation f_2 represents the sensor or measurement distortion, which is frequently encountered in practice. A slightly more restricted version of the model, in which the inner function, f_1, is also assumed to be invertible, was proposed in Zhang and Chan (2006) and applied to causal analysis of stock returns.[1]

The PNL causal model has the most general form among all well-defined functional causal models in which the causal direction has been shown to be identifiable under mild assumptions. Clearly, it contains the linear model and the nonlinear additive noise model as special cases. The multiplicative noise model, $y = x \cdot n$, where all involved variables are positive, is another special case: the multiplicative noise model can be written as $y = \exp(\log x + \log n)$, where $\log n$ is considered as a new noise term, $f_1(x) = \log(x)$, and $f_2(\cdot) = \exp(\cdot)$.

Next, we discuss the identifiability conditions of the causal direction for the PNL causal model, which naturally contain those for the linear model and nonlinear additive noise model as special cases.

[1] In Zhang and Chan (2006) both functions f_1 and f_2 are assumed to be invertible; this causal model, as a consequence, can be estimated by making use of post-nonlinear independent component analysis (PNL-ICA; Taleb and Jutten, 1999), which assumes that the observed data are componentwise invertible transformations of linear mixtures of the independence sources to be recovered.

8.3.2 Identifiability of Causal Direction

The identifiability conditions of the causal direction according to the PNL causal model were established by a proof by contradiction (Zhang and Hyvärinen, 2009b). We assume the causal model holds in both directions $x \to y$ and $y \to x$ and show that this implies some very strong conditions on the distributions and functions involved in the model. Therefore, if the data are generated according to the PNL causal model in settings not fulfilling those strong conditions, the backward direction does not follow the model, and the causal direction can be determined. We will next explain this in more detail.

Assume that the data (x, y) is generated by the PNL causal model with the causal relation $x \to y$ in (8.13). Moreover, let us assume (by contradiction) that the backward direction, $y \to x$, also follows the PNL causal model with independent noise. That is,

$$x = g_2(g_1(y) + \tilde{n}) \tag{8.14}$$

where y and $\tilde{n}$ are independent, g_1 is nonconstant, and g_2 is invertible.

Equations (8.13) and (8.14) define the transformation from (x, n) to $(y, \tilde{n})$; as a consequence, using the change-of-variable technique, $p_{y,\tilde{n}}$ can be expressed in terms of $p_{x,n} = p_x p_n$. The identifiability results were derived by making use of linear separability of the logarithm of the joint density of independent variables, that is, for a set of independent random variables whose joint density is twice differentiable, the Hessian of the logarithm of their density is diagonal everywhere (Lin, 1998). Since y and $\tilde{n}$ are assumed to be independent, $\log p_{y,\tilde{n}}$ then follows such a linear separability property. This implies that the second-order partial derivative of $\log p_{y,\tilde{n}}$ w.r.t. y and $\tilde{n}$ is zero. It then reduces to a differential equation of a bilinear form. Under certain technical assumptions (e.g., p_n is positive on $(-\infty, +\infty)$), the solution to the differential equation gives all cases in which the causal direction is *not* identifiable according to the PNL causal model. Table 1 in Zhang and Hyvärinen (2009b) summarizes all five nonidentifiable cases. The first one is the linear-Gaussian case, in which the causal direction is well known to be nonidentifiable. Roughly speaking, to make one of those cases true, one has to adjust the data distribution and the involved nonlinear functions very carefully.

In other words, in the generic case, the causal direction is identifiable if the data were generated according to the PNL causal model. Simulations results were further presented in Zhang and Hyvärinen (2009b) to verify the established identifiability results.

8.3.3 Determination of Causal Direction Based on the PNL Causal Model

The commonly used approach to distinguishing cause from effect with nonlinear functional causal models consists of two steps, which are similar for both the nonlinear additive noise model and post-nonlinear model. First, one fits the nonlinear regression model on the data for both hypothetical causal directions, obtaining estimates for f and g. The second step is to do a statistical analysis of the regressors and the residuals to determine the causal direction.

For the nonlinear additive noise model, the functions f and g are usually estimated by performing quite conventional Gaussian process (GP) regression (Hoyer *et al.*, 2009). (For details on GP regression, one may refer to Rasmussen and Williams, 2006) In contrast, estimation of the PNL causal model (8.13) has several indeterminacies: the sign, mean, and scale of the noise term, and accordingly, the sign, location, and scale of f, g are arbitrary. In the estimation procedure, one may impose certain constraints to avoid such indeterminacies in the estimate. However, we should note that in principle, we do not care about those indeterminacies in the causal discovery context, since they do not change the statistical independence or dependence property between the estimated noise and the hypothetical cause.

It is well known that for linear regression, the maximum likelihood estimator of the coefficient is still statistically consistent even if the noise distribution is erroneously assumed to be Gaussian. However, this may not be the case for general nonlinear models. As shown in (Zhang *et al.* 2016, Section 3.2), if the noise distribution is misspecified, the estimated PNL causal model (8.13) may not be statistically consistent, even when the indeterminacies in the estimate discussed earlier are properly tackled. Therefore, the noise distribution should be adaptively estimated from data, if the true one is not known *a priori*.

Regarding the statistical analysis of the regressor and the residuals, performing independence tests between the estimated noise and hypothetical cause is one approach (Hoyer *et al.*, 2009), (Zhang and Hyvärinen, 2009b). A commonly used option is the Hilbert-Schmidt information criterion (HSIC; (Gretton *et al.*, 2005)), although many others could be used. In fact, for nonlinear additive noise models, as we discussed in Section 8.2.2, following Hyvärinen and Smith (2013), the independence can be evaluated using one-dimensional entropy estimators as well.

Considering concrete implementations in the literature, Zhang and Hyvärinen (2009b) proposed to estimate the PNL causal model (8.13) by mutual information minimization with the involved nonlinear functions represented by multilayer perceptrons (MLPs). Later, Zhang *et al.* (2016) proposed to estimate the PNL causal model by making use of warped Gaussian processes with a flexible noise distribution, which is represented by a mixture of Gaussians. We call these two implementations PNL-MLP and PNL-WGP-MoG, respectively.

8.4 ON THE RELATIONSHIPS BETWEEN DIFFERENT PRINCIPLES FOR MODEL ESTIMATION

So far, we have mainly discussed the identifiability of the causal direction in the two-variable case, and it should be noted that the results can be readily extended to the case with an arbitrary number of variables, as shown in Peters *et al.* (2011). The basic idea is that no matter how many variables are involved in the system, when we are interested in a particular pair of directly connected variables, it becomes the two-variable case if the values of relevant variables are fixed.

Maximum likelihood is usually used to fit the functional causal model together with a directed acyclic graph (DAG) to the given data. Not surprisingly, the negative likelihood (with the distribution of the noise adaptively estimated from data) is equivalent to the mutual information between the estimated noise terms, as stated in Theorem 3 in Zhang *et al.* (2016). The higher the likelihood, the less dependent the estimated noise terms. (Note that the root variables in the DAG are also counted as noise terms.)

On the other hand, traditionally, it has been noted that under the causal Markov condition, which states that each variable is independent from its nondescendants conditioning on its parents, and the faithfulness assumption, one could recover an equivalence class of the underlying causal structure based on conditional independence relationships of the variables (Spirtes *et al.*, 2001, Pearl, 2000). This is known as the constraint-based approach to causal discovery. How are these principles, including mutual independence of the estimated noise terms and the causal Markov condition, related to each other? Next, we will answer this question, and the results in this section hold for an arbitrary number of variables.

In the following, we consider optimization over different DAG structures to find the causal structure. We assume we have infinite data, and we optimally fit the nonlinear functions f_i according to the DAG structure given, using some hypothetical method, which is statistically consistent. Then the question is how the statistical properties of the estimated noise terms (residuals) are related to the conditional independence properties of the variables x_i, for each particular DAG.

Suppose (first) that we fit the nonlinear additive noise model given the DAG structure, that is,

$$x_i = f_i(pa_i) + n_i \tag{8.15}$$

where pa_i represents the direct causes of x_i, to the data, that is, parents in the DAG. It has been shown that mutual independence of the estimated residuals and conditional independence between observed variables (together with the independence between n_i and pa_i) are equivalent; furthermore, they are achieved if and only if the total entropy of the disturbances is minimized (Zhang and Hyvärinen, 2009a). More specifically, when fitting the model (8.15) with a hypothetical DAG causal structure to the given variables $x_1, \cdots, x_K$, the following three properties are equivalent:

(i) The estimated noise terms n_i are mutually independent.
(ii) The total entropy of the estimated noise terms, that is, $\sum_i H(n_i)$, is minimized, with the minimum being equal to $H(x_1, \cdots, x_K)$.
(iii) The causal Markov condition holds (i.e., each variable is independent of its nondescendants in the DAG conditioning on its parents), and in addition, the noise term in x_i is independent of the parents of x_i.

Let us then consider the PNL causal model. When one fits the PNL causal model

$$x_i = f_{i2}(f_{i1}(pa_i) + n_i) \tag{8.16}$$

to the data, the scale of the noise terms as well as f_{i1} is arbitrary, since f_{i2} is also to be estimated. Consequently, unlike for the nonlinear additive noise model, in the PNL causal model context, it is not meaningful to talk about the total entropy of the noise terms (see condition (iii)). However, as shown in Zhang and Hyvärinen (2009b), when fitting the PNL causal model with a hypothetical DAG causal structure to the data, we still have the equivalence between conditions (i) and (iii).

The next question is how to estimate a functional causal model for more than two variables in practice. one approach is to use exhaustive search: for all possible causal orderings, fit functional causal models for all hypothetical effects separately and then do model checking by testing for independence between the estimated noise and the corresponding hypothetical causes. However, note that the complexity of this procedure increases superexponentially along with the number of variables. Smarter approaches are thus needed.

The aforementioned theorem suggests a simpler two-step method to find the causal structure implied by the PNL causal model. We use here the relationship between mutual independence of the noise terms and the causal Markov condition combined with the independence between each noise term and its associated parents. One first uses the constraint-based approach (Spirtes *et al.*, 2001), (Pearl, 2000) to find the Markov equivalent class from conditional independence relationships given by some nonparametric conditional independence tests; for instance, one can adopt the kernel-based test (Zhang *et al.*, 2011). This approach first finds the skeleton of the causal graph by removing the edge between a pair of variables, if there exists some subset of variables (including the empty set) given that they are conditionally independent. It then uses the orientation rules to find the causal directions of some edges. The PNL causal model is then used to identify the causal directions that cannot be determined in the first step: for each DAG contained in the equivalent class, we estimate the noise terms and determine whether this causal structure is plausible by examining whether the disturbance in each variable x_i is independent of the parents of x_i. Consequently, one avoids the exhaustive search over all possible causal structures and high-dimensional statistical tests of mutual independence of all noise terms.

8.5 REMARK ON GENERAL NONLINEAR CAUSAL MODELS

We have discussed several functional causal models, namely, the linear model, nonlinear additive noise model, and PNL causal model. Now let us discuss the possibility of doing causal discovery with the general form of functional causal models. A functional causal model represents the effect y as a function of the direct causes x and some unmeasurable noise (Pearl, 2000):

$$y = f(x, n; \theta_1) \tag{8.17}$$

where n is the noise that is assumed to be independent of x, the function $f \in \mathcal{F}$ explains how y is generated from x, $\mathcal{F}$ is an appropriately constrained functional class, and θ_1 is the parameter set involved in f. We assume that the transformation from

(x, n) to (x, y) is invertible, such that n can be uniquely recovered from the observed variables x and y.

In the functional causal model (8.17), the noise term is assumed to be independent of the cause. If for the reverse direction, one cannot find a noise term that is independent of the hypothetical cause (which is y), then we can determine the true causal direction or distinguish cause from effect. As discussed earlier, in general, this is the case for the PNL causal model, as well as for the linear and nonlinear models with additive noise. Unfortunately, this is not the case if we do not impose any constraint on the function f.

As discussed in Hyvärinen and Pajunen (1999), given *any* two random variables x and y with continuous support, no matter how they are related, one can always construct another variable, denoted by $\mathring{n}$, which is statistically independent of x. In Zhang *et al.* (2016), the class of functions to produce such an independent variable $\mathring{n}$ (or called independent noise term in our causal discovery context) was given, and it was shown that this procedure is invertible: y is a function of x and $\mathring{n}$.

This is also the case for the hypothetical causal direction $y \rightarrow x$: we can also always represent x as a function of y and an independent noise term, if the functional form is not properly constrained. That is, any two variables would be symmetric according to the functional causal model, if f is not constrained. Therefore, in order for the functional causal models to be useful to determine the causal direction, we have to introduce certain constraints on the function f such that the independence condition on noise and hypothetical cause holds for only one direction. Examples of such constraints include the linear model, the nonlinear additive noise model, and the PNL causal model discussed earlier. As we have already seen, under appropriate assumptions, the constraints of additive noise and the PNL data-generating model serve such a goal.

8.6 SOME EMPIRICAL RESULTS

Various nonlinear functional causal models have been used to distinguish cause from effect on the cause-effect pairs available at `http://webdav.tuebingen.mpg.de/cause-effect/`. They consist of different data pairs, for which the causal direction is believed to be known, for testing the performance of causal detection algorithms. They are from different scientific disciplines including climate analysis, finance, and computer science. Here let us summarize the results reported in Zhang *et al.* (2016). The exploited approaches include the PNL causal model estimated by mutual information minimization with nonlinear functions represented by MLPs (Zhang and Hyvärinen, 2009b), denoted by PNL-MLP for short, the PNL causal model estimated by warped Gaussian processes with Gaussian noise, denoted by PNL-WGP-Gaussian, the PNL causal model estimated by warped Gaussian processes with MoG noise, denoted by PNL-WGP-MoG, the additive noise model estimated by Gaussian process regression (Hoyer *et al.*, 2009), denoted by ANM, the approach based on the Gaussian process prior on the function f (Mooij *et al.*, 2010), denoted by GPI, and IGCI (Janzing *et al.*, 2012). The data set consists of 77 data pairs. To reduce computational load, we used at most 500 points for each cause-effect pair: if the original data set consists of more than 500 points, we randomly sampled 500 points from them; otherwise, we simply used the original data set.

The accuracy of different methods (in terms of the percentage of correctly discovered causal directions) is reported as follows:

PNL-MLP: 70%
PNL-WGP-Gaussian: 67%
PNL-WGP-MoG: 76%
ANM: 63%
GPI: 72%
IGCI: 73%

One can see that all results are better than chance, illustrating the effectivity of using functional causal models to distinguish cause from effect. Here PNL-WGP-MoG gives the best performance among these methods.

8.7 DISCUSSION AND CONCLUSION

We have given a survey of functional causal models that enable us to fully identify the causal structure from observational data. We focused on the two-variable case, where the task is to distinguish cause from effect. We have reviewed the linear non-Gaussian causal model, nonlinear additive noise model, and the post-nonlinear causal model, listed from the most to the least restrictive. We addressed the identifiability of the causal direction: for those three models, in the generic case, the backward direction does not admit an independent noise term, and as a consequence, it is possible to distinguish cause from effect. We have also briefly discussed the procedure to achieve so, which consists of fitting the functional model and performing an independence test between the estimated noise and the hypothetical cause. For nonlinear additive noise models, we have also presented a likelihood-ratio-based approach to determining the causal direction.

There are some open problems along this line of research. First, one can consider functional causal models as a way to represent the conditional distribution of the effect, given the cause. Can we then find hints as to the causal direction directly from the data distribution? Or, in other words, can we find a general way to characterize the causal asymmetry directly in terms of certain properties of the data distribution? An attempt to do so is to make use of the so-called exogeneity property of a causally sufficient causal system (Zhang *et al.*, 2015).

Secondly, note that nonlinear functional causal models are usually intransitive. That is, if both causal processes $x_1 \to x_2$ and $x_2 \to x_3$ admit a particular type of functional causal model, the process $x_1 \to x_3$ does not necessarily follow the same model. (Linear models are transitive.) This could be a potential issue of functional-causal-model-based causal discovery: it may fail to discover indirect causal relations. (Here, by direct causal relations, we mean the causal relations in which only a single noise variable is involved.) On the other hand, this may be a benefit of using functional causal models for causal discovery, in that it is possible to detect the existence of the causal intermediate variable and further recover it. But how to do so is currently unclear.

In this chapter, we were concerned with causal discovery in the continuous case. In the discrete case, if one knows precisely what model class generated the effect from cause, which, for instance, may be the noisy AND or noisy XOR gate, then under mild conditions, the causal direction can be easily seen from the data distribution. Consider binary variables and take the noisy AND gate as the causal process. Then the probability of the effect variable taking value 1 is smaller than (or equal to, if the noise only takes value 1) that for the cause variable. However, generally speaking, if the precise model class of the causal process is unknown, it is difficult to recover the causal direction from observed data in the discrete case, especially when the cardinality of the variables is small. As an illustration, consider the situation where the causal process first generates continuous data and discretizes such data to produce the observed discrete ones. It is then not surprising that certain properties of the causal process are lost due to discretization, making causal discovery more difficult.

Finally, developing efficient methods for causal discovery of more than two variables based on functional causal models is an important step toward solving large-scale real-world causal analysis problems in various domains including neuroscience and biology. To make causal discovery computationally efficient, one may have to limit the complexity of the causal structure, say, limit the number of direct causes of each variable. Even so, a smart optimization procedure instead of exhaustive search is still missing in the literature.

The package for estimating the post-nonlinear causal model and causal direction identification based on this model is available at `http://webdav.tuebingen.mpg.de/causality/CauseOrEffect_NICA.rar` (with nonlinear functions represented by MLPs) or `http://people.tuebingen.mpg.de/kzhang/warpedGP.zip` (estimated by warped Gaussian processes with mixture-of-Gaussian noise).

REFERENCES

Dodge, Y. and Rousson, V. (2000) Direction dependence in a regression line. *Communications in Statistics: Theory and Methods*, **29**, 1957–1972.

Dodge, Y. and Rousson, V. (2001) On asymmetric properties of the correlation coefficient in the regression setting. *The American Statistician*, **55**, 51–54.

Dodge, Y. and Rousson, V. (2016) Recent developments on the direction of a regression line, in *Statistics and Causality: Methods for Applied Empirical Research* (eds W. Wiedermann and A. von Eye), John Wiley & Sons, Inc.

Gretton, A., Bousquet, O., Smola, A., and Schölkopf, B. (2005) Measuring statistical dependence with Hilbert-Schmidt norms, in *Algorithmic Learning Theory: 16th International Conference* (eds S. Jain, H. Simon, and E. Tomita), Springer-Verlag, Berlin, pp. 63–78.

Hoyer, P.O., Janzing, D., Mooij, J.M., Peters, J., and Schölkopf, B. (2009) Nonlinear causal discovery with additive noise models, in *Advances in Neural Information Processing Systems 21*, Vancouver, BC, Canada, pp. 689–696.

Hyvärinen, A. (1998) New approximations of differential entropy for independent component analysis and projection pursuit, in *Advances in Neural Information Processing Systems 10*, MIT Press, Cambridge, MA, pp. 273–279.

Hyvärinen, A. and Pajunen, P. (1999) Nonlinear independent component analysis: existence and uniqueness results. *Neural Networks*, **12** (3), 429–439.

Hyvärinen, A. and Smith, S. (2013) Pairwise likelihood ratios for estimation of non-Gaussian structural equation models. *Journal of Machine Learning Research*, **14**, 111–152.

Janzing, D., Mooij, J., Zhang, K., Lemeire, J., Zscheischler, J., Daniuvsis, P., Steudel, B., and Schölkopf, B. (2012) Information-geometric approach to inferring causal directions. *Artificial Intelligence*, **182**, 1–31.

Karvanen, J. and Koivunen, V. (2002) Blind separation methods based on Pearson system and its extensions. *Signal Processing*, **82**, 663–573.

Lin, J. (1998) Factorizing multivariate function classes, in *Advances in Neural Information Processing Systems 10*, MIT Press, Cambridge, MA, pp. 563–569.

Mooij, J., Stegle, O., Janzing, D., Zhang, K., and Schölkopf, B. (2010) Probabilistic latent variable models for distinguishing between cause and effect, in Advances in Neural Information Processing Systems 23 (NIPS 2010), Curran, NY.

Pearl, J. (2000) *Causality: Models, Reasoning, and Inference*, Cambridge University Press, Cambridge.

Peters, J., Mooij, J., Janzing, D., and Schölkopf, B. (2011) Identifiability of causal graphs using functional models, in Proceedings of UAI 2011, pp. 589–598.

Pham, D. and Garat, P. (1997) Blind separation of mixture of independent sources through a quasi-maximum likelihood approach. *IEEE Transactions on Signal Processing*, **45** (7), 1712–1725.

Rasmussen, C. and Williams, C. (2006) *Gaussian Processes for Machine Learning*, MIT Press, Cambridge, MA.

Shimizu, S., Hoyer, P., Hyvärinen, A., and Kerminen, A. (2006) A linear non-Gaussian acyclic model for causal discovery. *Journal of Machine Learning Research*, **7**, 2003–2030.

Spirtes, P., Glymour, C., and Scheines, R. (2001) *Causation, Prediction, and Search*, MIT Press, Cambridge, MA.

Taleb, A. and Jutten, C. (1999) Source separation in post-nonlinear mixtures. *IEEE Transactions on Signal Processing*, **47** (10), 2807–2820.

Wiedermann, W. and Hagmann, M. (2015) Asymmetric properties of the Pearson correlation coefficient: correlation as the negative association between linear regression residuals. *Communications in Statistics: Theory and Methods*, in press.

Zhang, K. and Chan, L. (2006) Extensions of ICA for causality discovery in the Hong Kong stock market, in Proceedings of the 13th International Conference on Neural Information Processing (ICONIP 2006).

Zhang, K. and Hyvärinen, A. (2009a) Causality discovery with additive disturbances: An information-theoretical perspective, in Proceedings of European Conference on Machine Learning and Principles and Practice of Knowledge Discovery in Databases (ECML PKDD) 2009, Bled, Slovenia.

Zhang, K. and Hyvärinen, A. (2009b) On the identifiability of the post-nonlinear causal model, in Proceedings of the 25th Conference on Uncertainty in Artificial Intelligence, Montreal, Canada.

Zhang, K. and Hyvärinen, A. (2010) Distinguishing causes from effects using nonlinear acyclic causal models, in JMLR Workshop and Conference Proceedings, vol. 6, pp. 157–164.

Zhang, K., Peters, J., and Janzing, D., and Schölkopf, B. (2011) Kernel-based conditional independence test and application in causal discovery, in Proceedings of the 27th Conference on Uncertainty in Artificial Intelligence (UAI 2011), Barcelona, Spain.

Zhang, K., Wang, Z., Zhang, J., and Schölkopf, B. (2016) On estimation of functional causal models: general results and application to the post-nonlinear causal model. *ACM Transactions on Intelligent Systems and Technology (TIST)*, vol. **7** (2), ACM, New York.

Zhang, K., Zhang, J., and Schölkopf, B. (2015) Distinguishing cause from effect based on exogeneity, in Proceedings of the 15th Conference on Theoretical Aspects of Rationality and Knowledge (TARK 2015).

PART III

GRANGER CAUSALITY AND LONGITUDINAL DATA MODELING

9

ALTERNATIVE FORMS OF GRANGER CAUSALITY, HETEROGENEITY, AND NONSTATIONARITY

Peter C. M. Molenaar and Lawrence L. Lo

Quantitative Developmental Systems Methodology, Department of Human Development and Family Studies, The Pennsylvania State University, University Park, PA, USA

9.1 INTRODUCTION

Granger causality testing is an increasingly popular approach to determine causal relations among dynamic processes. It originated within econometrics (Granger, 1969), but now also has important applications especially in biophysics (e.g., Valdes-Sosa *et al.*, 2011) and neuroimaging (e.g., Goebel *et al.*, 2003). Perhaps part of the popularity of Granger causality testing is due to its straightforward implementation that in essence is similar to the implementation underlying causal inference from data obtained with cross-lagged experimental designs (Rogosa, 1980). The implementation of Granger causality testing can be summarized in its barest form as follows. Suppose a bivariate process $z(t)$ with univariate components $x(t)$ and $y(t)$. A causal time series model (to be explained shortly) is fitted to the observations of $z(t)$, including lagged cross-relations between $x(t)$ and $y(t-u)$, $u > 0$, and lagged cross-relations between $y(t)$ and $x(t-u)$, $u > 0$. If the lagged cross-relations between

Statistics and Causality: Methods for Applied Empirical Research, First Edition.
Wolfgang Wiedermann and Alexander von Eye.

$x(t)$ and $y(t - u)$ are substantial whereas the lagged cross-relations between $y(t)$ and $x(t - u)$ are negligible, then $y(t)$ is considered to be a Granger cause of $x(t)$. Alternatively, if the lagged cross-relations between $y(t)$ and $x(t - u)$ are substantial while the lagged cross-relations between $x(t)$ and $y(t - u)$ are negligible, then $x(t)$ is considered to be a Granger cause of $y(t)$.

This simple implementation of Granger causality testing will be generalized in various important ways. In particular, alternative ways to carry out Granger causality testing are considered, using frequency-domain representations of causal time series models of p-variate processes, where $p > 0$ is a finite but arbitrary natural number. Also generalizations to nonstationary and heterogeneous processes (to be explained in due course) are considered, including reference to an innovative and extremely successful way to obtain a valid common time series model for a given set of heterogeneous replications. But the main focus will be on an important theoretical ambiguity in current Granger causality testing for which (to the best of our knowledge) no solution has been proposed yet. The details of this ambiguity will be discussed at some length, after which a new data-driven solution is presented. It is expected that this solution may considerably change the current practices of Granger causality testing.

9.2 SOME INITIAL REMARKS ON THE LOGIC OF GRANGER CAUSALITY TESTING

Before embarking upon the main issues that predominantly are of a statistical nature, it may be helpful to first clarify somewhat the logical reasoning underlying Granger causality testing. The rationale why the implementation of Granger causality testing described in the previous section indeed can result in substantial conclusions about causal relations between $x(t)$ and $y(t)$ is based on the following counterfactual reasoning. The basic assumptions are that the occurrence of a cause happens prior to its effects. Therefore, *if* variation in $x(t)$ is a sufficient cause of variation in $y(t)$, *then* at least some of the lagged cross-relations between $y(t)$ and $x(t - u)$, $u > 0$, will be substantial because these cross-relations are compatible with the assumed cause-effect timeline (presumed effects of $x(t - u)$, $u > 0$, on $y(t)$ occur at later times t). In schema: if A and P then Q, where

A: the occurrence of a cause happens prior to its effects;

P: variation in $x(t)$ is a sufficient cause of variation in $y(t)$;

Q: some of the lagged cross-relations between $y(t)$ and $x(t - u)$, $u > 0$, are substantial.

Of course, a similar counterfactual reasoning scheme results with the alternative specifications:

P′: variation in $y(t)$ is a sufficient cause of variation in $x(t)$;

Q′: some of the lagged cross-relations between $x(t)$ and $y(t - u)$, $u > 0$, are substantial.

This yields

if A and P′ then Q′.

Representing the rationale underlying Granger causality testing in this way immediately makes clear that it involves a logical fallacy. In empirical applications it is Q or Q′ that is affirmed. Suppose that Q is affirmed. Then the Granger causality test implies affirmation of P. In schema:

if A and P then Q; Q; hence P.

This argument is logically invalid, called affirmation of the consequent. It only would be valid if P is the only sufficient condition for Q, but this assumption rarely if ever is true. To yield a valid argument in general, it therefore has to be assumed that all possible causes of variation in $x(t)$ and $y(t)$ are explicitly included in the time series model underlying Granger causality testing. To the extent that this is practically impossible, the results obtained in Granger causality testing may not be valid.

In an interesting monograph (Kleinberg, 2013, section 2.4.2) summarizes philosophical critique of bivariate Granger causality testing, including references to the pertinent literature. She acknowledges that p-variate Granger causality testing, where p is a sufficiently large integer, avoids many of the criticisms of its bivariate analogue, but mentions practical difficulties in dealing with causal dynamic time series modeling of high-dimensional time series. It will be argued that our data-driven approach to be introduced below can alleviate these practical difficulties to a substantial degree. Kleinberg also presents an interesting alternative approach based on temporal logic and compares this in simulation studies with bivariate Granger causality testing (Kleinberg, 2013, chapter 7). It should be noted that only standard Granger causality testing in the time domain is considered; not the alternative time-frequency domain analogues that are the main focus of this chapter.

9.3 PRELIMINARY INTRODUCTION TO TIME SERIES ANALYSIS

To reiterate, Granger causality testing is based on fitting causal time series models to observations of multivariate dynamic processes. This section defines some basic concepts of time series analysis that will figure in what follows. For a more complete discussion, the reader is referred to, for instance, Brillinger (1975), Gardiner (1983), and Grigoriou (2002). Schelter *et al.* (2006) contains many contributions directly relevant to Granger causality testing.

Henceforth, a time series is conceived of as a stochastic dynamic process; the distinctions between the two are unimportant for our purposes. The definition of a time series is simple: it is a collection of random variables indexed by time. Hence for our purposes, sufficient description of a time series is in terms of its finite-dimensional distributions on the product space of time-indexed random variables making up the process. This definition goes back to Kolmogorov, whose so-called extension theorem provides the analytical bridge between cylinder sets of finite-dimensional distributions and the existence of an underlying stochastic dynamic process (Kolmogorov, 1933). Wiener (1930) gives an alternative functional definition that can be proven to be equivalent to Kolmogorov's (cf. Brillinger, 1975, section 2.11).

In what follows, the following simplifying assumptions are made. Time is conceived of as discrete with equal intervals between consecutive time points. The finite-dimensional distributions characterizing a stochastic dynamic process will be considered to be Gaussian. In addition, only weakly stationary time series models (to be defined shortly) will be considered. These assumptions are made to ease the presentation but can be dropped.

The following notational conventions are used. Manifest variables are denoted by Roman letters and latent variables by Greek letters. Matrices are denoted by boldface uppercase letters and vectors by boldface lowercase letters. Vectors are column vectors; if y is a (column) vector then y', where the apostrophe denotes transposition, is the corresponding row vector. No notational distinction is made between fixed and random variables; this distinction is always made within the text.

Let $\mathbf{y}(t) = [y_1(t), y_2(t), \ldots, y_p(t)]'$ be a p-variate time series, $p \geq 1$. The mean of $\mathbf{y}(t)$ at each time point t is $\mathrm{E}[\mathbf{y}(t)] = \boldsymbol{\mu}(t)$, where E[.] stands for the expectation operator. Considered as function of t, $\boldsymbol{\mu}(t)$ denotes the p-variate mean function (trend) of $\mathbf{y}(t)$. If $\boldsymbol{\mu}(t) = \boldsymbol{\mu}$, that is, if the mean function is constant in time, then $\mathbf{y}(t)$ is said to have a stationary mean function. The sequential covariance of $\mathbf{y}(t)$ between each distinct pair of time points t_1 and t_2 is defined as $\boldsymbol{\Sigma}(t_1, t_2) = cov[\mathbf{y}(t_1), \mathbf{y}(t_2)']$, where cov[.] stands for covariance. Considered as function of two-dimensional time, that is, for $\forall t_1, \forall t_2, \boldsymbol{\Sigma}(t_1, t_2)$, denotes the (p, p)-variate covariance function of $\mathbf{y}(t)$. If $\boldsymbol{\Sigma}(t_1, t_2)$ only depends on the relative time difference, called lag, $t_1 - t_2 = u$, that is, $\boldsymbol{\Sigma}(t_1, t_2) = \boldsymbol{\Sigma}(t_1 - t_2) = \boldsymbol{\Sigma}(u), \forall u$, then $\mathbf{y}(t)$ has stationary covariance function depending only on lag u. If both the mean function and covariance function of $\mathbf{y}(t)$ are stationary, then $\mathbf{y}(t)$ is called a weakly stationary p-variate time series. Hence, the statistical characterization of a weakly stationary p-variate series consists of the specification of its p-variate mean level $\boldsymbol{\mu}$ and the sequence of (p, p)-dimensional covariance matrices $\boldsymbol{\Sigma}(u)$ specifying the sequential dependencies at all lags u. Because Gaussian series are completely characterized by the first two moment functions, weakly stationary Gaussian series are also strongly stationary in the sense that all their finite-dimensional distributions are invariant under time translations.

Granger causality testing almost always is based on the class of linear time series models for weakly stationary dynamic processes called vector autoregressive moving-average (VARMA) models. The VARMA(a, b) model for a p-variate series $y(t)$ is defined by

$$\mathbf{y}(t) = \boldsymbol{\mu} + \boldsymbol{\Phi}_1\mathbf{y}(t-1) + \ldots + \boldsymbol{\Phi}_a\mathbf{y}(t-a) + \varepsilon(t) + \boldsymbol{\Theta}_1\varepsilon(t-1) + \ldots + \boldsymbol{\Theta}_b\varepsilon(t-b) \tag{9.1}$$

where a and b are the natural numbers indicating, respectively, the order of the autoregressive component and the order of the moving average component. In what follows, we will, to ease the presentation, assume that the mean function equals zero: $\boldsymbol{\mu} = 0$.

The zero-mean p-variate $\varepsilon(t)$ process is the so-called white noise, implying that it lacks any sequential dependencies. Hence, the covariance function of $\varepsilon(t)$ is $cov[\varepsilon(t+u), \varepsilon(t)'] = \delta(u)\boldsymbol{\Sigma}_{\varepsilon}$, where the Kronecker delta equals 1 if $u = 0$ and equals 0 otherwise. The sequence of (p, p)-dimensional matrices $\boldsymbol{\Phi}_k, k = 1, \ldots, p$, contain along the diagonals the lagged autoregressive coefficients, while the off-diagonal elements are the lagged cross-regression coefficients. The sequence of (p, p)-dimensional matrices $\boldsymbol{\Phi}_k, k = 1, \ldots, q$, contains the coefficients of the lagged relationships among the elements of $\varepsilon(t)$.

Equation 9.1 is called a causal time series model because $\boldsymbol{y}(t)$ only depends upon previous realizations $\boldsymbol{y}(t-k), k > 0$, as well as on contemporaneous and previous realizations of $\varepsilon(t)$. Noncausal VARMA models also include dependencies of $\boldsymbol{y}(t)$ on future realizations $\boldsymbol{y}(t+k), k > 0$, as well as future realizations of $\varepsilon(t)$. Noncausal models, also called physically unrealizable models, are relevant in the mathematical statistical theory underlying time series analysis. For instance, the original version of the Wold decomposition theorem includes a noncausal VARMA (Wold, 1954). Aggregation of causal VARMA models may also yield a noncausal model for the average (cf. Forni and Lippi, 1997). But for our purposes, the most important observation is that causal models may become noncausal after statistical analysis in the frequency domain and consequent inverse discrete Fourier transformation of the results thus obtained (cf. Molenaar, 1987). Hinich (1984) presents an alternative frequency domain approach involving Hilbert transforms that guarantees causal models after inverse transformation.

Hannan and Deistler (1988) present a thorough discussion of the class of VARMA models and its relationship with the class of state space models (to be defined below). For our present purposes, it suffices to only consider the subclass of vector autoregressive models (VARs): *VAR(a)* = VARMA$(a, 0)$, because Granger causality testing almost always is based on VARs. Appealing to the so-called decomposition theorem of Wold (1954), it turns out that any weakly stationary process can be approximated to any degree by a *VAR(a)* if its order a is chosen to be sufficiently large.

Introducing the backshift operator B defined by $B\boldsymbol{y}(t) = \boldsymbol{y}(t-1)$, each *VAR(a)* can be represented as

$$\boldsymbol{y}(t) = \boldsymbol{\Phi}_1\boldsymbol{y}(t-1) + \ \ldots \ + \boldsymbol{\Phi}_a \ \boldsymbol{y}(t-a) + \varepsilon(t) = \boldsymbol{\Phi}_1 B\boldsymbol{y}(t) + \ \ldots \ + \boldsymbol{\Phi}_a B^a \boldsymbol{y}(t) + \varepsilon(t)$$

Introducing the polynomial in B defined by $\boldsymbol{\Phi}(B, a) = \boldsymbol{I} - \boldsymbol{\Phi}_1 B - \ \ldots \ - \boldsymbol{\Phi}_a B^a$, where $\boldsymbol{I}$ denotes the (p, p)-dimensional identity matrix, the *VAR(a)* can now be compactly written as $\boldsymbol{\Phi}(B, a)\boldsymbol{y}(t) = \varepsilon(t)$. The backshift operator fulfills a role akin to the discrete Laplace transform or z-transform (Papoulis, 1977), where z is in general a complex number. Keeping only the imaginary part of z, hence substituting $B = \exp\ [-2\pi \boldsymbol{i}\omega_k]$, where $i = \sqrt{-1}$ and $k = 0, 1, \ldots$, the discrete Fourier transformed representation of a *VAR(a)* at each frequency ω_k becomes

$$\boldsymbol{\Phi}(\omega_k)\boldsymbol{y}(\omega_k) = \varepsilon(\omega_k) \tag{9.2}$$

where $\boldsymbol{\Phi}(\exp\ [-2\pi \boldsymbol{i}\omega_k], a)$ is concisely written as $\boldsymbol{\Phi}(\omega_k)$; $\boldsymbol{y}(\omega_k)$ and $\varepsilon(\omega_k)$ are, respectively, the discrete Fourier transforms of $\boldsymbol{y}(t)$ and $\varepsilon(t)$ (e.g., Brillinger, 1975). For a finite stretch of time series data $\mathbf{y}(t)$, $t = 0, 1, \ldots, T-1$, the frequency ω_k in the discrete Fourier transform is defined as $\omega_k = k/T, k = 0, 1, \ldots, T-1$.

Expression (9.2) is the basis for the representation of the spectral density matrices associated with a VAR. The latter are obtained by taking the inverse of $\mathbf{\Phi}(\omega_k)$: $\mathbf{\Gamma}(\omega_k) = \mathbf{\Phi}(\omega_k)^{-1}$. Then the spectral density matrix $cov[\mathbf{y}(\omega_k), \mathbf{y}(\omega_k)^*] = \mathbf{\Psi}(\omega_k)$, where * denotes the complex conjugated transpose, is proportional to

$$\mathbf{\Psi}(\omega_k) \propto \mathbf{\Gamma}(\omega_k)\mathbf{\Sigma}_\varepsilon \mathbf{\Gamma}(\omega_k)^* \tag{9.3}$$

where $\mathbf{\Sigma}_\varepsilon$ is the full covariance matrix of the white process noise $\varepsilon(t)$. For each frequency $\omega_k \neq 0,\ 1/2$ the (p, p)-dimensional complex-valued spectral density matrix $\mathbf{\Psi}(\omega_k)$ is Hermitian with real values (the autospectra) along the diagonal and complex values (the cross-spectra) off-diagonal.

For weakly stationary time series, there exists an invertible 1–1 relation between its covariance function and its spectral density matrices. This implies that statistical analysis of a weakly stationary series can be based on the covariance function (so-called analysis in the time domain) or, equivalently, on the spectral density matrices (so-called analysis in the frequency domain). In what follows, Granger causality testing will be considered both in the time and frequency domains, but mainly in the latter because the spectral density matrices associated with weakly stationary series observed at $t = 0, 1, \ldots, T-1$ are asymptotically (i.e., if T increases indefinitely) independently complex-Wishart distributed (Brillinger, 1975, chapter 7). This unique feature of the complex exponentials spanning the frequency domain considerably eases statistical analysis because spectral density matrices at distinct frequencies ω_k, ω_m, $k \neq m$, can be analyzed independently of each other. In contrast, analysis in the time domain requires simultaneous consideration of the covariance function at all lags.

9.4 OVERVIEW OF GRANGER CAUSALITY TESTING IN THE TIME DOMAIN

Only weakly stationary time series are considered; generalization to nonstationary series is discussed in later sections. Also, it is assumed that a Granger cause precedes its effect by at least one time step. Therefore, contemporaneous Granger causality (Lütkepohl, 2007, section 2.3.1) will not be the prime focus, although it is addressed in what follows.

To convey the main ideas, we consider the special situation that the simple VAR(1) holds for a bivariate series: $\mathbf{y}(t) = \mathbf{\Phi}y(t-1) + \varepsilon(t)$, where $\mathbf{y}(t) = [y_1(t), y_2(t)]'$. This model can be expanded as

$$\begin{aligned} y_1(t) &= \phi_{11}y_1(t-1) + \phi_{12}y_2(t-1) + \varepsilon_1(t) \\ y_2(t) &= \phi_{21}y_1(t-1) + \phi_{22}y_2(t-1) + \varepsilon_2(t) \end{aligned} \tag{9.4}$$

Following Lütkepohl (2007, section 2.3), Granger causality testing in this simple model reduces to testing whether ϕ_{12} differs significantly from zero and ϕ_{21} does not, which would indicate that $y_2(t)$ is a Granger cause of $y_1(t)$. Or, alternatively, testing whether ϕ_{21} differs significantly from zero and ϕ_{12} does not, which would indicate that $y_1(t)$ is a Granger cause of $y_2(t)$.

Generalization from a VAR(1) to a VAR(p) for bivariate series is straightforward:

$$\begin{aligned} y_1(t) &= \phi(1)_{11}y_1(t-1) + \phi(1)_{12}y_2(t-1) + \ \dots \ + \phi(p)_{11}y_1(t-p) \\ &\quad + \phi(p)_{12}y_2(t-p) + \varepsilon_1(t) \\ y_2(t) &= \phi(1)_{21}y_1(t-1) + \phi(1)_{22}y_2(t-1) + \ \dots \ + \phi(p)_{21}y_1(t-p) \\ &\quad + \phi(p)_{22}y_2(t-p) + \varepsilon_2(t) \end{aligned} \tag{9.5}$$

It now is more practical not to test whether each of $\phi(k)_{12}$ or $\phi(k)_{21}, k = 1, \dots, p$, differ significantly from zero, but instead follow a more indirect approach. For instance, to determine whether in (9.5) $y_2(t)$ is a Granger cause of $y_1(t)$, the following steps are carried out. First fit (Eq. 9.5) to the bivariate series $\mathbf{y}(t)$ at hand. Then fit univariate AR(q) and AR(r) models to $y_1(t)$ and $y_2(t)$, respectively, where the orders q and r may differ from the order p in (9.5)

$$\begin{aligned} y_1(t) &= \phi(1)y_1(t-1) + \ \dots \ + \phi(q)y_1(t-q) + \zeta(t) \\ y_2(t) &= \gamma(1)y_2(t-1) + \ \dots \ + \gamma(r)y_2(t-r) + \xi(t) \end{aligned} \tag{9.6}$$

Let $\mathbf{\Sigma}_\varepsilon$ be the (2, 2)-dimensional covariance matrix of the white process noise $\varepsilon(t)$ in (9.5). Then the total interdependence F_{y_1,y_2} between $y_1(t)$ and $y_2(t)$ can be computed as (cf. Wen *et al.*, 2013):

$$F_{y_1,y_2} = \ln \frac{var[\zeta(t)]var[\xi(t)]}{\det[\mathbf{\Sigma}_\varepsilon]} \tag{9.7}$$

where ln is the natural logarithm, det[.] denotes the determinant and $var[\zeta(t)]$, $var[\xi(t)]$ denote, respectively, the variances of the white process noise $\zeta(t)$ and $\xi(t)$ in (9.6). Geweke (1982) shows that F_{y_1,y_2} can be decomposed as

$$F_{y_1,y_2} = F_{y_1\rightarrow y_2} + F_{y_2\rightarrow y_1} + F_{y_1*y_2} \tag{9.8}$$

where $F_{y_1\rightarrow y_2} = \ln\{var[\zeta(t)]/var[\varepsilon_1(t)]\}$, $var[\varepsilon_1(t)]$ being the (1,1) element of Σ; $F_{y_2\rightarrow y_1} = \ln\{var[\xi(t)]/var[\varepsilon_2(t)]\}$, $var[\varepsilon_2(t)]$ being the (2, 2) element of Σ_ε; $F_{y_1*y_2} = \ln\{var[\varepsilon_1(t)]var[\varepsilon_2(t)]/\det[\mathbf{\Sigma}_\varepsilon]\}$. If $F_{y_1\rightarrow y_2}$ is large positive and $F_{y_2\rightarrow y_1}$ small (close to zero) then $y_1(t)$ is a Granger cause of $y_2(t)$; if $F_{y_2\rightarrow y_1}$ is large positive and $F_{y_1\rightarrow y_2}$ small then $y_2(t)$ is a Granger cause of $y_1(t)$.

Generalization to p-variate time series, where $p > 2$, is unproblematic (see Wen *et al.*, 2013, for details). Basically, two nonoverlapping subseries are selected from the p-variate time series $\mathbf{y}(t) = [y_1(t), \dots, y_p(t)]'$: the p_1-variate series $\boldsymbol{y}_1(t)$ consisting of p_1 component series of $\boldsymbol{y}(t)$ and the p_2-variate series $\boldsymbol{y}_2(t)$ consisting of p_2 component series of $\boldsymbol{y}(t)$. For instance, for $p = 6$, $p_1 = 2$, and $p_2 = 3$: $\boldsymbol{y}_1(t) = [y_1(t), y_3(t)]$ and $\boldsymbol{y}_2(t) = [y_2(t), y_5(t), y_6(t)]'$. Then $\boldsymbol{y}_1(t)$ and $\boldsymbol{y}_2(t)$ are treated in the same way as the univariate series $y_1(t)$ and $y_2(t)$, that is, appropriate multivariate generalizations of (9.5)–(9.8) above are determined. For instance in the computation of $F_{y_1 \to y_2}$ and $F_{y_2 \to y_1}$ variances are replaced by determinants of (p_1, p_1)- and (p_2, p_2)-dimensional covariance matrices of the white process noise concerned (see Barnett and Seth, 2014).

In their influential paper introducing Granger causality testing in fMRI brain imaging research, Goebel *et al.* (2003) present a simulation example of a bivariate VAR(1) such as Equation 9.4 above in which $\phi_{11} = -0.8454$, $\phi_{12} = 0$, $\phi_{21} = -0.5$, and $\phi_{22} = -0.8454$. The (2, 2)-dimensional covariance matrix Σ of the white process noise is taken to be diagonal with $var[\varepsilon_1(t)] = var[\varepsilon_2(t)] = 0.2853$. Because $\phi_{12} = 0$ and $\phi_{21} = -0.5$ it is clear that $y_1(t)$ is a Granger cause of $y_2(t)$. Based on a simulated series of length $T = 1000$, it is found that $F_{y_1 \to y_2} = 0.5076$ and $F_{y_2 \to y_1} = 0.0007$.

The most important question regarding Granger causality testing in the time domain based on $F_{y_1 \to y_2}$ and $F_{y_2 \to y_1}$, where $\boldsymbol{y}_1(t)$ and $\boldsymbol{y}_2(t)$ are, respectively, nonoverlapping p_1- and p_2-variate subseries of a given p-variate series $\boldsymbol{y}(t)$, concerns the consistency of results obtained in a series of such tests obtained with different subseries. If $p > 2$, then several distinct subsets of nonoverlapping p_1- and p_2-variate subseries $\boldsymbol{y}_1(t)$ and $\boldsymbol{y}_2(t)$ can be considered, and it is not guaranteed that the results thus obtained converge into a consistent overall causal network.

9.5 GRANGER CAUSALITY TESTING IN THE FREQUENCY DOMAIN

A decomposition analogous to (9.8) also applies in the frequency domain (cf. Wen *et al.*, 2013). The focus of Granger causality testing in the frequency domain, however, is on a set of measures derived from the spectrum $\boldsymbol{\Psi}(\omega_k)$ given by (9.3), $\boldsymbol{\Phi}(\omega_k)$ in the frequency domain representation (9.2) of VARs, and its inverse $\boldsymbol{\Gamma}(\omega_k) = \boldsymbol{\Phi}(\omega_k)^{-1}$. There are quite a few of such measures; see Schlögl and Supp (2006) for an overview. In what follows only a single measure will be considered, namely, the partial directed coherence (PDC), to be defined below. But first an important issue has to be addressed.

9.5.1 Two Equivalent Representations of a VAR(a)

Until now, it was understood that a *VAR*(a) for a weakly stationary p-variate series $\boldsymbol{y}(t)$ is defined by

$$\boldsymbol{y}(t) = \boldsymbol{\Phi}_1 \boldsymbol{y}(t-1) + \ \dots \ + \boldsymbol{\Phi}_a \boldsymbol{y}(t-a) + \varepsilon(t) \tag{9.9}$$

where the white process noise $\varepsilon(t)$ has full covariance matrix $\boldsymbol{\Sigma}_\varepsilon$. There exists, however, an equivalent *VAR*(a) representation defined by

$$\boldsymbol{y}(t) = \boldsymbol{\Xi}_0 \boldsymbol{y}(t) + \boldsymbol{\Xi}_1 \boldsymbol{y}(t-1) + \ \dots \ + \boldsymbol{\Xi}_a \boldsymbol{y}(t-a) + \boldsymbol{\upsilon}(t) \tag{9.10}$$

where the white process noise $\boldsymbol{\upsilon}(t)$ has *diagonal* covariance matrix $\boldsymbol{\Sigma}_\upsilon$. That is, the univariate component processes $\upsilon_k(t)$, $k = 1, \ldots, p$, are mutually uncorrelated.

In agreement with the pertinent literature representation (9.9) will be called the *standard VAR* while representation (9.10) will be referred to as a *structural VAR*. Each standard VAR (9.9) can be transformed into the equivalent structural VAR (9.10) by decomposing $\boldsymbol{\Sigma}_\varepsilon$ as

$$\boldsymbol{\Sigma}_\varepsilon = (\boldsymbol{I} - \boldsymbol{\Xi}_0)^{-1}\boldsymbol{\Sigma}_\upsilon(\boldsymbol{I} - \boldsymbol{\Xi}_0)^{-T} \tag{9.11}$$

where the superscript $-T$ denotes transposition and inversion. It is noted that (9.11) formally constitutes a Choleski decomposition. It then follows that the coefficient matrices associated with the lagged regressions in both representations are related by (cf. Gates *et al.*, 2010)

$$\boldsymbol{\Xi}_k = (\boldsymbol{I} - \boldsymbol{\Xi}_0)\boldsymbol{\Phi}_k \quad \text{and} \quad \boldsymbol{\Phi}_k = (\boldsymbol{I} - \boldsymbol{\Xi}_0)^{-1}\boldsymbol{\Xi}_k, \quad k = 1, \ldots, a \tag{9.12}$$

It is clear from (9.12) that the choice of representation (9.9) or (9.10), while equivalent, can yield entirely different results in Granger causality testing!

The important difference between the two equivalent representations is that in the structural VAR (9.10) the contemporaneous relations among the univariate component series $y_k(t)$, $k = 1, \ldots, p$, are explicitly represented by unidirectional regressions with coefficients in $\boldsymbol{\Xi}_0$, whereas in the standard VAR (9.9) these contemporaneous relations are a function of the contemporaneous associations among the univariate component process noise series $\varepsilon_k(t)$, $k = 1, \ldots, p$. One possible way to interpret this difference is the following: in the structural VAR (9.10) contemporaneous relations are endogenously generated while in the standard VAR (9.9) contemporaneous relations are exogenously generated. Additional aspects of the difference concerned are addressed in a later section.

9.5.2 Partial Directed Coherence (PDC) as a Frequency-Domain Index of Granger Causality

The focus is first on the standard *VAR*(a) (9.9) for a p-variate observed time series $\boldsymbol{y}(t)$. Discrete Fourier transformation of the associated polynomial in the backshift operator B, $\boldsymbol{\Phi}(B, a) = \boldsymbol{I} - \boldsymbol{\Phi}_1 B - \ldots - \boldsymbol{\Phi}_a B^a$, yields at each frequency ω_k the (p, p)-dimensional complex-valued matrix $\boldsymbol{\Phi}(\omega_k)$ in (9.2). Let $\boldsymbol{\Delta}_\varepsilon$ denote the diagonal (p, p)-dimensional matrix with the inverse diagonal elements of $\boldsymbol{\Sigma}_\varepsilon$ along the diagonal and zero elements off-diagonal. That is, the mth diagonal element of $\boldsymbol{\Delta}_\varepsilon$ is $1/var[\varepsilon_m(t)], m = 1, \ldots, p$. Then the so-called generalized partial directed coherence gPDC $\pi_{im}(\omega_k)$ between univariate component series $y_i(t)$ and univariate component series $y_m(t)$ is defined as follows (Baccalá and Sameshima, 2001; Faes and Nollo, 2010):

$$\pi_{im}(\omega_k) = \frac{(\Phi_{im}(\omega_k)/var[\varepsilon_i(t)]}{\sqrt{\boldsymbol{\phi}_{:m}(\omega_k)^* \boldsymbol{\Delta}_\varepsilon \boldsymbol{\phi}_{:m}(\omega_k)}} \tag{9.13}$$

where $\sqrt{.}$ denotes the square root and $\boldsymbol{\phi}_{:m}(\omega_k)$ is the mth column of $\boldsymbol{\Phi}(\omega_k)$. Because $\pi_{im}(\omega_k)$ is complex-valued, one usually takes the absolute value $|\pi_{im}(\omega_k)|$, where because of the normalization it always holds that $1 \geq \pi_{im}(\omega_k) \geq 0$.

The Granger causality test based on the gPDC is now straightforward. If for univariate component series $y_i(t)$ and $y_m(t)$ it holds that $|\pi_{im}(\omega_k)|$ is large, whereas $|\pi_{mi}(\omega_k)|$ is about zero, then $y_m(t)$ is a Granger cause of $y_i(t)$ at frequency ω_k. For a p-variate series $\boldsymbol{y}(t)$ application of the gPDC-based Granger causality test, therefore, involves $p(p-1)/2$ of such pairwise comparisons at each frequency.

Next the gPDC analogue for the equivalent structural *VAR*(a) (9.10) is considered. Define the following polynomial in the backshift operator B: $\boldsymbol{\Psi}[B,a] = \boldsymbol{I} - \boldsymbol{\Xi}_1 B - \ldots - \boldsymbol{\Xi}_a B^a$. The discrete Fourier transform at each frequency ω_k of $\boldsymbol{\Psi}[B,a)$ is denoted by the (p,p)-dimensional complex-valued matrix $\boldsymbol{\Psi}(\omega_k)$. Let $\boldsymbol{\Delta}_\upsilon$ denote the inverse of the diagonal covariance matrix $\boldsymbol{\Sigma}_\upsilon$. Then the so-called instantaneous partial directed coherence (iPDC) $\chi_{im}(\omega_k)$ between univariate component series $y_i(t)$ and univariate component series $y_m(t)$ at each frequency ω_k is defined by (Faes and Nollo, 2010):

$$\chi_{im}(\omega_k) = \frac{(\Psi_{im}(\omega_k)/var[\upsilon_i(t)])}{\sqrt{\boldsymbol{\psi}_{:m}(\omega_k)^* \boldsymbol{\Delta}_\upsilon \boldsymbol{\psi}_{:m}(\omega_k)}} \tag{9.14}$$

where $\boldsymbol{\psi}_{:m}(\omega_k)$ is the mth column of $\boldsymbol{\Psi}(\omega_k)$. Because $\chi_{im}(\omega_k)$ is complex-valued, one usually takes the absolute value $|\chi_{im}(\omega_k)|$, where because of the normalization it always holds that $1 \geq |\chi_{im}(\omega_k)| \geq 0$.

9.5.3 Some Preliminary Comments

Before presenting illustrations of the application of the gPDC and iPDC, the following comments should be made. First, note that in the definitions of gPDC and iPDC in, respectively, (9.13) and (9.14) the numerator and denominator of each ratio are scaled. In (9.13), the scaling is based on the diagonal elements (variances) of the covariance matrix $\boldsymbol{\Sigma}_\varepsilon$ of the white process noise $\varepsilon(t)$ in (9.9), whereas in (9.14) the scaling is based on the diagonal elements of the covariance matrix $\boldsymbol{\Sigma}_\upsilon$ of the white process noise $\boldsymbol{\upsilon}(t)$ in (9.10). It is noted that analogues of the gPDC and iPDC have been considered without such scaling. Similar scaled and unscaled analogues also exist for other frequency-domain measures for Granger causality. We focus on the scaled versions of gPDC and iPDC because these can be expected to be more robust against differences in variability of univariate component series $y_i(t)$, $i = 1, \ldots, p$. Second, it is noted that in defining the gPDC analogue for the equivalent structural *VAR*(a) (9.10) the polynomial in the backshift operator B could have been defined differently, namely as $\boldsymbol{\Omega}[B,a) = \boldsymbol{\Xi}_0 - \boldsymbol{\Xi}_1 B - \ldots - \boldsymbol{\Xi}_a B^a$. In $\boldsymbol{\Omega}[B,a)$ the first element is $\boldsymbol{\Xi}_0$, containing the directed contemporaneous relations among the univariate component series $y_i(t)$, $i = 1, \ldots, p$. In contrast, in $\boldsymbol{\Psi}[B,a)$ underlying (9.14) the first element is the identity matrix while the elements at lags $k > 0$ are equal to those of $\boldsymbol{\Omega}[B,a)$. Hence the iPDC $\chi_{im}(\omega_k)$ based on $\boldsymbol{\Psi}[B,a)$ only quantifies lagged Granger causality. Faes and Nollo (2010) call the analogous PDC index based on $\boldsymbol{\Omega}[B,a)$ the extended PDC because it quantifies both contemporaneous and lagged Granger causality. We will not further

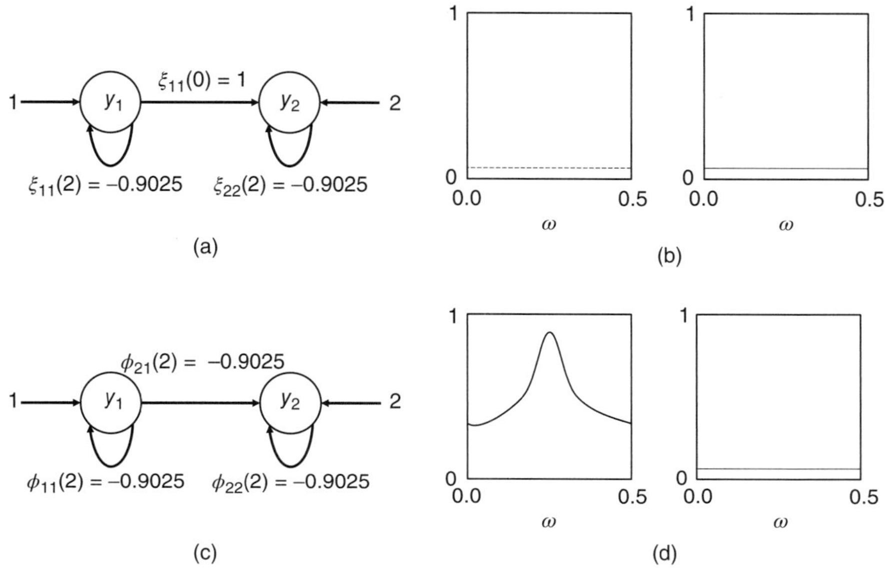

Figure 9.1 Representation of the model described in Equation 9.15. Time-domain representation of the model with instantaneous effects (a) and corresponding frequency domain representation (b). Time-domain representation of the equivalent VAR(2) model (c) with corresponding frequency-domain representation (d).

consider the latter extended PDC. Instead, an alternative test for contemporaneous Granger causality is introduced in what follows.

9.5.4 Application to Simulated Data

Faes and Nollo (2010) apply the gPDC and iPDC to data simulated by means of the following structural VAR(2):

$$
\begin{aligned}
y_1(t) &= -0.9025 y_1(t-2) + v_1(t) \\
y_2(t) &= y_1(t) - 0.9025 y_2(t-1) + v_2(t)
\end{aligned}
\tag{9.15}
$$

This model is depicted in Figure 9.1(a) (Fig. 9.1 is redrawn after Fig. 1 in Faes and Nollo, 2010, p. 392). The iPDCs $\chi_{21}(\omega_k)$ and $\chi_{12}(\omega_k)$ are depicted in, respectively, the left-hand and right-hand panels of Figure 9.1(b). Clearly the iPDCs are zero across all frequencies, yielding the correct conclusion that there is no lagged Granger causality between $y_1(t)$ and $y_2(t)$. Figure 9.1(c) shows the equivalent standard VAR(2). It has a cross-lagged relation directed from $y_1(t)$ to $y_2(t)$, suggesting that $y_1(t)$ is a lagged Granger cause of $y_2(t)$. The gPDCs $\pi_{21}(\omega_k)$ and $\pi_{12}(\omega_k)$ are depicted in, respectively, the left-hand and right-hand panels of Figure 9.1(d). Comparison of both panels clearly suggests that $y_1(t)$ is a lagged Granger cause of $y_2(t)$.

These results show that *if* the data are generated by a structural VAR *then* application of the gPDC based on a standard VAR yields incorrect results. It can similarly be shown that *if* the data are generated by a standard VAR *then* application of the iPDC based on a structural VAR yields incorrect results. Note that the latter implication is not addressed by Faes and Nollo (2010). We will return to the latter important observation in the following section.

9.6 A NEW DATA-DRIVEN SOLUTION TO GRANGER CAUSALITY TESTING

The application to simulated data suggests that the iPDC given by (9.14) is a valid and sensitive index to test for lagged Granger causality, that is, Granger causality not confounded by contemporaneous relations among the univariate component series of an observed vector-valued time series, *if the time series is generated by a structural VAR*. The iPDC is based on fitting the structural VAR (9.10) to the data, usually by means of a two-step procedure. In the first step, a standard *VAR*(a) (9.9) is fitted to the data. Then the estimated full covariance matrix of the white process noise thus obtained is subjected to the Choleski decomposition (9.11), yielding the coefficient matrix Ξ_0 of the equivalent structural VAR as well as the diagonal covariance matrix $\boldsymbol{\Sigma}_\upsilon$ of its white noise process. The remaining coefficient matrices Ξ_k, $k = 1, \ldots, a$, in the equivalent structural VAR then are obtained via (9.12).

There is, however, a major problem associated with this two-step procedure: the results obtained in the Choleski decomposition (9.11) depend upon the ordering of the p univariate component series $y_k(t)$, $k = 1, \ldots, p$ (cf. Lütkepohl, 2007, pp. 61–62). If this ordering is permuted, the results obtained in the Choleski decomposition of the associated permuted covariance matrix of the white process noise also change in that a different Ξ_0 is obtained. Because the ordering of the univariate component series in a vector-valued observed time series is arbitrary, this dependence of the coefficient matrices in the structural VAR obtained by means of the two-step procedure on the particular ordering chosen is unacceptable.

We have developed an alternative approach to fit structural VARs to the data in which the Choleski decomposition in the two-step procedure is not needed. This alternative approach, described in Gates *et al.* (2010), consists of rewriting the structural VAR as a structural equation model (SEM), called the unified SEM (uSEM), and using standard SEM software to fit the model directly to the data. Gates and Molenaar (2012) present the results of a large-scale simulation study validating this alternative approach. Henceforth, we will refer to this alternative approach as fitting a uSEM. The computer program implementing the fit of a uSEM can be freely accessed at `http://www.nitrc.org/projects/gimme/`.

While the availability of an alternative approach that sidesteps the problems with the usual two-step approach to fit structural VARs is in itself important, it also opens up a possibility to solve the problem of equivalent VAR representations in Granger causality testing. Remember that the standard VAR (9.9) and the structural VAR (9.10) are equivalent, but that the results of Granger causality testing based on either

(9.9) or (9.10) are different. In particular, as shown in the previous section, the results of applying the gPDC (9.13) based on the standard VAR or the iPDC (9.14) based on the structural VAR are different. Only if one knows the true stochastic dynamic system that generated the data (as one does in a simulation study) can one make the correct choice between the two equivalent representations. But for empirical data, the true dynamic model almost always is unknown. Our new approach to fit a uSEM, however, can be adapted to let the data decide what the appropriate representation is.

9.6.1 Fitting a uSEM

To describe the main steps in fitting a uSEM, the structural VAR(1) is taken as the most elementary example (see Gates *et al.*, 2010, for a complete description). Equation 9.10 then reduces to $\boldsymbol{y}(t) = \Xi_0\boldsymbol{y}(t) + \Xi_1\boldsymbol{y}(t-1) + \boldsymbol{v}(t)$. To fit this model to the data, the model is expanded for two consecutive time points in the following way:

$$\begin{aligned} \boldsymbol{y}(t) &= \boldsymbol{v}^{\circ}(t) \\ \boldsymbol{y}(t+1) &= \Xi_0\boldsymbol{y}(t+1) + \Xi_1\boldsymbol{y}(t) + \boldsymbol{v}(t+1) \end{aligned} \tag{9.16}$$

The first equation in (9.15) is simply an initial condition, where $\boldsymbol{v}^{\circ}(t)$ denotes p-variate pseudonoise equal to $\boldsymbol{y}(t)$. The $(2p, 2p)$-dimensional covariance matrix implied by (9.15) is a so-called block-Toeplitz matrix (Molenaar, 1985), the estimate of which serves as input to the SEM software. To fit a structural VAR/uSEM, the following steps are carried out:

(a) Fit (9.15) to the block-Toeplitz input covariance matrix, but fix all coefficients in Ξ_0 and Ξ_1 at zero. Only free up the diagonal elements of the (p, p)-dimensional subcovariance matrix of $\boldsymbol{v}(t+1)$. Free up all nonredundant elements of the (p, p)-dimensional subcovariance matrix of $\boldsymbol{v}^{\circ}(t)$. The goodness of fit of this model almost always will be bad. If so, go to step b. If not, stop.

(b) Perform the Lagrange multiplier tests of all fixed parameters in Ξ_0 and Ξ_1; each such value is asymptotically chi square distributed with one degree of freedom. Select the fixed parameter having the largest value of the Lagrange multiplier test.

(c) If this value is significant, free up this parameter. Refit the model thus extended and go to step b. If not, stop.

9.6.2 Extending the Fit of a uSEM

Note that in the procedure described in the previous section, the (p, p)-dimensional subcovariance matrix of $\boldsymbol{v}(t+1)$ is not changed. It is a diagonal covariance matrix during all steps of the procedure. Hence, the final result of the above procedure is a structural VAR/uSEM.

However, step b in the above procedure can be changed in such a way that the data decide whether a structural VAR, a standard VAR, or a hybrid of structural and standard VAR (hybrid VAR) is the appropriate representation. To accomplish this, step b is changed as follows:

(b′) Perform the Lagrange multiplier tests of all fixed parameters in Ξ_0, Ξ_1 and the (p,p)-dimensional subcovariance matrix of $\boldsymbol{\upsilon}(t+1)$. Each such value is asymptotically chi square distributed with one degree of freedom. Select the fixed parameter having the largest value of the Lagrange multiplier test.

We claim that with the stepwise procedure a, b′, c the best-fitting representation is obtained, whether it is a standard VAR, a structural VAR, or a hybrid VAR. The new procedure will be referred to as fitting a hybrid VAR. If only parameters in Ξ_0 and Ξ_1 are freed up, the best-fitting representation is a structural VAR/uSEM. If only parameters in Ξ_1 and the (p,p)-dimension subcovariance matrix of $\boldsymbol{\upsilon}(t+1)$ are freed up, the best-fitting representation is a standard VAR. If parameters in Ξ_0, Ξ_1 and the (p,p)-dimensional subcovariance matrix of $\boldsymbol{\upsilon}(t+1)$ are freed up, the best-fitting representation is a hybrid VAR.

9.6.3 Application of the Hybrid VAR Fit to Simulated Data

The following hybrid VAR(1) is used to generate data:

$$
\begin{aligned}
y_1(t) &= 0.7y_1(t-1) + \eta_1(t) \\
y_2(t) &= 0.7y_1(t) + 0.7y_2(t-1) + \eta_2(t) \\
y_3(t) &= 0.7y_3(t-1) + \eta_3(t) \\
y_4(t) &= 0.7y_4(t-1) + \eta_4(t)
\end{aligned} \tag{9.17}
$$

where $var[\eta_1(t)] = var[\eta_2(t)] = var[\eta_3(t)] = var[\eta_4(t)] = 1$. Unit variances were used for simplicity of this demonstration. All Gaussian white process noise component series are uncorrelated, save for $\eta_3(t)$ and $\eta_4(t)$: $\text{cov}[\eta_3(t), \eta_4(t)] = .7$. Hence, the simulation model is a hybrid VAR(1) involving a directed contemporaneous relation from $y_1(t)$ to $y_2(t)$ as well as a nonzero off-diagonal element of the $(4,4)$-dimensional covariance matrix of $\boldsymbol{\eta}(t)$ representing a contemporaneous association among $\eta_3(t)$ and $\eta_4(t)$. The hybrid VAR simulation model is depicted in Figure 9.2. The data were generated by means of the Fortran program (calling subroutines from IMSL that is part of the Fortran compiler) presented in the Appendix section.

Application of the hybrid VAR fit procedure to a time series of length $T = 900$ generated according to this model correctly recovers the true (hybrid) model structure, while all estimated parameters are within 95% confidence intervals about their true values. For instance, the estimate of the directed contemporaneous relation from $y_1(t)$ to $y_2(t)$ (true value 0.7) is 0.7 with standard error s.e. = 0.03. The estimate of the covariance between $\eta_3(t)$ and $\eta_4(t)$ (true value 0.7) is 0.66 ([s.e. = 0.04). Figure 9.3 shows the recovered hybrid VAR model.

The fit of a standard VAR to the same data yields a goodness of fit that is comparable to that of a hybrid VAR, but the set of estimated parameters contains two spurious

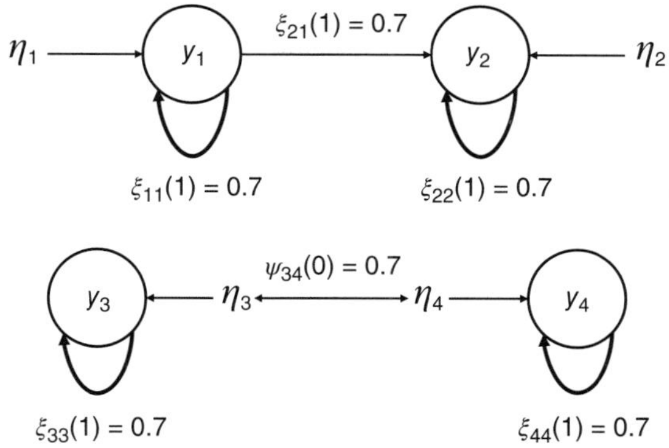

Figure 9.2 Time-domain representation of the hybrid VAR(1) model given by Equation 9.17.

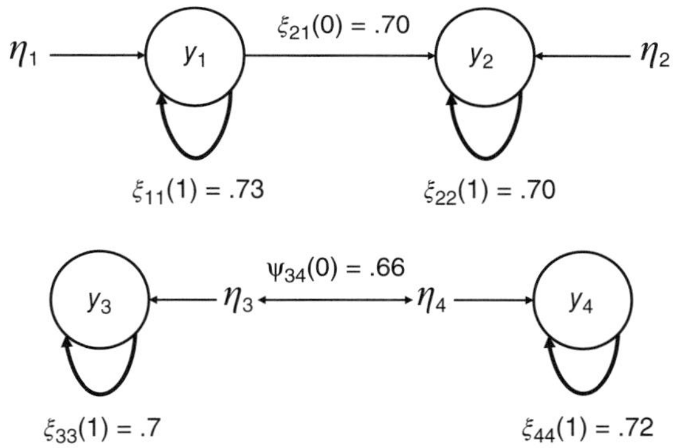

Figure 9.3 Hybrid VAR fit estimates for model given by Equation 9.17.

elements. While, like in the hybrid VAR fit, all autoregressive parameters are correctly recovered with values within 95% of the confidence intervals about their true values, there now also is a cross-lagged relationship in which $y_1(t-1)$ influences $y_2(t)$ with estimated value 0.5 (s.e. = 0.04). This significant cross-lagged relationship would incorrectly suggest that $y_1(t)$ is a lagged Granger cause of $y_2(t)$. Also, apart from the correct recovery of the covariance between $\eta_3(t)$ and $\eta_4(t)$ with estimated value 0.66 (s.e. = 0.04), there now is also a significant covariance between $\eta_2(t)$ and $\eta_1(t)$ with estimated value 0.7 (s.e. = 0.05). Figure 9.4 shows the estimated standard VAR.

Finally, the fit of a structural VAR (uSEM) to the same data again yields a goodness of fit that is comparable to that of a hybrid VAR, but the set of estimated parameters contains two spurious elements. Like in the hybrid VAR fit, all autoregressive parameters as well as the directed contemporaneous relation from $y_1(t)$ to $y_2(t)$ are correctly recovered with estimated values within 95% confidence interval about their true values. In addition, there is a contemporaneous directed relation from $y_4(t)$ to $y_3(t)$ with estimated value 0.69 (s.e. = 0.02). Obviously, this contemporaneous directed relation

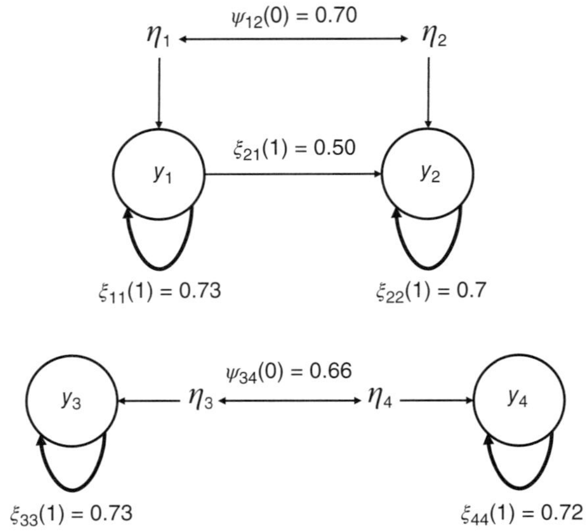

Figure 9.4 Standard VAR fit estimates for model given by Eq. 9.17.

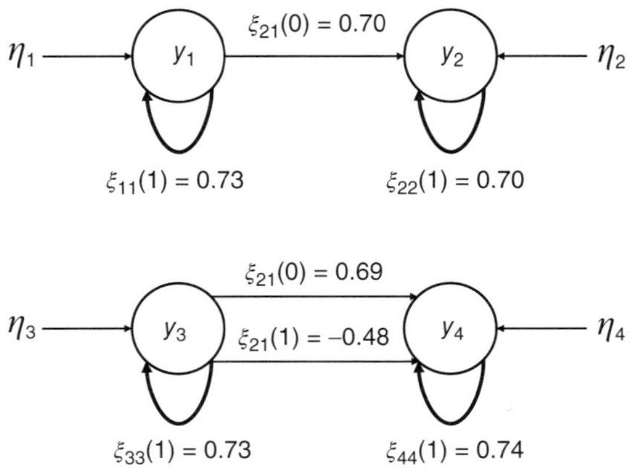

Figure 9.5 Structural VAR fit estimates for model given by Equation 9.17.

captures the contemporaneous covariance between $\eta_3(t)$ and $\eta_4(t)$. But there is also a significant directed lagged cross-relation from $y_4(t-1)$ to $y_3(t)$ with estimated value -0.48 (s.e. $= 0.03$). The latter spurious directed lagged relation suggests that $y_4(t)$ is a lagged Granger cause of $y_3(t)$. The estimated structural VAR is shown in Figure 9.5.

This initial illustration of the new hybrid VAR procedure is encouraging, showing its feasibility as well as the dangers involved in fitting a standard VAR or structural VAR when the true data generating mechanism is a hybrid VAR. These dangers generalize to Granger causality testing based on the fitted VAR models, either in the time or frequency domain. It is noted that the illustration is limited in important ways (one replication, one simulation model, etc.). What is needed is a large-scale simulation study in order to thoroughly investigate its statistical properties. This will be carried out in the near future.

9.7 EXTENSIONS TO NONSTATIONARY SERIES AND HETEROGENEOUS REPLICATIONS

Until now, it has been assumed that observed time series are weakly stationary. Also the focus has been on Granger causality testing for a single replication. In applied research, however, one often has to deal with multiple heterogeneous replications and/or nonstationary time series. In this section, some promising ways to deal with these are concisely discussed.

9.7.1 Heterogeneous Replications

The question to be addressed is how Granger causality testing can be carried out if heterogeneous replications are available. It is understood that heterogeneity implies that the time series data obtained with different cases in the set of replications obey different VARs. Of course, if each case is tested individually, then there is no problem. But it is supposed that the question to be addressed concerns how to carry out a Granger causality test that generalizes across the heterogeneous cases. For instance, Forni and Lippi (1997, Chapter 11) discuss some of the problems with Granger causality testing of aggregated heterogeneous data. Suffice it to state that aggregation, pooling, or concatenation (i.e., adding the series of the kth case at the end of the $(k-1)$th case, $k = 2, \ldots, N$, and treating the result as a single long series) is a bad idea because it generates spurious relations. Molenaar and Nesselroade (2014) present a simple example involving data that are simulated by means of a structural VAR for $N = 3$ replications. Two replications are homogeneous (same directed contemporaneous and lagged relations with identical parameter values) while the third replication is almost homogeneous, differing only in the direction of a single lagged relation. Yet the fit of a structural VAR to the pooled block-Toeplitz covariance matrix (cf. Molenaar, 1985) has several spurious relations that do not exist for any of the three replications.

An innovative alternative approach to arrive at a common structural VAR is called *Group Iterative Multiple Model Estimation* (GIMME; Gates and Molenaar, 2012) and is based on the following rationale. Suppose one has p-variate time series obtained with N possibly heterogeneous replications and, to ease the presentation, it can be assumed that each replication obeys a first-order structural VAR: $\mathbf{y}(t) = \Xi_0\mathbf{y}(t) + \Xi_1\mathbf{y}(t-1) + \mathbf{v}(t)$. Then GIMME first determines an instance of the structural VAR that has a common group structure across all N persons. That is, this so-called group model has exactly the same *pattern* of free parameters in Ξ_0 and Ξ_1 across all N replications, although the actual *values* of these parameters are allowed to differ arbitrarily between subjects. Hence in the first phase, GIMME determines a common dynamic network structure for the complete group of N replications, while allowing that the weights associated with each link are replication-specific.

GIMME determines the group model in an automatic data-driven way in which new parameters (directed links) in Ξ_0 and Ξ_1 are freed up sequentially, one by one, starting from a model that in default mode contains no free parameters. It is an option to start the sequential search with a model containing already free parameters in Ξ_0 and/or Ξ_1, where these parameters have been selected based on (theoretical) a priori knowledge. It also is an option to forbid that a subset of selected parameters in Ξ_0 and/or Ξ_1 is freed up during the sequential search, where this subset again is determined based on a priori knowledge. At each step in the sequential model search, it is determined which one of the subset of eligible parameters that have not yet been freed up will maximally improve the likelihood across the N subjects. If this improvement of the likelihood is significant for at least a fixed proportion P of the N persons, then this parameter (directed link) is added to the common dynamic network structure, the replication-specific values of the parameter are estimated and the sequential search moves to the next step. If not, the next phase of GIMME, to be described shortly, starts. Based on extensive simulation studies, the proportion P of the N replications for which the increase in the likelihood ratio should be significant has been fixed at 75%.

In the second phase of GIMME, the group network structure determined in the first phase constitutes the starting model in a sequential model search for each replication separately. In this search for each replication $k \in \{1, 2, \ldots, N\}$, replication-specific directed links are added one by one until no additional replication-specific directed links can be found that significantly improve the likelihood ratio for this replication. GIMME has been validated in extensive simulation studies, showing superb performance (Gates and Molenaar, 2012). GIMME is freely accessible at `http://www.nitrc.org/projects/gimme/`.

The group model obtained with the current implementation of GIMME constitutes a structural VAR, but this will be extended to a common group hybrid VAR in the near future. The group model thus obtained can serve as the starting point for Granger causality testing with heterogeneous replications because it has an identical pattern of contemporaneous and lagged relations across all replications, although the coefficient weights are replication-specific. Consequently, for each of the replications, a Granger causality test can be carried out in a way that is directly comparable across the N replications. No applications of this innovative approach are available yet, but research concerned is in progress.

9.7.2 Nonstationary Series

Taking as example a first-order standard VAR $\boldsymbol{y}(t) = \boldsymbol{\Phi}_1 \boldsymbol{y}(t-1) + \varepsilon(t)$, there are various possible causes of nonstationarity. For instance, the coefficient matrix $\boldsymbol{\Phi}_1$ can be time-varying. Or the covariance matrix $\boldsymbol{\Sigma}_\varepsilon$ of the white process noise is time-varying. Another possible cause of nonstationarity is that one or more eigenvalues of $\boldsymbol{\Phi}_1$ have modulus equal to or larger than 1 (Lütkepohl, 2007, Chapter 2). Finally, although throughout this chapter it is assumed that the mean function is stationary and zero, it is of course possible that this mean function is time-varying and hence nonstationary. In what follows, the focus will be on zero-mean observed processes, the nonstationarity of which is caused by time-varying coefficient matrices such as $\mathbf{\Phi}_1$ in the standard VAR.

The basic model is the so-called state-space model (SSM) for an observed p-variate time series $\boldsymbol{y}(t)$. First, the SSM for weakly stationary series is considered (Molenaar, 1985):

$$\begin{aligned} \boldsymbol{y}(t) &= \mathbf{\Lambda}\boldsymbol{\eta}(t) + \boldsymbol{\zeta}(t) \\ \boldsymbol{\eta}(t) &= \boldsymbol{B}\boldsymbol{\eta}(t-1) + \varepsilon(t) \end{aligned} \tag{9.18}$$

where $\boldsymbol{\eta}(t)$ is a q-variate latent state process, $\boldsymbol{\zeta}(t)$ is a p-variate white noise process representing measurement error, $\varepsilon(t)$ is q-variate white process noise, $\mathbf{\Lambda}$ is a (p, q)-dimensional matrix of so-called loadings, and $\boldsymbol{B}$ is a (q, q)-dimensional matrix with regression coefficients. The first equation making up the SSM defined by 9.16 has the form of a common factor model in which $\boldsymbol{\eta}(t)$ denotes the latent factor series. The second equation is a VAR(1) for the latent factor series. The zero lag (p, p)-dimensional covariance matrix of $\boldsymbol{\zeta}(t)$ is $\mathbf{\Sigma}_\zeta$; it is a diagonal matrix because the measurement error has to obey conditional independence. The zero lag (q, q)-dimensional full covariance matrix of $\varepsilon(t)$ is $\mathbf{\Sigma}_\varepsilon$.

The assumption that the latent state process $\boldsymbol{\eta}(t)$ obeys a VAR(1) is not restrictive because each *VAR*(a) can be rewritten as an equivalent VAR(1) by extending the dimension of the latent state process (Lütkepohl, 2007, Chapter 2). Clearly, the standard VAR is a special case of (9.16) in which $\mathbf{\Lambda} = I$, the (p, p)-dimensional identity matrix, and $\boldsymbol{\zeta}(t)$ is absent. Although the structural VAR also can be expressed as a special case of (9.16), it involves extending the dimension of the latent state process as well as the introduction of intricate nonlinear constraints, which is why we will not consider it. Similar remarks apply to the hybrid VAR as special case of (9.16). Durbin and Koopman (2012) and Shumway and Stoffer (2013) are excellent introductions to SSMs.

The SSM with time-varying parameters (TV-SSM) is defined as

$$\begin{aligned} \boldsymbol{y}(t) &= \mathbf{\Lambda}[\boldsymbol{\theta}(t)]\boldsymbol{\eta}(t) + \boldsymbol{\zeta}(t) \\ \boldsymbol{\eta}(t+1) &= \boldsymbol{B}[\boldsymbol{\theta}(t)]\boldsymbol{\eta}(t) + \varepsilon(t+1) \\ \boldsymbol{\theta}(t+1) &= \boldsymbol{\theta}(t) + \xi(t+1) \end{aligned} \tag{9.19}$$

The first equation of (9.17) shows that loadings in $\mathbf{\Lambda}[\boldsymbol{\theta}(t)]$ depend upon a time-varying parameter-vector $\boldsymbol{\theta}(t)$ and hence can change in time. The second equation describes the time evolution of the state process $\boldsymbol{\eta}(t)$; the autoregressive weights in $\boldsymbol{B}[\boldsymbol{\theta}(t)]$ depend upon $\boldsymbol{\theta}(t)$ and therefore can also be arbitrarily time-varying. The third equation in (9.17) describes the time-dependent variation of the unknown parameters. The r-variate parameter vector process $\boldsymbol{\theta}(t)$ obeys a so-called random walk with Gaussian white process noise $\xi(t)$. Other dynamic models for the parameter vector process $\boldsymbol{\theta}(t)$ are available (e.g., higher order random walks or autoregressive models).

To fit the TV-SSM to the data a special and intricate estimation method is required. A beta version of the computer program implementing this estimation method can be obtained from the first author. The method yields estimated trajectories of each parameter across the whole observation interval. Because the estimation algorithm has been developed only recently, a limited number of applications of the TV-SSM have been reported until now. A variant of the model was first considered in Molenaar (1994). The first application of (9.17) was to father–stepson interactions and is presented in Molenaar *et al.* (2009). Wang *et al.* (2014) present an application to patient-specific treatment of diabetes.

The TV-SSM can be straightforwardly extended in several ways, including the effects of external input and bilinear terms. But (9.17) suffices for the present purposes. At each time $t = 1, \dots, T$, the TV-SSM is subjected to discrete Fourier transformation, yielding a time-varying sequence of (p, p)-dimensional spectral density matrices and hence time-varying sequences of PDCs. The three-dimensional array of PDC values as function of time and frequency for each pair of distinct univariate component series of $\boldsymbol{y}(t)$ then can be treated as before to decide about Granger causality. To the best of our knowledge no applications have been published yet.

9.8 DISCUSSION AND CONCLUSION

Each VAR has two equivalent representations: a standard VAR and a structural VAR. The choice of which of these two equivalent representations is used for lagged Granger causality testing has in general a large effect on the outcomes thus obtained. For instance, using the gPDC (based on standard VAR modeling) will in general yield different conclusions about lagged Granger causality than the iPDC (based on structural VAR modeling). This raises the all-important question: Which of the two equivalent representations should one choose in Granger causality testing of a given empirical data set? The pertinent literature is mostly silent about this question (e.g., Barnett and Seth, 2014).

Faes and Nollo (2010) are a positive exception in that they explicitly recommend using the structural *VAR*(a). They distinguish the gPDC (9.13), the iPDC (9.14), and the extended PDC based on $\boldsymbol{\Omega}[B, a) = \Xi_0 - \Xi_1 B - \dots - \Xi_a B^a$. Faes and Nollo (2010) correctly indicate that results of lagged Granger causality testing based on the gPDC and the PDC derived from $\boldsymbol{\Omega}[B, a)$ will in general yield different results. They recommend using the iPDC to detect pure lagged Granger causality. Faes and Nollo (2010) do not discuss the problem associated with the commonly used two-step procedure to estimate structural VARs, that is, its dependence upon the ordering of the univariate component series in $\boldsymbol{y}(t)$.

We present an innovative alternative approach in which the decision whether a standard VAR, a structural VAR, or a hybrid VAR best describes the stochastic dynamic system underlying a given observed time series, is determined in a data-driven way. The new approach is invariant under permutations of the univariate component series in $\boldsymbol{y}(t)$. In an application to data generated by a hybrid VAR(1), incorporating both a directed contemporaneous relation and a contemporaneous correlation among a pair of univariate components of the white process noise, the new approach correctly recovers the true model and therefore also can yield the correct results about lagged Granger causality. In contrast, applications of the standard VAR as well as the structural VAR yield incorrect conclusions in this respect.

Having determined whether a standard VAR, a structural VAR, or a hybrid VAR best describes the stochastic dynamic system underlying a given observed time series, Granger causality testing in the frequency domain then can proceed as follows. If the selected model is a standard VAR, the gPDC is the index of choice for testing lagged Granger causality while it also is concluded that contemporaneous Granger causality is absent. If the structural VAR is selected, then the iPDC is the index of choice for testing lagged Granger causality. Moreover, Ξ_0 constitutes a straightforward indication of the presence of contemporaneous Granger causality. If $\xi_{ik}(0)$ is substantial while $\xi_{ki}(0)$ is close to zero, then $y_k(t)$ is a contemporaneous Granger cause of $y_i(t)$. If the selected model is a hybrid VAR, then again the iPDC is the index of choice for testing lagged Granger causality and Ξ_0 is scrutinized again in the same way as for the structural VAR to test for contemporaneous Granger causality.

This completely data-driven approach has been developed only very recently. It has to be properly implemented and then subjected to large-scale simulation studies. All this is in progress. We are confident that the new approach will prove to be of considerable value for Granger causality testing in both the time and frequency domains, solving several ambiguities as indicated.

REFERENCES

Baccalá, L.A. and Sameshima, K. (2001) Partial directed coherence: a new concept in neural structure determination. *Biological Cybernetics*, **84** (6), 463–474.

Barnett, L. and Seth, A.K. (2014) The MVGC multivariate Granger causality toolbox: a new approach to Granger-causal inference. *Journal of Neuroscience Methods*, **223**, 50–68.

Brillinger, D.R. (1975) *Time Series: Data Analysis and Theory*, Holt, Rinehart & Winston, New York.

Durbin, J. and Koopman, S.J. (2012) *Time Series Analysis by State Space Methods*, 2nd edn, Oxford University Press, Oxford.

Faes, L. and Nollo, G. (2010) Extended causal modeling to assess partial directed coherence in multiple time series with significant instantaneous interactions. *Biological Cybernetics*, **103** (5), 387–400.

Forni, M. and Lippi, M. (1997) *Aggregation and the Microfoundations of Dynamic Macroeconomics*, Clarendon Press, Oxford.

Gardiner, C. (1983) *Handbook of Stochastic Methods for Physics, Chemistry and the Natural Sciences*, Springer-Verlag, Berlin.

Gates, K.M. and Molenaar, P. (2012) Group search algorithm recovers effective connectivity maps for individuals in homogeneous and heterogeneous samples. *Neuroimage*, **63** (1), 310–319.

Gates, K.M., Molenaar, P., Hillary, F.G., Ram, N., and Rovine, M.J. (2010) Automatic search for fMRI connectivity mapping: an alternative to Granger causality testing using formal equivalences among SEM path modeling, VAR, and unified SEM. *Neuroimage*, **50** (3), 1118–1125.

Geweke, J. (1982) Measurement of linear dependence and feedback between multiple time series. *Journal of the American Statistical Association*, **77** (378), 304–313.

Goebel, R., Roebroeck, A., Kim, D.S., and Formisano, E. (2003) Investigating directed cortical interactions in time-resolved fMRI data using vector autoregressive modeling and Granger causality mapping. *Magnetic Resonance Imaging*, **21** (10), 1251–1261.

Granger, C.W. (1969) Investigating causal relations by econometric models and cross-spectral methods. *Econometrica*, **37**, 424–438.

Grigoriou, M. (2002) *Stochastic Calculus: Applications in Science and Engineering*, Birkhäuser, Boston, MA.

Hannan, E.J. and Deistler, M. (1988) *The Statistical Theory of Linear Systems*, John Wiley & Sons, Inc., New York.

Hinich, M.J. (1984) Estimating the gain of a linear filter from noisy data, in *Time Series in the Frequency Domain*. Handbook of Statistics, vol. **3** (eds D. Brillinger and P. Krishnaiah), Elsevier, Amsterdam, pp. 157–168.

Kleinberg, S. (2013) *Causality, Probability, and Time*, Cambridge University Press, Cambridge.

Kolmogorov, A. (1933) *Grundbegriffe der Wahrscheinlichkeitsrechnung*, Springer-Verlag, Berlin.

Lütkepohl, H. (2007) *New Introduction to Multiple Time Series Analysis*, Springer-Verlag, Berlin.

Molenaar, P.C. (1985) A dynamic factor model for the analysis of multivariate time series. *Psychometrika*, **50** (2), 181–202.

Molenaar, P.C. (1987) Dynamic factor analysis in the frequency domain: causal modeling of multivariate psychophysiological time series. *Multivariate Behavioral Research*, **22** (3), 329–353.

Molenaar, P.C. (1994) Dynamic latent variable models in developmental psychology, in *Analysis of Latent Variables in Developmental Research* (eds A. von Eye and C. Clogg), Sage Publications, Newbury Park, CA, pp. 155–180.

Molenaar, P. and Nesselroade, J. (2014) Systems methods for developmental research, in *Handbook of Child Psychology and Developmental Science*, Theory and Method, vol. **1**, 7th edn (eds W.F. Overton and P.C.M. Molenaar), John Wiley & Sons, Inc., Hoboken, NY, in press.

Molenaar, P., Sinclair, K., Rovine, M., Ram, N., and Corneal, S. (2009) Analyzing developmental processes on an individual level using non-stationary time series modeling. *Developmental Psychology*, **45**, 260–271.

Papoulis, A. (1977) *Signal Analysis*, McGraw-Hill, New York.

Rogosa, D. (1980) A critique of cross-lagged correlation. *Psychological Bulletin*, **88** (2), 245–258.

Schelter, B., Winterhalder, M., and Timmer, J. (eds) (2006) *Handbook of Time Series Analysis: Recent Theoretical Developments and Applications*, Wiley-VCH Verlag GmbH, Weinheim.

Schlögl, A. and Supp, G. (2006) Analyzing event-related EEG data with multivariate autoregressive parameters, in *Event-Related Dynamics of Brain Oscillations*, Progress in Brain Research, vol. **159** (eds C. Neuper and W. Klimesch), Elsevier, Amsterdam, pp. 135–147.

Shumway, R.H. and Stoffer, D.S. (2013) *Time Series Analysis and its Applications: With R Examples*, 3rd edn, Springer-Verlag, New York.

Valdes-Sosa, P.A., Roebroeck, A., Daunizeau, J., and Friston, K. (2011) Effective connectivity: influence, causality and biophysical modeling. *Neuroimage*, **58** (2), 339–361.

Wang, Q., Molenaar, P., Harsh, S., Freeman, K., Xie, J., Gold, C., Rovine, M., and Ulbrecht, J. (2014) Personalized state-space modeling of glucose dynamics for type 1 diabetes using continuously monitored glucose, insulin dose, and meal intake an extended Kalman filter approach. *Journal of Diabetes Science and Technology*, **8** (2), 331–345.

Wen, X., Rangarajan, G., and Ding, M. (2013) Multivariate Granger causality: an estimation framework based on factorization of the spectral density matrix. *Philosophical Transactions of the Royal Society of London, Series A: Mathematical, Physical and Engineering Sciences*, **371** (1997), 20110 610.

Wiener, N. (1930) Generalized harmonic analysis. *Acta Mathematica*, **55** (1), 117–258.

Wold, H. (1954) *A Study in the Analysis of Stationary Time Series*, 2nd edn, Almquist and Wiksell Book Co., Uppsala .

APPENDIX

Source Code for Program Generating Data for Model Presented in Equation 9.17

```
      program gen
      parameter(my=9,mm=999)
      implicit double precision (a-h,o-z)
      implicit integer (i-n)
      dimension par(1),lagar(1),pma(1),lagma(1)
      dimension a(my),wi(my),w(my),y(my,mm)
      dimension be(my,my),aa(my,my), ps(my,my)
      dimension d(my,my),dd(my),ds(my,my)
      open(1,file='gpar.doc')
      open(2,file='svar.doc')
      read(1,*)ny,nt
      do 10 i=1,ny
 10   read(1,*)(aa(i,j),j=1,ny)
      do 11 i=1,ny
 11   read(1,*)(be(i,j),j=1,ny)
      do 12 i=1,ny
 12   read(1,*)(ps(i,j),j=1,ny)
      do 30 i=1,ny
      do 31 j=1,ny
 31   d(i,j)=0.d0
 30   d(i,i)=1.d0
      do 32 i=1,ny
      do 32 j=1,ny
 32   d(i,j)=d(i,j)-aa(i,j)
      call dlinrg(ny,d,my,aa,my)
 99   format(6(1x,f10.4))
      do 34 i=1,ny
      do 34 j=1,ny
      d(i,j)=0.d0
      do 34 k=1,ny
 34   d(i,j)=d(i,j)+aa(i,k)*be(k,j)
      call dchfac(ny,ps,my,tol,irank,ds,my)
      do 35 i=1,ny
      do 35 j=1,ny
 35   ps(i,j)=0.d0
      do 36 i=1,ny
      do 36 j=1,i
 36   ps(i,j)=ds(j,i)
      do 37 i=1,ny
 37   write(6,99)(ps(i,j),j=1,ny)
      do 38 i=1,ny
      do 38 j=1,ny
      ds(i,j)=0.d0
      do 38 k=1,ny
 38   ds(i,j)=ds(i,j)+aa(i,k)*ps(k,j)
      do 91 jj=2,nt+50
      lagar(1)=1
```

```
      lagma(1)=0
      par(1)=0.d0
      pma(1)=0.d0
      wi(1)=0.d0
      call drnarm(ny,0.d0,1,par,lagar,0,pma,lagma,0,1.d0,a,wi,w)
      do 92 i=1,ny
      dd(i)=0.d0
      do 92 j=1,ny
92    dd(i)=dd(i)+ds(i,j)*w(j)
      do 93 i=1,ny
      y(i,jj)=dd(i)
      do 93 j=1,ny
93    y(i,jj)=y(i,jj)+d(i,j)*y(j,jj-1)
91    continue
      do 22 i=51,nt+50
22    write(2,*)(y(j,i),j=1,ny)
      stop
      end
```

Example Input for Generating Data Following Equation 9.17

```
 4   900
 0   0   0   0
.7   0   0   0
 0   0   0   0
 0   0   0   0

.7   0   0   0
 0  .7   0   0
 0   0  .7   0
 0   0   0  .7

 1   0   0   0
 0   1   0   0
 0   0   1  .7
 0   0  .7   1
```

10

GRANGER MEETS RASCH: INVESTIGATING GRANGER CAUSATION WITH MULTIDIMENSIONAL LONGITUDINAL ITEM RESPONSE MODELS

INGRID KOLLER

Institute for Psychology, Department of Developmental and Educational Psychology, Alpen-Adria-Universität Klagenfurt, Klagenfurt, Austria

CLAUS H. CARSTENSEN

Psychology and Methods of Educational Research, University of Bamberg, Bamberg, Germany

WOLFGANG WIEDERMANN

Department of Educational, School & Counseling Psychology, College of Education, University of Missouri, Columbia, MO, USA

ALEXANDER VON EYE

Department of Psychology, Michigan State University, East Lansing, MI, USA

10.1 INTRODUCTION

Longitudinal data analysis is an important topic in the field of assessing educational trajectories. For example, the aim of the interdisciplinary National Educational Panel

Statistics and Causality: Methods for Applied Empirical Research, First Edition.
Wolfgang Wiedermann and Alexander von Eye.

Study (NEPS) in Germany is to assess individual educational processes from early childhood to late adulthood (Blossfeld *et al.*, 2011). For this purpose, several tests to assess competencies, such as reading, mathematics, and science, are constructed and administered in a longitudinal design.

The ambitious goal to assess competencies across the life span results in several methodological challenges, for example, to measure the same competence across several phases of the life span. Unidimensional measurement in time is addressed by complex instrument development and study designs and is evaluated by scaling the observed data using item response theory models. In NEPS, the Rasch model (RM; Rasch, 1960) and unidimensional extensions (e.g., the Partial Credit Model; Masters, 1982) are used for scaling competence tests (Pohl and Carstensen, 2013).

Further challenges arise through the question "Which further competencies or context variables affect modifiability of the target competence?" To answer this question, one approach could include a two-step procedure, that is, scaling the competencies of interest separately and subsequently estimating the influence of a competence X on the change in Y using regression analysis. But this approach has one disadvantage. Analyzing data using manifest variables (e. g., the raw score or person ability parameters) without considering the estimation error of the measurement model may lead to attenuated covariance estimates (e. g., von Davier *et al.*, 2009). Therefore, it is recommended to scale and model the data in one step. Here, latent associations are modeled while considering potential measurement error. In this chapter, we propose a one-step approach that combines multidimensional longitudinal item response models and Granger causation techniques. The chapter is structured as follows: First, we introduce the basics of Granger causation, followed by an introduction in the RM and its multidimensional extension. Then, multidimensional longitudinal item response models and the investigation of Granger causation with multidimensional longitudinal item response models are described. An empirical example on the topic of science knowledge of children is given. In the discussion section, we address potential limitations of the approach and outline potential future research.

10.2 GRANGER CAUSATION

The concept of Granger causation enables researchers to estimate the causal relations between two series of observations (Granger, 1969; see also, Lütkepohl, 2007; von Eye *et al.*, 2015, von Eye *et al.*, 2014a,b). In essence, Granger causality analysis is based on a prediction error approach, which is, from a philosophical perspective, deeply rooted in Hume's conceptualization of causality, that is, the cause must precede the effect (Hume, 1777/1975). More recently, von Eye and Wiedermann (2015) integrated Granger causation in mechanistic concepts of causality that also acknowledge the existence of contemporary causes. Let Ω_t be the information set containing all the relevant information up to time point t. Further, let $\sigma^2_{Y_t}(h|\Omega_t)$ be the prediction error variance of the h-step predictor of Y_t based on the information Ω_t. Then, a variable X_t is said to "Granger-cause" a variable Y_t, if $\sigma^2_{Y_t}(h|\Omega_t) < \sigma^2_{Y_t}(h|\Omega_t \backslash \{X_s|s < t\})$. Note that Granger (1969) original formulation

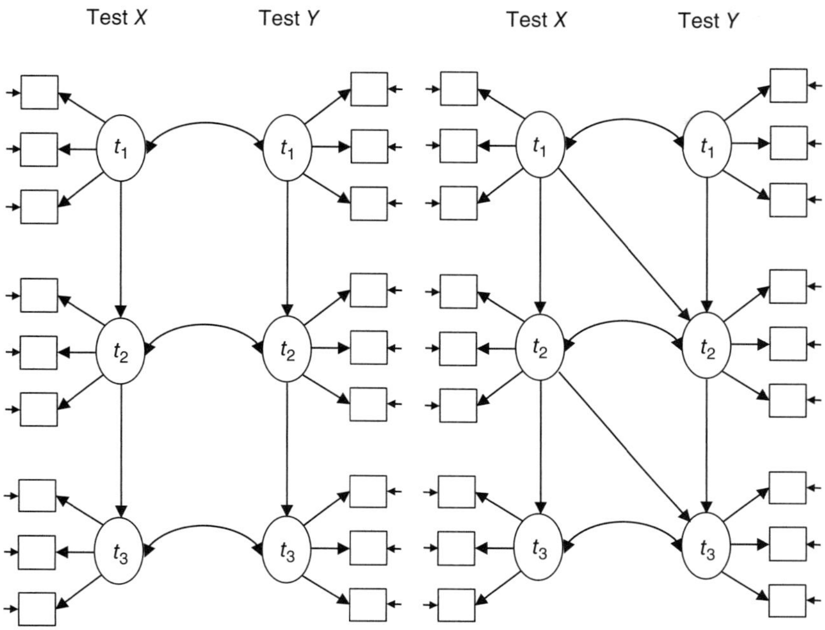

Figure 10.1 Path diagrams for two linear autoregressive models, the right panel assumes that X Granger-causes Y.

included contemporary causes. Thus, $\sigma^2_{Y_t}(h|\Omega_t\backslash\{X_s|s < t\})$ should be replaced with $\sigma^2_{Y_t}(h|\Omega_t\backslash\{X_s|s \leq t\})$. However, in practical applications researchers tend to ignore potential instantaneous effects because estimates of directional instantaneous effects are not uniquely identified without making strong assumptions concerning the direction of effects (von Eye and Wiedermann, 2015). Of course, the choice of the information set Ω_t is essential in Granger's conceptualization of causality. In practice, *all* relevant information up to time point t is not available. Thus, in the majority of cases, Ω_t reduces to the information in the past and present of the processes of interest. Then Granger causality statements simplify to the following: X_t is said to Granger-cause Y_t when lagged values of X_t provide significantly more information concerning future values of Y_t (see Fig. 10.1, right panel) than past values of Y_t alone (see Fig. 10.1, left panel).

Consider the following example of a first-order vector autoregressive process, VAR(1): $\boldsymbol{Y}(t) = \boldsymbol{\Phi Y}(t-1) + \boldsymbol{\varepsilon}(t)$ with $\boldsymbol{Y}(t) = [Y_t, X_t]'$, $\boldsymbol{Y}(t-1) = [Y_{t-1}, X_{t-1}]'$, and $\boldsymbol{\varepsilon}(t) = [\varepsilon_{1,t}, \varepsilon_{2,t}]'$, which expands to

$$\begin{aligned} Y_t &= \phi_{11} Y_{t-1} + \phi_{12} X_{t-1} + \varepsilon_{1,t} \\ X_t &= \phi_{21} Y_{t-1} + \phi_{22} X_{t-1} + \varepsilon_{2,t} \end{aligned} \tag{10.1}$$

In this simple case, Granger causality analysis involves testing the null hypotheses $H_0 : \phi_{12} = 0$ and $H_0 : \phi_{21} = 0$. When $H_0 : \phi_{12} = 0$ can be rejected and $H_0 : \phi_{21} = 0$ can be retained, one concludes that X_t Granger-causes Y_t. Alternatively, when $H_0 : \phi_{12} = 0$ is retained and $H_0 : \phi_{21} = 0$ is rejected, one has found empirical evidence that Y_t Granger-causes X_t.

Figure 10.1 depicts the principle of Granger causation in case of associated competencies (circles) together with the manifest indicators (rectangles). The latter constitute the measurement part of the model. Thus, the model involves three elements: (i) scaling the data using a measurement model, (ii) assessing modifiability of unidimensional competencies across time (autoregressive part), and (iii) assessing Granger causation in multidimensional functioning competencies through comparing the autoregressive model shown in the left panel of Figure 10.1 with the model in which cross-lagged effects are also considered (right panel of Figure 10.1).

10.3 THE RASCH MODEL

The Rasch model (RM) is a unidimensional item response model that offers several attractive mathematical properties (e. g., all items measure the same latent construct in the sense of unidimensionality) if the model holds for the data (see, e. g., Embretson and Reise, 2000; Fischer, 1995b; Molenaar, 2007). In the RM, the probability of solving an item $X_{vi} = 1$ given the ability θ of subject $v = 1, \ldots, V$ and the item difficulty β ($i = 1, \ldots, I$) is

$$P(X_{vi} = 1|\theta_v, \beta_i) = \frac{\exp(\theta_v - \beta_i)}{1 + \exp(\theta_v - \beta_i)} \tag{10.2}$$

The RM is a special case of the multidimensional random coefficient multinomial logit model (MRCML model; Adams *et al.*, 1997), which is a generalization of a wide class of Rasch type models, for example, the Partial Credit model (PCM; Masters, 1982) or the linear logistic test model (LLTM; Fischer, 1973). A detailed description and derivation of the multidimensional model is given in Adams *et al.* (1997). In the MRCML model, the probability of solving an item in category $k = 1, \ldots, K$ can be written as

$$P(\mathbf{X}_{vi} = 1; \mathbf{A}, \mathbf{B}, \boldsymbol{\beta}|\, \boldsymbol{\theta}) = \frac{\exp(\mathbf{b}_{ik}\boldsymbol{\theta} + \mathbf{a}'_{ik}\boldsymbol{\beta})}{\sum_{k=1}^{K_i} \exp(\mathbf{b}_{ik}\boldsymbol{\theta} + \mathbf{a}'_{ik}\boldsymbol{\beta})} \tag{10.3}$$

where $\boldsymbol{\beta} = 1, \ldots, I$ is the vector of item parameters, $\boldsymbol{\theta} = (\theta_1, \theta_2, \ldots, \theta_D)$ is the vector of person abilities across dimensions $d = 1, \ldots, D$, $\mathbf{A}$ is the design matrix with entries for each category k in item i, $\mathbf{A} = (a_{11}, a_{12}, \ldots, a_{1K_1}, a_{21}, \ldots, a_{2K_2}, \ldots, a_{IK_I})'$. The scoring matrix $\mathbf{B}$ includes $\mathbf{b}_{ik} = (b_{1k_1}, b_{1k_2}, \ldots, b_{iK_D})$, which are $D \times 1$ column vectors for each item i, category k, and dimension d. This term can be generalized to $\mathbf{B}_i = (b_{i1}, b_{i2}, \ldots, b_{iD})$, which is the scoring submatrix for each item i and dimension d, and $\mathbf{B} = (B'_1, B'_2, \ldots, B'_n)$, which is the scoring matrix for each item i. Considering,

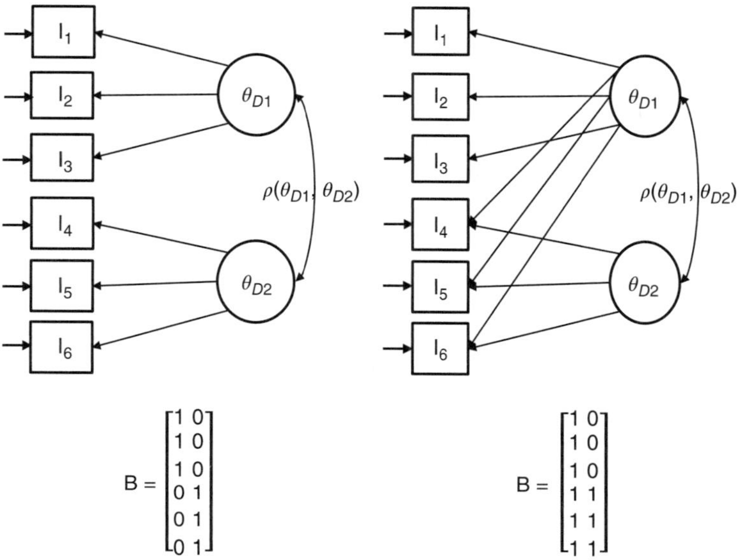

Figure 10.2 Path diagrams (upper panel) and the associated scoring matrices (lower panel) for the between item multidimensional model (left panel) and the within item multidimensional model (right panel).

for example, four items and two dimensions with two items per dimension, the design matrix (**A**) and the scoring matrix (**B**) can be written as

$$\mathbf{A} = \begin{bmatrix} 1 & 0 & 0 & 0 \\ 0 & 1 & 0 & 0 \\ 0 & 0 & 1 & 0 \\ 0 & 0 & 0 & 1 \end{bmatrix} \tag{10.4}$$

and

$$\mathbf{B} = \begin{bmatrix} 1 & 0 \\ 1 & 0 \\ 0 & 1 \\ 0 & 1 \end{bmatrix} \tag{10.5}$$

The design matrix defines the item parameter estimation, that is, one parameter for each item. The scoring matrix defines that the first two items are influenced by dimension one and the items three and four by dimension two.

In general, two types of multidimensionality models can be defined (see Adams *et al.*, 1997). First, if more than one dimension is included in item responses and each item is influenced by only one dimension, one refers to *between item multidimensionality* (see Fig. 10.2, left panel). Second, when a subset of items (or all items) is influenced by additional dimensions (see Fig. 10.2, right panel), one refers to *within multidimensionality*. Figure 10.2 shows the path diagrams (upper panel) and their associated scoring matrices (lower panel). The MRCML model is a generalized RM that allows modeling different types of multidimensionality including the evaluation of longitudinal data.

In the following section, we give a brief overview of longitudinal item response models and describe how to define the design and scoring matrices for the longitudinal MRCML model. After that, we show the definition of the model for the investigation of Granger causation.

10.4 LONGITUDINAL ITEM RESPONSE THEORY MODELS

Several longitudinal item response models are described in the literature (e. g., Bacci, 2012; Cho *et al.*, 2013; Fischer, 1976; Glück and Spiel, 1997, 2007; Hojitink, 2007; Hsieh *et al.*, 2013; von Davier *et al.*, 2011). Well-known models include, for example, the linear logistic test model for measuring change (LLTM; e. g., Fischer, 1974, 1989, 1995a), the linear logistic model with relaxed assumptions (LLRA; e. g., Fischer, 1974, 1989, 1995b), and associated hybrid forms of these models (e. g., Fischer and Ponocny-Seliger, 1998; Formann and Spiel, 1989). These models possess useful properties provided that certain preconditions are fulfilled (see, e. g., Koller *et al.*, 2015; Ponocny, 2002). For example, the models assume specific objectivity of change, which means only one change parameter is needed for all items. In other words, it is irrelevant which items are used for the investigation of change, and different changes are solely determined by different groups of subjects. Therefore, it is not possible to model associations between time points as well as individual differences in change, which hampers the application of these models to test hypotheses compatible with Granger causation.

Multidimensional models constitute a second class of models for measuring change. Two very popular models are the multidimensional Rasch model for learning and change (MRMLC; Embretson, 1991) and the multidimensional Rasch model for repeated testing (MRMRT; Andersen, 1985). These models offer the possibility to assess change on the individual level. In their classical formulation, the models are restricted to dichotomous items (an MRMLC for polytomous items is discussed in Fischer, 2001) and the RM has to hold within time points.

Comparatively less research has been done on multidimensional longitudinal item response models. Cho *et al.* (2013) discuss the analysis of change in multidimensional data using generalized explanatory item response models. However, as specified above, one can also use the MRCML model for the investigation of change (see Wang *et al.*, 1998). This class of models allows the analysis of dichotomous and polytomous items and enables researchers to assess more than one dimension or test within one time point. Further, as will be demonstrated, this class of models enables researchers (i) to formulate Granger causation hypotheses, (ii) to evaluate prespecified associations of latent variables, and (iii) to account for measurement error in a one-step approach. For these purposes, the MRCML model has to be combined with multidimensional models for measuring change. For a better understanding, the MRMLC and the MRMRT are introduced. Then we show how to define the model for

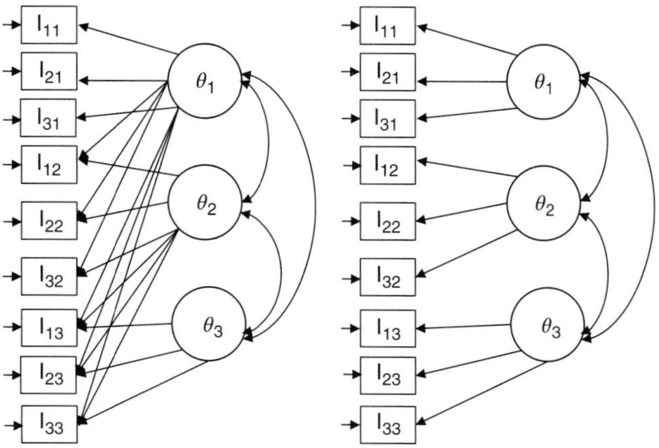

Figure 10.3 Path diagrams for Embretson's MRMLC (left panel) and for Andersen's MRMRT (right panel).

the investigation of change and further for empirically establishing Granger causation statements.

The MRMLC (Embretson, 1991) is a unidimensional multiparameter model (see Wang *et al.*, 1998) that combines the RM with a $\mathbf{T} \times \mathbf{D}$ matrix, where $\mathbf{T}$ are the time points ($t = 1, \ldots, T$) and $\mathbf{D}$ are the dimensions or abilities ($d = 1, \ldots, D$). The $\mathbf{T} \times \mathbf{D}$ matrix is a Wiener simplex with rows representing the time points and columns representing the dimensions, that is,

$$\mathbf{A} = \begin{bmatrix} 1 & 0 & 0 \\ 1 & 1 & 0 \\ 1 & 1 & 1 \end{bmatrix} \tag{10.6}$$

The MRMLC can be written as

$$P(X_{vi(t)} = 1|\boldsymbol{\theta}_v, \beta_i) = \frac{exp(\sum_{t=1}^{T} \theta_{vt} - \beta_i)}{1 + exp\ (\sum_{t=1}^{T} \theta_{vt} - \beta_i)} \tag{10.7}$$

where $\boldsymbol{\theta}_v$ is the vector of abilities with θ_{v1} being the initial ability at t_1 and $\theta_{v2}, \ldots, \theta_{vT}$ being the modifiabilities of further time points. For example, the ability at t_3 is given by $\boldsymbol{\theta}_{t3} = \theta_{t1} + \theta_{t2} + \theta_{t3}$. Item parameters β_i are also kept constant over time. The left panel of Figure 10.3 depicts the path diagram for three time points.

The second model is the MRMRT (Andersen, 1985; see Fig. 10.3, right panel), which is also a unidimensional multiparameter model. Here, the same items have to be administered across time points (see Wang *et al.*, 1998). The model equation can be written as

$$P(X_{vit} = 1|\theta_{vt}, \beta_i) = \frac{exp\ (\theta_{vt} - \beta_i)}{1 + exp(\theta_{vt} - \beta_i)}, \tag{10.8}$$

TABLE 10.1 Scoring Matrix B and Design Matrix A for the Joint MRMLC and MRCML Models (see also Wang *et al.*, 1998).

IT	*T*	*D*	Scoring Matrix **B**			Design matrix **A** (Saturated Model)								
1	1	1	1	0	0	−1	0	0	0	0	0	0	0	0
1	2	1	1	1	0	−1	0	0	−1	0	0	0	0	0
1	3	1	1	1	1	−1	0	0	0	−1	0	0	0	0
2	1	1	1	0	0	0	−1	0	0	0	0	0	0	0
2	2	1	1	1	0	0	−1	0	0	0	−1	0	0	0
2	3	1	1	1	1	0	−1	0	0	0	0	−1	0	0
3	1	1	1	0	0	0	0	−1	0	0	0	0	0	0
3	2	1	1	1	0	0	0	−1	0	0	0	0	−1	0
3	3	1	1	1	1	0	0	−1	0	0	0	0	0	−1

Note: For Model Identification You Have to Fix the Person Parameters or Remove One Parameter from A.

where θ_{vt} is the ability of person v at time point t. Parameter estimation is restricted to person abilities for each time point and their potential associations. The item parameters β_i are kept constant over time. In contrast to the MRMLC, θ_3 is the ability at t_3 and not the modifiability between t_2 and t_3. In their original formulation, both models, the MRMLC and MRMRT, lead to identical results (see von Davier *et al.*, 2011). Thus, the modifiability at t_2 in the MRMLC is given by the difference of change parameters $\theta_2 - \theta_1$ estimated using the MRMRT.

Wang *et al.* (1998) showed that the MRCML model is a generalization of the MRMLC and the MRMRT when the scoring matrix **B** and design matrix **A** are appropriately defined. In Table 10.1, an example of the MRMLC with three dichotomous items and three time points is given (for a more detailed description of the approach, see, Wang *et al.*, 1998). For each of the three items, a Wiener simplex scoring matrix **B** is defined. The design matrix **A** defines a model that estimates the maximum number of change parameters in the model (often called the *maximum model*; for models with other variations in item difficulties across time; see Wang *et al.*, 1998). Thus, one item-specific parameter together with one item-specific change parameter are estimated across all time points, that is, one change parameter for t_2 and one for t_3. No change parameter is estimated for the first time point because t_1 serves as the initial ability (the corresponding parameter being set to zero). Using this representation of **A**, the assumption of measurement invariance of items over time (by removing the last six columns of **A**) or the existence of item-specific change effects can be evaluated.

Integrating Andersen's MRMRT into the MRCML model is straightforward. Here, the scoring matrix **B** is defined not through t Wiener simplex matrices but through identity matrices for each time point (see Table 10.2).

Using the definitions of the scoring matrices and the design matrices stated above, it is possible to analyze change in the multidimensional case, irrespective of whether the change is assumed to be linear or nonlinear. In this chapter, only the case of

TABLE 10.2 Scoring Matrix B and Design Matrix A for the Combination of MRMRT with MRCML Model for the Investigation of Granger Causation.

IT	*T*	*D*	Scoring Matrix **B**						Design Matrix **A**					
1	1	1	1	0	0	0	0	0	−1	0	0	0	0	0
1	2	1	0	1	0	0	0	0	−1	0	0	0	0	0
1	3	1	0	0	1	0	0	0	−1	0	0	0	0	0
2	1	1	1	0	0	0	0	0	0	−1	0	0	0	0
2	2	1	0	1	0	0	0	0	0	−1	0	0	0	0
2	3	1	0	0	1	0	0	0	0	−1	0	0	0	0
3	1	1	1	0	0	0	0	0	0	0	−1	0	0	0
3	2	1	0	1	0	0	0	0	0	0	−1	0	0	0
3	3	1	0	0	1	0	0	0	0	0	−1	0	0	0
1	1	2	0	0	0	1	0	0	0	0	0	−1	0	0
1	2	2	0	0	0	0	1	0	0	0	0	−1	0	0
1	3	2	0	0	0	0	0	1	0	0	0	−1	0	0
2	1	2	0	0	0	1	0	0	0	0	0	0	−1	0
2	2	2	0	0	0	0	1	0	0	0	0	0	−1	0
2	3	2	0	0	0	0	0	1	0	0	0	0	−1	0
3	1	2	0	0	0	1	0	0	0	0	0	0	0	−1
3	2	2	0	0	0	0	1	0	0	0	0	0	0	−1
3	3	2	0	0	0	0	0	1	0	0	0	0	0	−1

Note: For Model Identification You Have to Fix the Person Parameters or Remove One Parameter from A.

linear effects is examined. Modeling approaches for nonlinear effects can be found in Wilson *et al.* (2012).

As we will show next, it is straightforward to define the scoring matrix **B** for the multidimensional case for constructing an IRT model to test Granger causation hypotheses. For this purpose, we integrate the Andersen MRMRT into the MRCML model. In this model, latent abilities are estimated for each time point. To account for the autoregressive structure, latent regression parameters are estimated instead of latent associations ($\text{MRMRT}_{\text{modified}}$). Next, the autoregressive models for *X* and *Y* and the associated latent associations within each time point (double-headed arrows in Fig. 10.1) are estimated in one step. The resulting model is given in the left panel of Figure 10.1. An example of a scoring matrix and design matrix for three time points, two dimensions (i. e., *X* and *Y*) while assuming measurement invariance of items across time is given in Table 10.2.

The above specification enables researchers to test whether earlier/past values of *X* provide more information about future values of *Y* than exclusive past values of *Y* by including additional regression parameters as displayed in Figure 10.1, right panel.

The model selection procedure consists of two modeling phases: First, more restrictive models are compared with a maximum model, for example, testing the assumption that one common change parameter is sufficient for all items (the case of item-specific change parameters is discussed in Wang *et al.*, 1998). Second, models with and without Granger causation parameters are compared using information criteria such as the Akaike Information Criterion (AIC; Akaike, 1974), the Bayesian Information Criterion (BIC; Schwarz, 1978), or the consistent Akaike Information Criterion (CAIC; Bozdogan, 1987). In addition, Granger causation parameters (i.e., cross-lagged effects) must statistically exist. In the following section, we demonstrate the application of the Granger-IRT model using an empirical example.

10.5 DATA EXAMPLE: SCIENTIFIC LITERACY IN PRESCHOOL CHILDREN

Empirical data used for illustration are part of a longitudinal study on the development of scientific literacy in preschool children in Germany (SNAKE [1]; Carstensen *et al.*, 2011, 2012; Steffensky *et al.*, 2012a,b). Goals of the project include the construction of a Rasch model conform scientific literacy test for 5-year-old children and to investigate the extent to which scientific literacy of children can be improved through instruction. The sample consisted of 257 ($n = 123$ girls) 5-year-old children who were tested at three time points within 7 months. The scientific literacy test developed for the study refers to the topics of water, its physical states, changes in state, and solutions in water (Carstensen *et al.*, 2011). The tasks were administered in structured interviews and the children's responses were scored by trained interviewers by using two- to four-category response formats. For scaling of responses, some response categories were collapsed in order to achieve discriminating response categories. An example task on melting and knowledge of science is presented in Figure 10.4.

The final test form at t_1 (initial ability) included 29 items that were administered and scored, at t_2 (after an intervention phase) 29 items, and at t_3 (follow-up phase) 31 items were scored. Twenty-four items were used as link items. The other items at t_2 and t_3 were on average slightly more difficult than the items used at t_1 and t_2. In addition, the subscale *Figural Analogies* of the Culture Fair test (CFT1;

When Paul and Jan walked to the kindergarten one morning, they find all puddles to be slippery and they can stand on them. Paul says "Look, the water has turned into ice". Jan asks "How did that happen?"

What do you think? Why has the water turned into ice?

Scoring:

(2 points) coldness, cold outside, low temperatures (responses which name coldness)

(0 points) other responses

(0 points) child says I don't know.

Figure 10.4 An example of melting and knowledge of science from the SNAKE study.

[1] Studie zur Naturwissenschaftlichen Kompetenzentwicklung im Elementarbereich.

Cattell *et al.*, 1997) that includes 12 dichotomous items was administered at t_1 to adjust for general cognitive abilities (reasoning).

Carstensen *et al.* (2012) analyzed possible change effects in SNAKE applying the MRMLC and the MRMRT using different numbers of link items. Based on the result from these analyses, Carstensen *et al.* (2012) concluded that the MRMLC and the MRMRT using 24 link items showed the best model fit compared to a model estimated with the maximum number of parameters ($BIC_{nolink} = 29846.9$; $BIC_{MRMLC} = 29756.4$; $BIC_{MRMRT} = 29756.3$). The modifiabilities of the MRMLC at t_2 had a mean of $\theta_{Y_2} = 0.49$ and were uncorrelated with the estimated ability parameters at t_1 (θ_{Y_1}); the modifiabilities at t_3 had a mean of $\theta_{Y_3} = 0.13$ and were moderately correlated with θ_{Y_1}. Thus, scientific literacy improved through the intervention. However, the development over a longer observational period was again correlated with the initial ability.

For illustrative purposes, we used data from t_1 and t_2, and dichotomized the items of SNAKE. Note that post hoc dichotomization of items may have a negative impact on results (e. g., loss of information, decreased correlation between variables). However, in our case of two- and three-categorical items, dichotomization resulted in no significantly different results than these reported in Carstensen *et al.* (2012). We used the $MRMRT_{modified}$ including 24 link items in SNAKE to test the hypothesis that CFT is Granger-causing SNAKE. Because the CFT was only assessed at t_1, we only analyzed whether CFT at t_1 is significantly related to SNAKE at t_2. Thus, this example is a very minimal version of Granger causation assessment. In addition, using the MRCML model, it is possible to assess item-specific intervention effects, but, for SNAKE, the assumption of measurement invariance holds across time (see Carstensen *et al.*, 2012). Thus, no item-specific intervention effects were assessed in this analysis. Two models were estimated and compared (see Fig. 10.5). The analyses were performed using MPlus version 7 (Muthén and Muthén, 2013). A code example for applying the $MRMRT_{modified}$ is given in the appendix.

The results show that CFT at t_1 did not Granger-cause SNAKE at t_2 ($MRMRT_{modified}$: LogLik = −9274.173, npar = 59, AIC = 18666.346, BIC = 18875; $MRMRT_{modified\ Granger}$: LogLik = −9273.981, npar = 60, AIC = 18667.962, BIC = 18880.906). The estimated correlation and regression parameters are given in Figure 10.5. Overall, results suggest a moderate latent relation between CFT and SNAKE at t_1; however, CFT results have no significant influence on the changed ability at t_2.

10.6 DISCUSSION

In this chapter, we proposed a new approach to assess Granger causation with multidimensional longitudinal item response models. For this purpose, the MRMRT was integrated in the MRCML model as described by Wang *et al.* (1998). This approach allows scaling of data, estimation of modifiabilities across time, and assessing Granger causation in a unified framework and has the advantage that measurement error is considered in the model.

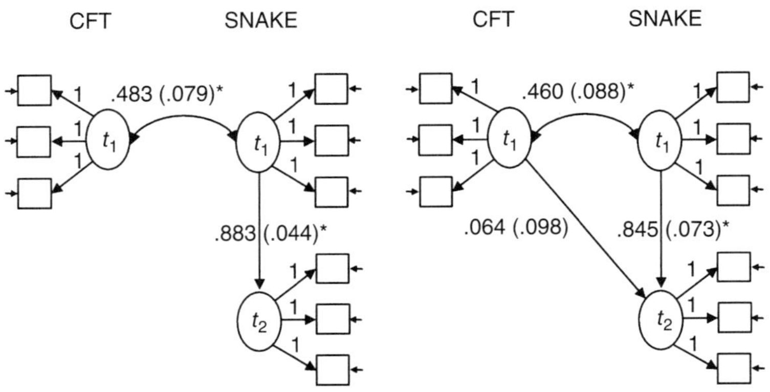

Figure 10.5 Path diagrams for the two linear autoregressive models. Correlation and regression parameters are displayed for each path (the standard errors are given in parentheses; $* = p < 0.05$.

Multidimensional longitudinal item response models offer the advantage that the assumption of Rasch model conformity within time points and across time points can be relaxed. Thus, these models allow the assessment of change in multidimensional data by defining the scoring and design matrices accordingly. In addition, one central precondition of several longitudinal item response models is measurement invariance of change across time, that is, all items of a test change in the same way. If this precondition is fulfilled, the property of specific objectivity of change holds for the data and it is, therefore, irrelevant which items are used to assess change (e. g., Koller *et al.*, 2015). The MRCML model can be used to evaluate this precondition by comparing different models (e. g., comparing the maximum model and a model with only one change parameter for all items), and to model violations of measurement invariance as item-specific change.

The data example serves as an illustration of the assessment of Granger causation with multidimensional longitudinal item response models. The example does not include two time series as depicted in Figure 10.1. The CFT was only administered at t_1. Results suggest that reasoning is moderately related to scientific literacy. However, past values of reasoning do not contribute to the prediction of scientific literacy at t_2. Note that changes in scientific literacy between t_1 and t_2 were not related with the literacy at t_1 either and were assumed to be due to the intervention. Because childcare groups were assigned to the treatment randomly, the intervention phase results should not covary with pretest intelligence.

It is important to note that complexity of model estimation increases with potential multidimensionality of observed data and, thus, may lead to inestimable parameters due to computational limitations. In this case, a more exploratory approach (e. g., Cho *et al.*, 2013; Stevenson *et al.*, 2013) in which, for example, the predictor X is only included as subject-covariate in analysis may be worthwhile. This approach can be seen as a two-step procedure in which, in the first step, the person parameters of the predictor X are estimated and, in a second step, the measurement model of Y

and Granger causation are assessed using explanatory item response models. However, this approach only accounts for error in the measurement model of Y, while measurement errors associated with X are not part of the model.

In this chapter, we focused on a Granger-IRT model that uses past information of series of observations. Contemporaneous effects are not considered. Granger (1969) formulated his causality principle for both, lagged and contemporaneous information (cf. von Eye and Wiedermann, 2015). While the former is generally accepted to provide valuable information to derive statements of Granger causality, contemporaneous effects are often modeled in terms of bidirectional associations. In other words, the direction of instantaneous effects must be derived from additional information about the relation of variables and cannot be established empirically (Lütkepohl, 2007; see, however, Wiedermann and von Eye, 2015a,b). Of course, the proposed Granger-IRT model can be reformulated to consider contemporaneous effects as well. To address the issue of potentially ambiguous interpretations of the direction of contemporaneous effects, recent advances in regression modeling may be considered. In cross-sectional data settings, recent studies of Dodge and Rousson (2001), Shimizu *et al.* (2006), or Wiedermann and von Eye (2015b) showed that contemporaneous effects can be unambiguously interpreted as directional provided that higher than second moments (i.e., skewness and kurtosis) of variables indicate deviations from the normal distribution. Time series applications of these insights are discussed by Hyvärinen *et al.* (2010). Combining these recent advances with the proposed $\text{MRMRT}_{\text{modified}}$ may enable researchers to account for potential contemporaneous effects within the IRT framework while solving the issue of ambiguity of contemporaneous effect interpretation. Such an approach, of course, would require latent ability distributions being nonnormal – a prerequisite that may be quite common in practice (Micceri, 1989).

Furthermore, when applying the classic principles of Granger causality, one must implicitly make the assumption of no hidden confounders. Various studies showed that results of Granger analyses can be seriously biased in the presence of unobserved confounders (Asghar, 2008, Hsiao, 1982, Peters *et al.*, 2013). Similarly, application of Granger-IRT models requires the assumption of the absence of latent confounders. This potential issue definitely warrants future studies to quantify robustness properties of the proposed model against confounder influences.

In sum, the possibility to estimate the measurement model and to assess Granger causation may be highly valuable for psychometric research. Future research in the field of multidimensional item response models should examine how many dimensions can be considered in one step and which effect complex linking designs can have on parameter estimation and interpretation.

REFERENCES

Adams, R., Wilson, M., and Wang, W.C. (1997) The multidimensional random coefficient multinomial logit model. *Applied Psychological Measurement*, **21** (1), 1–23.

Akaike, H. (1974) A new look at the statistical model identification. *IEEE Transactions on Automatic Control*, **19** (6), 716–723.

Andersen, E. (1985) Estimating latent correlations between repeated testings. *Psychometrika*, **50** (1), 3–16.

Asghar, Z. (2008) Simulation evidence on Granger causality in presence of a confounding variable. *International Journal of Applied Econometrics and Quantitative Studies*, **5** (2), Retrieved from `http://www.usc.es/economet/reviews/ijaeqs526.pdf`.

Bacci, S. (2012) Longitudinal data: different approaches in the context of item response theory models. *Journal of Applied Statistics*, **39** (9), 2047–2065.

Blossfeld, H.P., Rossbach, H.G., and von Maurice, J. (2011) *Education As a Lifelong Process: The German National Educational Panel Study (NEPS)*, Special Issue in Zeitschrift für Erziehungswissenschaft, Springer-Verlag, Wiesbaden.

Bozdogan, H. (1987) Model selection and Akaike's information criterion (AIC): the general theory and its analytical extensions. *Psychometrika*, **52** (3), 345–370.

Carstensen, C., Lankes, E.M., and Steffensky, M. (2011) Ein Modell zur Erfassung naturwissenschaftlicher Kompetenz im Kindergarten [a model for analyzing scientific literacy in pre-school children]. *Zeitschrift für Erziehungswissenschaft*, **14** (4), 651–669.

Carstensen, C., Lankes, E.M., and Steffensky, M. (2012) Modellierung von längsschnittlichen Daten am Beispiel einer quasi-experimentellen Studie zur Erfassung von naturwissenschaftlichen Kompetenzen im Kindergartenalter [modeling longitudinal data using an example of a quasi-experimental study to scientific literacy in preschool], in *Item-Response-modelle in der sozialwissenschaftlichen Forschung [Item-Response-Models in Social Science Research]* (eds W. Kempf and R. Langeheine), Verlag Irena Regener, Berlin.

Cattell, R., Weiss, R., and Osterland, J. (1997) *Grundintelligenztest Skala 1 [Basic intelligence test scale 1]*, Braunschweig, Westermann.

Cho, S.J., Athay, M., and Preacher, K. (2013) Measuring change for a multidimensional test using a generalized explanatory longitudinal item response model. *British Journal of Mathematical and Statistical Psychology*, **66** (2), 353–381.

von Davier, M., Gonzalez, E., and Mislevy, R. (2009) What are plausible values and why are they useful? in *IERI Monograph Series: Issues and Methodologies in Large Scale Assessments*, vol. **2** (eds M. von Davier and D. Hastedt), IEA-ETS Research Institute, Princeton, NJ, pp. 9–36.

von Davier, M., Xu, X., and Carstensen, C. (2011) Measuring growth in a longitudinal large-scale assessment with a general latent variable model. *Psychometrika*, **76** (2), 318336.

Dodge, Y. and Rousson, V. (2001) On asymmetric properties of the correlation coefficient in the regression setting. *The American Statistician*, **55** (1), 51–54.

Embretson, S. (1991) A multidimensional latent trait model for measuring learning and change. *Psychometrika*, **56** (3), 495–515.

Embretson, S. and Reise, S. (2000) *Item Response Theory for Psychologists*, vol. **4**, Lawrence Erlbaum, Mahwah, NJ.

von Eye, A. and Wiedermann, W. (2015) Manifest variable Granger causality models for developmental research: a taxonomy. *Applied Developmental Sciences*, **19** (4), 183–195.

von Eye, A., Wiedermann, W., and Koller, I. (2015) Granger causality–Linear regression and logit models, in *Dependent Data in Social Sciences Research: Forms, Issues, and Methods of Analysis* (eds M. Stemmler, A. von Eye, and W. Wiedermann), Springer, Switzerland, pp. 127–148.

von Eye, A., Wiedermann, W., and Mun, E.Y. (2014) Granger causality–statistical analysis under a configural perspective. *Integrative Psychological and Behavioral Science*, **48** (1), 79–99.

Fischer, G. (1973) The linear logistic test model as an instrument in educational research. *Acta Psychologica*, **37** (6), 359–374.

Fischer, G. (1974) *Einführung in die Theorie psychologischer Tests: Grundlagen und Anwendungen [Introduction to the theory of psychological tests: Basics and applications]*, H. Huber, Bern.

Fischer, G. (1976) Some probabilistic models for measuring change, in *Advances in Psychological and Educational Measurement* (eds D. de Gruijter and L. van der Kamp), John Wiley & Sons, Inc., New York, pp. 97–110.

Fischer, G. (1989) An IRT-based model for dichotomous longitudinal data. *Psychometrika*, **54** (4), 599–624.

Fischer, G. (1995a) Derivations of the Rasch model, in *Rasch Models: Foundations, Recent Developments, and Applications* (eds G. Fischer and I. Molenaar), John Wiley & Sons, Inc., New York, pp. 15–38.

Fischer, G. (1995b) Some neglected problems in IRT. *Psychometrika*, **60** (4), 459–487.

Fischer, G. (2001) Gain scores revisited under an IRT perspective, in *Essays on Item Response Theory* (eds A. Boomsma, M. van Duijn, and T. Snijders), Springer-Verlag, New York, pp. 43–68.

Fischer, G. and Ponocny-Seliger, E. (1998) *Structural Rasch Modeling. Handbook of the Usage of LPCM-WIN 1.0*, ProGAMMA, Groningen.

Formann, A. and Spiel, C. (1989) Measuring change by means of a hybrid variant of the linear logistic model with relaxed assumptions. *Applied Psychological Measurement*, **13** (1), 91–103.

Glück, J. and Spiel, C. (1997) Item response models for repeated measures designs: application and limitations of four different approaches. *Methods of Psychological Research Online*, **2** (1), 1–18. `http://www.dgps.de/fachgruppen/methoden/mpr-online/issue2/art6/article.html`.

Glück, J. and Spiel, C. (2007) Item response models for repeated measures designs: application and limitation of four different approaches, in *Oxford Handbook of Methods in Positive Psychology* (eds A. Ong and M. van Dulmen), Oxford University Press, Oxford, pp. 349–361.

Granger, C. (1969) Investigating causal relations by econometric models and cross-spectral methods. *Econometrica*, **37** (3), 424–438.

Hojitink, H. (2007) Linear and repeated measures models for the person parameters, in *Rasch Models: Foundations, Recent Developments, and Applications* (eds G. Fischer and I. Molenaar), Springer-Verlag, Oxford, pp. 203–214.

Hsiao, C. (1982) Autoregressive modeling and causal ordering of economic variables. *Journal of Economic Dynamics and Control*, **4**, 243–259.

Hsieh, C.A., von Eye, A., Maier, K., Hsieh, H.J., and Chen, S.H. (2013) A unified latent growth curve model. *Structural Equation Modeling: A Multidisciplinary Journal*, **20** (4), 592–615.

Hume, D. (1777/1975) *Enquiries Concerning Human Understanding and Concerning the Principles of Morals*, Clarendon Press, Oxford.

Hyvärinen, A., Zhang, K., Shimizu, S., and Hoyer, P.O. (2010) Estimation of a structural vector autoregression model using non-Gaussianity. *Journal of Machine Learning Research*, **11**, 1709–1731.

Koller, I., Wiedermann, W., and Glück, J. (2015) Item response models for dependent data: quasi-exact tests for the investigation of some pre-conditions for measuring change, in *Dependent Data in Social Sciences Research: Forms, Issues, and Methods of Analysis* (eds M. Stemmler, A. von Eye, and W. Wiedermann), Springer, Switzerland, pp. 263–280.

Lütkepohl, H. (2007) *New Introduction to Multiple Time Series Analysis*, Springer-Verlag, Berlin.

Masters, G. (1982) A Rasch model for partial credit scoring. *Psychometrika*, **47** (2), 149–174.

Micceri, T. (1989) The unicorn, the normal curve, and other improbable creatures. *Psychological Bulletin*, **105** (1), 156.

Molenaar, I. (2007) Some background for item response theory and the Rasch model, in *Rasch Models: Foundations, Recent Developments, and Applications* (eds G. Fischer and I. Molenaar), Springer-Verlag, Oxford, pp. 3–14.

Muthén, B. and Muthén, L. (2013) *Mplus User's Guide*, 7th edn, Muthen & Muthen, Los Angeles, CA.

Peters, J., Janzing, D., and Schölkopf, B. (2013) Causal inference on time series using restricted structural equation models, in *Advances in Neural Information Processing Systems*, vol. **26** (eds C.J.C. Burges, L. Bottou, M. Welling, Z. Ghahramani, and K.Q. Weinberger), Curran Associates, Inc., pp. 154–162, `http://papers.nips.cc/paper/5063-causal-inference-on-time-series-using-restricted-structural-equation-models.pdf`.

Pohl, S. and Carstensen, C. (2013) Scaling of competence tests in the national educational panel study–many questions, some answers, and further challenges. *Journal for Educational Research Online/Journal für Bildungsforschung Online*, **5** (2), 189–216.

Ponocny, I. (2002) On the applicability of some IRT models for repeated measurement designs: conditions, consequences, and goodness-of-fit tests. *Methods of Psychological Research Online*, **7** (1), 22–40.

Rasch, G. (1960) *Probabilistic Models for Some Intelligence and Attainment Tests*, vol. **1**, Danish Institute for Educational Research, Copenhagen.

Schwarz, G. (1978) Estimating the dimension of a model. *Annals of Statistics*, **6** (2), 461–464.

Shimizu, S., Hoyer, P.O., Hyvärinen, A., and Kerminen, A. (2006) A linear non-Gaussian acyclic model for causal discovery. *Journal of Machine Learning Research*, **7**, 2003–2030.

Steffensky, M., Lankes, E.M., and Carstensen, C. (2012a) Was bedeutet naturwissenschaftliche Kompetenz bei Fünfjährigen und wie kann man sie erfassen? [what is scientific literacy for five years old and how can one assess it?], in *Mixed Methods in der empirschen Bildungsforschung [Mixed methods in empirical educational research]* (eds M. Gläser-Zikuda, T. Seidel, C. Rohlfs, A. Gröschner, and S. Zeigelbauer), Waxmann, Münster, pp. 107–119.

Steffensky, M., Lankes, E.M., Carstensen, C., and Nölke, C. (2012b) Alltagssituationen und Experimente: was sind bessere Lerngelegenheiten für Kindergartenkinder? Ergebnisse aus dem SNaKE-projekt [Daily life and experiments: what are the appropriate learning settings for children in kindergarten? Findings from the SNaKE project]. *Zeitschrift für Erziehungswissenschaften*, **15** (1), 37–54.

Stevenson, C., Hickendorff, M., Resing, W., Heiser, W., and de Boeck, P. (2013) Explanatory item response modeling of children's change on a dynamic test of analogical reasoning. *Intelligence*, **41** (3), 157–168.

Wang, W.C., Wilson, M., and Adams, R. (1998) Measuring individual differences in change with multidimensional Rasch models. *Journal of Outcome Measurement*, **2** (3), 240–265.

Wiedermann, W. and von Eye, A. (2015a) Direction of effects in mediation analysis. *Psychological Methods*, **20** (2), 221–244, doi: 10.1037/met0000027.

Wiedermann, W. and von Eye, A. (2015b) Direction of effects in multiple linear regression models. *Multivariate Behavioral Research*, **50** (1), 23–40.

Wilson, M., Zheng, X., and McGuire, L. (2012) Formulating latent growth using an explanatory item response model approach. *Journal of Applied Measurement*, **13** (1), 1–22.

APPENDIX

The following Mplus code can be used to estimate the modified MRMRT to assess Granger causality assuming three items for each latent factor (*lat_u*, *lat_v*1, and *lat_v*2). Removing the path '*lat_v*1 on *lat_u*' in the MODEL command, estimates the baseline model given in the left panel of Figure 10.5. Missing values are assumed to be coded with 999.

```
TITLE: Modified MRMRT for testing Granger causality hypotheses;
DATA:
FILE = "C:\fakepath\mydata.dat";
VARIABLE:
NAMES = v11 v21 v31 v21 v22 v23 v31 v32 v33;
USEVARIABLES = v11 v21 v31 v21 v22 v23 v31 v32 v33;
MISSING = all (999);
CATEGORICAL = v11 v21 v31 v21 v22 v23 v31 v32 v33 u11 u21 u31;
ANALYSIS:
TYPE = missing;
ESTIMATOR = ML;
ITERATIONS = 2000;
CONVERGENCE = 0.00005;
COVERAGE = 0.10;
INTEGRATION = MONTECARLO (5000);
MODEL:
lat_u by u11@1 u21@1 u31@1; ! estimating the latent factor U
lat_v1 by v11@1 v21@1 v31@1; ! estimating the latent factor V at t1
[lat_v1@0];
[v11$1](01);
[v21$1](02);
[v31$1](03);
lat_v2 by v12@1 v22@1 v32@1; ! estimating the latent factor V at t2
[lat_v2*];
[v12$1](01);
[v22$1](02);
[v32$1](03);
lat_v1 with lat_v2; ! estimating the correlation between U and V at t1.
lat_v1 on lat_u; ! testing Granger causality
OUTPUT:
SAMPSTAT RESIDUAL STANDARDIZED TECH1 TECH3 TECH4;
```

11

GRANGER CAUSALITY FOR ILL-POSED PROBLEMS: IDEAS, METHODS, AND APPLICATION IN LIFE SCIENCES

KATEŘINA HLAVÁČKOVÁ-SCHINDLER

Department of Adaptive Systems, Institute of Information Theory and Automation, Academy of Sciences of the Czech Republic, Prague, Czech Republic

VALERIYA NAUMOVA

Center for Biomedical Computing, Simula Research Laboratory, Lysaker, Norway

SERGIY PEREVERZYEV JR.

Applied Mathematics Group, Department of Mathematics, University of Innsbruck, Innsbruck, Austria

11.1 INTRODUCTION

Causality describes the relation between a cause and its effect (its consequence). One can say that the *inverse problems*, where one would like to discover unobservable features of a cause from the observable features of an effect (Engl *et al.*, 1996), can be seen as causality problems. When several elements or phenomena are considered and the causal relationships among them are questioned, we talk about the so-called causality network. A *causality network* can be seen as a directed graph with nodes, which correspond to the variables $\{x^j, j = 1, \cdots, p\}$ and directed edges, which represent the causal influences between variables. The variables represent entities or

Statistics and Causality: Methods for Applied Empirical Research, First Edition.
Wolfgang Wiedermann and Alexander von Eye.

objects, for example, genes. We write $x^i \leftarrow x^j$ if the variable x^j has a causal influence on the variable x^i.

11.1.1 Causality Problems in Life Sciences

Causality networks arise in various scientific contexts. For example, in cell biology, one considers causality networks that involve sets of active genes of a cell. An active gene produces a protein. In biological experiments, it has been observed that the amount of the protein, which is produced by a given gene, may depend on or may be *causally* influenced by the amount of proteins produced by other genes. In this way, causal relations between genes and corresponding causality network arise. These causality networks are also called *gene regulatory networks*. In cell biology, these networks are used in the research of causes of genetic diseases.

In neuroscience, causality networks are widely used to express the temporal interactions between various regions of the brain. Knowledge of these interactions can help to understand the human cognition or neurological diseases (Paluš *et al.*, 2001, Seth, 2005, Marinazzo *et al.*, 2012).

In practice, the first important information that can be observed about a network is the temporal evolution (time series) of the involved variables $\{x_t^j, t = 1, \cdots, T\}$, where t is the index of time and j is the index of the concrete variable in the network. How can this information be used for inferring causal relations between variables?

The statistical approach to deriving causal relations between a target variable y and potential predictor variables $\{x^j, j = 1, \cdots, p\}$ using the known temporal evolution of their values $\{y_t, x_t^j, t = 1, \cdots, T, j = 1, \cdots, p\}$ consists of specifying a model of the relations between y and $\{x^j, j = 1, \cdots, p\}$. As a first step, one can consider a linear model for variable y_t:

$$y_t \approx \sum_{j=1}^{p} \beta^j x_t^j, t = 1, \cdots, T$$

The coefficients $\{\beta^j, j = 1, \cdots, p\}$, which can be estimated using the least-squares method, serve as indicators of causal relations. For instance, in statistics (Wikipedia, 2013) by fixing the value of a threshold parameter $\beta_{\text{tr}} > 0$, one says that there is a causal relationship $y \leftarrow x^j$ if $|\beta^j| > \beta_{\text{tr}}$.

The goal of this chapter is to overview existing approaches for the reconstruction of the causal relations and to present novel techniques, originating from regularization theory, that allow for a more accurate and robust reconstruction of causality networks.

11.1.2 Outline of the Chapter

In Section "Granger Causality and Multivariate Granger Causality", we continue our discussion of causality in general terms and introduce the notion of Granger Causality and Multivariate Granger Causality. We also discuss some methods for the reconstruction of causalities in gene regulatory networks. Consequently, we present the concept of gene regulatory networks and some recent approaches for their reconstruction. Since we consider causality problems as a special case of inverse problems, in

Section "Regularization of Ill-Posed Inverse Problems," we introduce the state of the art in the regularization theory for treating inverse ill-posed problems. We will mainly focus on approaches for treating problems with incomplete, high-dimensional, and noisy data, because of their high relevance to real-life applications. Section "Multivariate Granger Causality" describes the state of the art for its analysis. Further, we discuss quality measures, which are used in numerical experiments for checking the performance of the methods. Finally, we discuss novel regularization techniques for the reconstruction of causal relationships and present results of numerical experiments on gene regulatory network reconstruction using the classical approaches such as Lasso and our novel methods.

11.1.3 Notation

First, we introduce some standard notation that will be used in this paper. The entries of a matrix X are denoted by lowercase letters and the corresponding indices. We define the Frobenius norm of a matrix X as

$$\| X\|_F := \left(\sum_{i,j} |x_{i,j}|^2 \right)^{\frac{1}{2}}$$

where $x_{i,j}$ is the entry (i,j) of the matrix X. It is also convenient to introduce the ℓ_p^n vector norms

$$\| x\|_{\ell_p^n} :=\| x\|_p := \left(\sum_{i=1}^{n} |x_i|^p \right)^{1/p}, \quad 0 < p < \infty.$$

This notation will be used in Section "Multivariate Granger Causality Approaches Using ℓ_1 and ℓ_2 Penalties." More specific notations will be defined in the paper, where they turn out to be useful.

11.2 GRANGER CAUSALITY AND MULTIVARIATE GRANGER CAUSALITY

In 1969, the econometrist Clive Granger introduced a method to quantify temporal-causal relations among time series measurements (Granger, 1969), which gained great success across many scientific domains and in a variety of applications. In 2003, Granger was for his achievements awarded with the Nobel Prize in Economics. He introduced Wiener's concept of causality into the analysis of time series (Wiener, 1956) and the notion of the "computationally measurable" causality. His method is usually referred to as *Granger causality*. Granger causality is based on the statistical predictability of one time series using knowledge from one or more other time series. The basic idea of the method is straightforward.

Consider two simultaneously measured signals x and y, and examine two predictions of the values of y: the first one uses only the past values of the signal y, and the second one uses the past values of both signals y and x. If the second prediction is significantly better than the first one, then we call x to be *causal* to y (Wiener, 1956). Note that the contemporaneous effects are not considered. The standard test developed by Granger is based on linear regression models and leads to the two well-known alternative test statistics, the Granger–Sargent and the Granger–Wald (discussed in detail below) tests (Abramowitz and Stegun, 1972). The probabilistic nature of Granger causality leads to uncertainty concerning the relationship of cause and effect, which are fundamentally deterministic. The efficient applicability of the original Granger causality is impaired by several crucial problems of discovering latent confounding effect, missing counterfactual reasoning and capturing instantaneous and nonlinear causal relationships (Spirtes *et al.*, 2001, Pearl, 2009, Bahadori and Liu, 2013b). Nevertheless, due to its simplicity and scalability, Granger causality remains a popular method for uncovering temporal dependencies and for detecting interactions between time series.

Rather than referring to Granger causality as a causal analysis tool, we will define it in our paper as a temporal dependence discovery method. Being aware of the above-mentioned criticism, we will use the terms "G-causality," "G-causal" or "Granger causality" in terms of temporal dependency or inference.

11.2.1 Granger Causality

Granger causality, GC hereafter, characterizes the extent to which a process x_t influences another process y_t and builds upon the notion of incremental predictability. It is said that the *process x_t Granger causes another process y_t* if future values of y_t can be better predicted using the past values of x_t and y_t rather than only past values of y_t.

The standard test of Granger causality is based on a linear regression model

$$y_t = a_0 + \sum_{l=1}^{L} b_{1l} y_{t-l} + \sum_{l=1}^{L} b_{2l} x_{t-l} + \xi_t \tag{11.1}$$

where ξ_t are uncorrelated random variables with zero mean and variance σ^2, L is the time lag, which denotes the maximum number of the considered past values of a variable, and $t = L + 1, \cdots, N$. The null hypothesis that x_t does not Granger cause y_t is supported when $b_{2l} = 0$ for $l = 1, \cdots, L$, reducing (11.1) to

$$y_t = a_0 + \sum_{l=1}^{L} b_{1l} y_{t-l} + \xi_t'. \tag{11.2}$$

This model leads to the two well-known test statistics, the Granger–Sargent (GS) and the Granger–Wald (GW) test. The Granger–Sargent test is defined as

$$GS = N \frac{(R_2 - R_1)/L}{R_1/(N - 2L)} \tag{11.3}$$

where R_1 is the residual sum of squares in Equation (11.1) and R_2 is the residual sum of squares in Equation (11.2). The GS test statistic has an F-distribution with L and $N - 2L$ degrees of freedom. The Granger–Wald test is defined as

$$GW = N\frac{\hat{\sigma}^2_{\xi'_t} - \hat{\sigma}^2_{\xi_t}}{\hat{\sigma}^2_{\xi_t}} \tag{11.4}$$

where $\hat{\sigma}^2_{\xi'_t}$ is the estimate of the variance of ξ'_t from model (11.2) and $\hat{\sigma}^2_{\xi_t}$ is the estimate of the variance of ξ_t from model (11.1). The GW statistic follows the χ^2_L distribution under the null hypothesis of no causality.

11.2.2 Multivariate Granger Causality

The bivariate Granger causality can straightforwardly be extended to p-dimensional multivariate time series represented by $x_t \in R^{p\times 1}$.

Based on the intuition that the cause should precede its effect (i.e., following Hume's definition of causality), in multivariate Granger causality one states that a (vector) variable x^i can be potentially G-caused by the past versions of the involved variables $\{x^j, j = 1, \cdots, p\}$. Then, in the spirit of the statistical approach described above and using a (multivariate) vector autoregressive model (VAR) for the *G-causal relations among p (scalar)* variables x^j_t, we consider the following approximation problem for the scalar values:

$$x^i_t \approx \sum_{j=1}^{p}\sum_{l=1}^{L} \beta^j_l x^j_{t-l}, \quad t = L+1, \cdots, T \tag{11.5}$$

where L is the *maximal time lag*. The approximation problem (11.5) can be specified using the least-squares approach:

$$\sum_{t=L+1}^{T}\left(x^i_t - \sum_{j=1}^{p}\sum_{l=1}^{L} \beta^j_l x^j_{t-l}\right)^2 \to \min_{\beta^j_l}.$$

Then, the coefficients $\{\beta^j_l\}$ can be determined from a system of linear equations. In the following sections, we denote the coefficient matrix by $A^{\text{est}} = (\beta^1, \beta^2, \cdots, \beta^p)$, where the coefficients are obtained by an approximation method. Performing a statistical significance test on the value of coefficients, one identifies the Granger causes of the target series. As in the statistical approach, one can now fix the value of the threshold parameter (i.e., of the substantive cutoff) $\beta_{\text{tr}} > 0$ and say that

$$x^j \text{ Granger causes } x^i, \text{ denoted by } x^i \leftarrow x^j \quad \text{if} \quad \sum_{l=1}^{L} |\beta^j_l| > \beta_{\text{tr}}. \tag{11.6}$$

It is well known from the literature (see, e.g., Lozano *et al.*, 2009) that an application of Granger causality on gene regulatory networks with a large p may not lead to satisfactory results. This poor performance is reflected in the nonuniqueness of the solution of the corresponding minimization problem and the potentially large number of reconstructed spurious relations. Actually, in practice one would expect to have only a few causal relations for a given gene, which means that the vector (β_l^j) is sparse. In this case, the statistical significant tests are inefficient, while they lead to higher chances of spurious correlations. Moreover, the high dimensionality of biological data leads to further challenges. To address this issue, various *variable selection procedures* can be applied. Most of them are extensions of "classical" variable selection procedures such as Lasso (Tibshirani, 1996), LARS (Efron *et al.*, 2004), and elastic nets (Zou and Hastie, 2005).

Lasso (least absolute shrinkage and selection operator) is an alternative regularized version of least squares, which, in addition to the minimization of the residual sum of squares, imposes an ℓ_1 norm on the coefficients $\{\beta_l^j\}$. Due to the nature of ℓ_1 norm, Lasso shrinks the regression coefficients toward 0 and returns some coefficients that are exactly 0, implementing variable selection in this way. In the following, we will refer to *Lasso Granger* as to an algorithm for learning the temporal dependency among multiple time series based on variable selection using Lasso.

LARS (least angle regression) is a less greedy version of traditional forward selection method. A simple modification of the LARS algorithm is computationally less intensive compared to Lasso. The efficiency of the LARS algorithm makes it widely used in variable selection problems.

However, for highly correlated variables, Lasso tends to select only one variable instead of the whole group. To overcome this challenge, the elastic net method, which combines ℓ_2 and ℓ_1 penalties on the coefficients, was proposed. The convex function induced by the ℓ_2 penalty helps elastic net to achieve a grouping effect, where strongly correlated predictors tend to be in or out of the model together. Elastic net often outperformed Lasso in terms of prediction error for correlated data.

In the following sections, we continue our discussion on existing variable selection procedures with an emphasis on methods for discovering causal relations in gene regulatory networks.

11.3 GENE REGULATORY NETWORKS

Biomolecular interactions in a cell, called transcriptional regulation, show a complex nonlinear dynamics. Models of transcriptional regulation are commonly depicted in the form of networks, where directed connections between nodes represent regulatory interactions. The goal of these models is to infer on (or to reconstruct) the structure of gene regulatory networks from experimental data. Biological samples are usually profiled using the so-called gene expression microarrays, which correspond to the vector measurements and provide quantitative information to assess molecular control mechanisms. An experiment as sample, y, is a result of a single microarray experiment corresponding to a single column in the matrix of gene expressions, $y = (x_j^1, \cdots, x_j^n)'$ where n is the number of genes in the data set. A gene expression profile from microarrays has typically 5000–100,000 variables (genes) and just 15–100 measurements.

The detection of causality in a gene regulatory network from gene expression measurements is a challenging problem, being solved by various computational methods with various successes.

The most popular methods to model interactions in gene regulatory networks from experimental data are the so-called *Dynamic Bayesian networks* (see, for example, Yu *et al.*, 2004). The application of ordinary differential equations is also popular in biological modeling (see, for example, Cao and Zhao, 2008, Bansal *et al.*, 2007, Zou and Conzen, 2005). These methods are reliable for modeling the local kinetics among a small number of genes; however, for larger gene regulatory networks, these approaches are computationally intensive.

Several other methods for modeling interactions among genes have been recently proposed and applied to gene expression data, such as *Structural Equation Models, Probabilistic Boolean Networks* and *Fuzzy Controls* (see, for example, Cao and Zhao, 2008, Shmulevich *et al.*, 2002, just to mention a few). These methods are mainly applied to small genetic networks to study the dynamics of adjacent genes and will not be discussed in this paper.

Taking into account the increasing interest of biologists in investigating interactions among large number of genes together with the scalability and simplicity of Granger causality methods, we focus in this chapter on these methods together with various ℓ_1 and ℓ_2 penalties.

11.4 REGULARIZATION OF ILL-POSED INVERSE PROBLEMS

The problem of the reconstruction of a gene regulatory network belongs to the class of inverse problems with high-dimensional data set and sparse number of measurements. Recently, this problem attracted increasing attention from various scientific communities in inverse problems, machine learning, and approximation theory.

A general inverse problem (see, e.g., Engl *et al.*, 1996, Hofmann, 1999, Rieder, 2003, Kabanikhin, 2008, Lu and Pereverzev, 2013) can be seen as an operator equation

$$\mathbf{y} = A\boldsymbol{\beta} \tag{11.7}$$

where $\mathbf{y}$ represents the data obtained in observational experiments, in other words, the *effect*, $\boldsymbol{\beta}$ is the solution to be reconstructed, the *cause*, and the operator A represents the *model* between the cause and its effect. The approximation problem (11.5) can be seen as a problem of form (11.7).

In practice, one has to take into account that the data $\mathbf{y}$ in (11.7) are *noisy*. Ideally, one assumes that there is a hidden cause $\boldsymbol{\beta}^\dagger$ with corresponding *ideal* data $\mathbf{y}^\dagger$ such that $\mathbf{y}^\dagger = A\boldsymbol{\beta}^\dagger$. The data $\mathbf{y}$ deviate from $\mathbf{y}^\dagger$, and the norm $\delta := \| \mathbf{y}^\dagger - \mathbf{y} \|$ is referred to as noise. Typically, the sources of the noise are imperfect measurements and modeling errors.

Inverse problems are often *ill-posed*, which means that Equation (11.7) using the noisy data $\mathbf{y}$, may have no solution, or the solution of (11.7) may be arbitrarily far away from the expected cause $\boldsymbol{\beta}^\dagger$. So-called *regularization methods* are proposed to deal with the ill-posedness of inverse problems.

A well-known class of regularization methods is the so-called *Tikhonov-type* regularization (see, e.g., Tikhonov, 1963, Tikhonov and Arsenin, 1977, Tikhonov *et al.*, 1995), where the solution of (11.7) is constructed as the minimizer $\boldsymbol{\beta}(\lambda)$ of the following functional:

$$\| \mathbf{y} - A\boldsymbol{\beta}\|^2 + \lambda \ \rho(\boldsymbol{\beta}) \to \min_{\beta} . \tag{11.8}$$

In (11.8), λ is the so-called *regularization parameter*, and $\rho(\cdot)$ is a functional that is often similar to a norm functional. The methods (11.11)–(11.13), which are discussed below, have form (11.8).

The appropriate choice of the regularization parameter λ is very important for the successful application of regularization methods. The goal is to choose λ such that the reconstruction error $\| \boldsymbol{\beta}^{\dagger} - \boldsymbol{\beta}(\lambda) \|$ is minimal. This choice has to be made without the knowledge of $\boldsymbol{\beta}^{\dagger}$. From a theoretical viewpoint (Bakushinskii, 1984), the choice of λ has to be coupled to the noise level δ and to the data $\mathbf{y}$.

In the theoretical analysis of choice rules, one tries to obtain an estimate for the reconstruction error $\| \boldsymbol{\beta}^{\dagger} - \boldsymbol{\beta}(\lambda) \|$ that converges to zero as the noise level tends to zero. Also, the rate of convergence of $\| \boldsymbol{\beta}^{\dagger} - \boldsymbol{\beta}(\lambda) \|$ is of interest, and one tries to design choice rules such that the convergence rate of the error is optimal over a class of solutions $\boldsymbol{\beta}^{\dagger}$. In this respect, the so-called *balancing principle* (Mathé and Pereverzev, 2003, Pereverzev and Schock, 2005, Lazarov *et al.*, 2007) is highly important.

Although the knowledge of the noise level is important from the theoretical point of view, in practice it is either unknown or it is challenging to estimate its value reliably. This is the case, for example, for inverse problem (11.5). In this case, one uses *heuristic choice rules*. In this paper, we use the so-called *quasi-optimality criterion*, which we present in Section “Granger Causality with Multipenalty Regularization.” This choice rule has a close connection to the above-mentioned balancing principle, providing a certain reliability in its results.

In the context of causality detection, the concept of *consistency* of the reconstruction methods, which is discussed in the following section, seems to be similar to the concept of the above-mentioned *convergence* of regularization methods. However, clear links between these two concepts seem to be missing, and consideration of these links is an interesting subject for future research.

11.5 MULTIVARIATE GRANGER CAUSALITY APPROACHES USING ℓ_1 AND ℓ_2 PENALTIES

In Liu *et al.* (2009) and Song and Bickel (2011), statistical properties of the Lasso Granger methods were reviewed. Prior to these papers, Arnold *et al.* (2007) and Fujita *et al.* (2007) discussed the consistency of the Lasso Granger algorithm and proved that the learned temporal dependencies will converge to the ground truth exponentially fast, if the time series data are generated from linear Gaussian models.

Inspired by Liu *et al.* (2009), the common objectives in the analysis of Lasso Granger methods are showing that the method is consistent in terms of three performance metrics:

(1) *Prediction consistency* states that the estimation matrix A^{est} by Lasso Granger can be used to accurately predict the future values of the time series. Formally, an estimation A_T^{est} obtained from a time series of length T is a consistent estimator if

$$\frac{1}{T}\sum_{t=1}^{T}\left\|\sum_{l=1}^{L}(A_T^{\text{est},l} - A^{\text{true}}l)\mathbf{x}(t-l)\right\|_2^2 \rightarrow 0, \text{for} T \rightarrow \infty \tag{11.9}$$

where A^{true} is the true coefficient matrix.

(2) *Parameter estimation consistency* states that the estimated coefficients should be close to the true coefficients:

$$\| A_T^{\text{est},l} - A^{\text{true},l}\|_F \rightarrow 0, T \rightarrow \infty, \text{ for } l = 1, \cdots, L \tag{11.10}$$

(3) *Support recovery* states that the nonzero pattern of the estimate A_T^{est} matches the nonzero pattern of the true coefficient matrix with a high probability as $T \rightarrow \infty$.

The consistency of the Lasso Granger method in terms of the first two performance metrics under some additional assumptions on the matrix A^{true} has been shown in a recent paper (Song and Bickel, 2011). We refer to Song and Bickel (2011) for further discussion on consistency of the method and the corresponding error estimates. Unfortunately, no consistency for support recovery and asymptotic normality are ensured for the Lasso Granger method. These results, however, were derived for special modifications of Lasso, such as adaptive Lasso (Song and Bickel, 2011, Zou, 2006).

The multivariate Granger causality methods that apply Lasso to the problem of reconstruction of gene regulatory networks were first proposed by Arnold *et al.* (2007). This method and its variations belong to *Graphical Lasso Granger* (*GLG*) methods. The model of the GLG method has the form

$$\sum_{t=L+1}^{T}\left(x_t^i - \sum_{j=1}^{p}\sum_{l=1}^{L}\beta_l^j x_{t-l}^j\right)^2 + \lambda \parallel \boldsymbol{\beta}\|_1 \rightarrow \min_{\beta_l^j} \tag{11.11}$$

where $\lambda > 0$ and L denotes the lag of the time series.

The solution of (11.11) for each variable $\{x^i, i = 1, \cdots, p\}$ with the causality rule (11.6) defines an estimator of the causality network between the variables $\{x^i\}$. Although method (11.11) enjoys great computational advantages and excellent performance, it is a well-known fact that the Lasso has a tendency to overselect the variables, that is, reconstruct spurious causation.

In many situations, natural groupings exist between variables, and variables belonging to the same group should be either selected or eliminated as a group. Yuan and Lin (2006) proposed an extension of Lasso, the so-called *Group Lasso*, to address this issue. This approach was used in Lozano *et al.* (2009) to develop a novel methodology, termed *Graphical Group Lasso Granger* (GgrLG), which overcomes the limitations mentioned above for the detection of causal relations. In particular, given J groups of variables that partition the set of predictors, the so-called *group Lasso estimate* $\hat{\boldsymbol{\beta}}_{\text{group}}(\lambda)$ (Yuan and Lin, 2006) is defined as the minimizer of

$$\sum_{t=L+1}^{T}\left(x_t^i - \sum_{j=1}^{p}\sum_{l=1}^{L}\beta_l^j x_{t-l}^j\right)^2 + \lambda \sum_{j=1}^{J} \| \boldsymbol{\beta}_{G_j} \|_2 \tag{11.12}$$

where $\boldsymbol{\beta}_{G_j} = \{\boldsymbol{\beta}_k : k \in G_j\}$, $\lambda > 0$ is a regularization parameter. This form of the functional presupposes that the groups are of equal length, which is a quite natural assumption in this case since they correspond to the number of sampling points realized in regression. It is worthwhile to mention that the use of the ℓ_2−norm as a penalty norm enforces the coefficients $\boldsymbol{\beta}_{G_j}$ within a given group to be similar in amplitude (as opposed to using the ℓ_1 norm). A limitation of the group Lasso is that it requires a priori information of group structures, which is often unavailable. Moreover, the procedures of minimizing (11.12) are nonlinear and require the solution of $O(pL)$ equations on each iteration step. This can be computationally intensive for large numbers of genes.

By extending upon these results, Zeng and Xie (2012) proposed two new methods to select variables in correlated data, the so-called gLars and gRidge. These methods conduct grouping and selecting at the same time and therefore work well when prior information of group structures is not available. Simulations and real examples show that the proposed methods often outperform the existing variable selection methods, including least angle regression (LARS) and elastic net, in terms of both reducing prediction error and preserving sparsity of representation. Another method based on group Lasso penalty with a linear autoregressive model was proposed and applied to gene regulatory networks by Kojima *et al.* (2008).

Analysis of Granger causality between two groups of time series was also applied in brain functional connectivity analysis, where the functional connection between two regions of brain is investigated by analyzing multiple time series representing each region. Using the concept of canonical correlation (Soto *et al.*, 2010), canonical Granger causality is proposed to be calculated between two time series representing the groups of times series, which are linear combinations of the time series in each group (Ashrafulla *et al.*, 2012).

Rajapakse and Mundra (2011) experimentally tested the stability of multivariate vector autoregressive (MVAR) methods with ridge, Lasso, and elastic net penalties by simulation on synthetic data and on gene expression data sets gathered over the HeLa cell cycle. The stability of these MVAR methods with Lasso and with elastic net were comparable, and their accuracies were much higher than the MVAR with the ridge function.

Other methods to infer causal relationships, the so-called *Adaptive Thresholding Lasso Granger* (AtrLG; Shojaie *et al.*, 2012b) and *Graphical Truncating Lasso Granger* (GtrLG; Shojaie and Michailidis, 2010), were proposed by Shojaie and Michailidis and their consistency was proved in Shojaie *et al.* (2012a). Let $\mathbf{x}$ be the

$n \times p$ matrix of observations and let x_t denote the matrix corresponding to the t-th time point, and x_t^j be its j-th column. The truncating Lasso estimate of the graphical Granger model is found by solving the following estimation problem for $i = 1, \cdots, p$

$$\underset{\beta^t \in R^p}{\operatorname{argmin}} \frac{1}{n} \left\| x_T^i - \sum_{j=1}^{p} \sum_{l=1}^{L} x_{T-l}^j \beta_l^j \right\|_2^2 + \lambda \sum_{l=1}^{L} \psi^t \sum_{j=1}^{p} |\beta_l^j| w_j^l$$

$$\psi^1 = 1, \psi^l = M^{I\{\|a^{(l-1)}\|_0 < p^2 \beta/(T-l)\}}, l \geq 2 \tag{11.13}$$

where M is a large constant, β is the allowed false negative rate, determined by the user, and $a^{l-1} = (\beta_{l-1}^1, \cdots, \beta_{l-1}^p)$ is a vector of coefficients, estimated at $(l-1)$. In practice is selected $M = g \exp\ n$ for g a large positive number, see [65]. Selection of β can be based on the cost of false negatives in the specific problem at hand, as well as the sample size; as sample size increases, smaller values of β can be considered. A practical strategy for selecting β is to first find the Lasso (or adaptive Lasso) estimate and select β so that the desired false negative rate is achieved.

The truncating effect of the proposed penalty (imposed by ψ^l) is motivated by the rationale that the number of effects (edges) in the graphical model decreases as the time lag increases. Consequently, if there are fewer than $(p^2\beta/T - l)$ edges in the $(l-1)$ estimate, all the later estimates are forced to zero. Hence, the truncating Lasso penalty provides an estimate of the order of the underlying VAR model. In addition, by applying this penalty, the number of covariates in the model is reduced as the coefficients for effects of genes on each other after the estimated time lags are forced to zero. Shojaie and Michailidis (2010) showed that the resulting estimate is consistent for variable selection (i.e., the correct edges are estimated with increasing probability, as the sample size increases) in the high-dimensional sparse setting. With high probability, the signs of the effects are consistently estimated and the order of the underlying VAR model is correctly estimated.

Similar to GtrLG, AtrLG method attempts to simultaneously estimate the order of the VAR model and the structure of the network. While the truncating Lasso estimate is based on the assumption that the effects of genes on each other decay over time, the adaptively thresholded Lasso estimator relies on a less stringent structural assumption that sets a lower bound on the number of edges in the adjacency matrix of the graphical Granger model at each time point. The relaxation of the decay assumption allows the new estimator to correctly estimate the order of the time series in a broader class of models. The GtrLG may fail in situations where the decay assumption is violated. The method has two more drawbacks. First, the order of the VAR model d is often unknown and is, therefore, set to $T-1$, resulting in $p(T-1)$ covariates in the weighted Lasso estimation problem. Moreover, the weighted Lasso estimate may potentially include edges from different time points of variable x_j to any given variable x_i, $i \neq j$. We also refer the reader to the recent work of Shojaie, where reconstruction of gene regulatory networks by regularization techniques was addressed, for more detailed analysis of the above-presented methods and their extensions (Shojaie, 2013).

As another application of a Lasso Granger method, Bahadori and Liu (2013a) used the copula approach and proposed a semiparametric algorithm (*Granger nonparanormal* (*G-NPN*)) for dependency analysis of time series with non-Gaussian marginal distributions, called *Copula Granger* (CG) method. Modeling of the dependency relations requires p time series $O(p^2)$ parameters, which can lead to high dimensionality and inconsistency of the nonparametric methods. The goal of the copula approach is to separate the marginal properties of the data from their dependency structure. The marginal distribution of the data can efficiently be estimated using nonparametric techniques with exponential convergence rate. The ℓ_1 regularization technique could be used to estimate the dependency structure in high-dimensional settings.

The learning G-NPN model involves three steps:

(i) Find the empirical marginal distribution for each time series $\hat{F}_i$.

(ii) Map the observations into the copula space as $\hat{f}_i(x_t^i) = \hat{\mu}_i + \hat{\sigma}_i \Phi^{-1}(\hat{F}_i(x_t^i))$ where $\hat{\mu}_i$ and $\hat{\sigma}_i$ are the mean and standard deviation of the original time series. Φ^{-1} is the inverse cumulative distribution function of a standard normal.

(iii) Find the Granger causality among $\hat{f}_i(x_t^i)$.

In practice, the Winsorized estimator of the distribution function is used, to avoid the large numbers $\Phi^{-1}(0^+)$ and $\Phi^{-1}(1^-)$, (Bahadori and Liu, 2013a):

$$\tilde{F}_j = \begin{cases} \delta_n, & \text{if } \hat{F}(x^j) < \delta_n \\ \hat{F}(x^j) & \text{if } \delta_n \leq \hat{F}(x^j) < 1 - \delta_n \\ (1 - \delta_n) & \text{if } \hat{F}(x^j) > 1 - \delta_n. \end{cases}$$

The Winsorized estimator is the transformation of statistics by limiting extreme values in the statistical data with the goal of reducing the effect of possibly spurious outliers, see, for example, Hastings *et al.* (1947). Bahadori and Liu (2013a) proved that the convergence rate for Copula Granger method is the same as the one for Lasso.

The Copula Granger method was tested with respect to the Granger method and the Lasso Granger method on synthetic and experimental data (Twitter applications) with the best precision for Copula Granger Lasso method (Bahadori and Liu, 2013a). We compared the Copula Granger method to the Lasso Granger in Hlaváčková-Schindler and Bouzari (2013) on the network of 19 genes with better results for the Copula Granger method.

In our recent paper (Pereverzyev and Hlaváčková-Schindler, 2014), we focused on an important tuning possibility of the Lasso, namely, an appropriate choice of the *threshold parameter* β_{tr} and introduced the so-called *GLG method with two-level-thresholding*. This method is equipped with an appropriate *thresholding strategy* and an appropriate *regularization parameter choice rule*.

In Hlaváčková-Schindler and Pereverzyev (2015), we compared our method to other Lasso Granger methods for gene regulatory network reconstruction, namely, to the Lasso Granger method from Arnold *et al.* (2007), Graphical Truncating Lasso from Shojaie and Michailidis, to the Copula Granger method from Bahadori and Liu (2013a), and to a method not using Lasso, that is, a modification of a Bayesian network method from Äijö and Lähdesmäki (2009). As in Shojaie and Michailidis (2010) and Lozano *et al.* (2009), we used the gene expression data for the set of selected genes from the data basis of genes active in human cancer (HeLa), analyzed by Whitfield *et al.* (2002). Our method was superior in this comparison. Details are discussed in Section "Novel Regularization Techniques".

Despite the computational benefit and simplicity of the linear regression, model (11.5) could be too simple to appropriately match the underlying dynamics of the phenomena and may sometimes lead to misspecifications. A more realistic situation would be to assume that the target function depends nonlinearly on relevant variables. This situation is much less studied and in the vast majority of the literature is restricted to the so-called *additive model,* where the target function is assumed to be the sum

$$f(x) = \sum_{j=1}^{p} f_j(x_j) \tag{11.14}$$

of nonlinear univariate functions f_j in some Reproducing Kernel Hilbert Spaces (RKHS) $\mathcal{H}_j$ such that $f_j \equiv 0$ for $j \notin \{v_i\}_{i=1}^{l}$. For the sake of brevity, we omit the discussion on RKHS, and refer the reader to the seminal paper (Aronszajn, 1950) on a comprehensive theory of RKHS.

Several authors, for example, Bach (2009), Mosci *et al.* (2011), just to mention a few, observed that detection of relevant variables in the model (11.14) can be performed using multiparameter regularization with special regularization terms: partial derivatives, different regularization spaces, and so on. However, the application of the proposed multiparameter methods on the real-life problems can be a nontrivial task due to several important reasons. First of all, the authors do not address the issue of selecting regularization parameters, which is a challenging and tedious task when there are more than two or three parameters. Second, the above-mentioned approaches can be computationally demanding and, therefore, are not always suitable for problems with higher dimensions.

In the context of regularization theory, the multiparameter regularization has been broadly studied as a mechanism to achieve the theoretically optimal rate of reconstruction without an a priori knowledge of relevant information on the solution. We refer the interested reader to recent papers on multiparameter and multipenalty regularization (Lu and Pereverzev, 2011, Naumova and Pereverzyev, 2013, Fornasier *et al.*, 2014). Taking our inspiration from these recent works and the above-mentioned findings in learning theory community, we propose in Section "Granger Causality with Multipenalty Regularization" a novel multipenalty regularization approach for detecting relevant variables from a priori given high-dimensional data under the assumption that the input–output relation is described by a nonlinear function depending on few variables. Different than the above-mentioned work on detection of relevant variables, the method we propose is simple and fast to implement, that is, there is no need for any sophisticated parameter choice rules.

11.6 APPLIED QUALITY MEASURES

Let a causality network among n elements be given by a directed graph with the nodes given by these elements. A *graphical method* is a method that reconstructs the causality network with the variables $\{x^j\}$ by means of a directed graph. Graphical methods are frequently used in biology; see, for example, Khanin and Wit (2007), Zhang and Kim (2014).

Intuitively, the quality of a graphical method can be evaluated by the ability of the method to reconstruct the *known* causality network. It can be tested by various ways, for example, by using the adjacency matrix. An *adjacency matrix* $A = \{a_{i,j} \mid \{i,j\} \subset \{1, \cdots, p\}\ \}$ for the causality network has the following elements:

$$a_{i,j} = 1 \quad \text{if} \quad x^i \leftarrow x^j; \quad a_{i,j} = 0 \quad \text{otherwise.}$$

Assume that there is a *true* adjacency matrix A^{true} of the true causality network, and its *estimator* A^{est}, which is produced by a graphical method. The elements of the adjacency matrix A^{est} can be classified as follows:

- If $a^{\text{est}}_{i,j} = 1$ and $a^{\text{true}}_{i,j} = 1$, then $a^{\text{est}}_{i,j}$ is called true positive. The number of all true positives of matrix A^{est} will be denoted as TP.
- If $a^{\text{est}}_{i,j} = 0$ and $a^{\text{true}}_{i,j} = 0$, then $a^{\text{est}}_{i,j}$ is called true negative. The number of all true negatives of matrix A^{est} will be denoted as TN.
- If $a^{\text{est}}_{i,j} = 1$ and $a^{\text{true}}_{i,j} = 0$, then $a^{\text{est}}_{i,j}$ is called false positive. The number of all false positives of matrix A^{est} will be denoted as FP.
- If $a^{\text{est}}_{i,j} = 0$ and $a^{\text{true}}_{i,j} = 1$, then $a^{\text{est}}_{i,j}$ is called false negative. The number of all false negative of matrix A^{est} will be denoted as FN.

The following quality measures of the estimator A^{est} will be considered:

- *Precision* (also called positive predictive value) of A^{est}:

$$P = \frac{TP}{TP + FP}, 0 \leq P \leq 1 \tag{11.15}$$

- *Recall* (also called sensitivity) of A^{est}:

$$R = \frac{TP}{TP + FN}, 0 \leq R \leq 1 \tag{11.16}$$

Since it is possible to have a high precision and low recall, and vice versa, one considers also an average between these two measures.

The so-called F_1*-score* is defined as the harmonic mean of precision and recall:

$$\frac{1}{F_1} = \frac{1/P + 1/R}{2}. \tag{11.17}$$

The computational complexity of Lasso Granger methods (i.e., including the above-mentioned one) is $O(nd^2p^2)$, where n is the number of observations (i.e., the length of the time series), p is the number of genes, and d is the order of the corresponding VAR model. The computational complexity of Graphical Truncating Lasso is $O(n\hat{d}^2p^2)$, where $\hat{d}$ is the estimate of the order d of VAR model (i.e., the effective number of time lags in VAR, noted L elsewhere) from the truncated Lasso penalty (Shojaie and Michailidis, 2010).

11.7 NOVEL REGULARIZATION TECHNIQUES WITH A CASE STUDY OF GENE REGULATORY NETWORKS RECONSTRUCTION

11.7.1 Optimal Graphical Lasso Granger Estimator

Assume that the true network arise in various scientific causality network with the variables $\{x^j\}$ is given by the adjacency matrix A^{true}. Assume further that the observation data $\{x_t^j\}$ are given. The *best* reconstruction of A^{true} that can be achieved by the so-called *optimal* GLG estimator we proposed in Pereverzyev and Hlaváčková-Schindler (2014).

Let $\boldsymbol{\beta}_i(\lambda)$ denote the solution of the minimization problem (11.11) in the GLG-method, and $\boldsymbol{\beta}_i^j(\lambda) = (\beta_{1,i}^j, \cdots, \beta_{L,i}^j)$. Then, the GLG estimator $A^{\mathrm{GLG}}(\lambda, \beta_{\mathrm{tr}})$ of the adjacency matrix A^{true} is defined as follows:

$$A_{i,j}^{\mathrm{GLG}}(\lambda, \beta_{\mathrm{tr}}) = 1 \text{ if } \quad \| \boldsymbol{\beta}_i^j(\lambda)\|_1 > \beta_{\mathrm{tr}}$$
$$A_{i,j}^{\mathrm{GLG}}(\lambda, \beta_{\mathrm{tr}}) = 0 \quad \text{otherwise.}$$

Let $A_{i,*}^{\mathrm{GLG}}(\lambda, \beta_{\mathrm{tr}})$ denote the i-th row of the GLG estimator. For the given regularization parameter λ, let $\beta_{\mathrm{tr}}^i(\lambda)$ be the threshold parameter that minimizes the number of false entries in the row $A_{i,*}^{\mathrm{GLG}}(\lambda, \beta_{\mathrm{tr}})$, that is, the threshold parameter that solves the following minimization problem:

$$\| A_{i,*}^{\mathrm{true}} - A_{i,*}^{\mathrm{GLG}}(\lambda, \beta_{\mathrm{tr}})\|_1 \to \min_{\beta_{\mathrm{tr}}}. \tag{11.18}$$

Then, we consider the minimization of the number of false entries with respect to the regularization parameter λ, that is, let $\lambda_{\mathrm{opt},i}$ solve

$$\| A_{i,*}^{\mathrm{true}} - A_{i,*}^{\mathrm{GLG}}(\lambda, \beta_{\mathrm{tr}}^i(\lambda))\|_1 \to \min_{\lambda}. \tag{11.19}$$

In this way, we obtain, what we call, the *optimal* GLG estimator $A^{\mathrm{GLG,opt}}$ of the true adjacency matrix A^{true}:

$$A_{i,j}^{\mathrm{GLG,opt}} = A_{i,j}^{\mathrm{GLG}}(\lambda_{\mathrm{opt},i}, \beta_{\mathrm{tr}}^i(\lambda_{\mathrm{opt},i})).$$

Note that the optimal GLG estimator minimizes the following quality measure, which we call Fs-measure:

$$\mathrm{Fs} = \frac{1}{p^2} \parallel A^{\mathrm{true}} - A^{\mathrm{est}} \parallel_1, 0 \leq \mathrm{Fs} \leq 1. \tag{11.20}$$

Fs-measure represents the number of *false* elements in the estimator A^{est} that is scaled with the total number of elements in A^{est}.

In practice, the minimization problems (11.18) and (11.19) can be approximated by the corresponding minimization problems over finite sets of parameters β_{tr}, λ. If we consider a set with N_{tr} values for β_{tr}, and a set with N_λ values for λ, then, in order to determine $A^{\mathrm{GLG,opt}}$, one needs to use $N_{\mathrm{tr}} \cdot N_\lambda$ Lasso Granger solvers. The computational complexity of one Lasso Granger solver was discussed in the previous section.

In the networks created by nature, the true causal relations among selected genes are often unknown. One can use the detected relations from available genetic databases, for example, from frequently updated gene and protein interactions data base Biogrid [1]. The Biogrid tool "Genemania" is a graphical database of detected interactions among genes by experimenting in genetic laboratories all over the world. The biological experiments are expensive, and, therefore, the knowledge of a reliable computational method is of high importance.

To approach the problem of how close one can get to $A^{\mathrm{GLG,opt}}$ without the knowledge of A^{true}, let us first focus on the choice of the threshold parameter β_{tr}.

11.7.2 Thresholding Strategy

The purpose of the threshold parameter β_{tr} is to differentiate the relations $x^i \leftarrow x^j$ with *small* values of $\parallel \boldsymbol{\beta}_i^j(\lambda)\parallel_1$ as the noncausal ones. When can we say that $\parallel \boldsymbol{\beta}_i^j(\lambda)\parallel_1$ is small? We propose considering the following *guiding indicators* of smallness:

$$\begin{aligned}
\beta^i_{\mathrm{min}}(\lambda) &= \min\{\parallel \boldsymbol{\beta}_i^j(\lambda)\parallel_1,\ j = 1,\cdots,p \mid \ \parallel \boldsymbol{\beta}_i^j(\lambda)\parallel_1 \neq 0\},\\
\beta^i_{\mathrm{max}}(\lambda) &= \max\{\parallel \boldsymbol{\beta}_i^j(\lambda)\parallel_1,\ j = 1,\cdots,p\}.
\end{aligned} \tag{11.21}$$

In particular, we propose considering the threshold parameter of the following form:

$$\beta^i_{\mathrm{tr},\alpha}(\lambda) = \beta^i_{\mathrm{min}}(\lambda) + \alpha(\ \beta^i_{\mathrm{max}}(\lambda) - \beta^i_{\mathrm{min}}(\lambda)\ ) \tag{11.22}$$

It should be noted that $\beta^i_{\mathrm{min}}(\lambda)$ and $\beta^i_{\mathrm{max}}(\lambda)$ determine the interval of possible values for β_{tr}, namely $\beta_{\mathrm{tr}} \in [\beta^i_{\mathrm{min}}(\lambda) - \epsilon_1, \beta^i_{\mathrm{max}}(\lambda)]$, where $\epsilon_1 > 0$ is a small constant. Thus, with $\alpha \in [-\epsilon_2, 1]$, where $\epsilon_2 > 0$ is another small constant, $\beta^i_{\mathrm{tr},\alpha}$ covers the entire range of possible values for β_{tr}. The choice $\alpha = 1/2$ is the default. Also, it is worth noting that the choice of the threshold (11.22) is independent of the scaling of the data.

[1] 1 (n.d.) Biological General Repository for Interaction Datasets, Biogrid 3.2.

The optimal GLG-estimator with the threshold parameter $\beta^i_{\mathrm{tr},1/2}$ can be defined as follows. Let $\lambda^{\mathrm{tr},1/2}_{\mathrm{opt},i}$ solve the minimization problem:

$$\| A^{\mathrm{true}}_{i,*} - A^{\mathrm{GLG}}_{i,*}(\lambda, \beta^i_{\mathrm{tr},1/2}(\lambda))\|_1 \to \min_{\lambda}.$$

Then, the corresponding optimal GLG-estimator is

$$A^{\mathrm{GLG,opt}}_{\mathrm{tr},1/2}(i,j) = A^{\mathrm{GLG}}_{i,j}(\, \lambda^{\mathrm{tr},1/2}_{\mathrm{opt},i}, \beta^i_{\mathrm{tr},1/2}(\lambda^{\mathrm{tr},1/2}_{\mathrm{opt},i})\,).$$

The choice of the threshold parameter $\beta^i_{\mathrm{tr},1/2}$ raises the following issue. A gene receives always causal relations, unless the solution of (11.11) $\boldsymbol{\beta}_i(\lambda)$ is zero. But how strong are these causal relationships compared to each other? The norm $\| \boldsymbol{\beta}^j_i(\lambda)\|_1$ can be seen as an *indicator of the strength* of the causal relationship $x^i \leftarrow x^j$.

Let us now construct a matrix $A^{\mathrm{GLG,opt};\boldsymbol{\beta}}_{\mathrm{tr},1/2}$, similar to the adjacency matrix $A^{\mathrm{GLG,opt}}_{\mathrm{tr},1/2}$, in which the norm $\| \boldsymbol{\beta}^j_i(\lambda)\|_1$ is used instead of the value 1. That is,

$$A^{\mathrm{GLG,opt};\boldsymbol{\beta}}_{\mathrm{tr},1/2}(i,j) = \| \boldsymbol{\beta}^j_i(\lambda^{\mathrm{tr},1/2}_{\mathrm{opt},i})\|_1 \quad \text{if} \quad \| \boldsymbol{\beta}^j_i(\lambda^{\mathrm{tr},1/2}_{\mathrm{opt},i})\|_1 > \beta^i_{\mathrm{tr},1/2}$$

$$A^{\mathrm{GLG,opt};\boldsymbol{\beta}}_{\mathrm{tr},1/2}(i,j) = 0 \quad \text{otherwise}.$$

The false causal relations of the estimator $A^{\mathrm{GLG,opt}}_{\mathrm{tr},1/2}$ showed up in the experiments on a gene regulatory network in Pereverzyev and Hlaváčková-Schindler (2014) to be actually weak. This observation suggested to use a second thresholding that is done on the network, at the level of the adjacency matrix.

Thresholding on the network level is similar to thresholding on the gene level. Specifically, let us define the guide indicators of smallness on the network level in a similar way (11.21):

$$A_{\min} = \min_{i,j=1,\cdots,p} \{A^{\mathrm{GLG,opt};\boldsymbol{\beta}}_{\mathrm{tr},1/2}(i,j) \neq 0\}$$

$$A_{\max} = \max_{i,j=1,\cdots,p} \{A^{\mathrm{GLG,opt};\boldsymbol{\beta}}_{\mathrm{tr},1/2}(i,j)\}.$$

And, similarly (11.22), define the threshold on the network level as follows:

$$A_{\mathrm{tr},\alpha} = A_{\min} + \alpha(\, A_{\max} - A_{\min}\,). \tag{11.23}$$

We propose terming the described combination of two thresholdings on the gene and network levels *two-level thresholding*. The adjacency matrix obtained by this thresholding strategy is as follows:

$$A^{\mathrm{GLG,opt}}_{\mathrm{tr},1/2;\alpha_1}(i,j) = 1 \quad \text{if} \quad A^{\mathrm{GLG,opt};\boldsymbol{\beta}}_{\mathrm{tr},1/2}(i,j) > A_{\mathrm{tr},\alpha},$$

$$A^{\mathrm{GLG,opt}}_{\mathrm{tr},1/2;\alpha_1}(i,j) = 0 \quad \text{otherwise}.$$

It turned out that with $\alpha = 1/4$, in (11.23) the optimal GLG-estimator for the gene regulatory network in Pereverzyev and Hlaváčková-Schindler (2014) can be fully recovered.

11.7.3 An Automatic Realization of the GLG-Method

For an automatic realization of the GLG-method, that is, when the true adjacency matrix A^{true} is not known, one needs in addition to a thresholding strategy, a choice rule for the regularization parameter λ in (11.11). For such a choice, we proposed in Pereverzyev and Hlaváčková-Schindler (2014) using the so-called *quasi-optimality criterion* (Tikhonov and Glasko, 1965, Bauer and Reiß, 2008, Kindermann and Neubauer, 2008). In this criterion, one considers a set of regularization parameters

$$\lambda_k = \lambda_0 q^k, \ q < 1, \ k = 0, 1, \cdots, n_\lambda \tag{11.24}$$

and for each λ_k the corresponding solution of (11.11) $\boldsymbol{\beta}^i(\lambda_k)$ is computed. Then, the index of the regularization parameter is selected as follows:

$$k^i_{\text{qo}} = \text{argmin}_k \{ \| \boldsymbol{\beta}_i(\lambda_{k+1}) - \boldsymbol{\beta}_i(\lambda_k) \|_1 \}. \tag{11.25}$$

Let us note that the motivation for the choice of the set of possible regularization parameters as (11.24), and for the choice of the regularization parameter as (11.25) is discussed in Pereverzev and Schock (2005).

In the experimental part of this paper, we compare the GLG-method with an appropriate thresholding to other discussed methods on the network of 19 genes given by gene expressions from biological experiments of Whitfield *et al.* (2002).

11.7.4 Granger Causality with Multi-Penalty Regularization

The natural groupings between the values x^t_j of variables x_j can be introduced into multivariate regression by considering, instead of (11.5) the following form:

$$x^t_v \approx \sum_{j=1}^{p} f_j \left(\sum_{l=1}^{L} \beta^l_j x^{t-l}_j \right), \ t = L+1, L+2, \cdots, T \tag{11.26}$$

where f_j are univariate functions in some RKHS $\mathcal{H}_j$. Then, a conclusion that gene x_k causes the gene x_v can be drawn by determining that variable x_k is a relevant variable of a function of the form (11.14).

In this section, we present a novel method for variable selection in (11.14) using the multipenalty regularization. To our best knowledge, this is the first work in the field that describes an application of multipenalty regularization for inferring causal relations in gene regulatory networks.

An estimator of the target function (11.14) can be constructed as the sum $\sum_{j=1}^{p} f_j^{\lambda}(x_j)$ of the minimizers of the functional

$$T_\lambda(f_1, f_2, \cdots, f_p;\ Z_N) = \frac{1}{N}\sum_{i=1}^{N}\left(y^i - \sum_{j=1}^{p} f_j(x_j^i)\right)^2 + \sum_{j=1}^{p} \lambda_j \parallel f_j \parallel_{\mathcal{H}_j}^2 \tag{11.27}$$

where $\lambda = (\lambda_1, \lambda_2, \cdots, \lambda_p)$ is a vector of the regularization parameters, and $Z_N = \{(x_1^i, x_2^i, \cdots, x_p^i, y^i)\}_{i=1}^{N}$ denotes a data set of observed values y^i, $i = 1, 2, \cdots, N$, of a variable of interest y paired with simultaneously observed values x_ν^i, $\nu = 1, 2, \cdots, p$, of the variables $x_1, x_2, \cdots, x_p$ that possibly interact with y.

On first sight, it may be seen that the results of the minimization of the functional (11.27) do not systematically lead to sparsity as in the previously addressed approaches. The sparse structure can be reconstructed following the next three steps.

The first step consists of constructing the minimizers $f_j = f_j^{\lambda_j}(x_j)$ of the functionals $T_{\lambda_j}(f_j; Z_N)$ defined by (11.27) with $p = 1$, $\lambda_1 = \lambda_j$, $x_1^i = x_j^i$, $\mathcal{H}_1 = \mathcal{H}_j$, $j = 1, 2, \cdots$. By using classical results from approximation theory (Kimeldorf and Wahba, 1970, Schölkopf *et al.*, 2001), such minimization is reduced to solving systems of N linear equations. Then the minimizers $f_j^{\lambda_j}(x_j)$ are used to rank the variables x_j according to the values of the discrepancies

$$D(f_j^{\lambda_j}(x_j);\ Z_N) = \left(\frac{1}{N}\sum_{i=1}^{N}\left(y^i - f_j^{\lambda_j}(x_j^i)\right)^2\right)^{1/2}, j = 1, 2, \cdots$$

as follows: the smaller the value of $D(f_j^{\lambda_j}(x_j); Z_N)$, the higher the rank of x_j. This step can be seen as an attempt to interpret the data Z_N by using only a univariate function, and the variable with the highest rank is considered as the first relevant variable x_{ν_1}.

The next step consists of testing the hypothesis that a variable with the second highest rank, say x_μ, is also relevant. For such a test, we compute the minimizers $f_{\nu_1}^{\lambda_{\nu_1}}, f_\mu^{\lambda_\mu}$ of the functional

$$T_\lambda(f_{\nu_1}, f_\mu;\ Z_N) = \frac{1}{N}\sum_{i=1}^{N}(y^i - f_{\nu_1}(x_{\nu_1}^i) - f_\mu(x_\mu^i))^2 + \lambda_{\nu_1} \parallel f_{\nu_1} \parallel_{\mathcal{H}_{\nu_1}}^2 + \lambda_\mu \parallel f_\mu \parallel_{\mathcal{H}_\mu}^2 \tag{11.28}$$

Our idea is based on the observation (Naumova and Pereverzyev, 2014) that in multi-penalty regularization with a componentwise penalization, such as (11.28), one requires small as well as large values of the regularization parameters $\lambda_{\nu_1}, \lambda_\mu$, that is, both λ_{ν_1} and $\lambda_\mu \ll 1$, and $\lambda_\mu > 1$ respectively. Therefore, in the proposed approach, variable x_μ is considered relevant if for $\{\lambda_{\nu_1}, \lambda_\mu\} \subset (0, 1)$, the values of the discrepancy

$$D(f_{\nu_1}^{\lambda_{\nu_1}}, f_\mu^{\lambda_\mu};\ Z_N) = \left(\frac{1}{N}\sum_{i=1}^{N}\left(y^i - f_{\nu_1}^{\lambda_{\nu_1}}(x_{\nu_1}^i) - f_\mu^{\lambda_\mu}(x_\mu^i)\right)^2\right)^{1/2} \tag{11.29}$$

are essentially smaller than the ones for $\lambda_{\nu_1} \in (0,1)$, $\lambda_\mu > 1$. If it is not the case, the above hypothesis is rejected, and we test in the same way the variable with the third highest rank and so on. When the variable x_μ was accepted as the second relevant variable, that is, $x_{\nu_2} = x_\mu$, we proceed with testing whether the variable with the third highest rank, say x_ν, can be taken as the third relevant variable, that is, whether $x_{\nu_3} = x_\nu$. Thus, we compute the minimizers $f_{\nu_1}^{\lambda_{\nu_1}}, f_{\nu_2}^{\lambda_{\nu_2}}, f_\nu^{\lambda_\nu}$ of the functional

$$T_\lambda(f_{\nu_1}, f_{\nu_2}, f_\nu;\ Z_N) = \frac{1}{N}\sum_{i=1}^{N}(y^i - f_{\nu_1}(x^i_{\nu_1}) - f_{\nu_2}(x^i_{\nu_2}) - f_\nu(x^i_\nu))^2 + \lambda_{\nu_1} \| f_{\nu_1} \|^2_{\mathcal{H}_{\nu_1}} + \lambda_{\nu_2} \| f_{\nu_2} \|^2_{\mathcal{H}_{\nu_2}} + \lambda_\nu \| f_\nu \|^2_{\mathcal{H}_\nu} \tag{11.30}$$

where, with a little abuse of notation, we use the same symbols f_{ν_1}, $f_{\nu_1}^{\lambda_{\nu_1}}$ as in (11.28),(11.29). Then, as above, variable x_ν is considered relevant if for $\{\lambda_{\nu_1}, \lambda_{\nu_2}, \lambda_\nu\} \subset (0,1)$, the values of the discrepancy

$$D(f_{\nu_1}^{\lambda_{\nu_1}}, f_{\nu_2}^{\lambda_{\nu_2}}, f_\nu^{\lambda_\nu};\ Z_N) = \left(\frac{1}{N}\sum_{i=1}^{N}\left(y^i - f_{\nu_1}^{\lambda_{\nu_1}}(x^i_{\nu_1}) - f_{\nu_2}^{\lambda_{\nu_2}}(x^i_{\nu_2}) - f_\nu^{\lambda_\nu}(x^i_\nu)\right)^2\right)^{1/2} \tag{11.31}$$

are essentially smaller than the corresponding values of (11.31) for $\{\lambda_{\nu_1}, \lambda_{\nu_2}\} \subset (0,1)$, $\lambda_\nu > 1$. In our experiments, the small parameter was chosen from the interval $[0.00001, 0.3]$ and the large one from $[1, B]$, where B is some large constant, say $B = 10$.

Otherwise, the variable with the next highest rank is tested in the same way.

If the discrepancy (11.31) exhibits the above-mentioned property, then for testing the variable with the next highest rank in accordance with the proposed approach, we need to add to (11.30) one more penalty term corresponding to that variable, so that the functional $T_\lambda(f_1, f_2, \cdots, f_p; Z_N)$ of the form (11.27) containing the whole set of penalties may appear only at the end of the testing procedure.

For the sake of brevity, we omit the theoretical justification of the presented approach here and refer the interested reader to our recent paper (Hlaváčková-Schindler *et al.*, 2014) for a detailed mathematical description and theoretical justification. However, we note that the theoretical results do not require any strict assumptions neither on the distribution of the data points nor on the number of them.

It is important to mention that the choice of the regularization parameter(s) is not a tedious and tricky task for the proposed method, since we are not interested in the exact reconstruction of the given value y^i but in values of the discrepancies for small and large values of the regularization parameters. Simulations of Monte Carlo type are used to make such comparisons. Namely, if $x_{\nu_1}, x_{\nu_2}, \cdots, x_{\nu_{l-1}}$ have been already accepted as relevant variables, then the values of D for the randomly chosen $(\lambda_{\nu_1}, \lambda_{\nu_2}, \cdots, \lambda_{\nu_l}) \in (0,1)^l$ are compared to the ones for the randomly chosen $(\lambda_{\nu_1}, \lambda_{\nu_2}, \cdots, \lambda_{\nu_l}) \in (0,1)^{l-1} \times [1, B]$, $B > 1$, and x_{ν_l} is accepted as relevant if these values are essentially dominated by the ones for $(\lambda_{\nu_1}, \lambda_{\nu_2}, \cdots, \lambda_{\nu_l}) \in (0,1)^{l-1} \times [1, B]$.

The computational complexity of multi-penalty regularization is $O(Np^2)$, where N is the number of given points and p is the number of variables.

11.7.5 Case Study of Gene Regulatory Network Reconstruction

We used the databases of gene expression data from the biological experiments of Whitfield *et al.* (2002), as in our papers Hlaváčková-Schindler and Bouzari (2013) and Hlaváčková-Schindler and Pereverzyev (2015). We selected 19 genes that are active in human cancer cell line, whose gene regulatory network was reconstructed based on the biological experiments by Li *et al.* (2006). The causal structure for these genes was adapted from Lozano *et al.* (2009) and is presented in Figure 11.1. We take this causal network as a benchmark network for a comparison of the discussed methods. The 19 genes, which we consider, play a substantial role at the human cancer cell lines. They have the following names: PCNA, NPAT, E2F1, CCNE1, CDC25A, CDKN1A, BRCA1, CCNF, CCNA2, CDC20, STK15, BUB1B, CKS2, CDC25C, PLK1, CCNB1, CDC25B, TYMS, and DHFR. The gene expressions in the database

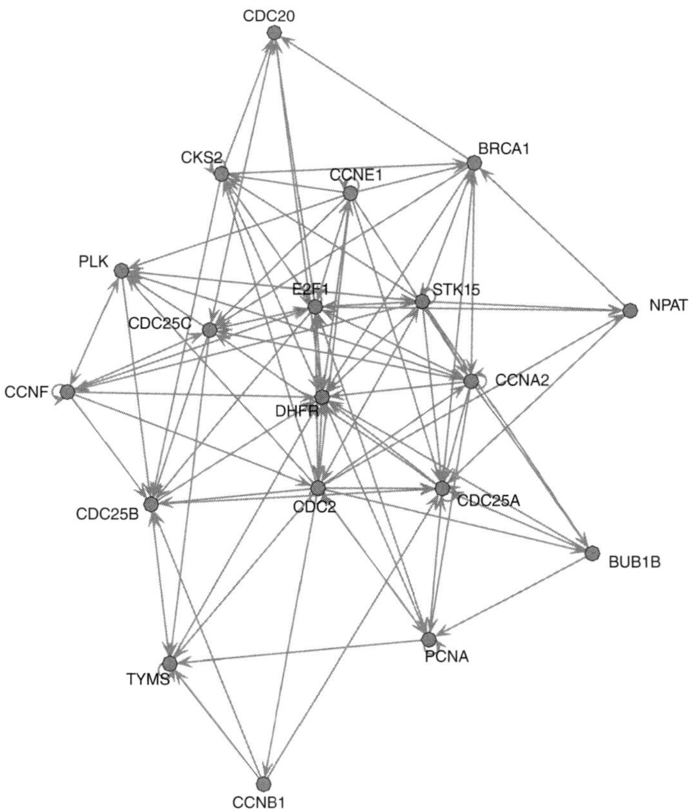

Figure 11.1 Causal structure from biological experiments for 19 selected genes. Reproduced from Lozano *et al.* (2009) with permission of Oxford University Press.

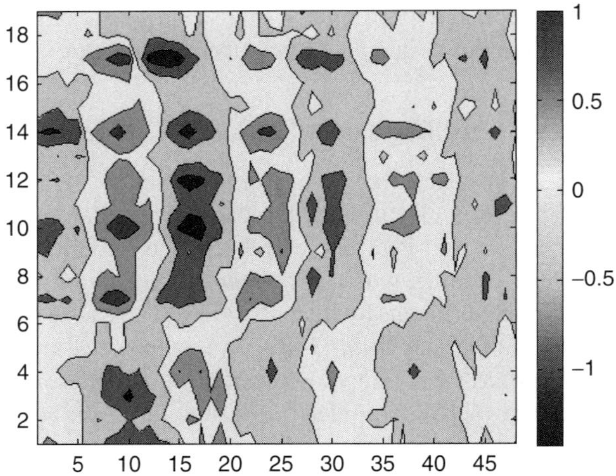

Figure 11.2 The horizontal coordinate x indicates the 48 time measurements and the vertical coordinate y indicates 19 genes ordered. The color of the pixel corresponds to the value determined by the color scale on the right-hand side.

from Whitfield *et al.* (2002) for these genes were given for 48 observations with one-hour intervals.

The data values are illustrated in Figure 11.2. The horizontal coordinate x indicates the 48 time measurements, the vertical coordinate y indicates the order of the 19 genes; the color of the pixel corresponds to the value determined by the color scale on the right-hand side.

We used the following MATLAB codes: our code for GLG-method with an appropriate thresholding that we extended with graphical outputs using MATLAB graphical software Graphviz4MATLAB Version 2.24. For experiments with Lasso Granger method, we used the MATLAB code from Bahadori written for the bivariate case, which we extended to the multivariate case. We extended this code also with the graphical outputs using Graphviz4MATLAB. Similarly, we extended the MATLAB code for Copula Granger method from Bahadori. These methods were compared to the method using dynamic Bayesian networks and ordinary differential equations from Äijö and Lähdesmäki (2009) in Hlaváčková-Schindler and Pereverzyev (2015). The latter method showed frequent overfitting with respect to the number of false positives and had high computationally costs. Here we compare the performance of the Lasso Granger methods with the Granger method with multi-penalty regularization. We developed the code for the multi-penalty regularization method in MATLAB.

As quality (performance) measures we considered the number of true positive outcomes denoted by *TP* and the classification accuracy $\mathrm{CA} = (\mathrm{TP} + \mathrm{TN})/(\mathrm{TP} + \mathrm{TN} + \mathrm{FP} + \mathrm{FN})$.

The Lasso Granger method was tested in four variations:

- Lasso Granger with zero threshold ($\beta_{\mathrm{tr}} = 0$ in (11.6)) and optimized regularization parameter λ in (11.11). We refer to this variation as LG.
- Lasso Granger with optimized regularization parameter and threshold, which is referred to as LG1.

TABLE 11.1 Quality Measures for the Considered Methods. The Number of the Causal Links in the Considered Gene Regulatory Network from Figure 11.1 Is 95. This Number Can Be Seen as the Maximal Possible Value for TP.

	CG	LG	LG1	LG2	LG3	MPR
CA	0.80	0.58	0.88	0.85	0.81	0.88
TP	58	38	63	51	42	53

- Lasso Granger with optimized regularization parameter and threshold given by formula (11.22) with $\alpha = 1/4$, LG2.
- And finally, Lasso Granger with regularization parameter chosen by quasioptimality criterion and threshold given by formula (11.22) with $\alpha = 1/4$, LG3. This is an automatic realization of the Lasso Granger method without the knowledge of the true adjacency matrix.

We call the Granger causality method with Multi-Penalty Regularization (MPR). Let us note that in MPR, the coefficients (β_j^l) in (11.26) have to be precomputed. For this purpose, one can use any (regularization) method for solving the approximation problem (11.5). In this case, of course, the results of MPR depend on the choice of this method. In Hlaváčková-Schindler *et al.* (2014), we used the l_2-regularized least-squares method for obtaining the coefficients (β_j^l). Here, we used the Lasso, which is the l_1-regularized least-squares method. The regularization parameter in both regularization methods was chosen by the quasi-optimality criterion.

The Copula Granger (CG) method , all mentioned variations of the Lasso Granger method, and the Granger Causality with Multi-Penalty Regularization required only a few seconds run at a PC workstation with 64-bit processor. CA and TP quality measures of the considered methods are summarized in Table 11.1. In Figure 11.3, we present the considered gene regulatory network and its reconstructions with LG3 and MPR in the circular layout.

One observes that although the CG gives the largest number of TPs among the automatic realizations of graphical methods, that is, CG, LG3, and MPR, it gives the lowest CA, while MPR gives the highest CA together with rather high TP. This makes MPR a very promising method for the reconstruction of gene regulatory networks.

11.8 CONCLUSION

The results of the reconstruction of the gene regulatory network in the experimental section emphasize the importance of the thresholding strategies for the variable selection regularization methods, such as the Lasso. The newly developed MPR technique (Hlaváčková-Schindler *et al.*, 2014) can be seen as an advanced thresholding, and our experimental results show its superior behavior with respect to our method with thresholding strategy.

As we noted earlier, the MPR requires a method that computes the coefficients (β_j^l) in (11.26). Currently, we tested the behavior of the MPR with l_2-regularization in Hlaváčková-Schindler *et al.* (2014), and here with l_1-regularization, which is the

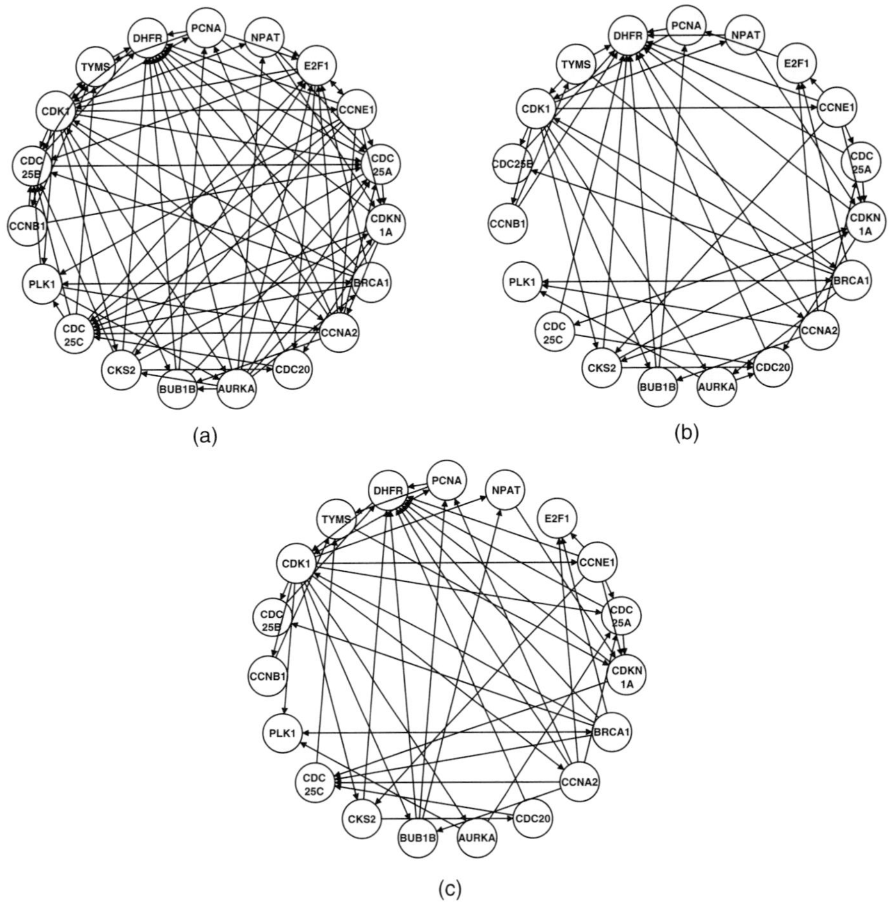

Figure 11.3 The considered gene regulatory network (a) and its reconstructions with LG3 (b) and MPR (c) in the circular layout.

Lasso. In our tests, the MPR gave superior results. Also, the coupling between these Lasso-modifications and MPR is interesting to realize. It would be also interesting to see the reconstruction of the gene regulatory network in Figure 11.1 by means of the discussed modifications of Lasso: truncating Lasso and adaptively thresholded Lasso.

The methods proposed in this paper are written in MATLAB and are available upon request.

ACKNOWLEDGMENTS

The first author gratefully acknowledges the partial support by the research grant GACR 13-13502S of the Grant Agency of the Czech Republic (Czech Science Foundation).

REFERENCES

Abramowitz, M. and Stegun, I.A. (1972) *Handbook of Mathematical Functions: With Formulas, Graphs, and Mathematical Tables*, 9th edn, Courier Corporation, New York.

Äijö, T. and Lähdesmäki, H. (2009) Learning gene regulatory networks from gene expression measurements using non-parametric molecular kinetics. *Bioinformatics*, **25** (22), 2937–2944.

Arnold, A., Liu, Y., and Abe, N. (2007) Temporal causal modeling with graphical Granger methods, in *Proceedings of the 13th ACM SIGKDD International Conference on Knowledge Discovery and Data Mining*, pp. 66–75.

Aronszajn, N. (1950) Theory of reproducing kernels. *Transactions of the American Mathematical Society*, **68** (3),337–404.

Ashrafulla, S., Haldar, J.P., Joshi, A.A., and Leahy, R.M. (2012) Canonical Granger causality applied to functional brain data, in *ISBI*, pp. 1751–1754.

Bach, F.R. (2009) Exploring large feature spaces with hierarchical multiple kernel learning, in *Advances in Neural Information Processing Systems 21*, pp. 105–112.

Bahadori, M.T. `http://www-scf.usc.edu/ mohammab/codes/codes.html` (accessed 13 January 2016).

Bahadori, T. and Liu, Y. (2013a) An examination of large-scale Granger causality inference, in *SIAM Conference on Data Mining*, ISBN 978-1-61197-262-7.

Bahadori, M.T. and Liu, Y. (2013b) An examination of practical Granger causality inference. Proceedings of the 2013 SIAM International Conference on data Mining, pp. 467–475.

Bakushinskii, A.B. (1984) Remarks on choosing a regularization parameter using the quasi-optimality and ratio criterion. *USSR Computational Mathematics and Mathematical Physics*, **24** (4), 181–182.

Bansal, M., Belcastro, V., Ambesi-Impiombato, A., and Di Bernardo, D. (2007) How to infer gene networks from expression profiles. *Molecular Systems Biology*, **3** (1), 78.

Bauer, F. and Reiß, M. (2008) Regularization independent of the noise level: an analysis of quasi-optimality. *Inverse Problems*, **24** (5), 055 009.

Cao, J. and Zhao, H. (2008) Estimating dynamic models for gene regulation networks. *Bioinformatics*, **24** (14), 1619–1624.

Efron, B., Hastie, T., Johnstone, I., and Tibshirani, R. (2004) Least angle regression. *Annals of Statistics*, **32** (2), 407–499.

Engl, H.W., Hanke, M., and Neubauer, A. (1996) *Regularization of Inverse Problems*, vol. **375**, Kluwer Academic Publishers, Dordrecht.

Fornasier, M., Naumova, V., and Pereverzyev, S.V. (2014) Parameter choice strategies for multi-penalty regularization. *SIAM Journal on Numerical Analysis*, **52** (4), 1770–1794.

Fujita, A., Sato, J.R., Garay-Malpartida, H.M., Yamaguchi, R., Miyano, S., Sogayar, M.C., and Ferreira, C.E. (2007) Modeling gene expression regulatory networks with the sparse vector autoregressive model. *BMC Systems Biology*, **1** (1), 39.

Granger, C.W. (1969) Investigating causal relations by econometric models and cross-spectral methods. *Econometrica*, **37**, 424–438.

Hastings, C. Jr., Mosteller, F., Tukey, J.W., and Winsor, C.P. (1947) Low moments for small samples: a comparative study of order statistics. *Annals of Mathematical Statistics*, **18**, 413–426.

Hlaváčková-Schindler, K. and Bouzari, H. (2013) Granger Lasso causal models in higher dimensions –application to gene expression regulatory networks. *The Proceedings of EVML/PKDD 2013, SCALE.*

Hlaváčková-Schindler, K. and Pereverzyev, S. Jr. (2015) Lasso Granger causal models: some strategies and their efficiency for gene expression regulatory networks, in *Decision Making: Uncertainty, Imperfection, Deliberation and Scalability*, Studies in Computational Intelligence, vol. **538** (eds T.V. Guy, M. Kárný, and D.H. Wolpert), Springer-Verlag, pp. 91–117.

Hlaváčková-Schindler, K., Naumova, V., and Pereverzyev, S., J. (2014) Multi-penalty regularization for detecting relevant variables. *Pre-print Nr 11, Leopold Franzens Universität Innsbruck.*

Hofmann, B. (1999) *Mathematik Inverser Probleme*, Teubner, Stuttgart.

Kabanikhin, S.I. (2008) Definitions and examples of inverse and ill-posed problems. *Journal of Inverse and Ill-Posed Problems*, **16** (4), 317–357.

Khanin, R. and Wit, E. (2007) Construction of malaria gene expression network using partial correlations, in *Methods of Microarray Data Analysis V* (eds P. McConnell, S.M. Lin, and P. Hurban), Springer-Verlag, Berlin, pp. 75–88.

Kimeldorf, G.S. and Wahba, G. (1970) A correspondence between Bayesian estimation on stochastic processes and smoothing by splines. *Annals of Mathematical Statistics*, **41**, 495–502.

Kindermann, S. and Neubauer, A. (2008) On the convergence of the quasi-optimality criterion for (iterated) Tikhonov regularization. *Inverse Problems and Imaging*, **2** (2), 291–299.

Kojima, K., Fujita, A., Shimamura, T., Imoto, S., and Miyano, S. (2008) Estimation of nonlinear gene regulatory networks via l1 regularized NVAR from time series gene expression data. *Genome Informatics*, **20**, 37–51.

Lazarov, R.D., Lu, S., and Pereverzev, S.V. (2007) On the balancing principle for some problems of numerical analysis. *Numerische Mathematik*, **106** (4), 659–689.

Li, X., Rao, S., Jiang, W., Li, C., Xiao, Y., Guo, Z., Zhang, Q., Wang, L., Du, L., Li, J., Li, L., Zhang, T., and Wang, Q.K. (2006) Discovery of time-delayed gene regulatory networks based on temporal gene expression profiling. *BMC Bioinformatics*, **7** (1), 26.

Liu, H., Lafferty, J., and Wasserman, L. (2009) The nonparanormal: semi-parametric estimation of high dimensional undirected graphs. *Journal of Machine Learning Research*, **10**, 2295–2328.

Lozano, A.C., Abe, N., Liu, Y., and Rosset, S. (2009) Grouped graphical Granger modeling for gene expression regulatory networks discovery. *Bioinformatics*, **25** (12), i110–i118.

Lu, S. and Pereverzev, S.V. (2011) Multi-parameter regularization and its numerical realization. *Numerische Mathematik*, **118** (1), 1–31.

Lu, S. and Pereverzev, S.V. (2013) *Regularization Theory for Ill-Posed Problems*, Selected Topics, Inverse and Ill-Posed Problems Series, vol. **58**, de Gruyter, Berlin.

Marinazzo, D., Pellicoro, M., and Stramaglia, S. (2012) Causal information approach to partial conditioning in multivariate data sets. *Computational and Mathematical Methods in Medicine*, **2012**, 1–8.

Mathé, P. and Pereverzev, S.V. (2003) Geometry of linear ill-posed problems in variable Hilbert scales. *Inverse Problems*, **19** (3), 789–803.

Mosci, S., Rosasco, L., Santoro, M., Verri, A., and Villa, S. (2011) Nonparametric sparsity and regularization, Technical Report MIT-CSAIL-TR-2011-041, vol. 41, MIT, CSAIL, Cambridge, USA.

Naumova, V. and Pereverzyev, S.V. (2013) Multi-penalty regularization with a component-wise penalization. *Inverse Problems*, **29** (7), 075 002.

Paluš, M., Komárek, V., Procházka, T., Hrnčír, Z., and Štěrbová, K. (2001) Synchronization and information flow in EEGs of epileptic patients. *Engineering in Medicine and Biology Magazine*, **20** (5), 65–71.

Pearl, J. (2009) *Causality: Models, Reasoning and Inference*, 2nd edn, Cambridge University Press, New York.

Pereverzev, S. and Schock, E. (2005) On the adaptive selection of the parameter in regularization of ill-posed problems. *SIAM Journal on Numerical Analysis*, **43** (5), 2060–2076.

Pereverzyev Jr., S. and Hlaváčková-Schindler K. (2014) Graphical Lasso Granger Method 2-Levels-Thresholding for Recovering Causality Networks: Chapter in the Book "System Modelling and Optimization", Editors C. Pötzsche et al., Springer, IFIP, **443**, pp. 220–229.

Rajapakse, J.C. and Mundra, P.A. (2011) Stability of building gene regulatory networks with sparse autoregressive models. *Bioinformatics*, **12** (Suppl. 13), S13.

Rieder, A. (2003) *Keine Probleme mit inversen Problemen. Eine Einführung in ihre stabile Lösung*, Vieweg, Wiesbaden.

Schölkopf, B., Herbrich, R., and Smola, A.J. (2001) A generalized representer theorem, in *Computational Learning Theory*, Lecture Notes in Computer Science, vol. **2111**, Springer-Verlag, Berlin, pp. 416–426.

Seth, A.K. (2005) Causal connectivity of evolved neural networks during behavior. *Network-Computation in Neural Systems*, **16** (1), 35–54.

Shmulevich, I., Dougherty, E.R., Kim, S., and Zhang, W. (2002) Probabilistic Boolean networks: a rule-based uncertainty model for gene regulatory networks. *Bioinformatics*, **18** (2), 261–274.

Shojaie, A. (2013) Link prediction in biological networks using multi-mode exponential random graph models, in *11th Workshop on Mining and Learning with Graphs*, Chicago, IL.

Shojaie, A., Basu, S., and Michailidis, G. (2012a) Adaptive thresholding for reconstructing regulatory networks from time-course gene expression data, Manuscript at `www.biostat.washington.edu` (accessed 23 December 2015).

Shojaie, A., Basu, S., and Michailidis, G. (2012b) Adaptively thresholded Lasso for gene regulatory networks. *Statistics in Biosciences*, **4** (1), 66–83.

Shojaie, A. and Michailidis, G. (2010) Discovering graphical Granger causality using the truncating Lasso penalty. *Bioinformatics*, **26** (18), i517–i523.

Song, S. and Bickel, P.J. (2011) Large vector auto regressions. *arXiv preprint arXiv:1106.3915*.

Soto, J., Pantazis, D., Jerbi, K., Baillet, S., and Leahy, R.M. (2010) Canonical correlation analysis applied to functional connectivity in MEG, in *ISBI*.

Spirtes, P., Glymour, C.N., and Scheines, R. (2001) *Causation, Prediction, and Search*, 2nd edn, MIT Press, Cambridge, MA.

Tibshirani, R. (1996) Regression shrinkage and selection via the Lasso. *Journal of the Royal Statistical Society. Series B (Statistical Methodology)*, **58** (1), 267–288.

Tikhonov, A.N. (1963) Solution of incorrectly formulated problems and the regularization method. *Soviet Mathematics*, **4**, 1035–1038.

Tikhonov, A.N. and Arsenin, V.Y. (1977) *Solutions of Ill-posed Problems*, Winston & Sons, Washington, DC.

Tikhonov, A.N. and Glasko, V. (1965) Use of the regularization method in non-linear problems. *USSR Computational Mathematics and Mathematical Physics*, **5** (3), 93–107.

Tikhonov, A.N., Goncharsky, A.V., Stepanov, V.V., and Yagola, A.G. (1995) *Numerical Methods for the Solution of Ill-posed Problems*, Kluwer Academic Publishers, Dordrecht.

Whitfield, M.L., Sherlock, G., Saldanha, A.J., Murray, J.I., Ball, C.A., Alexander, K.E., Matese, J.C., Perou, C.M., Hurt, M.M., Brown, P.O., and Botstein, D. (2002) Identification of genes periodically expressed in the human cell cycle and their expression in tumors. *Molecular Biology of the Cell*, **13** (6), 1977–2000.

Wiener, N. (1956) The theory of prediction, in *Modern Mathematics for Engineers* (ed. E.F. Beckenbach), McGraw-Hill, New York.

Wikipedia (2013) Causality—Wikipedia, the free encyclopedia, `http://en.wikipedia.org/w/index.php?title=Causality&oldid=575744857` (accessed 23 December 2015).

Yu, J., Smith, V.A., Wang, P.P., Hartemink, A.J., and Jarvis, E.D. (2004) Advances to Bayesian network inference for generating causal networks from observational biological data. *Bioinformatics*, **20** (18), 3594–3603.

Yuan, M. and Lin, Y. (2006) Model selection and estimation in regression with grouped variables. *Journal of the Royal Statistical Society. Series B (Statistical Methodology)*, **68** (1), 49–67.

Zeng, L. and Xie, J. (2012) Group variable selection for data with dependent structures. *Journal of Statistical Computation and Simulation*, **82** (1), 95–106.

Zhang, L. and Kim, S. (2014) Learning gene networks under SNP perturbations using eQTL datasets. *PLoS Computational Biology*, **10** (2), e1003 420.

Zou, H. (2006) The adaptive Lasso and its oracle properties. *Journal of the American Statistical Association*, **101** (476), 1418–1429.

Zou, M. and Conzen, S.D. (2005) A new dynamic Bayesian network (DBN) approach for identifying gene regulatory networks from time course microarray data. *Bioinformatics*, **21** (1), 71–79.

Zou, H. and Hastie, T. (2005) Regularization and variable selection via the elastic net. *Journal of the Royal Statistical Society. Series B (Statistical Methodology)*, **67** (2), 301–320.

12

UNMEASURED RECIPROCAL INTERACTIONS: SPECIFICATION AND FIT USING STRUCTURAL EQUATION MODELS

PHILLIP K. WOOD
Department of Psychological Sciences, University of Missouri, Columbia, MO, USA

12.1 INTRODUCTION

The opportunity to write this chapter is particularly well timed for me personally as it has been approximately 40 years since I first read Ludwig von Bertalanffy's (1968) *General Systems Theory*, the first book I encountered that used mathematics to describe mutual or reciprocal relationships in biological and physical systems. It also marks roughly 30 years since I have learned more about structural equation modeling (SEM) as a National Institute of Aging postdoctoral fellow at Penn State, where I learned how structural equation models can be used to specify models of growth and change over time. Since then I have been interested in using SEM to investigate possible causal models for longitudinal data with special emphasis on problem behaviors such as alcohol use.

Happily, this chapter provides occasion to address an issue that touches on all three topics. Specifically, I would like to focus on models with "instantaneous" reciprocal or autocausal effects which are characteristic of systems which involve reciprocal or self-referring loops. Given the other contributions to this book, it is necessary to first briefly contrast how these differ from cross-lagged panel correlation models and

Statistics and Causality: Methods for Applied Empirical Research, First Edition.
Wolfgang Wiedermann and Alexander von Eye.

Granger causality, which are more traditionally used. The effect of unincluded feedback loops on the parameters of causal models is qualitatively different in form than the usual biases in estimation that result from unincluded relevant variables, which involve traditional linear effects. Unincluded reciprocal effects are unusual in that they may result in under- or overestimation of unstandardized path coefficients but do not necessarily result in changes in the estimated amount of error variance in dependent variables. When reciprocal effects are estimated from cross-sectional data, it is often assumed that instrumental variables exist that are related to some but not all dependent variables of the system. Although instrumental variables make specification of a mathematically identified model easier, other alternatives are also possible when longitudinal data are considered. I will demonstrate a few of these examples using a small Monte Carlo study and actual data.

12.2 TYPES OF RECIPROCAL RELATIONSHIP MODELS

Many models have been proposed for characterizing reciprocal relationships, and, therefore, a brief review of proposed models is discussed in order to explain how the unmeasured reciprocal effects differ from other alternatives. Although methods based on the asymmetric properties of the correlation coefficient are taken up in several of the chapters in this book, the terms Granger causality and cross-lagged panel analysis are often mentioned in the context of longitudinal data. Although search heuristics have also been used to generate models in which the pattern of causal direction is based on search heuristics of patterns of conditional probability (e.g., Csardi and Nepusz, 2006; Epskamp *et al.*, 2012; Kalisch *et al.*, 2012; Pearl, 2009), they are not taken up here because these approaches do not permit specification of autocausal effects (described below), and their focus is generally on manifest as opposed to latent variable relationships. Some of the modules of the Tetrad program considered in Scheines *et al.* (1998) are exceptions to this, but even these modules only consider simple factor structures and not more complicated alternatives such as random intercept factors (Maydeu-Olivares and Coffman, 2006).

12.2.1 Cross-Lagged Panel Approaches

The cross-lagged panel approach for longitudinal data was originally proposed in the field of sociology by Lazarsfeld (1948). This approach involved use of two dichotomous variables measured over time in an analysis of the "two-attribute turn-overs" within a "16-fold table." This approach was extended in subsequent years to advance a theory of how class membership is related to voting over time (Lipset *et al.*, 1954) and later extended to cross-lagged associations for continuous data and experiments (e.g., Campbell, 1963). The general approach was then subsumed as part of the general analysis of parallel time series (Gottman *et al.*, 1969). The cross-lagged panel approach can be most simply illustrated by Figure 12.1, in which two variables are measured across three occasions. Bivariate associations between the two variables at a first occasion are modeled by a simple covariance ($\sigma_{x_1y_1}$), as are (often) the residual

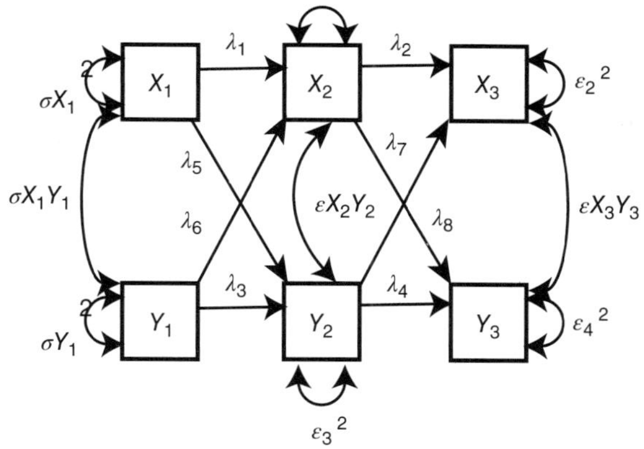

Figure 12.1 Cross-lagged panel correlation model.

effects ($\epsilon_{X_2Y_2}$, $\epsilon_{X_3Y_3}$). In addition, it is thought that each variable over time modeled by autoregressive components (λ_1, λ_2, λ_3, and λ_4). The important question, though, of the relative magnitude of causality between the constructs is assessed by the estimated regression weights from each construct at an earlier assessment to the other construct at the subsequent assessment (i.e., comparisons of λ_5 with λ_6 and λ_7 with λ_8).

Many limitations of cross-lagged panel models are not mentioned in applications of the technique, however. First, the cross-lagged model is not, strictly speaking, a "true" model of the kind of systems level instantaneous reciprocal causality proposed by von Bertalanffy (and discussed in more detail below). This aspect of the cross-lagged model is important because it is often referred to as a model of reciprocal effects (e.g., Selig and Little, 2011, p. 268) and latent variable extensions of the approach tend to only consider the basic logic of cross-lagged approach at the latent variable level (e.g., Sikora *et al.*, 2008). Assumptions of the cross-lagged model are often either ignored or overlooked. Although Kenny (1975) is clear that the cross-lagged panel approach makes the assumption of stationarity in the data, and others have noted that the approach is only valid for systems that are in equilibrium (e.g., (Coleman, 1968)), these limitations are frequently unmentioned in research applications (e.g., Grunberg *et al.*, 2006) or descriptions of the technique (e.g., Selig and Preacher, 2009). Although some (e.g., Rogosa, 1980) have argued that these limitations make the entire approach inadvisable, these cautions appear to be ignored in research applications.

12.2.2 Granger Causality

On first examination, the notion of Granger causality appears similar to the cross-lagged panel approach in that both models test for prospective associations of some other variable on a time series. The two approaches are also similar in that both require the system under examination to be at equilibrium and the measurement model of the constructs involved to be known. Tests of Granger causality are different, however, in that they first involve specification of the lagged structure of a series and then compared it to a model in which lagged values of a variable thought to be causally related to the series. The time series approach taken by Granger also permitted the accommodation of autoregressive effects within a time series model, permitting a different measurement model for the time series than the usual approach to cross-lagged models. Consider, for example, a variable y measured at time t. For each y_t, a regression of the lagged effects would have the following form:

$$y_t = a_0 + a_1 y_{t-1} + a_2 y_{t-2} + a_3 y_{t-3} \cdots (a_m y_{m-1}) + \text{error} \tag{12.1}$$

The second regression testing for Granger causality associated with a second variable x would have the following form:

$$\begin{aligned} y_t = a_0 + a_1 y_{t-1} + a_2 y_{t-2} + a_3 y_{t-3} + \cdots (a_m y_{m-1}) \\ + b_1 x_{t-1} + b_2 x_{t-2} + \cdots + b_m x_{t-m} + \text{error} \end{aligned} \tag{12.2}$$

where the subscript of b indicating lagged effects of a given length. The hypothesis that x does not Granger-cause y is rejected only if none of the lagged values of x are retained in the second regression. The basic idea of Granger causality appears to be extended to a wider variety of possible models, such as techniques not as sensitive to departures from normality of error variances (e.g., Hacker and Hatemi-J, 2012) or asymmetric causal models over time (e.g., Hatemi-J, 2012). Granger causality models, similar to cross-lagged models, however, do not allow for the specification of instantaneous reciprocal relationships such as those considered in systems theory.

12.2.3 Epistemic Causality

The models presented so far assume that causal relationships in the data manifest themselves over time in a stationary system at equilibrium. Causal models for any given system, however, can characterize a great variety of causal mechanisms and the time-delayed causal effects such as those considered in cross-lagged and Granger models can be used to specify and test a variety of causal structures (see von Eye and Wiedermann, 2015 for a taxonomy of several types of Granger causal models in data and their implications for model section and specification). These models, however, do not permit specification of feedback relationships that include instantaneous effects such as those related to chemical or cellular metabolism or thermoregulation, however.

The models considered in this paper differ from those often encountered in discussions of causal effects over time. This larger universe of entertained causal relationships is understood within the context of "epistemic causality" (Williamson, 2006a, 2006b, 2009). From this perspective, causality is a general term that describes several types of relationships between constructs and may include conditional probability, time-bound associations, propensity, or processes such as metabolic rates, for example. Across these various types of causality, however, a causality acts in conjunction with evidence and background knowledge to constrain one's beliefs about relationships between constructs and these constraints can in turn be represented as a causal graph. While some causal graphs may be acyclic and contain no feedback loops (as in the case of Granger causality models), other possible causal networks consist of directed cycle graphs (in which the variables are connected to form a polygon and all arrows are oriented in the same direction), or a reciprocal causal model (in which causal effects in both directions are estimated between at least two of the variables in the diagram. It is possible that competing theories may produce different causal models. These causal graphs are then used in conjunction with the data at hand to generate plausible structural models for the data.

The mapping from a causal graph to a causal model is not one-to-one: A given causal graph may imply more than one structural model, in which case the data at hand may be thought of as agnostic in deciding between equivalent candidate models. Parameters estimated under such equivalent models may, however, be useful in informing subsequent experiments regarding magnitude of effects. Subsequent experiments that include additional variables, longitudinal assessments, or experimental designs can permit more focused tests to enable adjudication between candidate models. In general, then, a given experiment may provide a useful test of an entire proposed causal structure, may be partially informative regarding parts of the model, or may be agnostic regarding the relative merits of candidate causal relationships.

12.2.4 Reciprocal Causality

Although cross-lagged panel correlations and Granger causality are models of reciprocal effects occur over the course of time, other models may contain reciprocal or circular effects that are instantaneous or so finely resolved in time as to be undetectable over time. Such feedback loops can be positive (in which a level of one variable causes more of that variable to be produced) or negative (in which the level of the variable is damped by other processes in the system). Positive feedback loop examples include audio feedback loops (the familiar sound produced when holding a microphone too near a speaker) and maternal oxytocin levels during labor (in which labor is initiated by a given level of oxytocin, which in turn causes further production of oxytocin until birth occurs). The thermostat example alluded to earlier is an example of a negative feedback loop, as well as many aspects of homeostasis such as control of blood sugar in the body or blood pressure regulation. Other examples from von Bertalanffy's (1968) General Systems Theory include predator–prey relationships and population control within ecological systems.

At the heart of all of these feedback systems is the notion of a signal transported within some closed circuit of the system. Within the field of electronic circuits, Mason (1953) addressed the mathematics of the transmission of electrical feedback loops by means of "flow graph" models, which enabled construction of a "gain formula" to model electrical transmission. It was quickly realized, however, that the mathematics underlying such feedback control systems had application to other mathematical systems involving circular transmission of a signal, such as some aspects of hydraulics. In the field of SEM, the transmission of a signal (in this case variability exogenous to the system) led Heise (2001) to extend such models to social science research questions. Although Heise (2001) and Mulaik (2009, p. 293) illustrate use of the gain formula within SEM to produce statements of the expected relationships between variables when the model contains feedback loops, generation of the predicted covariance matrix using standard SEM approaches is straightforward and produces models that correspond to unmeasured reciprocal effects under particular identification constraints.

Generally, feedback systems can consist of reciprocal effects in which arrows pointing in both directions appear as in the arrows are modeled as shown in the relationship between variables Y and Z in Figure 12.2a or as part of a larger model as shown in Figure 12.4a. Cycle relationships are another type of feedback model in which a chain of variables is both causal and antecedent to other variables as shown in Figure 12.4b. Under both reciprocal and cycle relationships, effects between

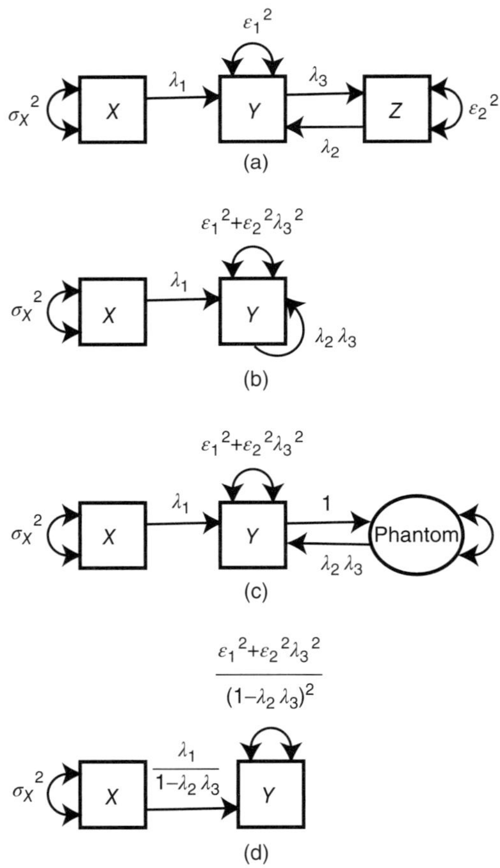

Figure 12.2 Reciprocal and autocausal effects models and bias of univariate regression. (a) Reciprocal effect with instrumental variable. (b) Autocausal effect. (c) Estimation of autocausal effect using phantom variable. (d) Bias of bivariate regression.

variables are referred to as "closed paths" in order to distinguish them from the "open paths" of directed effects more common in structural models.

Mason flow graph analysis is used to specify the total effect of signal transmission in the presence of feedback loops. This is useful because it can be used to describe the patterns of transmission in the circuit even when it was not practicable to directly assess the variable(s) responsible for the feedback but where the instantaneous feedback is determined from theory. Figure 12.2a is a frequently used example of a model containing a single reciprocal relationship between variables Y and Z. This model of reciprocal causation is mathematically identified due to the presence of variable X, which acts as an instrumental variable. The term "instrumental variable" was first introduced by economist Philip Wright (Wright, 1922), the father of geneticist Sewell Wright who was, in turn, the father of path analysis. Wright's original definition of an instrumental variable applied to path models and stated simply that the variable X be unrelated to the error term of the criterion variable, Z. Pearl (2009) proposed a more general definition of instrumental variables that relies on graph theory and requires only the criteria of conditional independence of the instrument with the criterion given the predictor. In this chapter, we will confine ourselves to instrumental variables within structural equation models and make only the minor adjustment that it is possible that the instrumental variable of interest may be a manifest variable or a latent variable posited by the researcher. Within the context of structural equation models, the requirement of conditional independence does not represent sufficient conditions for the model, as it is also necessary that the proposed model by mathematically and empirically identified (as discussed below), but at this point, any variable that is conditionally independent of an error term will be taken as a working definition for an instrumental variable, which can potentially enable estimation of reciprocal effects of the type considered here. The critical issue in specifying and estimating reciprocal effects concerns whether a model is mathematically identified, which can often be a somewhat difficult issue as discussed in what follows.

Returning to Figure 12.2a, we can think of the boxes labeled X, Y, and Z as representing points on an electrical circuit (which, in the case of Y and Z, not only consist of both speakers on a sound system but also microphones that produce a feedback loop). Suppose that the researcher is interested in the "total signal" being delivered to the component Y. In this diagram, a signal, with magnitude indicated by σ_x^2 , is transmitted from X to Y. The proportion of the signal sent is indicated by the path coefficient λ_1. A second unique signal is directly sent to component Y and its amount is indicated by ϵ_1^2. Component Y also transmits its (total) signal to variable Z in proportion to λ_3. Variable Z, in turn, receives a unique signal indicated by ϵ_2^2, and sends λ_2 of its total signal back to component Y, completing a feedback loop between variables Y and Z. Given this causal structure, reciprocal relationships can be modeled as structural equation models provided that the variables involved in the feedback loop, λ_2 and λ_3, satisfy certain properties to be described in the following section.

12.2.4.1 Predicted Covariance Matrix As with any structural equation model, the parameters of the model can be used to generate a statement of the predicted variance/covariance matrix based on the model. A standard way to express this predicted covariance matrix is $(I - A)^{-1}S(I - A)^{-1'}$, where S is a symmetric matrix of variances in the diagram and A is a matrix of path coefficients (McArdle and McDonald, 1984; McDonald, 1980). For these data, the exogenous variance of X, σ_X^2, and the errors of prediction associated with X and Z (ϵ_1^2 and ϵ_2^2, respectively) are expressed as

$$S = \begin{bmatrix} \sigma_X^2 & 0 & 0 \\ 0 & \epsilon_1^2 & 0 \\ 0 & 0 & \epsilon_2^2 \end{bmatrix} \tag{12.3}$$

while the matrix of path information,

$$A = \begin{bmatrix} 0 & 0 & 0 \\ \lambda_1 & 0 & \lambda_2 \\ 0 & \lambda_3 & 0 \end{bmatrix} \tag{12.4}$$

Although, for recursive models, the inverse of $I - A$ can be calculated as $I + A + AA + AAA$, the presence of a feedback loop in the diagram results in infinite sums, as the matrix of powers of A never become a null matrix. Specifically, the first several terms of $I + A + AA \cdots$ can be written as

$$\begin{bmatrix} 1 & 0 & 0 \\ \lambda_1 + \lambda_2\lambda_3\lambda_1 + \lambda_1\lambda_3^2\lambda_2^2 + \lambda_1\lambda_3^3\lambda_2^3 + \lambda_1\lambda_3^4\lambda_2^4 & \lambda_2\lambda_3 + \lambda_3^2\lambda_2^2 + \lambda_3^3\lambda_2^3 + \lambda_3^4\lambda_2^4 + \lambda_2^5\lambda_3^5 + 1 & \lambda_2 + \lambda_3\lambda_2^2 + \lambda_3^2\lambda_2^3 + \lambda_3^3\lambda_2^4 + \lambda_3^4\lambda_2^5 \\ \lambda_3\lambda_1 + \lambda_2\lambda_1\lambda_3^2 + \lambda_1\lambda_2^2\lambda_3^3 + \lambda_1\lambda_2^3\lambda_3^4 + \lambda_1\lambda_3^5\lambda_2^4 & \lambda_3 + \lambda_2\lambda_1\lambda_3^2 + \lambda_2^2\lambda_3^3 + \lambda_2^3\lambda_3^4 + \lambda_2^4\lambda_3^5 & \lambda_3\lambda_2 + \lambda_2^2\lambda_3^2 + \lambda_2^3\lambda_3^3 + \lambda_2^4\lambda_3^4 + \lambda_3^5\lambda_2^5 + 1 \end{bmatrix} \tag{12.5}$$

which can be rewritten as

$$\begin{bmatrix} 1 & 0 & 0 \\ \lambda_1 \sum\limits_{k=0}^{\infty} \lambda_2^k\lambda_3^k & \sum\limits_{k=0}^{\infty} \lambda_2^k\lambda_3^k & \lambda_2 \sum\limits_{k=0}^{\infty} \lambda_2^k\lambda_3^k \\ \lambda_1\lambda_3 \sum\limits_{k=0}^{\infty} \lambda_2^k\lambda_3^k & \lambda_3 \sum\limits_{k=0}^{\infty} \lambda_2^k\lambda_3^k & \sum\limits_{k=0}^{\infty} \lambda_2^k\lambda_3^k \end{bmatrix} \tag{12.6}$$

This infinite series converges (as explained below) provided that $|\lambda_2\lambda_3| < 1$. The predicted covariance matrix then is

$$\begin{bmatrix} 1 & 0 & 0 \\ \lambda_1 \sum\limits_{k=0}^{\infty} \lambda_2^k\lambda_3^k & \sum\limits_{k=0}^{\infty} \lambda_2^k\lambda_3^k & \lambda_2 \sum\limits_{k=0}^{\infty} \lambda_2^k\lambda_3^k \\ \lambda_1\lambda_3 \sum\limits_{k=0}^{\infty} \lambda_2^k\lambda_3^k & \lambda_3 \sum\limits_{k=0}^{\infty} \lambda_2^k\lambda_3^k & \sum\limits_{k=0}^{\infty} \lambda_2^k\lambda_3^k \end{bmatrix} \begin{bmatrix} \sigma_X^2 & 0 & 0 \\ 0 & \epsilon_1^2 & 0 \\ 0 & 0 & \epsilon_2^2 \end{bmatrix}$$

$$\begin{bmatrix} 1 & \lambda_1 \sum_{k=0}^{\infty} \lambda_2^k \lambda_3^k & \lambda_1 \lambda_3 \sum_{k=0}^{\infty} \lambda_2^k \lambda_3^k \\ 0 & \sum_{k=0}^{\infty} \lambda_2^k \lambda_3^k & \lambda_3 \sum_{k=0}^{\infty} \lambda_2^k \lambda_3^k \\ 0 & \lambda_2 \sum_{k=0}^{\infty} \lambda_2^k \lambda_3^k & \sum_{k=0}^{\infty} \lambda_2^k \lambda_3^k \end{bmatrix} \tag{12.7}$$

which, upon multiplying out and factoring yields

$$\begin{bmatrix} \sigma_X^2 & \lambda_1 \sigma_X^2 \sum_{k=0}^{\infty} \lambda_2^k \lambda_3^k & \lambda_1 \lambda_3 \sigma_X^2 \sum_{k=0}^{\infty} \lambda_2^k \lambda_3^k \\ \lambda_1 \sigma_X^2 \sum_{k=0}^{\infty} \lambda_2^k \lambda_3^k & (\sigma_X^2 \lambda_1^2 + \epsilon_1^2 + \epsilon_2^2 \lambda_2^2) \sum_{k=0}^{\infty} \lambda_2^k \lambda_3^{k^2} & (\lambda_3 \sigma_X^2 \lambda_1^2 \epsilon_1^2 \lambda_3 + \epsilon_2^2 \lambda_2) \sum_{k=0}^{\infty} \lambda_2^k \lambda_3^{k^2} \\ \lambda_1 \lambda_3 \sigma_X^2 \sum_{k=0}^{\infty} \lambda_2^k \lambda_3^k & (\lambda_3 \sigma_X^2 \lambda_1^2 + \epsilon_1^2 \lambda_3 + \epsilon_2^2 \lambda_2) \sum_{k=0}^{\infty} \lambda_2^k \lambda_3^{k^2} & (\sigma_X^2 \lambda_1^2 \lambda_3^2 + \epsilon_1^2 \lambda_3^2 + \epsilon_2^2) \sum_{k=0}^{\infty} \lambda_2^k \lambda_3^{k^2} \end{bmatrix}$$

Since $\sum_{k=0}^{\infty} \lambda_2^k \lambda_3^k = \frac{1}{1-\lambda_2 \lambda_3}$ (provided, again, that $|\lambda_2 \lambda_3| < 1$), we can rewrite this as

$$\begin{bmatrix} \sigma_X^2 & \dfrac{\lambda_1 \sigma_X^2}{1 - \lambda_2 \lambda_3} & \dfrac{\lambda_1 \lambda_2 \sigma_X^2}{1 - \lambda_2 \lambda_3} \\ \dfrac{\lambda_1 \sigma_X^2}{1 - \lambda_2 \lambda_3} & \dfrac{\epsilon_1^2 + \epsilon_2^2 \lambda_3^2 + \sigma_X^2 \lambda_1^2}{(1 - \lambda_2 \lambda_3)^2} & \dfrac{\lambda_2 \sigma_X^2 \lambda_1^2 + \epsilon_1^2 \lambda_2 + \epsilon_2^2 \lambda_3}{(1 - \lambda_2 \lambda_3)^2} \\ \dfrac{\lambda_1 \lambda_2 \sigma_X^2}{1 - \lambda_2 \lambda_3} & \dfrac{\lambda_2 \sigma_X^2 \lambda_1^2 + \epsilon_1^2 \lambda_2 + \epsilon_2^2 \lambda_3}{(1 - \lambda_2 \lambda_3)^2} & \dfrac{\sigma_X^2 \lambda_1^2 \lambda_2^2 + \epsilon_1^2 \lambda_2^2 + \epsilon_2^2}{(1 - \lambda_2 \lambda_3)^2} \end{bmatrix}.$$

This model is, then estimable, provided, then, that the researcher knows in advance the causal structure of the system the relative magnitude of particular reciprocal effects, that the product of λ_2 and λ_3 is less than 1, and the actual and predicted variance/covariance matrices are positive semidefinite. There is reason, in some situations, to believe that this is the case. For example, Pearl (2009) considers the case of a randomized experiment in which a coin flip (the X variable in the diagram) determines assignment to experimental condition (the Y variable) in order to assess the magnitude of the causal relationship of an independent variable (Y) on the dependent variable (variable Z variable in the diagram). In Pearl's presentation, this structural model is presented with the goal of arguing that such a structural model yields an unambiguous direction of causality, as the randomized variable X (the coin flip) cannot be correlated with the dependent variable.

As an example of a study corresponding to Figure 12.2a, suppose that a researcher decides on a randomized experimental manipulation in which the treatment consists assigning individuals to drinking one alcoholic beverage per day for a month and the control group is to assign a placebo drink. Suppose further that the researcher is interested in knowing whether alcohol consumption affects individuals' perceptions of the effects of alcohol consumption (say, the expectation that alcohol consumption make social interactions easier). If, on completion of the experiment, an association is found between alcohol consumption and the social expectation variable, this is taken as evidence of the causal association of alcohol on expectancy. Although Pearl (2009) used this experimental design to determine the direction of the effect between the two variables, from a structural modeling perspective such an experiment permits joint estimation of both coefficients, thereby permitting an estimate of a feedback effect. The feedback loop given in the top of Figure 12.2a can be used to simultaneously estimate two causes of alcohol consumption: one based on the randomized variable X, but another due to the social lubrication variable Z. This model could then be compared with unidirectional models of effect to determine whether a reciprocal relationship actually obtains between the variables of alcohol consumption and social lubrication.

In similar manner, one can consider the case of a directed cycle system in which variables are arranged in a connected graph in which causal direction is the same, as shown in Figure 12.3b. This system does not have individual instrumental variables usually encountered in discussions of reciprocal effects, but these can be estimated. For example, von Bertalanffy (1968) considers the case of a feedback loop involving a thermostat, a furnace, and a (heated) room as an example of a feedback loop. Such a cycle involving three variables is just identified and, provided that $|\lambda_1\lambda_2\lambda_3| < 1$, the model may be estimated. The predicted covariance matrix is

$$\begin{bmatrix} \dfrac{\epsilon_1^2+\epsilon_2^2\lambda_2^2\lambda_3^2+\epsilon_3^2\lambda_2^2}{(1-\lambda_1\lambda_2\lambda_3)^2} & \dfrac{\epsilon_1^2\lambda_1+\epsilon_2^2\lambda_2\lambda_3+\epsilon_3^2\lambda_1\lambda_2^2}{(1-\lambda_1\lambda_2\lambda_3)^2} & \dfrac{\epsilon_1^2\lambda_1\lambda_3+\epsilon_2^2\lambda_2\lambda_3^2+\epsilon_3^2\lambda_2}{(1-\lambda_1\lambda_2\lambda_3)^2} \\ \dfrac{\epsilon_1^2\lambda_1+\epsilon_2^2\lambda_2\lambda_3+\epsilon_3^2\lambda_1\lambda_2^2}{(1-\lambda_1\lambda_2\lambda_3)^2} & \dfrac{\epsilon_1^2\lambda_1^2+\epsilon_2^2+\epsilon_3^2\lambda_1^2\lambda_2^2}{(1-\lambda_1\lambda_2\lambda_3)^2} & \dfrac{\epsilon_1^2\lambda_3\lambda_1^2+\epsilon_2^2\lambda_3+\epsilon_3^2\lambda_1\lambda_2}{(1-\lambda_1\lambda_2\lambda_3)^2} \\ \dfrac{\epsilon_1^2\lambda_1\lambda_3+\epsilon_2^2\lambda_2\lambda_3^2+\epsilon_3^2\lambda_2}{(1-\lambda_1\lambda_2\lambda_3)^2} & \dfrac{\epsilon_1^2\lambda_3\lambda_1^2+\epsilon_2^2\lambda_3+\epsilon_3^2\lambda_1\lambda_2}{(1-\lambda_1\lambda_2\lambda_3)^2} & \dfrac{\epsilon_1^2\lambda_1^2\lambda_3^2+\epsilon_2^2\lambda_3^2+\epsilon_3^2}{(1-\lambda_1\lambda_2\lambda_3)^2} \end{bmatrix} \tag{12.8}$$

and it is seen that all terms in the predicted variance/covariance matrix are adjusted by $(1-\lambda_1\lambda_2\lambda_3)^2$. Note that circular systems differ from models with instrumental variables in that the direction of causality can be reversed without loss of fit. Specifically, a model in which X causes Y which causes Z which in turn causes X fits just as well as a model in which the causal direction of these arrows is reversed. If circular effects are specified in the presence of at least one instrumental variable, however, the direction of the effects makes the direction of causal effect distinguishable.

12.3 UNMEASURED RECIPROCAL AND AUTOCAUSAL EFFECTS

It seems unlikely, however, that only a single reciprocal effect or even single cycle is present in the data as many additional unmeasured variables may be present in any self-regulatory system process. Alcohol expectations, for example, may involve many other attributions in addition to social lubrication effects. As mentioned, in

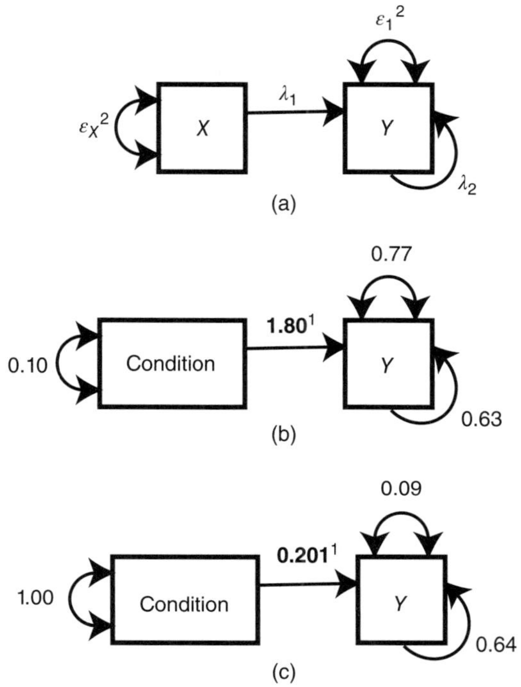

Figure 12.3 Autocausal model and estimates for drinking data. (a) Model parameters. (b) Unstandardized ([1]fixed parameter). (c) Standardized ([1]fixed parameter).

some situations it appears that a variable may seem to "cause itself" by increasing or decreasing values in direct proportion to their current levels. Such effects are often discussed as damping or stimulative effects (as in the thermal regulation or oxytocin examples mentioned above). Given, however, that technically such effects are due to other components of the system, one may alternatively think of such autocausal effects as unmeasured reciprocal effects.

As an example of this, consider now a slightly different experiment in which individuals are again enrolled in either the alcohol or placebo conditions but are asked (or assessed) regarding their actual level of alcohol consumption. Assuming that individuals would have some nonzero base rate of alcohol consumption in absence of the experimental manipulation, reports of alcohol consumption in the experimental condition that are higher or lower than that associated with the control condition may be thought of as representing satiation (in the case of lower-than-expected baseline consumption patterns) or promotion effects (in the case of elevated alcohol consumption).

As an example of an unincluded reciprocal effect, in Figure 12.2b the variable Z (alcohol's social lubrication effects in our example) is omitted. This model contains six estimated parameters, and, based on only bivariate information, is not identified. If, however, the problem can be reduced to one of estimation of three parameters, the problem can be specified as a just-identified model and the model shown in Figure 12.2b can be estimated. If the statistical software used does not allow for the specification of autocausal effects, they can be programmed using a phantom variable with zero variance as shown in Figure 12.2c. Consider, however, the case of a quasi-randomized design in which the coin flip variable is defined as earlier, but unmeasured reciprocal interactions are also present. The reduction in the number of freely estimated parameters can be accomplished in the case of a randomized assignment variable such as the coin flip due to the fact that the researcher also knows that exact value of the regression weight assigned in addition to knowing the causal direction. For example, if the coin flip is coded as a dummy vector, the regression weight λ_1 corresponds to the number of points on the variable Y associated with a one point gain in the coin flip variable. In our example above, for example, if the X variable consists of the number of drinks consumed in a day, λ_1 will be known and fixed to 1. If the free parameters of the model consist then of the freely estimated variance of the coin flip variable, the residual variance (which is the sum of $\epsilon_1^2 + \lambda_3\epsilon_2^2$), and the net effect of the unmeasured reciprocal interaction $(\lambda_2\lambda_3)$ the model is mathematically identified. In order to estimate this net effect, the model can be rewritten using phantom variables (Rindskopf, 1984). The biasing effects of unmeasured, but correlated predictors is well understood, but in the case of the randomizing variable X, there can be no such correlated predictors given the construction of the variable. Although the "usual" unmeasured predictors can still introduce bias in standardized coefficients (due to the failure to correctly model the error term associated with X), the only remaining source of bias in the Y variable is due to the net effect of the unmeasured reciprocal effect, $\lambda_2\lambda_3$, which will cause values greater or less than the true value of λ_1 depending on whether $\lambda_2\lambda_3$ is positive (a positive feedback loop) or negative (indicating a negative feedback loop).

12.3.1 Bias in Standardized Regression Weight

Suppose that the researcher models the relationship between X and Y as a standard regression. By further rewriting the structural model to remove the presence of the unmeasured reciprocal relationship, the parameters of the regression model will correspond to those shown in Figure 12.2d. For example, the researcher calculates a regression model assuming only a unidirectional regression model, the regression weight predicting Y from X alone, this yields

$$\dot{\lambda}_1 = \frac{(\lambda_1\sigma_X^2/1 - \lambda_2\lambda_3)}{\sigma_X^2} = \frac{\lambda_1}{1 - \lambda_2\lambda_3}$$

From this we can see that the estimated regression weight will be an overestimate (in absolute value) of the true effect if λ_2 is the same sign as λ_3, and will be an underestimate (in absolute value) if λ_2 is opposite in sign from λ_3. In terms of bias in the standardized regression weight, note that the estimated standardized regression weight, β_1, can then be calculated as

$$\dot{\beta}_1 = \left(\frac{\lambda_1}{1 - \lambda_1\lambda_2}\right)\frac{\sigma_X}{(\sqrt{\epsilon_1^2 + \epsilon_2^2\lambda_3^2 + \sigma_X^2\lambda_1^2}/1 - \lambda_2\lambda_3)}$$

whereas the true standardized regression weight is

$$\beta_1 = \lambda_1 \frac{\sigma_X}{\sqrt{\epsilon_1^2 + \epsilon_2^2\lambda_3^2 + \sigma_X^2\lambda_1^2/(1-\lambda_2\lambda_3)}}$$

Now the variance in Y accounted for the reduced model is $((\lambda_1\sigma_X)^2/(1-\lambda_2\lambda_3)^2) = (\lambda_1^2\sigma_X^2/\lambda_2^2\lambda_3^2 - 2\lambda 2\lambda 3 + 1)$ leaving the error variance to be $(\epsilon_1^2 + \epsilon_2^2\lambda_3^2 + \sigma_X^2\lambda_1^2/(1-\lambda_2\lambda_3)^2) - (\lambda_1^2\sigma_X^2/(1-\lambda_2\lambda_3)^2) = (\epsilon_1^2 + \epsilon_2^2\lambda_3^2/(1-\lambda_2\lambda_3)^2)$ as shown in doubled-headed arrow associated with Y in Figure 12.2d.

That the estimated proportion of error is larger than the true value due to the failure to include the proportion of variance due to the unique effects of variable $Z(\epsilon_2^2\lambda_3^2/(1-\lambda_2\lambda_3)^2)$ is unsurprising, but the fact that the magnitude of the reciprocal effect not involving variability due to Z (as expressed in the term $(1-\lambda_2\lambda_3)^2$), means that the estimated error variance can also be larger than the true variance ϵ_1^2 if the product of λ_2 and λ_3 is positive but can produce smaller values if the product $\lambda_2\lambda_3$ is negative. It is, for example, possible to calculate the difference between the true error associated with Y and that based on the standard single predictor model and the true value of ϵ_2^2 as $(\epsilon_2^2\lambda_3^2/(1-\lambda_2\lambda_3)^2) + (\epsilon_1^2/(1-\lambda_2\lambda_3)^2) - \epsilon_1^2 = (\epsilon_2^2\lambda_3^2/(1-\lambda_2\lambda_3)^2) + (\epsilon_1^2\lambda_2\lambda_3(-\lambda_2\lambda_3+2)/(1-\lambda_2\lambda_3)^2)$. It is unsurprising that the magnitude of the estimated error variance, ϵ_2^2, is larger due to the failure to include the variable Z in the model. The fact, though, that a term exists that involves only ϵ_1^2 and that can cause the estimated error of the single predictor variable to be larger or smaller than the true value of ϵ_1^2 is, however, somewhat surprising. Specifically, the estimated error variance due to the unmodeled reciprocal relationship error is bigger than ϵ_1^2 when $\lambda_2\lambda_3$ is positive, but less than ϵ_1^2 when $\lambda_2\lambda_3$ is negative. The unexplained variability due solely to unmodeled reciprocal relationships constitutes an additional source of variation distinct from variances, covariances, and path coefficients usually considered in path diagrams. Although such variabilities may be of interest, it is essential that such effects be estimable from the data at hand. It is to this topic that we now turn.

For a simple model with one predictor such as the one shown in Figure 12.2b, the value of λ_1 must be fixed to some known value in order to identify the model and estimate the total signal associated with it via the reciprocal feedback term. It is not possible to identify the separate values of λ_2 and λ_3, given that variable Z is not assessed, but the net effect of reciprocal effects can be estimated. Given that the "self-referring arrows" of feedback effects are not possible in some SEM software packages, such effects can be modeled using phantom variables that contain "signal" or variance components set to zero (Rindskopf, 1984), as shown in Figure 12.2c. Finally, if it is the case that the researcher believes the structural model relating X to Y contains no feedback terms, the resulting parameters which be combinations of the true values as shown in Figure 12.2d.

12.3.2 Autocausal Effects

It may be the case that no specific unmeasured variable exists, but that deviations from the predicted amount of the dependent variable are a structural aspect of the system.

In the case of alcohol, for example, a researcher may be able to titrate the amount of alcohol administered to produce a desired blood alcohol level based on body mass but, on measuring the amount of alcohol actually in the blood stream, finds that this is significantly lower than predicted for some individuals. It would seem reasonable to believe that the physiological systems of these individuals appear able to more rapidly metabolize alcohol, an autocausal effect.

Taken together flow diagrams or path diagrams with reciprocal relationships highlight the qualitatively different ways in which reciprocal causality affect causal structure. These effects have not been frequently addressed, perhaps due to the belief that such unmodeled sources of variability are covered under the more general topic of unincluded variables. The behavior of unincluded open-ended loops results in only two effects: overestimation of error variability and bias in parameter estimates if the unincluded term is correlated with other exogenous variables in the model. The situation is somewhat different for models with reciprocal relationships, however. If the magnitude of the feedback effect is known, it becomes possible to estimate some parameters of the structural model without bias. In addition, as shown in Figure 12.2d, estimated path coefficients associated with variables unrelated to the feedback loop can be either over- or underestimates of their true values, depending on whether the composite effect of the feedback loop is positive or negative. In the text that follows, however, the argument is made that unincluded reciprocal causal effects are qualitatively different from the case of unincluded variables because reciprocal effects constitute an additional source of unmodeled variability in the variables that are included in the model as well.

In order to demonstrate estimation of unmeasured reciprocal effects in a real-world example, follow-up data from an alcohol challenge study are considered, which were collected by Heath (2015). In this study, 193 college students were administered on average 1.78 ounces of alcohol, while 186 students participated in a control condition and 176 were given a placebo nonalcoholic drink. The following day, participants were asked how many alcoholic beverages (beer, wine, hard liquor) they consumed the next day. For purposes of the analysis, individuals in the control and placebo conditions were combined into a single group and reports of next day drinks of alcohol were converted into the corresponding ounces of alcohol consumed. These ounces of alcohol were then combined with the previous day's consumption (when the experiment occurred) to produce a single alcohol composite. The research question of interest explored whether the intervention of consuming alcohol reduced or enhanced individuals' base rates of alcohol consumption. The model consisted of a single dummy variable (indicating the random assignment to the alcohol or no-alcohol condition) with the regression weight from the dummy variable fixed to the amount of ounces of alcohol administered in the study and the dependent variable consisting of total ounces of alcohol consumed over the 2 days.

Unfortunately, this particular data set appears to show that participation in the condition of consuming alcohol had no effect whatever on alcohol consumption patterns of the students on the next day, with no evident differences in alcohol consumption at any level between the two groups. To illustrate what such an effect could look like, however, I deleted the data from individuals in the alcohol condition who did not consume alcohol the next day in order to simulate a condition in which individuals who consumed alcohol in the study had a higher base rate of alcohol consumption than those in the control groups. For this modified data set, a single predictor regression is fit in which the regression weight is fixed to the known value of alcohol administration (in this case equal to 1.8, the known number of ounces of alcohol administered) as shown in Figure 12.2b. This model has freely estimated variances for the dummy coded condition variable (0.102, $p < 0.001$), error variance (0.767, $p < 0.001$), and autocause (0.634, $p < 0.001$). (As this model is just identified, no traditional measures of goodness-of-fit measures are estimable for the data.) This autocause effect, then, could represent a property of the system (i.e., "promotion" or "satiety") or could represent the net effect of one or more feedback terms involving unmeasured variables in the system. This example, although contrived, however, illustrates how conditions of experimental control can also be used to produce a test for the presence of autocausal or unmeasured reciprocal interactions in the data. It highlights the fact that autocausal or unmeasured reciprocal effects model the discrepancy between the observed covariance between an exogenous variable and an endogenous variable and that predicted under the structural model.

12.3.3 Instrumental Variables

The estimation of reciprocal effects in general is made possible by such discrepancies between the observed covariance of an exogenous variable with some other endogenous variables in the system.. Discussions of reciprocal effects for structural models appear to assume that the exogenous variables of the system take the form of "instrumental variables" related to one, but not all endogenous variables in the system. For example, Heise (2001) notes: "Coefficients in non-recursive relations can also be identified and estimated provided that the specification of the system includes some variables with certain restricted features. These "instrumental variables"... may be part of the original specification of the system or they may be added to the system specification merely as a matter of research design. Regardless of whether instrumental variables are considered practically relevant, their conceptualization at the time of theorizing is a matter of utmost importance. The structural coefficients in a nonrecursive system can be estimated from cross-sectional data only if adequate instruments are available." As noted above, the number of instrumental variables and their relationship to variables in the model have important implications for the types of relationships that can be modeled between variables in a feedback loop. For example, in Figure 12.4a, the presence of two indicator variables in a path diagram permits the estimation of both reciprocal relationships between the dependent variables Y and Z and the estimation of the covariation between measurement errors, a constellation of effects that was not estimable if only one instrumental variable was present as shown in Figure 12.1. This model, however, is just identified, meaning that it involves the estimation of as many parameters as there are unique elements of the variance/covariance matrix for the data.

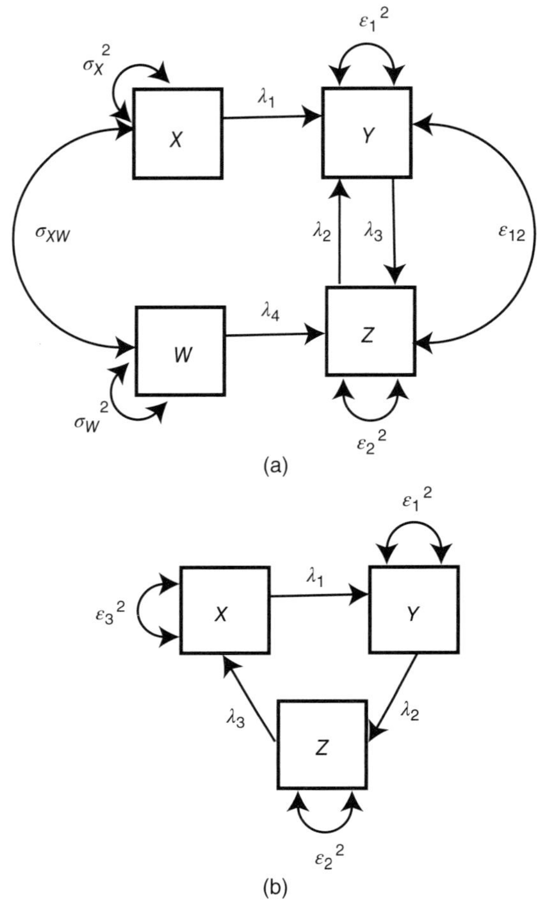

Figure 12.4 Examples of reciprocal and circular models. (a) Reciprocal model. (b) Circular model.

The same algebra can be used to generate the predicted variance/covariance matrix for other models that involve feedback loops and correlated error terms. Using the same algebra for the structural model shown in Figure 12.4a, the predicted covariance matrix is

$$\left[\begin{array}{ccc} \sigma_X^2 & \sigma_{WX} & \frac{\lambda_1\sigma_X^2+\lambda_2\lambda_4\sigma_{WX}}{1-\lambda_2\lambda_3} \\ \sigma_{WX} & \sigma_W^2 & \frac{\lambda_1\sigma_{WX}+\lambda_2\lambda_4\sigma_W^2}{1-\lambda_2\lambda_3} \\ \frac{\lambda_1\sigma_X^2+\lambda_2\lambda_4\sigma_{WX}}{1-\lambda_2\lambda_3} & \frac{\lambda_1\sigma_{WX}+\lambda_2\lambda_4\sigma_X^2}{1-\lambda_2\lambda_3} & \frac{\lambda_2^2\sigma_W^2\lambda_4^2+\sigma_X^2\lambda_1^2+\lambda_2(\lambda_4\lambda_1\sigma_{WX}+\epsilon_2^2)+2\epsilon_{12}+\epsilon_1^2}{(1-\lambda_2\lambda_3)^2} \\ \frac{\lambda_3\lambda_1\sigma_X^2+\lambda_4\sigma_{WX}}{1-\lambda_2\lambda_3} & \frac{\lambda_3\lambda_1\sigma_{WX}+\lambda_4\sigma_W^2}{1-\lambda_2\lambda_3} & \frac{\lambda_3\sigma_X^2\lambda_1^2+\lambda_1\sigma_{WX}(\lambda_4+\lambda_2\lambda_4\lambda_3)+\lambda_2\sigma_W^2\lambda_4^2+\lambda_3\epsilon_1^2+\epsilon_{12}(\lambda_2\lambda_3+1)+\lambda_2\epsilon_2^2}{(1-\lambda_2\lambda_3)^2} \end{array}\right.$$

$$\left.\begin{matrix} \dfrac{\lambda_3\lambda_1\sigma_X^2+\lambda_4\sigma_{WX}}{1-\lambda_2\lambda_3} \\ \dfrac{\lambda_3\lambda_1\sigma_{WX}+\lambda_4\sigma_W^2}{1-\lambda_2\lambda_3} \\ \dfrac{\lambda_3\sigma_X^2\lambda_1^2+\lambda_1\sigma_{WX}(\lambda_4+\lambda_2\lambda_4\lambda_3)+\lambda_2\sigma_W^2\lambda_4^2+\lambda_3\epsilon_1^2+\epsilon_{12}(\lambda_2\lambda_3+1)+\lambda_2\epsilon_2^2}{(1-\lambda_2\lambda_3)^2} \\ \dfrac{\sigma_X^2\lambda_3^2\lambda_1^2+\sigma_W^2\lambda_4^2+2\lambda_3\lambda_1\lambda_4\sigma_{WX}+\lambda_3(\lambda_3\epsilon_1^2+2\epsilon_{12})+\epsilon_2^2}{(1-\lambda_2\lambda_3)^2} \end{matrix}\right] \tag{12.9}$$

As can be seen, covariance terms between an exogenous variable with variables involved in the reciprocal loop are adjusted in the denominator by the magnitude of the reciprocal effect $(1 - \lambda_2\lambda_3)$ while covariance and variance terms between the variables involved in the reciprocal effect (even covariances between error terms) are adjusted by the square of this effect.

12.4 LONGITUDINAL DATA SETTINGS

12.4.1 Monte Carlo Simulation

When considering longitudinal data, it seems reasonable to think that repeated measurements of behavior could also constitute a chain meeting the requirement for instrumentality. If, for example, behavior is measured on three different occasions, and it is assumed that only autoregressive paths of length 1 are assumed, the discrepancy between the observed covariance between the initial and final measurement occasions would make it possible to estimate the magnitude of an autocausal effect for the third time of measurement or, if the magnitude of the autocausal effect is constrained to equality across either the second and third measurement occasions or across all three measurement occasions. An example of estimation of an autocausal model for three times of measurement is shown in Figure 12.5a and b, for example. Note that although this model is identified based on the number of estimated parameters, it would not be identified if the researcher had included a lagged effect of length two and that other models such as those that view measurements as indicative of a single trait over time constitute an alternative explanation of the data. In practice, it often becomes necessary to impose the constraint that the autocausal path not exceed 1 in absolute magnitude to avoid improper solutions and the use of large sample sizes or other model constraints may be necessary to secure a converged, proper solution. These issues are discussed further in the Monte Carlo simulation later.

When more than three measurement occasions are present, however, it becomes possible to consider autocausal effects even when the repeated measurements of behaviors under investigation are often thought to be due to one or more latent variables that span time. Although it is attractive to consider that the latent variable might itself constitute an unmeasured instrumental variable, it is not possible to use the congeneric factor model as a basis for the identification of autocausal or reciprocal effects between all manifest variables, as the solution is not uniquely identified. If, however, one specifies a single manifest variable as not containing

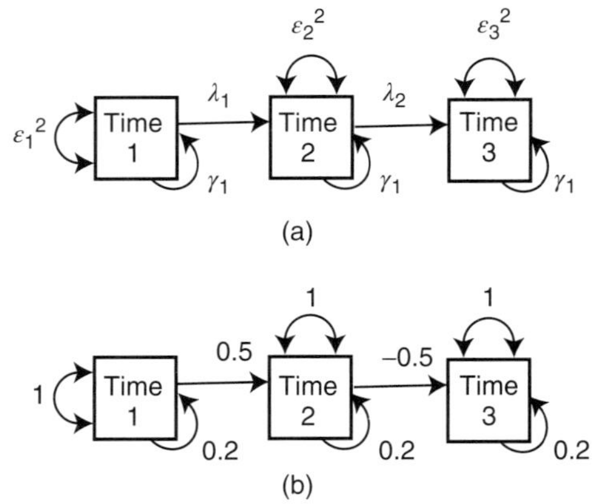

Figure 12.5 Autocausal effects for longitudinal data. (a) Model parameters. (b) Hypothetical values (autocauses set to equality).

autocausal effects, such unmeasured reciprocal effects are estimable, as they then model, as before, the discrepancy between the observed covariance between manifest variables and that predicted under the remaining structural model.

The number of longitudinal assessments necessary for a mathematically identified solution increases markedly, however. The number of estimated factor loadings, estimated error variances, and possible reciprocal effects at the manifest variable level requires a minimum of five measurement occasions to produce a model that is just identified. For those situations in which autoregressive paths between adjacent measurement occasions are also required, six measurement occasions are needed to produce a model with nonzero degrees of freedom.

In practice, however, a single underlying factor is often insufficient to characterize a general trait over time. In many situations, for example, a random intercept component is estimated as a latent variable (often represented as a latent variable with unit loadings and freely estimated factor variance, Maydeu-Olivares and Coffman, 2006). When measurement occasions span long periods of time, the postulated latent variables are estimated with nonzero means in order to model patterns of growth over time. Such growth curve models may also include autoregressive terms as well (e.g., Wood et al., under review). Such extensions, however, do not alter the conceptual status of the postulated latent variables as potential instrumental variables, but may increase the number of measurement occasions required to produce an identified model.

The question arises then, as to whether reciprocal effects can be detected when researchers consider longitudinal factor models that contain the necessary number of measurement occasions for identification. The path diagrams involved in specifying feedback loops often lead researchers to believe, for example, that unmeasured reciprocal effects, if present, would be subsumed in the error terms associated with manifest variables in the model or that the estimation difficulties in estimating such models preclude their use. To this end, I would like to present two reasonably realistic Monte Carlo simulations to make the following points: (i) Unmeasured reciprocal effects can be successfully recovered in factor models that contain sufficient measurement occasions to identify the model; (ii) misspecified models that do not include unmeasured reciprocal effects do not fit the data as well and, further, produce substantially different patterns of estimated factor loadings depending on whether the unincluded reciprocal effects are positive or negative; and (iii) when the data do not include unmeasured reciprocal effects, the estimated reciprocal effects are zero.

In the service of conducting a Monte Carlo study that better reflects the complexity of longitudinal models often encountered, I considered a Gompertz growth curve model as representative of nonlinear growth. The Gompertz growth curve considered had the following functional form as a function of time t:

$$y_t = e^{-e^{-\alpha(t-\lambda)}} \tag{12.10}$$

with λ indicating the time at which maximum growth occurs and α indicating a rate of change. Grimm and Ram (2009) provide a more detailed discussion of this curve and compare its substantive merits relative to other parametric curves. It is worth noting here, however, that the Gompertz curve contains an inflection point, indicating the maximum rate of growth over time and that the curve is asymmetric about this inflection point with $1/e$ of the total growth (about 37%) occurring before the inflection point with the remainder occurring afterward. The curve has found application, for example, in growth populations with confined space or limited nutrients. Studies of human stature, for example, appear to show a similar "biphasic" pattern with some individuals demonstrating an initially slower rate of growth accompanied by more rapid growth at later ages and vice versa (e.g., Masuyama, 1979). For these data, λ (the horizontal displacement) was chosen to be 60 and α (the growth rate) to be 6. Parameter values for the simulation are shown in the first column of Table 12.1. To make this simulation realistic, unequally spaced times were chosen for assessment at times 1, 5, 6, 8, 9, 10, 11, and 13, a random intercept factor was also included, with means and variances of the intercept and growth factors were estimated at unity. The magnitude of change over the course of time is similar to the amount of change observed in some studies (such as cognitive outcomes of higher education), while markedly other studies (such as change in verbal ability during childhood (McArdle and Epstein, 1987), or stature (Masuyama, 1979). Error variances were chosen so that the internal consistency of the model (in the absence of autoregressive and reciprocal effects) was 0.85, which was taken as a reasonably representative degree of measurement reliability. Positive and negative autoregressive effects between measurement occasions were also estimated as shown in Table 12.1, as well as a variety of positive and negative reciprocal effects. Note also that for one measurement occasion in the simulation no autoregressive effects were modeled (between measurement occasions 7 and 8) and for one occasion, no reciprocal effect was modeled (measurement occasion 1). In order to mirror the fact that researchers may not have any a priori reason to anticipate the functional form of growth over time, the models fit to the data were

TABLE 12.1 Monte Carlo Results for Detection of Unincluded Reciprocal Effects.

	True Model					No Reciprocal Effects			
	Factor Loadings	Parameter Estimate	95% Coverage	% Sign.	Bias	Estimate	95% Coverage	% Sign.	Bias
T1	0.00	0.00	0.00	1.00	n.a.	0.48	0.00	1.00	n.a.
T2	0.05	0.05	0.92	0.57	−0.02	0.46	0.00	1.00	8.18
T3	0.19	0.20	0.92	0.90	0.03	0.53	0.00	1.00	1.75
T4	0.61	0.61	0.93	1.00	0.00	0.95	0.00	1.00	0.56
T5	0.76	0.77	0.94	1.00	0.00	1.60	0.00	1.00	1.10
T6	0.86	0.87	0.93	1.00	0.01	1.93	0.00	1.00	1.23
T7	0.92	0.93	0.92	1.00	0.00	2.17	0.00	1.00	1.35
T8	0.98	0.98	0.93	1.00	0.01	2.33	0.00	1.00	1.38
		Reciprocal Effects							
T1	0.00	0.00	1.00	0.05	n.a.				
T2	0.10	0.12	0.92	0.06	0.20				
T3	−0.20	−0.22	0.93	0.11	0.12				
T4	−0.30	−0.30	0.93	0.66	0.00				
T5	0.30	0.30	0.95	0.99	−0.01				
T6	0.40	0.40	0.94	1.00	−0.01				
T7	0.46	0.45	0.93	0.99	−0.02				
T8	0.40	0.39	0.95	0.86	−0.02				
				Autoregressive Effects					
T2 on T1	0.10	0.08	0.92	0.21	−0.24	0.31	0.00	1.00	
T3 on T2	−0.10	−0.09	0.93	0.36	−0.10	−0.14	0.35	1.00	0.41
T4 on T3	0.20	0.19	0.94	0.53	−0.03	−0.28	0.00	1.00	−2.41

T5 on T4	−0.20	−0.20	0.95	0.91	−0.01	−0.28	0.67	0.99	0.38
T6 on T5	0.40	0.40	0.94	1.00	0.01	0.74	0.00	1.00	0.85
T7 on T6	0.20	0.21	0.93	0.98	0.03	0.44	0.01	1.00	1.18
T8 on T7	0.00	0.00	0.95	0.05		−0.03	0.88	0.12	
					Factor Means				
S	1.00	1.00	0.93	1.00	0.00	1.34	0.00	1.00	0.34
I	1.00	1.00	0.94	1.00	0.00	0.35	0.00	1.00	−0.65
					Factor Variances				
S	1.00	1.00	1.00	0.00		1.00	1.00	0.00	
I	1.00	1.01	0.94	1.00	0.01	0.62	0.00	1.00	−0.38
					Residual Variances				
T1	0.18	0.18	0.92	0.90	−0.03	0.27	0.00	1.00	0.51
T2	0.18	0.18	0.92	0.90	−0.03	0.27	0.00	1.00	0.51
T3	0.18	0.19	0.91	0.65	0.08	0.06	0.00	1.00	−0.67
T4	0.24	0.24	0.92	1.00	0.01	0.13	0.00	1.00	−0.46
T5	0.28	0.28	0.95	1.00	0.02	0.60	0.00	1.00	1.16
T6	0.31	0.31	0.93	1.00	0.02	0.90	0.00	1.00	1.91
T7	0.24	0.26	0.91	1.00	0.06	0.85	0.00	1.00	2.50
T8	0.28	0.30	0.93	0.58	0.08	0.71	0.00	1.00	1.54

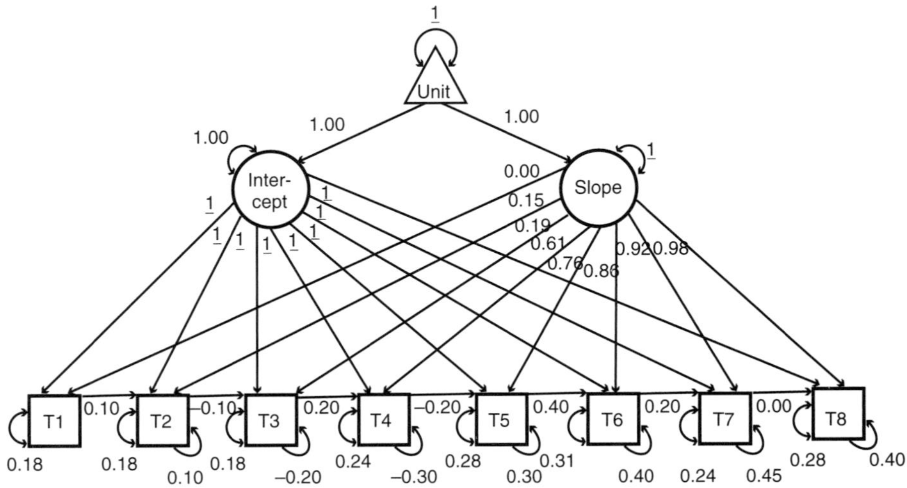

Figure 12.6 Longitudinal growth model with autocausal effects Monte Carlo values.

fit as free curve growth models (FCSI; Wood *et al.*, 2015). The path diagram corresponding to this model is shown in Figure 12.6. The Mplus model corresponding to this simulated data is shown in Table 12.2.

12.4.1.1 Convergence Issues In order to assure convergence during minimization, the magnitude of the reciprocal effects was constrained to be between −1 and 1 using the Model constraint command in Mplus. Although a large sample size of 1000 was considered for the analysis given that the real-world data considered below are also from a large-scale assessment, it should be noted that when the sample size is decreased to 500, models converged in only 97.3% of the simulations and when a sample size of 250 is considered, models converged in only 93% of the replications, suggesting that the issue of nonconvergence may be more significant for models that test for unincluded reciprocal effects.

12.4.1.2 Results As can be seen from the second column of Table 12.1, when the correct model is fit to these data, coverage (the proportion of times that the confidence interval of the simulation includes the true population value) occurs for all parameters, although a slight bias is noted for the smaller reciprocal effects (Times 1 and 2, 19.50% and 11.50%, respectively), and the autoregressive effect from Time 1 to Time 2 (−24.20%). As would be expected, observed χ^2 model fit statistics closely approximated the degrees of freedom for the model (average $\chi^2(19) = 18.78$) and the percentiles of the distribution appeared similar to their values based on normal theory. The average RMSEA was 0.003 across replications.

When, however, these same data are modeled as a standard growth curve model, however, fit of the resulting model is appreciably worse (average $\chi^2(25) = 63.43$), and the average RMSEA was 0.05. Failure to include the unincluded reciprocal effects

TABLE 12.2 Example Mplus Model for Estimated Autocausal Effects.

```
s by t1*0.500000 (L1)
t2*0.050427
t3*0.194092
t4*-0.610312
t5*0.762620
t6*0.861806
t7*0.921620
t8*0.975716;
i by t1@1
t2@1
t3@1
t4@1
t5@1
t6@1
t7@1
t8@1;
t2 on t2*0.10000(r2);
t3 on t3*-0.20000(r3);
t4 on t4*-0.30000(r4);
t5 on t5*0.30000(r5);
t6 on t6*0.40000(r6);
t7 on t7*0.46081(r7);
t8 on t8*0.40000(r8);
t2 on t1*0.1000;
t3 on t2*-0.1000;
t4 on t3*0.2000;
t5 on t4*-0.2000;
t6 on t5*0.4000;
t7 on t6*0.2000;
t8 on t7*0.0000;
[s*1];
[i*1];
s@1;
i*1;
t1*0.18(e1);
t2*0.18(e1);
t3*0.18(e2);
t4*0.2422;
t5*0.2791;
t6*0.3075;
t7*0.2422;
t8*0.2791;
```

produced markedly different loadings, which ranged from 0.48 to 2.33 across measurement occasions, resulting in biased estimates ranging between 55% and 800%. Bias in estimated factor means and the random intercept variance was also observed. Estimates of error variance were also underestimates in those measurement occasions associated with an unmeasured negative feedback loop (measurement occasions 3 and 4) but were overestimates for the remaining measurement occasions.

TABLE 12.3 Estimated Parameters Associated with Longitudinal Data on Alcohol Consumption and General Distress.

	QPLUS5				QHIGH				QDRUNK				BSI-GSI			
Loadings	Unstandardized		Standardized		Unstandardized		Standardized		Unstandardized		Standardized		Unstandardized		Standardized	
	Factor	Factor	Factor	Factor	Factor	Factor	Factor	Factor	Factor	Factor	Factor	Factor	Factor	Factor	Factor	Factor
Wave	1	2	1	2	1	2	1	2	1	2	1	2	1	2	1	2
1	**0.7**	**0.5**	**0.55**	**0.39**	**0.7**	**0.46**	**0.52**	**0.34**	**0.58**	**0.47**	**0.49**	**0.39**	**0.31**	**0.39**	**0.53**	**0.65**
2	**0.64**	**0.19**	**0.55**	**0.16**	**0.79**	**0.4**	**0.68**	**0.35**	**0.5**	**0.22**	**0.51**	**0.22**	**0.42**	**0.36**	**0.76**	**0.65**
3	**0.71**	**0.31**	**0.65**	**0.28**	**0.81**	**0.53**	**0.72**	**0.47**	**0.65**	**0.39**	**0.63**	**0.37**	**0.35**	**0.06**	**0.63**	**0.11**
4	**0.82**	**0.3**	**0.74**	**0.27**	**0.98**	**0.56**	**0.93**	**0.53**	**0.83**	**0.53**	**0.83**	**0.54**	**0.32**	**0.07**	**0.58**	**0.12**
5	**1.06**	**0.37**	**0.96**	**0.34**	**0.95**	**0.39**	**0.84**	**0.35**	**0.6**	**0.44**	**0.63**	**0.46**	**0.39**	0.01	**0.7**	0.02
6	**1.11**	−0.1	**1.05**	−0.09	**0.98**	0	**0.92**	0	**0.58**	−0.08	**0.64**	−0.09	**0.42**	0	**0.8**	0.01
7	0.07	**0.52**	0.06	**0.45**	**1.09**	0	**0.96**	0	−0.18	**0.55**	−0.18	**0.56**	**0.44**	0	**0.78**	0
8	**0.57**	0	**0.51**	0	**1.11**	0	**0.99**	0	**0.42**	0	**0.44**	0	**0.36**	0	**0.66**	0
							Reciprocal Autocausal Paths									
Reciprocal Effect	**−0.13**		**−0.13**		**−0.29**		**−0.29**		—		—		—		—	
							Autoregressive Paths									
t2 on t1	**0.25**		**0.27**		**0.17**		**0.2**		**0.19**		**0.22**		**−0.25**		**−0.28**	
t3 on t2	**0.2**		**0.22**		**0.15**		**0.16**		**0.09**		**0.08**		**0.15**		**0.14**	
t4 on t3	**0.15**		**0.15**		−0.05		−0.05		**−0.17**		**−0.17**		**0.19**		**0.2**	
t5 on t4	**−0.1**		**−0.1**		0.06		0.05		0		0		**0.08**		**0.08**	

t6 on t5	−0.07	−0.07	**0.11**	**0.12**	**0.23**	**0.24**	−0.01	−0.01
t7 on t6	**0.78**	**0.72**	0.05	0.05	**0.78**	**0.72**	−0.05	−0.05
t8 on t7	**0.41**	**0.42**	0.01	0.01	**0.4**	**0.42**	**0.1**	**0.1**
Factor Means	**0.6** 0.09		**1.07** 0.09		**0.6** **0.25**		**1.08** **0.49**	
Shift	**0.23**		0		**0.16**		**0.02**	
Factor Variance	1 1	1 1	1 1	1 1	1 1	1 1	1 1	1 1
Wave				Residual Variance				
1	**0.91**	**0.55**	**1.15**	**0.62**	**0.86**	**0.61**	**0.11**	**0.31**
2	**0.91**	**0.67**	**1.15**	**0.85**	**0.5**	**0.5**	**0.11**	**0.37**
3	**0.62**	**0.52**	**0.91**	**0.71**	**0.41**	**0.38**	**0.14**	**0.45**
4	**0.55**	**0.44**	**0.65**	**0.58**	**0.26**	**0.26**	**0.13**	**0.44**
5	**0.49**	**0.39**	**0.96**	**0.75**	**0.36**	**0.39**	**0.13**	**0.43**
6	**0.3**	**0.27**	**0.72**	**0.64**	**0.3**	**0.36**	**0.1**	**0.37**
7	**0.7**	**0.53**	**0.87**	**0.67**	**0.32**	**0.33**	**0.14**	**0.44**
8	**0.7**	**0.56**	**0.82**	**0.66**	**0.45**	**0.49**	**0.14**	**0.46**

The bolded estimates are significant at $p < 0.05$.

Taken together, this simulation suggests that the failure to model unincluded reciprocal effects can result in markedly poorer fit for growth curve models even though the unstandardized estimates of loadings may appear markedly higher than their true values.

12.4.1.3 False Detection of Unincluded Reciprocal Effects Although the simulation described in the previous section makes the case that unincluded reciprocal effects may be a source of model misfit and may substantially affect estimated parameters, it is possible that modeling unincluded reciprocal effects may also result in false positives, that is, estimation of unincluded reciprocal effects where, in fact, none exist. To address this question, the simulation described above was rerun except that the true model now included no reciprocal effects. When this is done, average estimated reciprocal effects were all very close to zero (ranging from −0.02 to 0.02). Statistically significant reciprocal effects were found at levels close to the 0.05 alpha level of significance (ranging between 0.05 and 0.08 across measurement occasions). Parameter estimates for the remaining parts of the model appeared close to their population parameter values. On the basis of this initial Monte Carlo exploration, then, it seems that the estimation of unincluded reciprocal effects when they are not present does not appear to result in a necessarily high rate of false positive identification of unincluded reciprocal effects. This observation, however, must be tempered by the fact that the false-positive rate may be present if the data fail to meet other assumptions of the factor model, such as conditional normality of error variance or nonlinearity of parameter estimates.

12.4.2 Real-World Data Examples

Although the results of the Monte Carlo simulations provide some evidence that it is possible to estimate unincluded reciprocal interactions and that failure to do so may result in substantial bias in parameter estimates, it is unclear whether it is possible to do so in practice, given that real-world data may not possess the ideal properties of randomly generated data and patterns of missing data present in the data may make estimation of such parameters problematic. Accordingly, this chapter concludes with analyses of four constructs drawn from a study of collegiate problem behaviors in which individuals were assessed across eight measurement occasions. Because the focus of the study was to explore patterns of onset of and desistance in problem behaviors and related constructs during the college years, the structural models that were explored were models that explored latent variables with estimated mean levels. The IMPACTS data set is based on assessments of 3720 first-time college students at a large Midwestern University who were assessed twice a year for 4 years. Further details of the study are given in Sher and Rutledge (2007). Although the initial sample of students constituted a near-complete assessment of all eligible freshmen, over the course of the 4 years missing data occurred between 69% and 60% of the original sample. In order to explore whether unmeasured reciprocal effects could be assessed with these data, three constructs were considered: general distress, as assessed by the Brief Symptom Inventory (BSI; Derogatis, 1975) report of the number of times in the past 3 months that the individual felt drunk (QDRUNK), felt high from alcohol (QHIGH), or consumed five or more drinks in one sitting (PLUS5). Because patterns of change for these four constructs are quite likely to be qualitatively different, the factor structure of each construct was first explored by means of confirmatory factor models in an attempt to "right size" the model before fitting growth curve models to the data (Wood *et al.*, 2015). Specifically, an exploratory factor analysis was first conducted on the data to determine the dimensionality of the data and then, based on this, models were considered that explored whether the dimensionality could be best recovered using congeneric factors (i.e., factors in which freely estimated loadings were estimated) or using a random intercept factor (i.e., a factor in which loadings were set to equality). For these variables, however, it was found that two-factor solutions fit the data well based on exploratory factor analyses, and, given that all four variables are characterized by a general pattern of desistance over the college years, it was decided to identify the factor structure of each construct by fixing the loading at the last measurement occasion to zero for the second extracted factor. Corresponding growth curve models were then explored for the data in which factor means were estimated while constraining manifest variables to zero or to equality (a constraint that results in McDonald's (1967) linear factor model. See Wood *et al.* (under review) for a more complete description).

12.4.2.1 QPLUS5 For the report of the number of times that the individual reported drinking five or more drinks in one sitting during the previous 3 months, a two-factor solution fit the data best based on an exploratory factor analysis of the data. Given that the normative pattern of alcohol use during undergraduate years shows a normative pattern of desistence over time, it was decided to identify this factor structure by freely estimating factor loadings across all measurement occasions for the first factor and to mathematically identify the second factor by constraining the loading associated with the first measurement occasion to zero. Because this resulted in a factor solution in which the factor loading for the second wave was nonsignificant, it was decided to fix the second loading to zero as well, making the factor representative of behavior after the first year of college. When this model is fit with intercepts constrained to zero and freely estimated factor means, the resulting fit is quite good, and, based on this, it was decided to explore whether unincluded reciprocal effects could be present. Although the χ^2 statistic from the resulting model is statistically significant ($\chi^2(13) = 39.25; p = 0.0002$), the fit of the model is quite good based on other fit statistics (CFI = 0.99; TLI = 0.99; RMSEA = 0.02, CI = [0.02 – 0.03]). In addition, the χ^2 difference test (also known as $\Delta\chi^2$) comparing the model with a reciprocal effect with the model containing no reciprocal effects is statistically significant ($\chi^2 = 31.31, p < 0.0001$).

12.4.2.2 QHIGH For the reports of the number of times the person reported being "high" or "buzzed" while using alcohol, a similar pattern of results was found as for the QPLUS5 variable. Overall fit for the model with common reciprocal effects appeared good ($\chi^2(14) = 49.74$; CFI = 0.99, TLI = 0.99; RMSEA = 0.03[0.02 – 0.03]), and a model in which no unmeasured reciprocal effects were included fit significantly worse ($\chi^2(1) = 15.07, p < 0.0001$). A model in which the magnitude of unincluded reciprocal relationships was equal across time appeared to be a parsimonious alternative to modeling different magnitudes of reciprocal relationships at each measurement occasion ($\chi^2(6) = 0.01, p < 0.0001$).

12.4.2.3 QDRUNK No reciprocal effects were found for the QDRUNK variable, and, the estimated value of the unmeasured reciprocal effect, when estimated, was positive (0.17) rather than negative as found for the other alcohol consumption variables. The $\Delta\chi^2$ test comparing the reciprocal model with a model with no reciprocal effects was not statistically significant ($\chi^2(1) = 1.60, p < 0.21$).

12.4.2.4 BSI-GSI The pattern of results for BSI-GSI was different than that for the alcohol consumption variables. Specifically, the model with magnitudes of reciprocal effects across measurement occasions appeared to fit the data the best.

For these data, reports of the quantity of drinking five or more drinks in a sitting and quantity of times that the person drank until "high" were associated with a negative autocausal effect, suggesting that some "satiety" or unmeasured reciprocal effect was present, which caused covariances between a measurement occasion and remaining occasions to be lower than expected under the growth curve models estimated. The other two variables relating to general psychological distress and quantity of times reported drunk were not associated with autocausal or unmeasured reciprocal effects. The finding regarding the number of times drunk variable is somewhat surprising, given that the autocausal effects regarding five or more drinks and "feeling high" appeared interpretable as reflecting "satiety" effects. It should be noted, however, that the factor loadings associated with the QDRUNK variable were, however, much lower than those for the other two alcohol variables, suggesting that the failure to find autocausal effects may be due to a lack of reliable assessment of the construct. Although a more detailed discussion of the conceptual interpretation of these relationships cannot be taken up in the interests of space, it appears that these data show that it is possible to estimate autocausal effects and/or unmeasured reciprocal effects in real-world longitudinal data with several measurement occasions.

12.5 DISCUSSION

The identification of autocausal, reciprocal, and cycle effects presented in this chapter is based on the discrepancies between observed variables and based on the predicted covariance under the remaining structural model assumed. The effect of unincluded reciprocal effects in structural models appears qualitatively different from the usual bias terms associated with the simple failure to include relevant predictor variables to the model in that it deals with unexplained covariance and not differences in amount of explained variance in the variables under study. Put another way, unmeasured reciprocal effects do not appear to be absorbed into the error terms or autoregressive components of longitudinal models.

The specification of models that can test for the presence of such unincluded reciprocal effects appears both feasible and not overly prone to false positives, given the preliminary Monte Carlo work presented here. It also appears that estimation of such effects can be done with real-world data, given the analyses based on the longitudinal study presented here. Results from the collegiate data suggest that feedback loops associated with alcohol consumption indicate that, when present, the unincluded feedback loops are negative, which suggests that when such reciprocal effects are not included in the model that estimated error variances are underestimates.

The possible detection of unincluded reciprocal effects causes speculation as to the specific nature of such reciprocal relationships. It may well be, for example, that such unincluded reciprocal effects are the result of feedback relationships between other variables in the system. For example, it may be that peer effects of alcohol consumption affect one's consumption of alcohol (either positively or negatively) and the example of one's drinking in turn affects the alcohol consumed by one's peers. Alternatively, however, it may be that the feedback loops associated with a given behavior are "autoreciprocal" or "autochthonous" and arise from the behavior itself. It may be, for example, that consumption of one alcoholic beverage may lead to another one or, alternatively, that some appetitive aspect of alcohol consumption has been fulfilled, resulting in a negative feedback loop in alcohol consumption.

As McArdle and Nesselroade (2014) have noted, the estimation of reciprocal (and presumably autocausal and cycle effects) can be quite difficult in real-world applications. As we have seen, however, imposition of nonlinear constraints during estimation (such as those necessary to ensure that autocausal or the product of reciprocal effects does not exceed the absolute value of 1) seem to improve the estimation of such models considerably. When feedback loops are present in the data, the standard errors associated with estimation appear somewhat larger than in acyclic systems, and, for this reason, models in which more equality constraints are present have appeared in practice to improve the convergence rates of these models. Clearly, however, there is additional work to be done in improving the design, estimation, and interpretation of models with feedback loops. It is important to recall, however, that a researcher who tests for and finds possible unincluded reciprocal effects in data has not necessarily proven their existence. A variety of alternative explanations for an observed feedback loop may exist, which have nothing to do with autocausal or reciprocal causation. These effects rely strongly on the assumption that the remaining parts of the structural model have been correctly specified and to the extent, for example, that relationships between variables in the model are not linear but condition or nonlinear, estimated reciprocal effects can be ephemeral. Ad hoc modifications to scoring or the assumed measurement model of manifest variables explore the reasonableness of such counterarguments, however. To the extent, however, that such effects may be present in the data, their detection serves as an initial conjecture regarding the dynamics of the behavior of interest over time.

ACKNOWLEDGMENTS

The college alcohol data presented in this chapter was collected under grant NIH grant R37AA07231 to Kenneth J. Sher.

REFERENCES

von Bertalanffy (1968) *General Systems Theory: Foundations, Development, Applications*, Braziller, New York.

Campbell, D.T. (1963) From description to experimentation: interpreting trends as quasi-experiments, in *Problems in Measuring Change* (ed. C.W. Harris), University of Wisconsin Press, Madison, WI.

Coleman, J.S.D. (1968) The mathematical study of change, in *Methodology in Social Research* (eds H.M. Blalock and A.B. Blalock), McGraw-Hill, New York, pp. 428–478.

Csardi, G. and Nepusz, T. (2006) The igraph software package for complex network research. *Interjournal Complex Systems*, p. 1695–.

Derogatis, F.R. (1975) *The Brief Symptom Inventory*, Clinical Psychometric Research, Baltimore, MD.

Epskamp, S., Cramer, A., Waldorp, L., Schmittmann, V., and Borsboom, D. (2012) Qgraph: network visualizations of relationships in psychometric data. *Journal of Statistical Software*, **48**, 1–18.

von Eye, A. and Wiedermann, W. (2015) Manifest variable granger causality models for developmental research: a taxonomy. *Applied Developmental Science*, doi: 10.1080/10888691.2014.1001512.

Gottman, J.M., McFall, R.M., and Barnett, J.T. (1969) Design and analysis of research using time series. *Psychological Bulletin*, **72** (4), 299–306.

Grimm, K.J. and Ram, N. (2009) Nonlinear growth models in Mplus and SAS. *Structural Equation Modeling*, **16** (4), 676–701.

Grunberg, L., Moore, S., Sikora, P., and Greenberg, E. (2006) Downsizing and alcohol use: a cross-lagged longitudinal analysis.. *Political and economic change program. Working Paper PEC2006-0002. Institute of Behavioral Science, Boulder, CO.*

Hacker, S. and Hatemi-J, A. (2012) A bootstrap test for causality with endogenous lag length choice: theory and application in finance. *Journal of Economic Studies*, **39** (2), 144–160.

Hatemi-J, A. (2012) Asymmetric causality tests with an application. *Empirical Economics*, **43** (1), 447–456.

Heath, A. (2015) Alcohol effects on executive cognitive function: specifying component processes, Multi-disciplinary Alcoholism Research Center Grant (P60 AA011998), Project 8 (Sub-Project ID: 5979).

Heise, D.R. (2001) *Causal Analysis*, Author, Bloomington, IN, electronic edn. Original publication: 1975, New York: Wiley. Retrieved 7/30/2014 from `http//www.-indiana.edu/socpsy/public_files/CausalAnalysis.zip`.

Kalisch, M., Maechler, M., Colombo, D., Maathuis, M., and Buehlmann, P. (2012) Causal inference using graphical models with the R package PcAlg. *Journal of Statistical Software*, **47**, 1–26.

Kenny, D.A. (1975) Cross-lagged panel correlation: a test for spuriousness. *Psychological Bulletin*, **82** (6), 887–903.

Lazarsfeld, P.F. (1948) The use of panels in social research. *Proceedings of the American Philosophical Society*, **92**, 405–410.

Lipset, S.M., Lazarsfeld, P.F., Barton, A.H., and Linz, J. (1954) The psychology of voting: an analysis of political behavior, in *Handbook of Social Psychology*, vol. **2** (ed. G. Lindzey), Addison-Wesley, Reading, MA.

Mason, S.J. (1953) Feedback theory –some properties of signal flow graphs, in *Proceedings of the IRE*, pp. 1144–1156.

Masuyama, M. (1979) *Human Biochemical Individual Variabilities and their Quasi-Constancy*, Sogo Printing Company, Tokyo.

Maydeu-Olivares, A. and Coffman, D. (2006) Random intercept item factor analysis. *Psychological Methods*, **11**, 344–362.

McArdle, J. and Epstein, D. (1987) Latent growth curves within developmental structural equation models. *Developmental Psychology*, **58**, 110–133.

McArdle, J.J. and McDonald, R.P. (1984) Some algebraic properties of the reticular action model for moment structures. *British Journal of Mathematical and Statistical Psychology*, **37** (2), 234–251.

McArdle, J. and Nesselroade, J.R. (2014) *Longitudinal Data Analysis Using Structural Equation Models*, APA, Washington, DC.

McDonald, R.P. (1967) *Nonlinear Factor Analysis*, Psychometric Monograph, vol. **15**, Byrd Press, Richmond, VA.

McDonald, R.P. (1980) A simple comprehensive model for the analysis of covariance structures: some remarks on applications. *British Journal of Mathematical and Statistical Psychology*, **33** (2), 161–183.

Mulaik, S.A. (2009) *Linear Causal Modeling with Structural Equations*, CRC Press, Boca Raton, FL.

Pearl, J. (2009) *Causality: Models, Reasoning and Inference*, Cambridge University Press, London.

Rindskopf, D. (1984) Using phantom and imaginary latent variables to parameterize constraints in linear structural models. *Psychometrika*, **49** (1), 37–47.

Rogosa, D. (1980) A critique of cross-lagged correlation. *Psychological Bulletin*, **88** (2), 245–258.

Scheines, R., Spirtes, P., Glymour, C., Meek, C., and Richardson, T. (1998) The TETRAD project: constraint-based aids to causal model specification. *Multivariate Behavioral Research*, **33** (1), 65–117.

Selig, J.P. and Little, T.D. (2011) Panel and cross-lag models, in *Handbook of Developmental Research Methods* (eds B. Laursen, T.D. Little, and N.A. Card), Guilford, New York.

Selig, J.P. and Preacher, K.J. (2009) Mediation models for longitudinal data in developmental research. *Research in Human Development*, **6**, 144–164.

Sher, K.J. and Rutledge, P.C. (2007) Heavy drinking across the transition to college: predicting first-semester heavy drinking from precollege variables. *Addictive Behaviors*, **32** (4), 819–835.

Sikora, P., Moore, S., Greenberg, E., and Grunberg, L. (2008) Downsizing and alcohol use: a cross-lagged longitudinal examination of the spillover hypothesis. *Work & Stress*, **22** (1), 51–68.

Williamson, J. (2006a) Causal pluralism versus epistemic causality. *Philosophica*, **77**, 69–96.

Williamson, J. (2006b) Dispositional versus epistemic causality. *Minds and Machines*, **16**, 259–276.

Williamson, J. (2009) Probabilistic theories, in *The Oxford Handbook of Causation* (eds H. Beebee, C. Hitchcok, and P. Menzies), Oxford University Press, Oxford, pp. 185–212.

Wood, P.K., Jackson, K., and Sher, K.J. (under review) Alternative state-trait models of comorbidity and their implications for psychopathology.

Wood, P.K., Steinley, D., and Jackson, K. (2015) Right-sizing statistical models for longitudinal data. *Psychological Methods*, **20** (4), 470–488, doi: 10.1037/met0000037.

Wright, S. (1922) Coefficients of inbreeding and relationship. *American Naturalist*, **56**, 330–338.

PART IV

COUNTERFACTUAL APPROACHES AND PROPENSITY SCORE ANALYSIS

13

LOG-LINEAR CAUSAL ANALYSIS OF CROSS-CLASSIFIED CATEGORICAL DATA

KAZUO YAMAGUCHI

Department of Sociology, University of Chicago, Chicago, IL, USA

13.1 INTRODUCTION

This chapter describes a method for the analysis of two-way and three-way tables of *adjusted* cross-classified frequency data that reflect the effects of categorical variable X on categorical variable Y that would be realized after eliminating selection bias in the effects of X on Y generated by confounding covariates $\boldsymbol{V}$. It is also a method for applying semiparametric logit and multinomial logit regression models that do not specify the effects of the confounding covariates $\boldsymbol{V}$ on Y.

Sociological researchers' interests in the use of log-linear and log-multiplicative models of contingency tables seem to have been largely lost in the past two decades mainly because of the limitations of the analyses for handling many control variables. Although log-linear models that are in the regression form, such as logit, multinomial logit, and ordered logit models, continue to be used frequently, the effectiveness of a multivariate regression form in controlling confounding variables for causal analysis has gradually been called into question, starting from the Rosenbaum and Rubin's (1983; 1984) pioneering work on the use of propensity score adjustments.

This chapter responds to these trends by relying on a new log-linear causal analysis of a categorical dependent variable (Yamaguchi, 2012). Semiparametric methods based on the use of the propensity score, including propensity score matching

Statistics and Causality: Methods for Applied Empirical Research, First Edition.
Wolfgang Wiedermann and Alexander von Eye.

methods and the inverse-probability weighting based on the propensity score (Rosenbaum and Rubin, 1983, 1984; Rubin, 1985), have been widely used. However, the use of these methods for models with a logit or multinomial logit link function has serious limitations. The propensity score tries to attain independence between the treatment variable X and the confounding covariates $\boldsymbol{V}$ in the *pseudopopulation* by conditioning semiparametrically on the propensity score. However, even under statistical independence between X and $\boldsymbol{V}$, we cannot omit $\boldsymbol{V}$ from the regression equation for the logit and multinomial logit models in order to obtain unbiased estimates of X on Y (Gail *et al.*, 1984). This is known as the lack of *collapsibility* for regressions with a logit link function. Since the primary aim of log-linear causal analysis for contingency table data is to generate an adjusted two-way contingency table of X by Y frequencies that retain the causal effects of X on Y when summed across the categories of $\boldsymbol{V}$, the lack of collapsibility is a major problem. Since this issue is not widely recognized, we explain the problem formally and with illustrative examples later.

The collapsibility problem can be solved if we eliminate selection bias in the effect of X on Y due to confounding variables $\boldsymbol{V}$ by eliminating the effects of $\boldsymbol{V}$ on Y, instead of attaining independence between X and $\boldsymbol{V}$. In particular, standardization methods for the odds, such as Xie's (1989) "CD-purging" method, which eliminates the effects of $\boldsymbol{V}$ on Y, should work if $p(y|x, \boldsymbol{v}) > 0$ always holds for the sample estimate of $p(y|x, \boldsymbol{v})$. When $\boldsymbol{V}$ includes many variables, however, $p(y|x, \boldsymbol{v}) = 0$ usually occurs for some cases of $\boldsymbol{V}$ if we employ nonparametric estimates for $p(y|x, \boldsymbol{v})$. However, an estimate $p(y|x, \boldsymbol{v}) = 0$ for a particular $(x, \boldsymbol{v})$ makes the estimate for the conditional log odds ratio involving the $(x, \boldsymbol{v})$ either nonfinite or indeterminate, and makes known standardization methods that eliminate the effects of $\boldsymbol{V}$ on Y ineffective. Although the use of a parametric regression model for $p(y|x, \boldsymbol{v})$ can assure $p(y|x, \boldsymbol{v}) > 0$, this alternative makes a strong assumption that we wish to avoid. In this chapter, we solve this problem by employing semiparametric models for $p(y|x, \boldsymbol{v})$.

The main aim of this article is thus to describe a new standardization method (Yamaguchi, 2012) to solve the collapsibility problem and the problem of zero estimates for conditional probabilities $p(y|x, \boldsymbol{v})$ in order to obtain the *adjusted* conditional probabilities $P(y|x)$, where a capital P indicates an *expected probability* from a given model, that eliminate selection bias in X generated by the effects of confounding covariates $\boldsymbol{V}$ on the association between X and Y. The method is also extended to obtain $P\left(y|x, v_i\right)$ that retains odds ratios that characterize the causal effects of X on Y and the dependence of those effects on a particular covariate V_i, when interaction effects of X and V_i on Y exist on the log odds of $p(y|\boldsymbol{v}, x)$. From those adjusted conditional probabilities, we can generate two-way or three-way contingency tables of adjusted frequencies whose odds ratios reflect the causal effects of X on Y and the possible dependence of those effects on a covariate. The chapter also describes how to estimate the standard errors of parameter estimates in the analysis of association and conditional association between X and Y based on the use of those adjusted frequency data.

13.2 PROPENSITY SCORE METHODS AND THE COLLAPSIBILITY PROBLEM FOR THE LOGIT MODEL

The causal analysis introduced by Rosenbaum and Rubin (1983) is based on a distinction between the treatment assignment characterized by observed variable X and the treatment that yields that outcome and can be counterfactual. When we have a dichotomous distinction of the treatment group for whom $X = 1$ and the control group for whom $X = 0$, we assume a pair of potential outcome variables, Y_1 and Y_0, for each subject for an outcome under the treatment and an outcome under no treatment, respectively. One of those two that is counterfactual is considered missing because we observe only Y_1 for the treatment group and only Y_0 for the control group. The observed outcome variable is defined as $Y_{\text{obs}} = xY_1 + (1-x)Y_0$, and for the *linear* model, the *average* treatment effect is given as $E\left(Y_1\right) - E\left(Y_0\right)$.

In order to obtain a consistent estimate of the average treatment effect, Rosenbaum and Rubin (1983) assume that the *strongly ignorable treatment assignment* (SITA) condition holds. The SITA assumption posits that both Y_1 and Y_0 are conditionally independent of the treatment assignment when conditioned by a set of covariates $\boldsymbol{V}$, that is, $\{Y_1, Y_0\} \perp X | \boldsymbol{V}$. Then, under this and assuming all covariates to be categorical, we obtain

$$\begin{aligned} E(Y_1) &= \sum_{\boldsymbol{v}} E(Y_1|\boldsymbol{v})p(\boldsymbol{v}) = \sum_{\boldsymbol{v}} E(Y_1|x=1,\boldsymbol{v})p(\boldsymbol{v})(\text{since } E(Y_1|\boldsymbol{v},x) = E(Y_1|\boldsymbol{v})) \\ &= \sum_{\boldsymbol{v}} E(Y_{\text{obs}}|x=1,\boldsymbol{v})p(\boldsymbol{v}) = \sum_{\boldsymbol{v}} \omega_1(\boldsymbol{v})E(Y_{\text{obs}}|x=1,\boldsymbol{v})p(\boldsymbol{v}|x=1), \end{aligned} \tag{13.1}$$

where $p\,(.)$ indicates the probability measure, $E\,(.)$ indicates the expectation, and

$$\begin{aligned} \omega_1\left(\boldsymbol{v}\right) &\equiv \frac{p\left(\boldsymbol{v}\right)}{p\left(\boldsymbol{v}|x=1\right)} = \frac{p\left(\boldsymbol{v}\right)}{p\left(\boldsymbol{v},x=1\right)/p\left(x=1\right)} = \frac{p\left(\boldsymbol{v}\right)}{p\left(x=1|\boldsymbol{v}\right)p\left(\boldsymbol{v}\right)/p\left(x=1\right)} \\ &= p\left(x=1\right)/p\left(x=1|\boldsymbol{v}\right) \end{aligned} \tag{13.2}$$

Equations (13.1) and (13.2) indicate that an unbiased estimate of $E\left(Y_1\right)$ is obtained by a weighted average of Y_{obs} for the treatment group with weights inversely proportional to the propensity score. Similarly, an unbiased estimate of $E\left(Y_0\right)$ is obtained by a weighted average of Y_{obs} for the control group with weights $\omega_0\left(\boldsymbol{v}\right) \equiv p\left(x=0\right)/p\left(x=0|\boldsymbol{v}\right)$. The inverse probability weighting (IPW) relies on consistent sample estimates of $p\left(x|\boldsymbol{v}\right)$ for weighting the sample. For the treatment effects that are defined in terms of log odds ratios, as in the case of the logit or multinomial logit model, however, the above-described procedure does not work for obtaining a consistent parameter estimate for the average treatment effect, as explained in what follows.

Equations (13.1) and (13.2) indicate that the weights make treatment variable X independent of covariates $\boldsymbol{V}$ in the pseudopopulation under the condition

$p(x|\mathbf{v}) > 0$ because the weighted joint probability $p^*(x, \mathbf{v})$ of X and $\mathbf{V}$ satisfies $p^*(x, \mathbf{v}) \equiv p(x, \mathbf{v})(p(x)/p(x|\mathbf{v})) = p(x)p(\mathbf{v})$. In this pseudopopulation that satisfies $X \perp \mathbf{V}$, $E^*(Y_{\text{obs}}|x = 1) = E(Y_1)$ and $E^*(Y_{\text{obs}}|x = 0) = E(Y_0)$ both hold. $E^*(.)$ here and henceforth indicates the expectation in the pseudopopulation with adjusted probability measure $p^*(x, \mathbf{v}) = p(x)p(\mathbf{v})$. It follows that the *marginal treatment effect*, which we obtain as the treatment effect for a given link function $l(.)$ of the generalized linear model (McCullagh and Nelder, 1989) by collapsing data across states of $\mathbf{V}$ after attaining independence between X and $\mathbf{V}$, can be expressed as

$$l(E(Y_1)) - l(E(Y_0)) = l(E^*(Y_{\text{obs}}|x = 1)) - l(E^*(Y_{\text{obs}}|x = 0)) \tag{13.3}$$

On the other hand, under the SITA assumption, we can express the general model of *observed* outcomes at the individual level for individual i by the following semi-parametric regression equation for a given link function $l(.)$:

$$l(E(Y_{\text{obs},i}|x_i, \mathbf{v}_i)) = \alpha + \beta x_i + \phi(\mathbf{v}_i|\boldsymbol{\theta}_1) + \varphi(\mathbf{v}_i|\boldsymbol{\theta}_2)x_i \tag{13.4}$$

where $\boldsymbol{\theta}_1$ and $\boldsymbol{\theta}_2$ are parameter sets included in functions $\phi(.)$ and $\varphi(.)$, respectively. Without loss of generality, we can assume $\sum_{\mathbf{v}}\phi(\mathbf{v}|\boldsymbol{\theta}_1)p(\mathbf{v}) = \sum_{\mathbf{v}}\varphi(\mathbf{v}|\boldsymbol{\theta}_1)p(\mathbf{v}) = 0$ in order to make α and β represent the average effects as described below. Since $E(Y_1|\mathbf{v}) = E(Y_1|X = 1, \mathbf{v}) = E(Y_{\text{obs}}|X = 1, \mathbf{v})$ and $E(Y_0|\mathbf{v}) = E(Y_0|X = 0, \mathbf{v}) = E(Y_{\text{obs}}|X = 0, \mathbf{v})$ hold under the SITA assumption, Equation (13.4) implies$l(E(Y_{1i}|\mathbf{v}_i)) = l(E(Y_{\text{obs},i}|x_i = 1, \mathbf{v}_i) = \alpha + \beta + \phi(\mathbf{v}_i|\boldsymbol{\theta}_1) + \varphi(\mathbf{v}_i|\boldsymbol{\theta}_2)$and $l(E(Y_{0i}|\mathbf{v}_i) = l(E(Y_{\text{obs},i}|x_i = 0, \mathbf{v}_i)) = \alpha + \phi(\mathbf{v}_i|\boldsymbol{\theta}_1)$, and thus the treatment effect for individual i with covariates $\mathbf{v}_i$ is given as $\beta + \varphi(\mathbf{v}_i|\boldsymbol{\theta}_2)$, and parameter β indicates the *average treatment effect*.

A problem of causal reasoning for logit and multinomial logit regression models here is that the average treatment effect based on Equation (13.4) differs from the marginal treatment effect defined by Equation (13.3). This can be readily inferred from the fact that given Equation (13.4), the marginal treatment effect is given as

$$\begin{aligned}
&l(E^*(Y_{\text{obs}}|x = 1)) - l(E^*(Y_{\text{obs}}|x = 0)) \\
&\quad = l(E^*_{\mathbf{v}}(E(Y_{\text{obs}}|x = 1, \mathbf{v}))) - l(E^*_{\mathbf{v}}(E(Y_{\text{obs}}|x = 0, \mathbf{v}))) \\
&\quad = l\left(\sum_{\mathbf{v}} l^{-1}(\alpha + \beta + \phi(\mathbf{v}|\boldsymbol{\theta}_1) + \varphi(\mathbf{v}|\boldsymbol{\theta}_2))p(\mathbf{v})\right) - l\left(\sum_{\mathbf{v}} l^{-1}(\alpha + \phi(\mathbf{v}|\boldsymbol{\theta}_1))p(\mathbf{v})\right).
\end{aligned}$$

The disagreement between the marginal treatment effect and the average treatment effect yields the lack of *collapsibility*.

Prior to a more formal analysis of this collapsibility problem and its solution, we first illustrate the problem by presenting simple examples with a dichotomous dependent variable Y, a dichotomous treatment variable X, and a single dichotomous covariate V. Data sets 1, 2, and 3 in Table 13.1 present distinct hypothetical 2-by-2-by-2 cross-classified frequencies, where X is made independent of V for each

TABLE 13.1 Three Hypothetical Data Sets.

Data set 1

	$V = 1$		$V = 0$		Total	
	$X = 1$	$X = 0$	$X = 1$	$X = 0$	$X = 1$	$X = 0$
$Y = 1$	88	80	80	75	168	155
$Y = 0$	72	80	20	25	92	105

Data set 2

	$V = 1$		$V = 0$		Total	
	$X = 1$	$X = 0$	$X = 1$	$X = 0$	$X = 1$	$X = 0$
$Y = 1$	64	60	80	75	144	135
$Y = 0$	56	60	20	25	76	85

Data set 3

	$V = 1$		$V = 0$		Total	
	$X = 1$	$X = 0$	$X = 1$	$X = 0$	$X = 1$	$X = 0$
$Y = 1$	80	70	80	75	160	145
$Y = 0$	60	70	20	25	80	95

data set. They, respectively, represent situations where the interaction effect of X and V on Y is absent for the linear probability model (data set 1), the log-linear probability model (data set 2), and the logit model (data set 3), for the following specifications of the three models.

$$P(Y = 1|x, v) = \alpha + \beta_1 X + \gamma V + \beta_2 XV \tag{13.5a}$$

$$\log (P(Y = 1|x, v)) = \alpha + \beta_1 X + \gamma V + \beta_2 XV \tag{13.5b}$$

$$\log \left(\frac{P(Y = 1|x, v)}{P(Y = 0|x, v)} \right) = \alpha + \beta_1 X + \gamma V + \beta_2 XV \tag{13.5c}$$

The issue of collapsibility is concerned with whether, given the statistical independence between X and V, we can retain the *average treatment effect* if we omit the effect of V from the regression models –or equivalently, if we analyze data summed across the states of V –by applying the following models with the same link function.

$$P(Y = 1|x) = a + bX \tag{13.6a}$$

$$\log (P(Y = 1|x)) = a + bX \tag{13.6b}$$

$$\log \left(\frac{P(Y = 1|x)}{P(Y = 0|x)} \right) = a + bX \tag{13.6c}$$

TABLE 13.2 Analysis of Data Sets in Table 13.1.

		β_1	γ	β_2	Average β	b
Linear probability	Data set 1	0.050	−0.250	0.000	0.050	0.050
	Data set 2	0.050	−0.250	−0.017	0.041	0.041
	Data set 3	0.050	−0.250	0.021	0.063	0.063
Log-linear probability	Data set 1	0.065	−0.405	0.031	0.084	0.081
	Data set 2	0.065	−0.405	0.000	0.065	0.065
	Data set 3	0.065	−0.495	0.069	0.105	0.098
Logit	Data set 1	0.288	−1.099	−0.087	0.234	0.213
	Data set 2	0.288	−1.099	−0.154	0.204	0.176
	Data set 3	0.288	−1.099	0.000	0.288	0.270

Table 13.2 shows the estimates of regression coefficients β_1 and β_2 obtained by applying regression models (13.5a), (13.5b), and (13.5c) to each of the three data sets, and those of regression coefficient b obtained similarly by applying regression models (13.6a), (13.6b), and (13.6c). The "average β" in Table 13.2 is a weighted average of the treatment effects $\beta_1 + \beta_2 V$ from models (13.5a), (13.5b), and (13.5c), defined as $\sum_{v} p(V)\left(\beta_1 + \beta_2 V\right)$, with weights $p(V)$ equal to the proportion of the covariate states.

The results of Table 13.2 show that

(a) the linear probability model preserves the average treatment effect regardless of the presence of interaction effects of X and V on Y;

(b) the log-linear probability model preserves the average treatment effect only if there exist no interaction effects of X and V on Y; and

(c) the logit model does not preserve the average treatment effect regardless of the presence of interaction effects of X and V on Y.

The following section describes a theorem that shows that characteristics (a), (b), and (c) given above generally hold for any data. Finding (c) implies the lack of collapsibility for the logit model.

13.3 THEOREM ON STANDARDIZATION AND THE LACK OF COLLAPSIBILITY OF THE LOGIT MODEL

We formalize observations we made in the analysis of Table 13.2 by introducing a theorem on standardization. We assume here that covariates $\boldsymbol{V}$ affect X, and both $\boldsymbol{V}$ and X affect Y. We first define two functions. The first function is the *link function*, $l(x)$, as the term is defined for generalized linear models (McCullagh and Nelder, 1989). When the effects of X and V on Y are additive, we express the model as

$l(P(y|\boldsymbol{v},x)) = \beta^{Y} + \beta^{Y|V} + \beta^{Y|X}$, and when interaction effects of $\boldsymbol{V}$ and X on Y exist, as $l(P(y|\boldsymbol{v},x)) = \beta^{Y} + \beta^{Y|V} + \beta^{Y|X} + \beta^{Y|VX}$, where β^{Y}, $\beta^{Y|V}$, $\beta^{Y|X}$, and $\beta^{Y|VX}$ indicate the Y-intercept, the effects of $\boldsymbol{V}$ on Y, the effects of X on Y, and the interaction effects of $\boldsymbol{V}$ and X on Y, respectively, and capital P indicates the expected probability from a given model.

We refer to the second function, which is assumed to be a smooth strongly monotonically increasing function, as the *standardization function*, $s(x)$. With this function, the standardized conditional probabilities $P(y|x)$ are defined as follows.

$$s(P(y|x)) = \sum_{\boldsymbol{v}} w(\boldsymbol{v})\, s(P(y|\boldsymbol{v},x)) \tag{13.7}$$

where $w(\boldsymbol{v})$ is the *standard distribution* of $\boldsymbol{V}$ that satisfies $w(\boldsymbol{v}) \geq 0$ and $\sum_{\boldsymbol{v}} w(\boldsymbol{v}) = 1$. Then, the following theorem holds[1]:

Theorem

(1) When the effects of $\boldsymbol{V}$ and X on Y are additive, such that $l\left(P\left(y_k|\boldsymbol{v},x_j\right)\right) = \beta_k^{Y} + \beta_k^{Y|V} + \beta_{k|j}^{Y|X}$, the standardized conditional probability preserves the effects of X on Y in $P(y|\boldsymbol{v},x)$ such that $l\left(P\left(y_k|x_{j_1}\right)\right) - l\left(P\left(y_k|x_{j_2}\right)\right) = \beta_{k|j_1}^{Y|X} - \beta_{k|j_2}^{Y|X}$ if either $s(x) = l(x)$ or $s(x) = \exp\ (l(x))$, and those effects are independent of the standard distribution $w(\boldsymbol{v})$.

(2) Even when the effects of $\boldsymbol{V}$ and X are additive, the standardized conditional probabilities do not preserve the effects of X on Y when the link function is the logit or multinomial logit function and the standardization function is the identity function $s(x) = x$.

(3) When interaction effects of $\boldsymbol{V}$ and X on Y exist such that $l\left(P\left(y_k|\boldsymbol{v},x_j\right)\right) = \beta_k^{Y} + \beta_k^{Y|V} + \beta_{k|j}^{Y|X} + \beta_{k|j}^{Y|VX}$, then the standardized conditional probability preserves the *average* effects of X on Y if $s(x) = l(x)$ and $w(\boldsymbol{v}) = p(\boldsymbol{v})$, but not if $s(x) = \exp\ (l(x))$.

A proof of the theorem is given in Appendix A. In particular, part (1) of the theorem means if the standardization function is either a multinomial logit function, such as

$$\log\left(\frac{P\left(y_k|x_j\right)}{P\left(y_1|x_j\right)}\right) = \sum_{\boldsymbol{v}} w(\boldsymbol{v}) \log\left(\frac{P\left(y_k|\boldsymbol{v},x_j\right)}{P\left(y_1|\boldsymbol{v},x_j\right)}\right) \tag{13.8}$$

or, a multinomial odds function, such as

$$\frac{P\left(y_k|x_j\right)}{P\left(y_1|x_j\right)} = \sum_{\boldsymbol{v}} w(\boldsymbol{v}) \left(\frac{P\left(y_k|\boldsymbol{v},x_j\right)}{P\left(y_1|\boldsymbol{v},x_j\right)}\right) \tag{13.9}$$

[1] Little and Pullum (1979) proved a part of this theorem for the case where the link function is equal to the standardization function. As far as the author knows, no proof was presented prior to a study by Yamaguchi (2012) in a general form with the standardization function for the case with $s(x) = \exp\ (l(x))$.

then, if no interaction effects of X and $\boldsymbol{V}$ on Y exist, the effects of X on Y, in terms of conditional odds ratios between X and Y for a given $\boldsymbol{V}$ in $P(y|\boldsymbol{v},x)$, are retained in the standardized probability $P(y|x)$, and those odds ratios do not depend on weights $w(\boldsymbol{v})$.

Part (2) of the theorem means that if the standardized conditional probability is defined as

$$P\left(y_k|x_j\right) = \sum_{\boldsymbol{v}} w(\boldsymbol{v}) P\left(y_k|\boldsymbol{v},x_j\right) \tag{13.10}$$

it does not preserve the effects of X on Y in $p(y|\boldsymbol{v},x)$ for the multinomial logit link function. This yields the lack of the collapsibility of the logit and multinomial logit models.

Part (3) of the theorem suggests that the use of the standardization Equation (13.8) combined with $w(\boldsymbol{v}) = p(\boldsymbol{v})$ works best if interaction effects of X and $\boldsymbol{V}$ on Y exist. However, this method is not effective if we employ the nonparametric estimation for $p(y|\boldsymbol{v},x)$ and the estimates $p(y|\boldsymbol{v},x)$ include a value of zero, as explained in the following section.

13.4 THE PROBLEM OF ZERO-SAMPLE ESTIMATES OF CONDITIONAL PROBABILITIES AND THE USE OF SEMIPARAMETRIC MODELS TO SOLVE THE PROBLEM

13.4.1 The Problem of Zero-Sample Estimates of Conditional Probabilities

We use shorter notations $P_{k|ij}^{Y|VX}$ and $P_{k|j}^{Y|X}$, respectively, for conditional probability $P\left(y_k|\boldsymbol{v}_i,x_j\right)$ and $P\left(y_k|x_j\right)$. Suppose that the multinomial logit regression with interaction effects of X and $\boldsymbol{V}$ on Y is specified as $\log\left(P_{k|ij}^{Y|VX}/P_{1|ij}^{Y|VX}\right) = \beta_k^Y + \beta_{k|i}^{Y|V} + \beta_{k|j}^{Y|X} + \beta_{k|ij}^{Y|VX}$, where $\beta_1^Y = \beta_{1|i}^{Y|V} = \beta_{1|j}^{Y|X} = \beta_{1|ij}^{Y|VX} = 0$, and $\beta_{k|1}^{Y|X} = \beta_{k|i1}^{Y|VX} = 0$, with the first category for each variable as the baseline category. Then the average effect of $X = j$ versus $X = 1$ on $Y = k$ versus $Y = 1$ would be obtained by taking a weighted average of $\beta_{k|j}^{Y|X} + \beta_{k|ij}^{Y|VX}$, with weights equal to $p(\boldsymbol{v})$. The average effect is given as

$$\sum_i p\left(\boldsymbol{v}_i\right)\left(\beta_{k|j}^{Y|X} + \beta_{k|ij}^{Y|VX}\right) = \sum_i p\left(\boldsymbol{v}_i\right)\log\left\{\frac{P_{k|ij}^{Y|VX} P_{1|i1}^{Y|VX}}{P_{k|i1}^{Y|VX} P_{1|ij}^{Y|VX}}\right\}. \tag{13.11}$$

Hence, the average treatment effects are the weighted averages of conditional log odds ratios between X and Y in $P(y|\boldsymbol{v},x)$, and this is what we would obtain for log odds ratios in the set of $P(y|x)$ by applying the standardization Equation (13.8) and setting $w(\boldsymbol{v}) = p(\boldsymbol{v})$.

If, however, covariates $\boldsymbol{V}$ are categorical and we characterize $p_{k|ij}^{Y|VX}$ nonparametrically for the combined states of $\boldsymbol{V}$, then it is very likely that the *sample estimate* for the component odds ratio of Equation (13.11) will take a value of zero for the numerator, the denominator, or both, for some states of $\boldsymbol{V}$, and thereby make the average of log odds ratios meaningless.

Hence, the method of retaining the average treatment effect by using Equation (13.11) is not effective when we employ nonparametric estimates for $p_{k|ij}^{Y|VX}$ with many covariates. Although this problem can be solved if we specify a parametric model for $p_{k|ij}^{Y|VX}$ to assure $P_{k|ij}^{Y|VX} > 0$, we wish to make the assumption of the model as weak as possible and use instead *semiparametric* models that do not specify the effects of $\boldsymbol{V}$ on Y, as described in the following section.

13.4.2 Method for Obtaining Adjusted Two-Way Frequency Data for the Analysis of Association between X and Y

Let us assume that all covariates $\boldsymbol{V}$ are categorical, or categorized, variables and let us denote by C a categorical variable whose categories are combined states of covariates $\boldsymbol{V}$, by excluding states of C with no sample observations. We indicate by f_{ijk}^{CXY} the observed joint frequencies of C, X, and Y, and also express the one-way and two-way marginal frequencies by indicating the corresponding variables in superscripts. For the reason to be explained in Section 13.5, the method introduced here requires that the joint marginal frequencies of the focal variables X and Y, $f_{jk}^{XY} = \sum_i f_{ijk}^{CXY}$, have no sampling zeros, that is, $f_{jk}^{XY} > 0$.

If the interaction effects of C and X on Y are not significant, we can purge the interaction effects from $p_{k|ij}^{Y|CX}$ by the maximum likelihood estimation. That is the same as applying the following semiparametric multinomial logit model, which we refer to as the *additive effect* model of the treatment variable and covariates.

$$\log\left(\frac{P_{k|ij}^{Y|CX}}{P_{1|ij}^{Y|CX}}\right) = \beta_k^Y + \beta_{k|i}^{Y|C} + \beta_{k|j}^{Y|X} \tag{13.12}$$

where $\beta_1^Y = \beta_{1|i}^{Y|C} = \beta_{1|j}^{Y|X} = 0$ and $\beta_{k|1}^{Y|X} = 0$. Since parameter $\beta_{k|i}^{Y|C}$ is used for each state of C, a state of C that satisfies $f_{ij}^{CX} > 0$ for just one state of X in the sample does not provide information in estimating $\beta_{k|j}^{Y|X}$, because the outcome of Y is completely explained as the effect of $\beta_{k|i}^{Y|C}$ for those cases. In particular, if every sample subject has a distinct category of C for cross-sectional survey data, no information remains in data in assessing the effects of X on Y. Hence, researchers should make sure that the proportion of such noninformative sample to be small for a given set of covariates in order not to lose efficiency greatly.

The maximum likelihood (hereafter ML) estimates of $p_{k|ij}^{Y|CX}$ that satisfy Equation (13.12) can be obtained by using the iterative proportional adjustment procedure described in Appendix B. As the appendix shows, the estimates for $p_{k|ij}^{Y|CX}$ are always given under the condition of $f_{jk}^{XY} > 0$.

After the elimination of the interaction effects of C and X on Y, the covariate-state-specific conditional probability $P_{k|ij}^{Y|CX}$ reflects the same effects of X on Y, expressed by conditional odds ratios, at each state of C, *except for cases of C with $f_{ik}^{CY} = 0$*. When $f_{ik}^{CY} = 0$, the odds ratio $\left(P_{k|ij}^{Y|CX} P_{1|i1}^{Y|CX}\right) / \left(P_{1|ij}^{Y|CX} P_{k|i1}^{Y|CX}\right)$ involves zero for both the numerator and the denominator and, therefore, becomes indeterminate. This fact leads to a requirement in the application of Equation (13.8) or (13.9) that we set weights $w(c) = 0$ for the cases of C for which $f_{ik}^{CY} = 0$ for at least one state of Y.

With such a specification for $w(c)$, we can obtain by using Equation (13.8) or (13.9) the standardized conditional probability $P(y|x)$ that satisfies

$$\log\left(\frac{\left(P_{k|j}^{Y|X} P_{1|1}^{Y|X}\right)}{\left(P_{k|1}^{Y|X} P_{1|j}^{Y|X}\right)}\right) = \beta_{k|j}^{Y|X} \tag{13.13}$$

Without loss of generality, we may set weights of Equations (13.8) or (13.9) to be equal to the *relative* proportion of C's states among those that satisfy $f_{ik}^{CY} > 0$ for all Y's states. Note that the theorem indicates that the standardized conditional probabilities do not depend on weights when no interaction effects of C and X on Y exist and, therefore, we are justified in manipulating weights to exclude cases involving indeterminate odds ratios, without loss of generality, by setting zero weights $w(\mathbf{v}) = 0$ for those cases.

Since only the odds ratios, and not the conditional probabilities, preserve the effects of X on Y, we may further adjust the adjusted two-way frequencies obtained initially as $f_j^X P_{k|j}^{Y|X}$ to retain the same marginal distributions of X and Y as the observed frequencies f_{jk}^{XY} by iterative proportional adjustments starting from $f_j^X P_{k|j}^{Y|X}$, because then the adjusted two-way frequencies are determined uniquely regardless of the particular specification we choose for weights $w(\mathbf{v})$, and the estimates of parameters pertaining to log odds ratios and their standard errors in log-linear analysis are not affected by this readjustment of marginal frequencies, as will be shown later.

13.4.3 Method for Obtaining an Adjusted Three-Way Frequency Table for the Analysis of Conditional Association

Now we consider cases where we can expect significant interaction effects of X and a particular covariate Z on Y. We consider this as an extension of the semiparametric model, for which we do not impose any constraints on the covariate effects, including Z, on Y. We can easily extend the method for cases with two or more interacting covariates by treating the combined states of those covariates as Z. Let us denote by C the set of combined categories of covariates other than Z. For the reason explained in Section 13.5, we assume that $f_{jkm}^{XYZ} > 0$.

Then, we obtain an extension of the model of Equation (13.12) by including the interaction effects of C and Z on Y as well as X and Z on Y such that

$$\log\left(\frac{P_{k|ijm}^{Y|CXZ}}{P_{1|ijm}^{Y|CXZ}}\right) = \beta_k^Y + \beta_{k|j}^{Y|X} + \beta_{k|m}^{Y|Z} + \beta_{k|jm}^{Y|XZ} + \beta_{k|i}^{Y|C} + \beta_{k|im}^{Y|CZ} \tag{13.14}$$

where the β parameters take a value of zero for the first category of each variable. We call this model a *conditionally additive effect model* of the treatment variable and covariates because the effects of X on Y and the effects of C on Y are additive for each given state of Z.

The ML estimates of Equation (13.14) can be obtained, without applying the regression model, by using the iterative proportional adjustment procedure described in Appendix B. As the appendix shows, the estimates for $P_{k|ijm}^{Y|CXZ}$ are always given under the condition of $f_{jkm}^{XYZ} > 0$.

Then, applying the standardization method of Equation (13.9), we obtain

$$\begin{aligned}
\log\left(P_{k|jm}^{Y|XZ} / P_{1|jm}^{Y|XZ}\right) &= \sum_i w\left(c_i\right) \log\left(\frac{P_{k|ijm}^{Y|CXZ}}{P_{1|ijm}^{Y|CXZ}}\right) \\
&= \sum_i w\left(c_i\right)\left(\beta_k^Y + \beta_{k|j}^{Y|X} + \beta_{k|m}^{Y|Z} + \beta_{k|jm}^{Y|XZ} + \beta_{k|i}^{Y|C} + \beta_{k|im}^{Y|CZ}\right) \\
&= \beta_k^Y + \beta_{k|j}^{Y|X} + \beta_{k|m}^{Y|Z} + \beta_{k|jm}^{Y|XZ} + \sum_i w\left(c_i\right)\left(\beta_{k|i}^{Y|C} + \beta_{k|im}^{Y|CZ}\right)
\end{aligned} \tag{13.15}$$

It follows that the *conditional* log odds ratios between X and Y at each level of Z in the adjusted conditional probabilities become

$$\log\left(\frac{P_{k|jm}^{Y|XZ} P_{1|1m}^{Y|XZ}}{P_{1|jm}^{Y|XZ} P_{k|1m}^{Y|XZ}}\right) = \beta_{k|j}^{Y|X} + \beta_{k|jm}^{Y|XZ} \tag{13.16}$$

and, therefore, preserve parameters that characterize the effects of X on Y and the interaction effects of X and Z on Y.

On the other hand, the conditional log odds ratios between Z and Y at each level of X in the adjusted conditional probabilities become

$$\log\left(\frac{P_{k|jm}^{Y|XZ} P_{1|j1}^{Y|XZ}}{P_{1|jm}^{Y|XZ} P_{k|j1}^{Y|XZ}}\right) = \beta_{k|m}^{Y|Z} + \beta_{k|jm}^{Y|XZ} + \sum_i w\left(c_i\right) \beta_{k|im}^{Y|CZ} \tag{13.17}$$

These log odds ratios reflect weighted averages of the interaction effects of C and Z on Y with weights $w(c)$ and since we cannot set $w(c) = p(c)$ to make them reflect the average effects because we have to rely only on the subset of C's categories for the estimation of $P^{Y|XZ}_{k\lfloor jm}$, the effects of Z on Y become a function of an arbitrary specification for weights $w(c)$.

Equations (13.16) and (13.17) indicate that when applying conditional association models to adjusted three-way contingency table data, we should keep in mind that the estimates for parameters that characterize the association between X and Y, and those that characterize dependence of the X–Y association on Z, can be interpreted as causal effects, but those that characterize the association between Z and Y cannot be.

With the ML estimates for $P^{Y|CXZ}_{k|ijm}$, we can obtain $P^{Y|XZ}_{k\lfloor jm}$ by using Equation (13.9) if we set weights to be $w(c) = p\left(c|f^{CYZ} > 0\right)$ and $w(c) = 0$ for cases of C involving $f^{CYZ} = 0$ for at least one combined state of Y and Z. Again, we may readjust $f^{XZ}_{im}P^{Y|XZ}_{k\lfloor jm}$ to make readjusted frequencies to have the same XZ-marginal frequencies and Y-marginal frequencies as those of observed frequencies f^{XYZ}_{jkm}, so that adjusted three-way frequencies are uniquely identified regardless of the use of particular weights $w(c)$.

13.5 ESTIMATION OF STANDARD ERRORS IN THE ANALYSIS OF ASSOCIATION WITH ADJUSTED CONTINGENCY TABLE DATA

It is well known that the use of weighted data with which we obtain consistent estimates for the parameters of interest to us still generates bias in the estimates of the standard errors of the parameter estimates, because standard errors depend on unweighted sample counts. Clogg and Eliason (1987) introduced a method that attempted to solve this problem in the log-linear analysis by fitting unweighted frequencies by the model while characterizing weighted frequencies by the model's parameters. Skinner and Vallet (2010), however, have recently shown that the standard errors estimated by the Clogg–Eliason method have systematic downward bias and are inconsistent *unless weights do not vary within each cell of the contingency table data*. Since sampling weights vary with sample individuals within each cell, the Skinner–Vallet study invalidated the Clogg–Eliason method for contingency-table data weighted by sampling weights.

However, we can employ the Clogg–Eliason method for the present analysis. We wish to estimate parameters that characterize the adjusted frequencies in which causal relationship among variables is embedded, while fitting unadjusted sample counts for estimating standard errors. Unlike standard applications of log-linear models to contingency-table data, we do not need to take into account sampling weights, because if the semiparametric model we assume for the population is correct, the causal relationship between X and Y and its dependence on Z do not depend on the sampling variability. Hence, we only need to apply the adjustment weights for standardization, but unlike the sampling weights, the adjustment weights are cell-specific and do not vary with sample individuals. Hence, as shown by Skinner and Vallet, the Clogg–Eliason method becomes valid in this case. We denote below by f^{XY}_{jk} and f^{XYZ}_{jkm} the cross-classified *unweighted* observed frequencies, by

g_{jk}^{XY} and g_{jkm}^{XYZ} the cross-classified adjusted frequencies based on the application of the semiparametric models to unweighted data and the standardization method described in the preceding sections. For the two-way table analysis of adjusted cross-classified frequencies g_{jk}^{XY}, let us set "cell weight" to be $w_{jk}^{XY} = f_{jk}^{XY}/g_{jk}^{XY}$ for observed frequency f_{jk}^{XY}. Then the log-linear or log-multiplicative model of association between X and Y for the adjusted two-way frequencies g_{jk}^{XY} can be specified by using the Clogg–Eliason method as

$$\log\left(\frac{F_{jk}^{XY}}{w_{jk}^{XY}}\right) = \lambda + \lambda_j^X + \lambda_k^Y + \rho_{jk}^{XY} \tag{13.18}$$

where F_{jk}^{XY} is the unweighted expected frequency from the model, and ρ_{jk}^{XY} characterizes the log odds ratios between X and Y. The model makes the unweighted expected frequencies F_{jk}^{XY} fit with the unweighted observed frequencies f_{jk}^{XY}, while the parameters of the model characterize adjusted frequencies g_{jk}^{XY}. The association component can be characterized by various log-linear or log-multiplicative association parameters (Goodman, 1979 1986; Clogg and Shihadeh, 1994).

13.6 ILLUSTRATIVE APPLICATION

13.6.1 Data

We employ the multiyear data of the General Social Survey 1972–2008 with the population of people aged 20–79 at the time of the survey. A particular trichotomous variable Happiness ("not happy," "pretty happy," and "very happy") is employed as the dependent variable with 41,200 samples. As the treatment variable, marital status with five categories (married, widowed, divorced, separated, never married) are employed. Seven categorical covariates are employed: (i) age (12 categories: 20–24, 25–29, 30–34, 35–39, 40–44, 45–49, 50–54, 55–59, 60–64, 65–69, 70–74, and 75–79), (ii) education (5 categories: less than 12 years, 12 years, 13–15 years, 16 years, and more than 16 years), (iii) gender, (iv) race (2 categories: African American and other), (v) family intactness at age 16 (2 categories: intact and nonintact), (vi) region of residence at age 16 (3 categories: foreign, Pacific, and other), (vii) period (4 categories: 1972–1979, 1980–1989, 1990–1999, and 2000–2008). The number of possible combined states of those seven covariates is 5280, including 1966 states that have no corresponding observations in the sample. Among those with corresponding observations, covariate states for which observations are made just with one category of marital status, and are, therefore, not informative for the estimation of the effects of marital status on the dependent variable include 1106 states and 1764 sample observations, with the average of 1.54 observations per state. On the other hand, informative covariate states for which observations are made with two or more states of marital status include 3208 states and 39,436 observations, with the average of 12.29 observations per state. As a result, the proportion of observations in the sample that are not informative turned out to be very small (4.3%).

Those seven covariates are significantly associated with both marital status and the dependent variable. Although the results are not presented because they are not central subjects of this chapter, the multinomial logit model that assumes the additive effects of seven covariates indicates the following. All seven covariates showed strong effects, but marital status, especially "married" versus other marital statuses is the strongest predictor of being "very happy" rather than "not happy." Education and race also have impacts greater than the other four predictors, with lower education and being African American leading to lower odds of being "very happy" rather than "not happy."

13.6.2 Software

An executable program, `LLCAUSAL.EXE`, that the author developed for the estimation of two-way and three-way tables of adjusted frequencies and information about the proportion of noninformative samples is available from the author by a request.

13.6.3 Analysis

We employed three-way table analysis because significant interaction of the treatment variable (marital status) and a covariate (race) are found in the preliminary analysis. Hence, we employ race as the conditioning variable in the models described in Section 13.4.3. Table 13.3 presents (i) adjusted frequencies and (ii) unweighted sample counts for the three-way table.

Table 13.4 presents the results from four multinomial logit models: (i) the multinomial logit model with marital status, race, and their interaction effects as the only

TABLE 13.3 Unweighted and Adjusted Frequencies of Happiness.

	Other Ethnicities			African Americans		
	Very Happy	Pretty Happy	Not Happy	Very Happy	Pretty Happy	Not Happy
			Adjusted frequencies			
Married	9,462.58	11,627.88	1,397.55	751.34	1,180.53	296.13
Widowed	593.57	1,854.47	600.96	95.75	367.54	156.71
Divorced	928.24	2,931.59	782.17	110.19	485.89	219.92
Separated	175.52	548.86	247.62	91.13	336.37	175.51
Never married	937.33	2,868.07	695.59	238.33	662.81	379.85
			Unweighted sample counts			
Married	9,316	11,632	1,540	693	1,220	315
Widowed	732	1,723	594	120	353	147
Divorced	937	2,916	789	120	501	195
Separated	168	542	262	86	341	176
Never married	996	2,896	609	216	740	325

TABLE 13.4 Multinomial Logit Models.

	The Effects of Marital Status (vs. Never Married)					
	(a) Among African Americans (vs. Very Happy)		(b) Among Other Races (vs. Very Happy)		Racial Difference: (a)–(b)	
	Unhappy	Pretty Happy	Unhappy	Pretty Happy	Unhappy	Pretty Happy
(i) Multinomial logit with marital status as the single predictor						
Married	−1.197**	−0.666***	−1.308***	−0.845***	0.111	0.180
Windowed	−0.206	−0.152	0.283***	−0.211***	−0.489**	0.059
Divorced	0.077	0.198	0.320***	0.068	−0.243	0.130
Separated	0.308	0.146	0.936***	0.104	−0.629***	0.042
(ii) Multinomial logit with marital status and seven other predictors						
Married	−1.292***	−0.611***	−1.461***	−0.829***	0.170	0.219*
Windowed	−0.160	0.109	0.365***	0.071	−0.525**	0.038
Divorced	0.072	0.262*	0.208**	0.078	−0.137	0.184
Separated	0.169	0.185	0.687***	0.068	−0.519**	0.117
(iii) Model 2 plus the interaction effects of race and age						
Married	−1.084***	−0.482***	−1.497***	−0.842***	0.413**	0.360***
Widowed	0.220	0.306*	0.294***	0.052	−0.074	0.254
Divorced	0.331*	0.435***	0.170*	0.061	0.161	0.374**
Separated	0.372*	0.317*	0.664***	0.058	−0.293	0.259
(vi) Semiparametric multinomial logit model						
Married	−1.397***	−0.571***	−1.614***	−0.912***	0.217	0.341***
Windowed	0.027	0.322*	0.311***	−0.021	−0.284	0.301*
Divorced	0.225	0.461***	0.127	0.032	0.098	0.429**
Separated	0.189	0.283*	0.642***	0.022	−0.453*	0.261

Note: * $p < 0.05$, ** $p < 0.01$, *** $p < 0.001$;

predictors; (ii) the multinomial logit model with additional additive effects of the six other covariates; (iii) a third model that adds the interaction effects of age and race to the second model; and (iv) the semiparametric multinomial logit model of Equation (13.14). Table 13.4 omits the presentation of the main effects of race from all models, as well as the effects of the six covariates for models 2 and 3 because we are interested only in the effects of marital status and their dependence on race.

The most conspicuous finding in Table 13.4 is the great difference in results between models 2 and 3. While the effects of marital status on Happiness change only a little among other races when the interaction effects of race and age are added, the results change greatly among African Americans. Since marital status and age are strongly associated, the interaction effects of race and age, if they are present, strongly affect the interaction effects of marital status and race, and a failure to include the former effects greatly distorts the latter effects.

Note that by including the interaction effects of age and race, the results of model 3 become closer, though not uniformly, to the results from the semiparametric multinomial logit model (model 4) than the results of model 2 are to those of model 4. Those findings indicate that the results from multinomial logit models that characterize the effects of covariates parametrically may depend heavily on a particular specification of the model regarding the covariates' effects. On the other hand, the semiparametric model does not suffer from that problem.

13.7 CONCLUSION

Although log-linear and log-multiplicative association models are useful for the analysis of cross-classified frequency data, their major limitations have been (i) their inability to use many control variables and (ii) their lack of association with causal analysis. A multivariate regression analysis of a categorical dependent variable, such as multinomial logit regression model, provided only a partial remedy for those limitations because it usually makes a strong assumption that there are linear additive effects of control variables, and because the type of association that can be tested is limited. This article introduced a new method that greatly reduced limitations (i) and (ii).

The log-linear causal analysis introduced in this chapter enables the analysis of association between two focal variables and the possible dependence of that association on a third variable, by controlling for confounding effects of many covariates without the parametric modeling of covariate effects as employed in multivariate regression models.

The major limitation of the method introduced in this chapter for causal analysis, however, is the assumed absence of *unobserved* confounding variables, as the same limitation applies to any use of propensity score methods with the SITA assumption in causal analysis.

REFERENCES

Clogg, C.C. and Eliason, S.R. (1987) Some common problems in loglinear analysis. *Sociological Methods & Research*, **16**, 8–44.

Clogg, C.C. and Shihadeh, E.S. (1994) *Statistical Models for Ordinal Variables*, Sage Publications, Thousand Oaks, CA.

Gail, M.H., Weiand, W., and Piantadosi, S. (1984) Biased estimates of treatment effect in randomized experiments with nonlinear regressions and omitted covariates. *Biometrika*, **71**, 431–444.

Goodman, L.A. (1979) Simple models for the analysis of association in cross-classifications having ordered categories. *Journal of the American Statistical Association*, **74**, 537–552.

Goodman, L.A. (1986) Some useful extensions for the usual corresponding analysis approach and the usual loglinear approach in the analysis of contingency tables. *International Statistical Review*, **54**, 243–270.

Little, R.J.A. and Pullum, T.W. (1979) The general linear model and the direct standardization: a comparison. *Sociological Methods & Research*, **7**, 475–501.

McCullagh, P. and Nelder, J.A. (1989) *Generalized Linear Models*, 2nd edn, Chapman & Hill, New York.

Rosenbaum, P.R. and Rubin, D.B. (1983) The central role of the propensity score in observational studies for causal effects. *Biometrika*, **70**, 41–55.

Rosenbaum, P.R. and Rubin, D.B. (1984) Reducing bias in observational studies using sub-classification on the propensity scores. *Journal of the American Statistical Association*, **79**, 516–524.

Rubin, D.B. (1985) The use of propensity scores in applied Bayesian inference, in *Bayesian Statistics*, vol. **2** (eds J.M. Bernardo, M.H. De Groot, D.V. Lindley, and A.F.M. Smith), Elsevier, North-Holland, pp. 463–472.

Skinner, C.J. and Vallet, L.A. (2010) Fitting log-linear models to contingency tables from surveys with complex sampling designs: an investigation of the Clogg–Eliason approach. *Sociological Methods & Research*, **39**, 83–108.

Xie, Y. (1989) An alternative purging methods: controlling the composition-dependent interaction in the analysis of rates. *Demography*, **26**, 711–716.

Yamaguchi, K. (2012) Loglinear causal analysis. *Sociological Methodology*, **42**, 257–285.

APPENDIX A: A PROOF OF THE THEOREM

For part (1) of the theorem, when $s(x) = l(x)$,

$$l\left(P_{k|j}^{Y|X}\right) = \sum_i w(\boldsymbol{v}_i)\, l\left(P_{k|ij}^{Y|\boldsymbol{V}X}\right) = \sum_i w(\boldsymbol{v}_i)\left(\beta_k^Y + \beta_{k|i}^{Y|\boldsymbol{V}} + \beta_{k|j}^{Y|X}\right)$$
$$= \beta_k^Y + \beta_{k|j}^{Y|X} + \sum_i w(\boldsymbol{v}_i)\, \beta_{k|i}^{Y|\boldsymbol{V}} \tag{A1}$$

It follows that $l\left(P_{k|j_1}^{Y|X}\right) - l\left(P_{k|j_2}^{Y|X}\right) = \beta_{k|j_1}^{Y|X} - \beta_{k|j_2}^{Y|X}$.

When $s(x) = \exp\ (l(x))$,

$$l\left(P_{k|j}^{Y|X}\right) = \log\left(s\left(P_{k|j}^{Y|X}\right)\right) = \log\left\{\sum_i w(\boldsymbol{v}_i)\, s\left(P_{k|ij}^{Y|\boldsymbol{V}X}\right)\right\}$$
$$= \log\left\{\sum_i w(\boldsymbol{v}_i) \exp\ \left(l\left(P_{k|ij}^{Y|\boldsymbol{V}X}\right)\right)\right\}$$
$$= \log\left(\sum_i w(\boldsymbol{v}_i) \exp\ \left(\beta_k^Y + \beta_{k|i}^{Y|\boldsymbol{V}} + \beta_{k|j}^{Y|X}\right)\right)$$
$$= \beta_k^Y + \beta_{k|j}^{Y|X} + \log\left(\sum_i w(\boldsymbol{v}_i) \exp\ \left(\beta_{k|i}^{Y|\boldsymbol{V}}\right)\right) \tag{A2}$$

It also follows that $l\left(P_{k|j_1}^{Y|X}\right) - l\left(P_{k|j_2}^{Y|X}\right) = \beta_{k|j_1}^{Y|X} - \beta_{k|j_2}^{Y|X}$. This ends the proof of part (1).

For part (2) where $\log\left(P_{k|ij}^{Y|\boldsymbol{V}X} \big/ P_{1|ij}^{Y|\boldsymbol{V}X}\right) = \beta_k^Y + \beta_k^{Y|\boldsymbol{V}} + \beta_{k|j}^{Y|X}$ and $s(x) = x$, we obtain

$$P_{k|j}^{Y|X} = \sum_i w(\boldsymbol{v}_i)\, P_{k|ij}^{Y|\boldsymbol{V}X} = \sum_i w(\boldsymbol{v}_i) \frac{\exp\ \left(\beta_k^Y + \beta_{k|i}^{Y|\boldsymbol{V}} + \beta_{k|j}^{Y|X}\right)}{1 + \sum_{k=2}^K \exp\ \left(\beta_k^Y + \beta_{k|i}^{Y|\boldsymbol{V}} + \beta_{k|j}^{Y|X}\right)}$$
$$= \exp\ \left(\beta_k^Y + \beta_{k|j}^{Y|X}\right)\left(\sum_i w(\boldsymbol{v}_i) \frac{\exp\ \left(\beta_{k|i}^{Y|\boldsymbol{V}}\right)}{1 + \sum_{k=2}^K \exp\ \left(\beta_k^Y + \beta_{k|i}^{Y|\boldsymbol{V}} + \beta_{k|j}^{Y|X}\right)}\right) \tag{A3}$$

and

$$\frac{P_{k|j}^{Y|X}}{P_{1|j}^{Y|X}} = \exp\ \left(\beta_k^Y + \beta_{k|j}^{Y|X}\right)\left(\frac{\sum_i w(\boldsymbol{v}_i) \dfrac{\exp\ \left(\beta_{k|i}^{Y|\boldsymbol{V}}\right)}{1 + \sum_{k=2}^K \exp\ \left(\beta_k^Y + \beta_{k|i}^{Y|\boldsymbol{V}} + \beta_{k|j}^{Y|X}\right)}}{\sum_i w(\boldsymbol{v}_i) \dfrac{1}{1 + \sum_{k=2}^K \exp\ \left(\beta_k^Y + \beta_{k|i}^{Y|\boldsymbol{V}} + \beta_{k|j}^{Y|X}\right)}}\right) \tag{A4}$$

and, therefore,

$$\log\left(\frac{P^{Y|X}_{k|j_1}}{P^{Y|X}_{1|j_1}}\right) - \log\left(\frac{P^{Y|X}_{k|j_2}}{P^{Y|X}_{1|j_2}}\right) = \beta^{Y|X}_{k|j_1} - \beta^{Y|X}_{k|j_2}$$

$$+\log\left[\frac{\left\{\sum_i w(\boldsymbol{v}_i)\dfrac{\exp\left(\beta^{Y|V}_{k|i}\right)}{1+\sum_{k=2}^{K}\exp\left(\beta^{Y}_{k}+\beta^{Y|V}_{k|i}+\beta^{Y|X}_{k|ij_1}\right)}\right\}\left\{\sum_i w(\boldsymbol{v}_i)\dfrac{1}{1+\sum_{k=2}^{K}\exp\left(\beta^{Y}_{k}+\beta^{Y|V}_{k|i}+\beta^{Y|X}_{k|ij_2}\right)}\right\}}{\left\{\sum_i w(\boldsymbol{v}_i)\dfrac{\exp\left(\beta^{Y|V}_{k|i}\right)}{1+\sum_{k=2}^{K}\exp\left(\beta^{Y}_{k}+\beta^{Y|V}_{k|i}+\beta^{Y|X}_{k|ij_2}\right)}\right\}\left\{\sum_i w(\boldsymbol{v}_i)\dfrac{1}{1+\sum_{k=2}^{K}\exp\left(\beta^{Y}_{k}+\beta^{Y|V}_{k|i}+\beta^{Y|X}_{k|ij_1}\right)}\right\}}\right] \tag{A5}$$

Hence, the adjusted conditional probability does not preserve the effects of X on Y in $p(y|\boldsymbol{v},x)$.

For part (3), when $s(x) = l(x)$ and $w(\boldsymbol{v}) = p(\boldsymbol{v})$,

$$\begin{aligned} l\left(P^{Y|X}_{k|j}\right) &= \sum_i p(\boldsymbol{v}_i)\, l\left(P^{Y|VX}_{k|ij}\right) \\ &= \sum_i p(\boldsymbol{v}_i)\left(\beta^{Y}_{k}+\beta^{Y|V}_{k|i}+\beta^{Y|X}_{k|j}+\beta^{Y|VX}_{k|ij}\right) \\ &= \beta^{Y}_{k}+\beta^{Y|X}_{k|j}+\sum_i p(\boldsymbol{v}_i)\left(\beta^{Y|V}_{k|i}+\beta^{Y|VX}_{k|ij}\right) \end{aligned} \tag{A6}$$

It follows that

$$l\left(P^{Y|X}_{k|j_1}\right) - l\left(P^{Y|X}_{k|j_2}\right) = \beta^{Y|X}_{k|j_1} - \beta^{Y|X}_{k|j_2} + \sum_i p(\boldsymbol{v}_i)\left(\beta^{Y|VX}_{k|ij_1} - \beta^{Y|VX}_{k|ij_2}\right) \tag{A7}$$

This indicates that adjusted conditional probability retains the average of the effects of X on Y in $p(y|\boldsymbol{v},x)$ regarding the effects of $X = j_1$ versus $X = j_2$.

On the other hand, when $s(x) = \exp\ (l(x))$ and $w(\boldsymbol{v}) = p(\boldsymbol{v})$,

$$\begin{aligned} l\left(P^{Y|X}_{k|j}\right) &= \log\left(\sum_i p(\boldsymbol{v}_i)\exp\ \left(\beta^{Y}_{k}+\beta^{Y|V}_{k|i}+\beta^{Y|X}_{k|j}+\beta^{Y|VX}_{k|ij}\right)\right) \\ &= \beta^{Y}_{k}+\beta^{Y|X}_{k|j}+\log\left(\sum_i p(\boldsymbol{v}_i)\exp\ \left(\beta^{Y|V}_{k|i}+\beta^{Y|VX}_{k|ij}\right)\right) \end{aligned} \tag{A8}$$

It follows that

$$l\left(P^{Y|X}_{k|j_1}\right) - l\left(P^{Y|X}_{k|j_2}\right) = \beta^{Y|X}_{k|j_1} + \beta^{Y|X}_{k|j_2} + \log\left\{\frac{\sum_i p(\boldsymbol{v}_i)\exp\ \left(\beta^{Y|V}_{k|i}+\beta^{Y|VX}_{k|ij_1}\right)}{\sum_i p(\boldsymbol{v}_i)\exp\ \left(\beta^{Y|V}_{k|i}+\beta^{Y|VX}_{k|ij_2}\right)}\right\} \tag{A9}$$

This quantity is not equal to the average effects of X on Y unless $\beta^{Y|VX}_{k|ij} = 0$.

APPENDIX B: ITERATIVE PROPORTIONAL ADJUSTMENT PROCEDURES THAT PURGE THE INTERACTION EFFECTS OF V AND X ON Y

The conditional probabilities that purge the interaction effects of C and X on Y are given as

$$P_{k|ij}^{Y|CX} = \gamma_k^Y \gamma_{ik}^{CY} \gamma_{jk}^{XY} \theta_{ij}^{CX} \tag{B1}$$

where θ_{ij}^{CX} is an adjustment factor to give $\sum_k P_{k|ij}^{Y|CX} = 1$ and, therefore, $\theta_{ij}^{CX} \equiv 1/\sum_k \gamma_k^Y \gamma_{ik}^{CY} \gamma_{jk}^{XY}$.

For the iterative proportional adjustments, we use the fact that the ML estimates for $P_{k|ij}^{Y|CX}$ preserve the XY and CY marginal frequencies such that

$$\sum_i P_{k|ij}^{Y|CX} f_{ij}^{CX} = f_{jk}^{XY} \quad \text{for every } (j,k) \text{ and}$$
$$\sum_j P_{k|ij}^{Y|CX} f_{ij}^{CX} = f_{ik}^{CY} \quad \text{for every } (i,k) \tag{B2}$$

As the conditional probability $p_{k|ij}^{Y|CX}$ also satisfies

$$\sum_k P_{k|ij}^{Y|CX} = 1 \quad \text{for every } (i,j) \tag{B3}$$

By starting from initial values, such as $P_{k|ij,0}^{Y|CX} = f_{ik}^{CY} f_{jk}^{XY} / f_k^Y$, that impose the absence of the interaction effects of C and X on Y, we alternate the proportional adjustments of satisfying Equations (B2) and (B4), and the proportional adjustments for satisfying Equations (B3) and (B4), for each round t of iteration steps starting from $t = 0$, where each round takes two steps, until Equations (B2)–(B4) all converge. The rounds and steps of iterations are as follows:

$$Q_{k|ij,2t+1}^{Y|CX} = P_{k|ij,2t}^{Y|CX} \frac{f_{jk}^{XY}}{\sum_i P_{k|ij,2t}^{Y|CX} f_{ij}^{CX}} \quad \text{and} \quad \left(\gamma_k^Y \gamma_{jk}^{XY} \text{ adjustment stage}\right) \tag{B4}$$

$$P_{k|ij,2t+1}^{Y|CX} = \frac{Q_{k|ij,2t+1}^{Y|CX}}{\sum_k Q_{k|ij,2t+1}^{Y|CX}} \quad \left(\theta_{ij}^{XY} \text{ adjustment stage}\right) \tag{B5}$$

We skip the following steps when $f_{ik}^{CY} = 0$:

$$Q_{k|ij,2t+2}^{Y|CX} = P_{k|ij,2t+1}^{Y|CX} \frac{f_{ik}^{CY}}{\sum_j P_{k|ij,2t+1}^{Y|CX} f_{ij}^{CX}} \quad \text{and} \quad \left(\gamma_k^Y \gamma_{ik}^{CY} \text{ adjustment stage}\right) \tag{B6}$$

$$P_{k|ij,2t+2}^{Y|CX} = \frac{Q_{k|ij,2t+2}^{Y|CX}}{\sum_k Q_{k|ij,2t+2}^{Y|CX}} \quad \left(\theta_{ij}^{XY} \text{ adjustment stage}\right) \tag{B7}$$

Note that the formula $\theta_{ij}^{CX} \equiv 1/\sum_k \gamma_k^Y \gamma_{ik}^{CY} \gamma_{jk}^{XY}$ indicates that $\sum_k f_{ik}^{CY} f_{jk}^{XY} > 0$ needs to be satisfied in order for $P_{k|ij}^{Y|CX}$ to be estimable. This condition is always satisfied, given the assumption that $f_{jk}^{XY} > 0$, because since we need to estimate $P_{k|ij}^{Y|CX}$ only for cases for which $f_i^C > 0, f_{ik}^{CY} > 0$ holds for at least one value of $Y = k$.

In the case of generating the adjusted three-way table of X, Y, and Z that purges the interaction effects of X and C on Y, but retains the interaction effects of X and Z on Y, and the interaction effects of Z on C on Y, the model can be written as

$$P_{k|ij}^{Y|CXZ} = \gamma_k^Y \gamma_{jk}^{XY} \gamma_{km}^{YZ} \gamma_{jkm}^{XYZ} \gamma_{ik}^{CY} \gamma_{ikm}^{CYZ} \theta_{ijm}^{CXZ} \tag{B8}$$

where θ_{ijm}^{CXZ} is an adjustment factor to give $\sum_k P_{k|ijm}^{Y|CXZ} = 1$ and, therefore, $\theta_{ijm}^{CXZ} \equiv 1/\sum_k \gamma_k^Y \gamma_{jk}^{XY} \gamma_{km}^{YZ} \gamma_{jkm}^{XYZ} \gamma_{ik}^{CY} \gamma_{ikm}^{CYZ}$. The ML estimates of $P_{k|ijm}^{Y|CXZ}$ for this model retain XYZ and CYZ marginal frequencies. Hence, the following three equations need to be satisfied:

$$\sum_i P_{k|ijm}^{Y|CXZ} f_{ijm}^{CXZ} = f_{jkm}^{XYZ} \qquad \text{for every } (j, k, m) \tag{B9}$$

$$\sum_j P_{k|ijm}^{Y|CXZ} f_{ijm}^{CXZ} = f_{ikm}^{CYZ} \qquad \text{for every } (i, k, m) \text{ and} \tag{B10}$$

$$\sum_k P_{k|ijm}^{Y|CXZ} = 1 \qquad \text{for every } (i, j, m) \tag{B11}$$

Starting from such estimates as $P_{k|ijm,0}^{Y|CXZ} = f_{ikm}^{CYZ} f_{jkm}^{XYZ} / f_{km}^{YZ}$ that impose the absence of (CXY) interactions, the iterative alternating procedure of satisfying Equations (B10) and (B11) in every odd round, and Equations (B10) and (B11) in the every even round will attain convergence.

14

DESIGN- AND MODEL-BASED ANALYSIS OF PROPENSITY SCORE DESIGNS

PETER M. STEINER

Department of Educational Psychology, School of Education, University of Wisconsin-Madison, Madison, WI, USA

14.1 INTRODUCTION

The popularity of propensity score (PS) techniques for estimating causal treatment effects from observational data has been constantly increasing during the last decades. One reason for the increased usage of PS matching, PS stratification, and inverse-propensity weighting lies in their design advantages in comparison to standard regression approaches (Rubin, 2007, 2008). First, in mimicking the design of a randomized experiment, PS techniques try to remove baseline differences between the treatment and control group by forming comparable groups on the estimated PS. Baseline differences in observed outcomes can be removed without looking at the outcome or even before the outcome is measured. Estimating a balancing PS – a score that balances the treatment and control group's covariate distribution – does not require any outcome data. Thus, in estimating the PS, researchers can be blinded from the outcome data, precluding that their expectations regarding the treatment effect can influence the analysis. Second, PSs provide a convenient way for assessing the treatment and control group's comparability, that is, the balance in the observed covariate distributions. The two groups might be too heterogeneous for a meaningful comparison if the treated and untreated subjects lack considerable overlap on the estimated PS. Lacking overlap, the treatment effect might not be estimable without

Statistics and Causality: Methods for Applied Empirical Research, First Edition.
Wolfgang Wiedermann and Alexander von Eye.

relying on strong extrapolations. Third, PS techniques require weaker assumptions than outcome-modeling regression approaches. In conducting a PS analysis, both the PS and the treatment effect can be estimated nonparametrically without relying on functional form and distributional assumptions. Because of the PS techniques' design advantages and intentions to establish comparable treatment and control groups as in a randomized trial, we refer to PS matching, PS stratification, and inverse-propensity weighting as PS designs. In analogy to experimental designs and survey sampling designs (Cox, 2006, Kish, 1987), it is possible to distinguish between design- and model-based formulations of PS designs. Design-based formulations assume nonstochastic outcomes and rely on randomization distributions for inferential purposes – no probabilistic assumptions about the outcome are required. In contrast, model-based formulations rely on stochastic outcomes and a probabilistic model in order to derive point and variance estimators and corresponding hypothesis tests.

The distinction between design- and model-based formulations of PS designs is neither explicitly stressed in methodological PS articles nor in reviews of PS techniques (e.g., Guo and Fraser, 2010, Imbens, 2004, Morgan and Winship, 2007, Schafer and Kang, 2008, Steiner and Cook, 2013, Stuart, 2010). Almost all publications implicitly discuss PS techniques from a model-based point of view. The sole exemptions are publications by Rosenbaum (2002, 2009) who advocates a strict design-based formulation of PS designs and Rubin (2006, 2008) who strongly emphasizes design advantages of PS analyses. In this chapter, we explicitly contrast design- and model-based formulations of PS designs from a frequentist perspective and argue that the distinction between the two formulations is helpful with regard to theoretical and practical questions about the generalizability of results, the selection of an adequate PS model, and the choice of a variance estimator.

The chapter is organized as follows. The next section briefly reviews the Rubin Causal Model (RCM) and defines the major causal estimands of interest. The subsequent section motivates design- and model-based formulations by means of a completely randomized experiment. Then, after contrasting design- and model-based PS matching, stratification, and inverse-propensity weighting estimators, we discuss the statistical issues involved in implementing a PS design from a design- and model-based point of view. The issues include finite sample properties of PS estimators, PS model selection procedures, and the choice of variance estimators. The discussion section summarizes the differences between the three PS designs and their analytic formulations.

14.2 CAUSAL MODELS AND CAUSAL ESTIMANDS

The RCM (Rubin, 1974, Holland, 1986) and its potential outcomes notation provides a convenient way of formulating causal estimands and stating assumptions required for identifying causal effects. The standard RCM formalizes the effect of a single manipulable cause, that is, the effect of a treatment ($Z_i = 1$) as compared to a control or alternative treatment condition ($Z_i = 0$). Each subject i of a target population of $i = 1, \dots, N$ subjects then has two potential outcomes: the potential control outcome, Y_i^0, which is observed if subject i is exposed to the control condition ($Z_i = 0$), and the potential treatment outcome, Y_i^1, which is observed if subject i is exposed to the treatment condition ($Z_i = 1$). In standard RCM, both potential outcomes (Y_i^0, Y_i^1) are unknown prior to treatment and assumed to be nonstochastic. However, since we never observe both potential outcomes simultaneously, we cannot directly infer the *individual* treatment effect $\tau_i = Y_i^1 - Y_i^0$ from observed data. Thus, RCM typically focuses on the *average* treatment effect (ATE) asserting that the counterfactual situation is more validly inferred for groups of subjects rather than individual subjects. For a sample or population of N subjects, the ATE is defined as the difference in the average potential treatment and control outcomes:

$$\text{ATE} = \overline{Y}^1 - \overline{Y}^0 = \frac{1}{N}\sum_{i=1}^{N} Y_i^1 - \frac{1}{N}\sum_{i=1}^{N} Y_i^0 \tag{14.1}$$

In order to emphasize the nonstochastic character of potential outcomes, we use simple population averages of potential outcomes instead of corresponding expectations (which typically imply a random variable). The ATE may either refer to the population average treatment effect (PATE) of a well-defined target population or the sample average treatment effect (SATE) for the sample in hand (Imbens, 2004).

If we allow for stochastic potential outcomes, the definition of individual and average causal effects involves a probabilistic model (Steyer *et al.*, 2000a,b, Steyer, 2005). Then, the *individual* treatment effect is defined as difference in individual expectations, $\tau_i = E(Y_i^1) - E(Y_i^0)$ and the *average* treatment effect is given by the average of individual effects across the population or sample of N subjects:

$$\text{ATE} = \frac{1}{N}\sum_{i=1}^{N} \tau_i = \frac{1}{N}\sum_{i=1}^{N} E(Y_i^1) - \frac{1}{N}\sum_{i=1}^{N} E(Y_i^0) \tag{14.2}$$

One may even go a step further and impose a full probability model on $(Y_i^0, Y_i^1, Z, \mathbf{X})$, where $\mathbf{X}$ is a vector of covariates used in identifying and estimating the treatment effect. Rubin (2006) considers a full probabilistic formulation as an optional part of RCM and advocates Bayesian inference for analyzing the data (Rubin, 1978, 2010). Structural causal models as laid out by Pearl (2009, 2010; see also Heckman, 2005, and Spirtes *et al.*, 1993) also rely on a full probabilistic formulation. Corresponding potential outcomes and treatment effects are identified on the basis of a hypothesized structural model and its probabilistic formulation. Thus, the substantive meaning of the treatment effect directly derives from the hypothesized *structural model* rather than an *experimental* or *quasi-experimental design*. The structural and design-based ATEs are identical only if the structural model correctly represents the data-generating process in the population under investigation (Steiner *et al.*, 2014). In any case, the full probabilistic formulation requires defining the average treatment effect in terms of expectations:

$$\text{ATE} = E(Y^1) - E(Y^0) \tag{14.3}$$

As we discuss later, defining the ATE in terms of deterministic potential outcomes as in Equation (14.1) directly suggests a design-based analysis of the treatment effect, whereas the definitions (14.2) and (14.3) in terms of stochastic outcomes imply a model-based analysis. Researchers are frequently also interested in causal quantities other than ATE, for instance, in effect ratios, quantiles, or conditional average treatment effects like the average treatment effect for the treated population (ATT). For deterministic potential outcomes, ATT is defined as

$$\text{ATT} = \overline{Y}^1_{Z=1} - \overline{Y}^0_{Z=1} = \frac{1}{\sum_i Z_i} \sum_{i:Z_i=1} Y_i^1 - \frac{1}{\sum_i Z_i} \sum_{i:Z_i=1} Y_i^0 \tag{14.4}$$

The ATT for stochastic outcomes is analogously defined but with expectations.

Both ATE and ATT cannot directly be computed from observed data because the potential treatment and control outcomes are never observed simultaneously. Depending on the treatment status, either the potential treatment or control outcome is observed: $Y_i = Y_i^1 Z_i + Y_i^0(1 - Z_i)$, where Y_i without any superscript denotes the observed outcome. Using the observed outcomes Y_i, we can define the prima facie effect (PFE) which represents the difference in the treatment and control group's average outcomes,

$$\text{PFE} = \frac{1}{N_T} \sum_{i \in T} Y_i - \frac{1}{N_C} \sum_{i \in C} Y_i = \overline{Y}_T - \overline{Y}_C$$

where $T = \{i : Z_i = 1\}, C = \{i : Z_i = 0\}$, $N_T = \sum_{i=1}^N Z_i$ and $N_C = \sum_{i=1}^N (1 - Z_i)$. Note that the PFE does not represent a causal effect per se. If subjects self-select into treatment or are selected by third persons, the PFE confounds selection effects with causal effects.

However, an unbiased estimation of the average treatment effect from observed data is possible if (i) the selection or assignment mechanism is ignorable (Rosenbaum and Rubin, 1983) and (ii) the stable-unit-treatment-value assumption (SUTVA) is met (Rubin, 1990a,b).[1] The ignorability of a selection mechanism mainly depends on the design of a study. In a randomized experiment, assignment is ignorable due to the randomization of subjects into the treatment and control condition. In an observational study, selection is ignorable only if the confounding part of the selection mechanism is reliably measured. In the following, we briefly review the design- and model-based analysis of randomized experiments, which sets the stage for the discussion of analytic strategies for PS designs.

14.3 DESIGN- AND MODEL-BASED INFERENCE WITH RANDOMIZED EXPERIMENTS

One can distinguish between two instances of randomizing subjects to treatments. First, *random assignment* of *deliberately* selected subjects into treatment and

[1] Strong ignorability is a sufficient but unnecessary condition for identifying average treatment effects (see Steyer *et al.*, 2000a, for a discussion of other conditions).

control conditions. Second, *random sampling* of subjects from a well-defined *target population* into treatment and control conditions (cf. Cox, 2006, Kish, 1987, Neyman, 1990, Rubin, 1990b, 2010). The two instances of randomization differ with respect to the underlying target population. Random sampling of n_T treatment and n_C control subjects from a well-defined target population allows us to generalize the estimated treatment effect to the underlying target population of N subjects (with $N > n_T + n_C = n$). For experimental designs with deliberately chosen subjects, such a generalization is in general not warranted – at least not on statistical grounds.

Inferences based on random assignment or random sampling may either be formalized in a design- or model-based way (Cox, 2006, Kish, 1987, Särndal, 1978). The *design-based* formulation considers the potential outcomes as nonstochastic and relies on randomization distributions for inferential purposes. The *model-based* formulation assumes stochastic outcomes with a well-defined probabilistic data-generating model.

14.3.1 Design-Based Formulation

First, consider an RCT where a *deliberate sample of subjects* is randomly assigned into a treatment or control condition. A deliberate sample is characterized by the lack of randomly sampling subjects – for example, a convenience sample of volunteers or a quota sample. Assuming nonstochastic potential outcomes and a perfectly implemented randomization procedure, location differences in the treatment, and control group's outcome distribution can be tested using the randomization distribution under the sharp null hypothesis (i.e., treatment has a null effect on each subject). Though nonparametric and parametric significance test can be derived from randomization distributions, we only discuss parametric analyses in this article (for an excellent discussion of nonparametric analyses see Rosenbaum, 2002).

In pursuing a parametric approach, the PFE, $\overline{Y}_T - \overline{Y}_C$, is an unbiased estimator of the average treatment effect as defined in Equation (14.1): $E_R(\overline{Y}_T - \overline{Y}_C) = \text{ATE}$, where the subscript R indicates that the expectation is taken over all possible permutations of treatment assignments (with fixed group sizes; Cox, 2006, Rubin, 1974). The PFE has an asymptotic variance of $v^2 = s_T^2/n_T + s_C^2/n_C$, where s_T^2 and s_C^2 are the treatment and control group's variance estimates of the outcome, respectively. Using a version of the central limit theorem, the parametric formulation directly leads to the conventional two-sample t-test with an asymptotically t-distributed test statistic $t = (\overline{Y}_T - \overline{Y}_C)/v$, with $n_T + n_C - 2$ degrees of freedom. Note that the asymptotic t-distribution is obtained despite the nonstochastic character of the potential outcomes. The asymptotic results solely derive from the randomization of subjects to treatments – no probabilistic assumptions are involved (Cox, 2006, Cox and Hinkley, 1974, Freedman, 2008). However, group sizes need to be sufficiently large for a reasonable approximation of asymptotic results; otherwise, nonparametric tests are preferable.

A drawback of standard RCTs with deliberately selected subjects is that the results do not automatically generalize to a broader target population since inferences are based on the randomization distribution derived from hypothetical replications of randomly assigning study participants to treatment conditions. However, if the treatment and control subjects were *randomly drawn from a well-defined target population*, effect estimates obtained from RCTs generalize to the underlying target population – first discussed by Neyman in 1923 (Neyman, 1990). This type of randomization can be characterized as a "repeated sampling randomization-based" approach since the randomness comes from randomly sampling and assigning subjects into treatment and control conditions (Rubin, 1990b). As before, the PFE is an unbiased estimator of the target population's average treatment effect, that is, $E_{RR} = (\overline{Y}_T - \overline{Y}_C) = \text{ATE}$. The subscript RR indicates that the expectation is taken over repeated random samples drawn from the underlying target population. Thus, the expectation refers to the average across all possible permutations of choosing n_T treatment and n_C control subjects from the target population of N subjects. Then, the same asymptotic two-sample t-test as before results, though the rationale is rather different since the randomization distribution is now derived from random sampling rather than random treatment assignment. The treatment and control groups' average outcomes and respective variances required for the t-test represent special cases of Horvitz–Thompson estimates (Lohr, 1999). Asymptotic results are based on large target populations with sampling ratios for the treatment and control group approaching zero (Cox, 2006, Lohr, 1999, Neyman, 1990).

14.3.2 Model-Based Formulation

The completely randomized design can also be formalized within a model-based framework where the potential outcomes represent random variables (cf. Cox, 2006):

$$Y_i = \upsilon_0 + \tau_R Z_i + \epsilon_i \tag{14.5}$$

where Y_i is the observed outcome, υ_0 the population mean of the potential control outcome, and Z_i the treatment indicator. The error term ϵ_i is assumed to be independent and identically distributed according to a normal distribution with expectation $E(\epsilon_i) = 0$ and variance $V(\epsilon_i) = \sigma_\epsilon^2$. We use these standard assumptions throughout the article, though the normality and homoscedasticity assumptions may be relaxed. Depending on the probabilistic formulation of the causal model, τ_R represents ATE as defined either in Equation (14.2) or (14.3).

The model-based formulation directly allows a regression-based estimation and testing of the treatment effect. In contrast to the design-based formulation, the test statistic and its distribution derive from probabilistic assumptions – independence, homoscedasticity, normality – rather than randomization (Cox, 2006, Freedman, 2008). While the design- and model-based formulations lead to the same point estimate, they differ in their variances. The regression-based variance estimator relies on the homoscedasticity assumption, which is not required for the design-based formulation. However, the more stringent probabilistic assumptions of the model-based formulation allow the generalization of results to a larger target- or hypothetical superpopulation, provided that Equation (14.5) correctly represents the data-generating process.[2] In contrast to design-based generalizations of results obtained from "repeated sampling randomization-based" designs, model-based generalizations are warranted by the validity of the data-generating model and its probabilistic assumptions rather than the validity of the RCT's design (Cox, 2006, Spanos, 1999).

14.4 DESIGN- AND MODEL-BASED INFERENCES WITH PS DESIGNS

Now consider the case where treatment selection is based on self-selection, administrator-, or third-person selection instead of randomization. The bias induced by differential selection into the treatment and control groups can be completely removed if treatment selection is strongly ignorable. According to Rosenbaum and Rubin (1983), treatment assignment or selection is said to be strongly ignorable if potential treatment and control outcomes are independent of treatment Z given an observed vector of covariates $\mathbf{X}$:

$$(Y^0, Y^1) \perp Z|\mathbf{X}, \text{with } 0 < P(Z = 1|\mathbf{X}) < 1$$

Rosenbaum and Rubin (1983) also showed that treatment selection is strongly ignorable if we condition on the propensity score $e(\mathbf{X}) = P(Z = 1|\mathbf{X})$ alone:

$$(Y^0, Y^1) \perp Z|e(\mathbf{X}), \text{with } 0 < e(\mathbf{X}) < 1$$

If the strong ignorability assumption is met, the average treatment effect can be written as an average of conditional PFEs, that is,

$$\tau = E_{e(\mathbf{X})}\{E(\overline{Y}_T - \overline{Y}_C|e(\mathbf{X}))\}$$

where the inner expectation is taken over the design- or model-based distribution of the mean difference, conditional on the propensity score $e(\mathbf{X}) = e$. The outer expectation, $E_{e(\mathbf{X})}\{.\}$, is taken over the frequency or probability distribution of $e(\mathbf{X})$. ATE is obtained if the PS distribution with respect to the overall study population is used, and ATT results if the treated population PS distribution is used.

Given a strongly ignorable selection mechanism, subjects with identical PSs can be considered as randomly assigned to treatment conditions. This is equivalent to random assignment based on a covariate (Rosenbaum, 2002, Rubin, 1974, 1977). Thus, matching or stratifying subjects on the PS suffices to remove all the selection bias. The same parametric and nonparametric analytic procedures as for a randomized matched pair or randomized block design can then be used.

[2] A superpopulation is a hypothetical target population from which the observed subjects would have been drawn if they were randomly sampled. Thus, the covariate distribution of the underlying superpopulation is assumed to resemble the distribution of the realized sample.

14.4.1 Propensity Score Designs

In order to illustrate the basic ideas of PS matching, PS stratification, and inverse-propensity weighting (IPW), we begin our discussion of PS designs with an unrealistic but simple example, where the ignorability establishing propensity scores $e(\mathbf{X}_i)$ are known and take on a restricted range of Q values, $e_i \in \{e_1, e_2, \dots, e_q, \dots, e_Q\}$ with $q = 1, \dots, Q$. Moreover, assume that both treatment and control subjects are observed for each of the Q distinct propensity scores – that is, for each treatment subject there exists at least one control subject with the same PS and vice versa (i.e., the PS distributions of the treatment and control group completely overlap). This guarantees that the treatment effect is identified and estimable for each PS. Such a situation occurs in practice when a small number of categorical covariates determines treatment selection. In discussing the different PS designs, we focus on parametric analyses, though nonparametric randomization tests might sometimes be preferable, particularly when sample sizes are small (Rosenbaum, 2002, 2009).

14.4.1.1 PS Matching Design The simplest PS matching design is 1:1 matching (without replacement) that matches each treated subject to a control subject with exactly the same PS. Since 1:1 PS matching mimics a randomized matched pair design, the matched data structure needs to be reflected in the analysis of the outcome (i.e., the analysis is conducted conditional on the matched pairs). In using a *design-based* formulation of the matched pair design, we estimate the treatment effect as the average difference in matched pairs, and test the effect using a one-sample t-test, given sample sizes are large enough such that the test statistic is approximately t-distributed. Table 14.1 formally states the estimator and its analytic requirements. Note that using the outcome differences in the matched pairs removes the between-pairs variance.

In relying on a *model-based* formulation (i.e., potential outcome are treated as random variables) treatment effects can be directly estimated within a regression framework: $Y_{ir} = \mu + \tau_M Z_{ir} + \pi_r + \epsilon_{ir}$, where τ_M is the treatment effect, π_r represents the variation between the $r = 1, \dots, R$ matched pairs, either in form of fixed effects (dummies) or random effects, and μ is the intercept whose interpretation depends on the coding scheme chosen (for instance, with dummy coded effects the intercept represents the predicted control outcome for the first (or last) matched pair). It is important to note that the model-based analysis relies on the probabilistic assumptions imposed on the error term ϵ_{ir} (independence, homoscedasticity, and normality). Given a strongly ignorable selection mechanism, the regression estimator of the PS-matched pair design, $\hat{\tau}_M$, is an unbiased estimator of the average treatment effect for the treated (ATT). The design- and model-based matching estimators estimate ATT instead of ATE since a control subject is matched to each treatment subject, thereby preserving the population of treated subjects.[3]

[3]Some authors argue that the one-sample t-test or the inclusion of fixed or random effects for the matched pairs (in the model-based formulation) is inappropriate (Hill, 2008, Hill and Reiter, 2006, Stuart, 2008, Schafer and Kang, 2008). Others (Austin, 2008, Rosenbaum, 2002) defend the one-sample t-test or the inclusion of matched pairs effects. Here we provide a clear rationale why the matched pairs structure should at least be considered in the design-based formulation of the matching design (because the drawn inferences are conditional on the matched pairs).

TABLE 14.1 Parametric Design-Based Analysis of PS Designs.

PS Design	Estimators (Treatment Effect and Variance)	Estimand	Analytic Requirements for Estimation and Hypothesis Testing
PS pair-matching	$\hat{\tau}_M = \sum_r D_r / R$ where $D_r = Y_{Tr} - Y_{Cr}$ is the difference score of the $r = 1, \ldots, R$ matched pairs $\hat{\upsilon}_M^2 = s_D^2/R$	ATT	• PS is a balancing design • Large samples[1]
PS stratification	$\hat{\tau}_S = \sum_q w_q \hat{\tau}_q$ where $\hat{\tau}_q = \overline{Y}_{Tq} - \overline{Y}_{Cq}$ are the stratum-specific effect estimates and $w_q = n_q / \sum_q n_q$ (ATE) or $w_q = n_{Tq} / \sum_q n_{Tq}$ (ATT) are the weights $\hat{\upsilon}_S^2 = \sum_{q=1}^{Q} w_q^2 \hat{\upsilon}_q^2$ with $\hat{\upsilon}_q^2 = s_{Cq}^2 / n_{Cq} + s_{Tq}^2 / n_{Tq}$	ATE or ATT	• PS is a balancing score • Large samples[1] • Large number of strata
Inverse propensity weighting	$\hat{\tau}_{IPW} = \overline{Y}_T^W - \overline{Y}_C^W$ where $\overline{Y}_T^W$ and $\overline{Y}_C^W$ are the weighted averages of treatment and control cases with ATE weights $W_i = Z_i / e_i + (1 - Z_i)/(1 - e_i)$ or ATT weights $W_i = Z_i + (1 - Z_i) e_i / (1 - e_i)$ $\hat{\upsilon}_{IPW}^2 = s_{WC}^2 / n_c + s_{WT}^2 / n_T$ with weighted variances s_{WC}^2 and s_{WT}^2 of the control and treatment group	ATE or ATT	• PSs correctly represent the probabilities of treatment selection • Strict overlap • Large samples / sampling ratios close to zero[1]

[1] Large samples are required because of the parametric approximation of the randomization distribution and the finite sample bias. If samples are small nonparametric methods can be used.

One-to-one matching is only one possible matching design. Alternative matching designs include k:1 matching, k:m matching (Imbens, 2004), and optimal full matching that allows for a flexible matching with group-specific matching ratios k_g:m_g (Hansen, 2004, Hansen and Klopfer, 2006). However, since optimal full matching groups all subjects into homogeneous strata of varying sizes, it can be classified as a PS stratification design that we discuss next.

14.4.1.2 PS Stratification Design Using our simple example with complete overlap on the PS distribution, we can define Q strata according to the distinct values of $e(\mathbf{X})$ such that the propensities of selecting the treatment are constant within each stratum: $e_{qi} = e_{qj}$ for two different subjects i and j within stratum q. Thus, the PS stratification design mimics a randomized block design where subjects are randomized to treatment conditions within blocks (with varying assignment probabilities across blocks but constant assignment probabilities within blocks). Analytic methods used for analyzing a randomized block design directly apply to the PS stratification design.

In the *design-based* formulation of a PS stratification design, ATE is estimated by averaging stratum-specific effect estimates. Let n_q denote the number of subjects in stratum $q = 1, \ldots, Q$, with $\sum_q n_q = n$, and $\hat{\tau}_q = \overline{Y}_{Tq} - \overline{Y}_{Cq}$ denote the treatment effect in stratum q, where $\overline{Y}_{Tq}$ and $\overline{Y}_{Cq}$ represent the outcome means of the treated and control subjects in stratum q. The average treatment effect is then given by the weighted average of stratum-specific treatment effects: $\hat{\tau}_S = \sum_q w_q \hat{\tau}_q$ with weights $w_q = n_q / \sum_q n_q$. $\hat{\tau}_S$ is an unbiased estimator of ATE as defined in Equation (14.1). Similarly, we get the treatment effect's variance: $\upsilon_S^2 = \sum_{q=1}^{Q} w_q^2 \hat{\upsilon}_q^2$ with stratum-specific variances $\hat{\upsilon}_q^2 = s_{Cq}^2 / n_{Cq} + s_{Tq}^2 / n_{Tq}$, where s_{Cq}^2 and s_{Tq}^2 are the sample variances of the control and treatment group, respectively. For large samples, the resulting test statistic $\hat{\tau}_S / \hat{\upsilon}_S$ is asymptotically normally distributed (Lohr, 1999, Neyman, 1990). Note that these results do not rely on any distributional assumptions regarding the potential outcomes, they are solely derived from the strong ignorability assumption and the resulting random selection property within homogeneous strata (Rosenbaum, 2002). For the average treatment effect for the treated (ATT), stratum weights w_q are derived from the stratum distribution of treated subjects: $w_q = n_{Tq} / \sum_q n_{Tq}$, where n_{Tq} is the number of treated subjects in stratum q. If the analytic sample represents a random sample drawn from a well-defined target population (with sampling ratios approaching zero) the parametric stratification estimator and its variance can be considered as a special case of the Horvitz–Thompson estimator, where subjects are stratified according to their unequal selection probabilities (Lohr, 1999).

The *model-based* formulation assumes a data-generating process that depends on an intercept μ, a treatment effect τ_S, stratum-specific fixed or random effects π_q, and independent and normally distributed error terms with expectation zero and constant variance: $Y_{iq} = \mu + \tau_S Z_{iq} + \pi_q + \epsilon_{iq}$. The regression estimator $\hat{\tau}_S$ is in general a biased estimator of ATE as defined in Equations (14.2) and (14.3) because regression adjustments in estimating $\hat{\tau}_S$ involve a variance-of-treatment weighting (Angrist and

Pischke, 2009). In order to illustrate the variance-of-treatment weighting, consider a model-based formulation with fixed PS-stratum effects, that is, stratum-dummies are included in a standard regression model: $Y_i = \mu + \tau_S Z_i + \boldsymbol{D}_i' \boldsymbol{\pi} + \epsilon_i$, with $\boldsymbol{D}_i$ being the vector of dummy variables representing the strata. One can then show that the treatment effect from the stratum-dummy regression is given by $\hat{\tau}_S = (1/\sum_q v_q w_q) \sum_q v_q w_q \hat{\tau}_q$, where $\hat{\tau}_q = \overline{Y}_{Tq} - \overline{Y}_{Cq}$ and $w_q = n_q / \sum_q n_q$ are the stratum-specific effects and frequency weights as defined before. $v_q = p_q(1 - p_q)$ are the stratum-specific Bernoulli variances of the treatment with $p_q = \sum_{i:e_i=e_q} z_i / n_q$ representing the proportion of treated subjects within stratum q. Multiplying frequency weights and variances results in $v_q w_q = n_q p_q (1 - p_q) / \sum_q n_q$, where the numerator represents each stratum's treatment variance (binomial variance). Thus, the regression estimator $\hat{\tau}_S$ involves two weighting schemes, the frequency distribution across strata (just like the design-based estimator) but also the treatment variance across strata. Consequently, $\hat{\tau}_S$ does not estimate ATE except if the stratum-specific treatment effects are constant across strata ($\tau_1 = \cdots = \tau_Q$) or the proportion of treatment subjects is constant ($p_1 = \cdots = p_Q$). Since none of the two conditions typically holds in practice, the stratum-dummy regression fails to estimate ATE as defined in Equations (14.2) and (14.3). But this does not imply that the variance-of-treatment weighted average treatment effect is outside the scope of RCM – it only requires defining the variance-of-treatment weighted effect as another causal estimand of interest. However, one can still use the stratum-dummy regression to estimate ATE, but it requires the inclusion of (treatment × stratum)-interaction terms in the regression model and the computation of ATE as a weighted linear combination of parameter estimates.

14.4.1.3 Inverse-Propensity Weighting (IPW) In addition to PS matching and stratification designs, IPW is another design-based PS approach, though its rationale is quite different. IPW is directly related to inverse-probability weighting as suggested by Horvitz and Thompson (1952), but instead of one sample, two samples are drawn without replacement from a target population of N subjects: one sample for the treatment condition with selection probabilities e_i and another sample for the control condition with selection probabilities $1 - e_i$. Selection probabilities are directly derived from propensity scores e_i and must be strictly greater than zero and less than one.

In our example with the small number of known propensity scores $e_i \in \{e_1, e_2, \ldots, e_q, \ldots, e_Q\}$, the weights for treated subjects are given by $1/e_i$ and for the untreated subjects by $1/(1 - e_i)$. In using more efficient normalized IPW weights, $W_i = (1/\sum_{k=1}^{n} Z_k/e_k) \times (1/e_i)$ for treated subjects and $W_i = (1/\sum_{k=1}^{n} (1 - Z_k)/(1 - e_k)) \times (1/1 - e_i)$ for control subjects, we obtain *design-based* Horvitz–Thompson estimates of the two group means (Lohr, 1999, Lumley, 2010). Then, ATE of Equation (14.1) is estimated by the difference between the treatment and control group's weighted averages (see Table 14.1). The very same weights are used to obtain the Horvitz–Thompson variance estimates for the group means. Note that the exact finite sample variance is rarely estimable since the selection probabilities' dependence structure induced by sampling without replacement is unknown. Therefore, propensities e_i and $1 - e_i$ represent *average* selection probabilities, that is, the average across the probabilities of being selected in the first draw, the second draw, the third draw, and so on. While the average selection probabilities are sufficient for getting an unbiased treatment effect, the corresponding variance estimator is only consistent. However, the bias in the variance estimator is negligibly small if the sampling ratios n_T/N and n_C/T are close to zero. The corresponding z-test statistic for testing the treatment effect is asymptotically normally distributed (Lohr, 1999). Substituting ATT weights for ATE weights allows us to estimate ATT: Treated subjects receive a nonnormalized weight of one and untreated subjects a weight of $e_i/(1 - e_i)$; weights can be normalized as indicated above.

In the *model-based* formulation, weighted-least-squares (WLS) regression is used to estimate the average treatment effect: $Y_i = \mu + \tau_{IPW} Z_i + \epsilon_i$, with IPW weights for the treated and control subjects as defined above (Busso *et al.*, 2009a, 2009b). Conceptually, the WLS estimator $\hat{\tau}_{IPW}$ and its regression-based variance estimator do not represent Horvitz–Thompson estimators because the finiteness of the population and the selection probabilities' dependence structure is not taken into account. This highlights that the estimators are derived from a probabilistic data-generating model rather than the sampling design. The efficiency of regression-based variance estimators depends on the calibration of weights. By switching from normalized weights to stabilized or augmented weights, the estimator's efficiency can be increased (Lunceford and Davidian, 2004, Robins, 1999b). Note that marginal structural models (Robins, 1999a) and marginal mean weighting based on PS stratification (Hong, 2010) belong to the same class of model-based weighting estimators. See Robins *et al.* (1995) and Lunceford and Davidian (2004) for a discussion of a more general class of IPW estimators.

14.4.2 Design- versus Model-Based Formulations of PS Designs

The discussion of design- and model-based formulations of PS designs revealed that the choice of a formulation has direct implications for the analysis and interpretation of results. While the causal interpretation of design-based estimates is warranted by design, model-based analyses of PS designs derive their warrant from a correctly specified probabilistic model. Tables 14.1 and 14.2 summarize the design- and model-based estimators.

The rationale for *design-based* formulations hinges on the claim that subjects with the same PS can be considered as being randomized into the treatment and control condition. If the PS establishes a strongly ignorable selection mechanism, design-based formulations of PS designs directly mimic the design and analysis of a randomized matched pair design, randomized block design, or a survey sampling design with unequal selection probabilities. Thus, design-based formulations estimate the treatment effect and its variance *conditional* on the matched, stratified, or weighted data. This allows one to rely on a minimal set of two strong assumptions – strong ignorability and SUTVA. These two conditions suffice for establishing

TABLE 14.2 Model-Based Analysis of PS Designs.

PS Design	Basic Regression Estimators[1] (Treatment Effect and Variance)	Estimand	Analytic Requirements for Estimation and Hypothesis Testing
PS pair-matching	$Y_{ir} = \mu + \tau_M Z_{ir} + \pi_r + \epsilon_{ir}$ where π_r are the fixed or random effects of the $r = 1, \ldots, R$ matched pairs	ATT	• Correct specification of PS model • Large samples[2] • Probabilistic assumptions[3]
PS stratification	$Y_{iq} = \mu + \tau_S Z_{iq} + \pi_q + \epsilon_{iq}$ where π_q are the fixed or random effects of the PS strata	Variance-of-treatment weighted ATE	• Correct specification of PS model • Large samples[2] • Large number of strata • Probabilistic assumptions[3]
Inverse propensity weighting	$Y_i = \mu + \tau_{IPW} Z_i + \epsilon_i$ with ATE weights $W_i = Z_i/e_i + (1 - Z_i)/(1 - e_i)$ or ATT weights $W_i = Z_i + (1 - Z_i)e_i/(1 - e_i)$	ATE or ATT	• PSs correctly represent the probabilities of treatment selection • Strict overlap • Large samples / sampling ratios close to zero[2] • Probabilistic assumptions[3]

[1] Treatment effects and corresponding standard errors are directly obtained from the regression analysis. Alternative model specifications may result in different estimands (e.g., if PS stratification is implemented with an interaction term between the treatment and the strata, or with marginal mean weights instead of fixed or random stratum effects then ATE or ATT can be estimated).

[2] Large samples are required because of the finite sample bias.

[3] The probabilistic assumptions about the error term can be relaxed. With large samples or bootstrapped standard errors the normality of the error term is not required. Heteroscedasticity can be addressed via sandwich estimation.

a valid PS design and conducting inferential analyses conditional on the matched, stratified, or weighted data. No probabilistic assumptions with respect to the potential outcomes are required. The generalizability of results is also directly determined by the design. Analytic results obtained from a deliberately selected sample do not generalize beyond the sample in hand. If the sample was randomly drawn from a well-defined target population results generalize to the target population.

In contrast to design-based formulations, *model-based* formulations require probabilistic assumptions about the error term's distribution, dependence structure, and scedasticity in addition to the strong ignorability assumption and SUTVA. Treatment effect estimates generalize to the underlying target population or a superpopulation if the validity of the estimated model can be reasonably postulated for the entire population. Besides specifying the probabilistic model, a thoughtful specification of the outcome regression model is also of importance, particularly if treatment effects are heterogeneous. Design- and model-based point estimates of a given PS design coincide only if the data-generating model is correctly specified (e.g., by including (treatment × PS-stratum)-interaction terms in order to account for treatment heterogeneity). Design- and model-based variance estimates differ in general, even if the data-generating models are correctly specified. This is so because design-based analyses do not rely on probabilistic assumptions, particularly the homoscedasticity assumption. However, assumptions of the model-based formulations may be relaxed such that variance estimates and parametric tests of model- and design-based formulations are essentially identical – but the formal generalizability of results still differs.

14.4.3 Other Propensity Score Techniques

The three basic PS designs outlined do not represent all PS techniques available to researchers. Regression estimation using PS-related predictors is another frequently used technique in practice (Rosenbaum and Rubin, 1983, Schafer and Kang, 2008). Even more popular are mixed methods, also called doubly robust methods, which combine model-based formulations of PS designs with an additional regression adjustment. The rationale behind mixed methods is that the estimators are robust against the misspecification either of the PS model or the outcome model (Robins *et al.*, 1995, Robins and Rotnitzky, 1995, Rubin and Thomas, 2000). But if both models are misspecified, then the doubly robust estimators might perform worse than a pure PS estimator or regression adjustment (Kang and Schafer, 2007).

However, as already discussed for the PS stratification design, standard regression adjustments do not estimate ATE or ATT but a variance-of-treatment weighted average treatment effect. This is of particular importance for mixed methods since they combine a PS design that estimates ATE or ATT and a regression adjustment that aims at a variance-of-treatment weighted average treatment effect. Thus, in the case of heterogeneous treatment effects, a mixture of estimands results (similar results hold for covariance-adjusted RCTs; Freedman, 2008, Schochet, 2010). We can circumvent this problem by first estimating a separate regression model for the treatment and control group, and then by averaging the differences in predicted treatment and control outcomes. This analytic technique if frequently referred to as regression estimation or G-computation based on marginal structural models (Cochran and Rubin, 1973, Rubin, 1973, Schafer and Kang, 2008, Snowden *et al.*, 2011, Robins, 1986).

It is important to note that we do not consider these techniques as PS designs since (i) they involve selecting an adequate outcome model that violates the design requirement of not looking at any outcome data when removing selection bias; and (ii) they preclude design-based formulations because including PS-related predictors or additional covariates in the outcome regression directly implies stochastic potential outcomes. This does not mean that these PS techniques should not be used in practice. To the contrary, mixed methods frequently help in removing residual bias left by the PS design (e.g., if perfect balance cannot be achieved) and in increasing the efficiency of model-based variance estimators. However, we argue that PS regression estimation and mixed methods should only be conducted and reported *in addition* to the pure design- or model-based analysis of a PS design. This corresponds to the practice of analyzing RCTs where covariance-adjusted treatment effects are reported in addition to unadjusted mean differences.

14.5 STATISTICAL ISSUES WITH PS DESIGNS IN PRACTICE

So far we discussed all three basic PS designs under the assumption that the PS is actually known and that the observed treatment and control subjects perfectly overlap (i.e., both treatment and control subjects are observed for each realized PS). In this case, the treatment effect estimators of all three PS designs are unbiased but the estimators' efficiency differs across PS designs. However, in practice the true PS is unknown and the treatment and control groups frequently lack perfect overlap (i.e., with finite sample sizes and a large set of covariates it is hard to find a treatment and control subjects with exactly the same estimated PS). Thus, with unknown PSs and finite samples we face two sources of error in addition to randomization or sampling errors: First, estimation error because of estimating the unknown PS and, second, bias due to lack of overlap, inexact matches, or heterogeneous PS strata. Both sources of error directly affect the finite sample properties of variance estimators. Estimators that are optimal in case of perfect overlap, for instance, might perform worse in comparison to other estimators if the treatment and control groups' PS distributions do not fully overlap.

In the following sections, we discuss from a design- and model-based perspective the implications of (i) substituting the estimated PS for the true but unknown PS and (ii) lacking overlap. Throughout the discussion, we assume that the strong ignorability assumption and SUTVA are met and that the PS has been correctly estimated. We first briefly review the PS estimators' finite sample properties in order to provide guidance for choosing a specific PS design in practice. Then we elaborate on balance, overlap, and model selection issues involved in selecting an adequate PS model and discuss implications on variance estimation.

14.5.1 Choice of a Specific PS Design

In practice, the choice of a PS design and estimator matters because they have different asymptotic and finite sample properties. PS matching and IPW estimators are $\sqrt{N}$-*consistent* given strict overlap and a correctly specified PS model (Heckman *et al.*, 1997, Hirano *et al.*, 2003, Robins *et al.*, 1995). The strict overlap condition requires that the propensity scores are strictly bound away from zero and one: $\xi < e(\boldsymbol{X}) < 1 - \xi$, for some $\xi > 0$ and almost every $\boldsymbol{X} = \boldsymbol{x}$ (Busso *et al.*, 2009a). Strict overlap ensures that the treatment and control group completely overlap such that all treatment and control subjects share the common support region on the PS. This implies that control subjects are available even for treated subjects with a PS close to one (analogous for treatment or control subjects with PSs close to zero). Note that the strict overlap condition is defined in terms of the true PS and, thus, does not depend on the sample size (i.e., increasing the sample size cannot compensate for a violated strict overlap condition). The consistency of PS stratification estimators depends on an additional qualification: the number of strata needs to increase simultaneously with the sample size.

The $\sqrt{N}$-consistency of PS estimators implies that all PS estimators are in general biased for small samples because treatment and control subjects with identical or very similar PSs might not be available. ATT can still be estimated with minimal bias from small samples of treated subjects if the treatment group is comparatively much larger (Kolar *et al.*, 2014). Assuming *strict overlap* and a *correctly specified PS model*, Busso *et al.* (2009a) showed that IPW is the least-biased estimator for finite samples. Using a simulation study, they claim that IPW is "approximately unbiased" even for samples with not more than 100 subjects. However, with sample sizes of 500 subjects or more, Busso *et al.* demonstrate that 1:1 matching performs basically as well as IPW. PS stratification is slightly more biased due to the roughness of the strata (such that within each stratum the treatment and control groups' PS distributions still differs). Under some regularity conditions, Rosenbaum and Rubin (1983) showed that PS stratification with five strata removes approximately 90% of the selection bias (see also Cochran, 1968). For large data sets, it is frequently possible to use 10 or more strata such that the residual bias reduces to a negligibly small size. Thus, for data sets with at least 500 subjects, all three PS estimators – 1:1 matching, stratification on a sufficiently large number of strata, and IPW – can be considered as approximately unbiased, given strict overlap and a correctly specified PS model.

However, it is important to note that IPW estimators are more sensitive to *misspecifications* of the PS model than PS matching and PS stratification estimators. While PS model misspecifications that only result in a strictly monotonic transformation of the correctly estimated PSs (e.g., due to omitting a quadratic term or a variable, or due to choosing an incorrect link function) do not affect PS matching and stratification estimators but the IPW estimator will be biased (Waernbaum, 2010). PS matching and stratification are insensitive to monotonic misspecifications because it does not change the PSs ranking and, thus, has no influence on the determination of the strata and only a negligible effect on matching. Some of the matched pairs might differ because the monotonic misspecification of the PS model results in slightly different PS distances between treatment and control subjects. In contrast, IPW is rather sensitive to model misspecifications because the inverse propensity weights are directly affected, even by minor misspecifications (Horvitz–Thompson estimators are unbiased only if the weights correctly reflect the selection probabilities). Therefore, IPW designs rely on stronger assumptions than PS matching or stratification designs.

If treatment and control groups do *not strictly overlap*, all PS estimators are biased. The extent of bias depends on the degree of overlap. Strong selection processes usually result in a weak overlap and considerably biased PS estimators. However, in deleting nonoverlapping subjects, one can remove the bias caused by the lack of overlap. But note that the deletion distorts the composition of the sample in hand and, thus, restricts the generalizability of effect estimates (unless constant treatment effects can be assumed). Though deleting nonoverlapping cases establishes the approximate unbiasedness of PS matching and stratification estimators with respect to the trimmed target population, IPW estimators remain biased. This is so because IPW weights refer to the overall study population and, therefore, are invalid for the trimmed study population. Again, this highlights IPW's more stringent requirements: the PSs need to represent unbiased estimates of the propensity score in the (sub)population of interest. Though we can obtain more valid inverse-propensity weights by reestimating the PS model with the trimmed data, the new set of weights commonly faces a lack of overlap again – though a less severe one than before. Thus, whenever treatment and control subjects show a considerable lack of overlap, PS matching and stratification designs are preferable to IPW designs.

Regarding the PS designs' *efficiency*, there is no conclusive evidence about which PS estimator is the most efficient one, though the IPW estimator almost achieves the semiparametric efficiency bound given strict overlap (Busso *et al.*, 2009a,b, Hahn, 1998). Busso *et al.* (2009a) show that with sample sizes of at least 500 cases PS stratification, 1:k and optimal full matching also come close to the semiparametric efficiency bound. This does not hold for 1:1 matching because dropping unmatched control subjects may drastically reduce the effective sample size. However, in their comparison of PS estimators, Busso *et al.* used model-based analyses that did not take the matched or stratified data structure into account (i.e., they implemented all estimators as weighting estimators without modeling the effects of matched pairs or strata). PS matching and stratification estimators that include the effects of matched pairs and strata (as discussed in the previous section) typically outperform IPW estimators if the matched pairs and strata explain a significant portion of the outcome's variance (Lunceford and Davidian, 2004). Moreover, if the strict overlap condition is violated IPW regularly results in increased standard errors due to extreme weights. Importantly, the relative efficiency of PS estimators might also depend on the heterogeneity of the treatment effect and the functional form of the response surfaces (Hill and Reiter, 2006).

To summarize, the estimators of all three PS designs – matching, stratification, and IPW – are approximately unbiased if (i) strong ignorability holds, (ii) the PS model is (approximately) correctly specified, (iii) the sample size is large enough, and (iv) the treatment and control groups strictly overlap. This is true for both design- and model-based estimators since point estimates are identical. IPW tends to be the least-biased estimator if the PS model is correctly specified and the strict overlap condition is met. If the treatment and control group lack strict overlap, all estimators are significantly biased unless the sample is restricted to the overlapping cases. With nonoverlapping cases removed, PS matching and stratification estimators tend to be less biased and more efficient than IPW estimators. However, PS stratification requires a sufficiently large number of strata, otherwise a nonnegligible residual bias may result (e.g., approximately 10% remaining bias with 5 strata). Again, note that dropping nonoverlapping cases restricts the estimate's generalizability to the empirically trimmed population with potentially unclear characteristics.

14.5.2 Estimation of Propensity Scores

Since the PS is rarely known in practice, the first step in implementing a PS design requires the estimation of a PS that balances pretreatment group differences on observed covariates. Though different binomial regression models and statistical learning algorithms (Berk, 2008, McCaffrey *et al.*, 2004) are available for estimating the unknown PS, in the following sections we assume a logistic PS model.

14.5.2.1 Criteria for Specifying PS Models: Balance The ultimate goal of PS designs is to establish a strongly ignorable selection mechanism such that all the selection bias can be removed. If the estimated PS fails to adequately equalize pretreatment group differences in observed covariates an invalid PS design results. Thus, in selecting a PS model, balance metrics help to probe the equivalence of the treatment and control group's covariate distribution. Goodness-of-fit criteria like Akaike's Information Criterion (AIC) are also useful but a correspondingly selected PS model also needs to establish balance in observed covariates. Nonetheless, the main goals of the two criteria are different: While selecting a PS model according to balance criteria aims at estimating a PS that establishes identical treatment and control groups with respect to the observed covariate distribution, goodness-of-fit criteria prioritize the correct specification of the unknown selection process. Thus, from a design point of view (without the aim to generalize results beyond the sample in hand), balance criteria are more important than goodness-of-fit criteria since getting comparable treatment and control groups matters most – even random imbalances should be removed. As with randomized matched pair or randomized block designs where the matching or blocking variables are perfectly balanced by design, we want to make sure that the treatment and control groups' covariate distributions are (almost) identical in the PS-matched or stratified sample. Since balance is a sample characteristic rather than a characteristic of an underlying target population, aiming at perfect balance makes particularly sense for PS designs with no intent to generalize beyond the sample in hand.

Achieving close to perfect balance in observed covariates is somewhat less important if one wants to generalize the estimated treatment effect to an underlying target or superpopulation. Balance in the sample does not necessarily imply that the corresponding PS model is correctly specified with respect to the target or superpopulation. Since valid generalizations require a correctly specified selection mechanism, criteria that prioritize a correct model specification over balance are more appropriate for PS designs that aim at generalizing the treatment effect. However, specifying the PS model according to goodness-of-fit criteria does not imply that balance is irrelevant. Significant imbalances in some covariates indicate an incorrectly specified PS model such that the model needs to be respecified.

The literature on PS techniques covers a broad range of balance criteria for assessing balance in observed covariates. Though balance with respect to the multivariate distribution should be investigated, each covariate's balance is frequently probed separately because checking the equivalence of the treatment and control group's multidimensional covariate distribution would require huge data sets and is computationally rather challenging. Like misspecification tests in regression analysis, balance tests include graphical methods, descriptive measures, and significance tests. Visual inspections include histograms, kernel density estimates, boxplots, and *QQ*-plots that compare the treatment and control groups' covariate distribution (Rosenbaum and Rubin, 1984, Sekhon, 2011). Descriptive measures for assessing balance include raw and standardized mean differences and variance ratios for probing the balance of distributions' second moment (Rubin, 2001). Significance tests like *t*-tests, Kolmogorov–Smirnov tests or regression tests, which regress each covariate on the treatment indicator (according to the PS-design's model-based formulation), help in assessing the significance of distribution imbalances. However, significance tests are not directly relevant for assessing balance because, as mentioned above, balance is a sample characteristic rather than a population characteristic (see Imai *et al.*, 2008, for a discussion on the balance test fallacy).

In checking covariate balance of a PS matched, stratified, or weighted data set, it is advisable to perform a variety of balance tests since any indication of considerable imbalance in observed covariates is sufficient to invalidate the PS design. Balance tests should not be restricted to the originally observed covariates, it is also useful to check balance on transformed covariates (e.g., quadratic, cubic, or interaction terms, but also log or other transformation) since they might directly indicate balance improving covariate transformations. Importantly, the more thoroughly one probes balance with nonredundant balance checks, the stronger a PS design's credibility for causal inference. Using a broad range of balance checks has the advantage that the weakness of each single balance test can be overcome by one of the other balance tests. For instance, power issues due to small sample sizes are a concern with significance tests but less so for descriptive and visual inspections. Or checking balance only with respect to the mean differences might fail to indicate imbalances on higher order moments, which can be detected by investigating variance ratios or mean differences within strata. Finally, it is notable that even perfect balance does not imply that all the selection bias is removed. All the selection bias is removed only if the strong ignorability assumption actually holds. Balance tests only check for overt bias due to observed covariates but not for hidden bias that is due to unobserved covariates.

Balance tests are most effective when implemented in accordance with the PS design chosen for estimating the treatment effect. For instance, if a 1:1 matching design is used, test balance on the matched data set. In case of PS stratification, probe differences in covariate means by applying the PS stratification estimator to each covariate separately. For IPW designs, use the IPW weights for plotting and computing balance metrics. Balance checks that are conducted in accordance with the chosen PS design help in establishing a more valid PS design than balancing procedures that are unrelated to the specific PS design. Conducting design-specific balance tests also implies an alignment with the causal estimand of interest. If ATE is the estimand of interest, balance tests need to refer to the entire sample of treated and untreated subjects. In case of ATT, balance with respect to the treated subjects matters, thus, for a PS stratification or IPW design, balance checks require ATT weights instead of ATE weights.

Such design-specific balance procedures make sense for design-based formulations but not necessarily for model-based formulations of PS designs. From a *modeling* point of view, PS model selection criteria that depend on the PS design and causal estimand have no justification since a population's data-generating selection mechanism never depends on the PS design and estimand chosen – the selection model has to be the same for all PS designs and causal estimands.

14.5.2.2 Overlap Another aspect of balance is overlap, that is, whether the treatment and comparison group's PS distribution share the same region of common support. Groups do not completely overlap if a nonnegligible portion of subjects at the lower or upper tail of the PS distribution – or even between the tails – does not have corresponding subjects with similarly low or high scores in the respective other group. Lack of overlap at the tails typically results when strong selection processes are at work. Overlap is effectively assessed by plotting the treatment and control group's distribution of the PS-logit using histograms or kernel density estimates (the logit of the PS is used instead of the PS since it better reveals the lack of overlap at the tails of the distribution).

In the case of a severe lack of overlap (i.e., the strict overlap condition is clearly violated), balance on observed covariates cannot be achieved, indicating that the two groups are too heterogeneous to be successfully equated in a PS design. As discussed earlier, lacking complete overlap results in biased PS estimators. In practice, we frequently face situations where strict overlap is not given but treatment and control subjects overlap at least partially – at least a large portion of subjects share the common support region on the PS. In order to achieve balance with partially overlapping groups, researchers frequently delete nonoverlapping subjects. In order to maintain a meaningful population, subjects are preferably discarded on the basis of observed covariates instead of the PS-logit itself. If balance statistics indicate satisfactory balance for the overlapping subjects, the treatment effect can be estimated without bias for the truncated subpopulation. But the estimated effect only generalizes to the correspondingly truncated target population – unless the assumption of constant treatment effects is reasonable. Also note that ATE and ATT involve different overlap requirements. In estimating ATE, both the treatment and control group need to share the same region of common support, while for ATT only the treated subjects need to be within the range of common support.

14.5.2.3 PS Model Selection PS model selection is rather challenging because the selection mechanism that generates the nonequivalent treatment and control group is typically unknown and frequently produces partially nonoverlapping groups. A feasible iterative strategy for selecting a PS model according to balance metrics was suggested by Steiner and Cook (2013): (i) Estimate an initial PS model; (ii) check overlap on the estimated PS-logit and each covariate separately and then delete nonoverlapping cases, preferably on the basis of the single covariates rather than the PS-logit; and (iii) check balance on observed covariates and the PS-logit – if balance is not satisfactory go back to (i), include the previously deleted cases, respecify the PS model using information obtained from the balance test, and then do the overlap and balance checks again; proceed until satisfactory balance is achieved.

But when is satisfactory balance achieved? The ideal scenario is that all balance tests suggest that the treatment and control group's covariate distributions are identical – as it would be the case for an exact matching on observed covariates. However, achieving perfect balance is nearly impossible in practice. But do we need perfect balance in practice? One may think of balance in analogy to the balance established by a completely randomized design. That is, the treatment and control group's distributions of observed covariates are equivalent in expectation but slightly differ for single realizations of the randomization (i.e., assuming independent covariates, 95% of the covariate distributions should not significantly differ between the two groups). But such an analogy is not appropriate for PS designs since even slight imbalances in covariates might still indicate systematic selection bias due to model misspecification and not just random imbalance as obtained from random assignment or random sampling. Imbalances in covariate distributions are attributable to chance only if we would know that we correctly specified the PS model. But without knowing this we cannot rule out that even small imbalances are still due to a misspecified functional form. In any case, one should always try to remove random imbalances in order to increase the design's validity (again, balance is a sample characteristic). From this point of view, significance tests for checking balance do not have a strong warrant. Moreover, an insignificant t-test with respect to a covariate's group means does not necessarily indicate the absence of systematic bias. The bias might only be too small to be detected by the test procedure, particularly, if sample sizes are not large. More important than an insignificant balance test is the magnitude of residual imbalance in substantive terms. For instance, a standardized mean difference of 0.1 standard deviations or greater in an observed covariate is unacceptable if the effect size of the treatment is expected to be as small as 0.1 standard deviations – particularly so for covariates that are assumed to be highly predictive of the outcome (e.g., a pretest of the outcome).

14.5.3 Estimating and Testing the Treatment Effect

14.5.3.1 Treatment Effect Once a balancing PS is obtained, the PS design is frozen and the treatment effect estimated. Freezing the PS design guarantees that the choice of the PS design and the specification of the PS model is unaffected by the researcher's knowledge of the observed outcome or the estimated treatment effect. Then, the *design-based* estimation and testing of the treatment effects is straightforward because the estimator and corresponding test statistic directly derive from the PS design. *Model-based* analyses require an additional testing of the probabilistic assumptions' adequacy, which is crucial for a valid model-based inference. If the data indicate violations of the independence, homoscedasticity and normality assumptions of the model's probabilistic structure need to be adapted.

14.5.3.2 ***Variance of the Treatment Effect*** Besides the point estimate, a reasonable estimate of the treatment effect's standard error for interval estimation and significance testing is required. In estimating the variance, we have to pay attention to the fact that the PS had to be estimated. It is well known that the estimated PS regularly produces better balance in observed covariates than the true PS because the estimated PS also removes random imbalances in addition to systematic imbalances (Rosenbaum, 1987, Rosenbaum and Rubin, 1983, Rubin and Thomas, 1996). Thus, the sampling error of the treatment effect is typically smaller when the estimated instead of true PS is used. Since the variance estimators discussed above do not reflect the uncertainty in the estimated PS an adjustment seems to be necessary – otherwise, biased variance estimates and significance tests might result. However, whether a variance adjustment is actually called for depends on the PS design and its formulation.

For *design-based* formulations of PS designs with no intent to generalize the effect estimates, estimating the unknown PS does not require an adjusted variance estimator because the randomization distribution of design-based formulations is obtained *conditional* on the matched, stratified, or weighted data. Given selection is strongly ignorable, subjects *within* matched pairs or strata are considered as being randomized into treatment and control conditions. The randomization distribution reflects all possible assignments *within* matched pairs or strata and does not depend on the variance between matched pairs or strata. Thus, how we arrive at matched pairs or strata – by matching on the true or estimated PS – does not matter. Similar arguments hold for the IPW design.

Whenever we wish to generalize the effect estimates beyond the sample in hand, the uncertainty in the estimated PS needs to be taken into account because in hypothetical replications of the study the reestimation of the PS model would always result in different parameter estimates and, thus, different propensity scores. Since the estimated PS tends to remove chance imbalances, matches and strata derived from the estimated PS are usually more homogeneous than matches or strata obtained from the true PS. Thus, across repeated samples with newly reestimated PS models, the estimated treatment effect shows less variation than a PS design's conventional variance estimator indicates.

In order to achieve a more accurate variance estimate, we can pursue two main strategies: either use adjusted variance estimators or bootstrapped estimators. Assuming that the PS is estimated via logistic regression, Schafer and Kang (2008) present formulas of adjusted variance estimators for different PS techniques (see also Imbens, 2004, Robins *et al.*, 1995). These formulas take the logistic estimation step of the PS into account. Alternatively, in bootstrapping the variance we actually mimic the whole process of respecifying the PS model as it actually would occur with repeated samples. Ideally, the PS model is entirely respecified according to balance or goodness-of-fit criteria for each bootstrap sample. Since such model specifications cannot always be fully automatized, at least the PS model should be reestimated (without further model respecifications). Note that despite the bootstrap's generality, it does not necessarily work for matching designs because of violated continuity requirements (Abadie and Imbens, 2002).

14.6 DISCUSSION

In this chapter, we discussed the strengths and weaknesses of design- and model-based formulations of three basic PS designs – PS matching, stratification, and IPW. We only focused on PS designs that are fully implementable without reference to any outcome data. Thus, we did not consider PS techniques that involve a direct modeling of the outcome, such as PS regression estimation or mixed methods that combine a PS design with an additional covariance adjustment. Nonetheless, in practice, it is advisable to use a mixed methods approach in addition to the analysis of the PS design chosen since it frequently increases efficiency and reduces residual bias in case of an imperfectly balanced PS design. However, there is no guarantee that an estimate resulting from a mixed method is less biased than the simple design-based estimate (Freedman, 2008, Kang and Schafer, 2007).

In practice, the PS and the treatment effect can be easily estimated using standard statistical software. Only PS matching is more challenging since it requires an efficient matching algorithm. Almost all standard statistical software tools offer matching procedures. For instance, in R the packages *optmach* (Hansen and Klopfer, 2006), *MatchIt* (Ho *et al.*, 2004), and *matching* (Sekhon, 2011) provide efficient algorithms for different matching approaches including optimal full and pair matching. Stata offers *match* (Abadie *et al.*, 2004), *psmatch2* (Leuven and Sianesi, 2003), and *pscore* (Becker and Ichino, 2002). The macros *Greedy* (Parsons, 2001), *Gmatch* and *Vmatch* (Kosanke and Bergstralh, 2004) are available in SAS (also *proc assign* and *proc netflow* can be used for optimal matching). Also for SPSS, a PS matching module is available (Thoemmes, 2012). Several of these packages also offer tools for checking balance and overlap.

The distinction between design- and model-based formulations of PS designs revealed that the choice of a specific design and formulation strongly depends on a researcher's intent to generalize the estimated treatment effect and willingness to rely on probabilistic assumptions. We also showed that design-based formulations are in accordance with the standard RCM since both rely on nonstochastic potential outcomes. Model-based formulations are based on stochastic outcomes and, thus, directly correspond to Steyer's generalization of RCM or Pearl's probabilistic structural causal model.

Design-based formulations of PS matching, stratification, or IPW designs do not require any probabilistic assumption. The sole assumptions required for a valid PS design and analysis are the strong ignorability assumption and SUTVA. If the estimated PS establishes balance on observed covariates, design-based PS estimators effectively estimate ATE or ATT for the sample in hand and the estimators do not need to account for the estimation step of the unknown PS. The purpose of design-based formulations of PS matching and stratification designs is to establish a strongly ignorable selection mechanism for the sample in hand. Thus, specifying the PS model according to balance criteria ensures that the matched or stratified treatment and control group are as similar as possible. IPW designs require a PS that accurately represents the selection probabilities in addition to the balancing property.

Model-based formulations of PS designs rely on probabilistic assumptions in addition to the strong ignorability assumption and SUTVA. By virtue of the probabilistic assumptions, results are generalizable to larger superpopulations – at least hypothetically. Thus, goodness-of-fit criteria might be more appropriate for PS model selection than balance metrics but a correctly specified model also implies balance on observed covariates. For model-based formulations without any intent to generalize, one may proceed as with design-based formulations, but still needs to probe the probabilistic assumptions.

Despite the significant conceptual differences between the three basic PS designs and their design- and model-based formulations, the estimates obtained from the different designs and formulations are frequently (but not always) very similar. The similarity of effect estimates was demonstrated by simulation studies (e.g., Busso *et al.*, 2009a, Schafer and Kang, 2008) and within-study comparisons that compare within the same study estimates from PS designs to estimates of a corresponding randomized experiment (e.g., Cook and Steiner, 2010, Pohl *et al.*, 2009, Shadish *et al.*, 2008). Not surprisingly, these studies also reveal that the availability and reliability of ignorability-establishing covariates is of much greater importance for getting approximately unbiased estimates than the choice of a specific PS design and formulation. While failing to measure all confounding covariates results in remaining selection bias, the choice of a specific PS technique has frequently no significant effect on effect estimates (Cook *et al.*, 2008, Glazerman *et al.*, 2003, Steiner *et al.*, 2010). Moreover, if subjects select into treatment on the basis of latent constructs, the unreliable measurement of these constructs may result in considerable remaining bias (Steiner *et al.*, 2011). Thus, the reliable measurement of all confounding covariates is by far much more important than the choice of a specific PS design and formulation.

Though the choice of a specific PS design does usually not result in significant different estimates, the distinction between different PS designs and their design- and model-based formulations helps in conceptualizing and implementing a PS study since it forces researchers to clearly think about their intent to generalize results and about their willingness to rely on probabilistic assumptions. If the results obtained from different PS designs and formulations significantly differ for a given data set, then determining the most credible design and formulation depends on the researcher's critical assessment of the underlying design and analytic assumptions. All these issues are even more important as the complexity of PS designs and analyses increases, as it is the case with clustered data or time-varying treatment regimes.

ACKNOWLEDGMENTS

The author thanks Coady Wing, Ron Serlin, Jee-Seon Kim, Yongnam Kim, David Kaplan, Bryan Keller, and Kelly Hallberg for helpful discussions. The research was partially supported by grants R305D100033 and R305D120005 from the Institute of Education Sciences, U.S. Department of Education and a grant from the W.T. Grant Foundation.

REFERENCES

Abadie, A., Drukker, D., Herr, J.L., and Imbens, G.W. (2004) Implementing matching estimators for average treatment effects in stata. *Stata Journal*, **4**, 290–311.

Abadie, A. and Imbens, G.W. (2002) Simple and bias-corrected matching estimators for average treatment effects, *NBER Technical Working Paper 283*, National Bureau of Economic Research, Cambridge.

Angrist, J.D. and Pischke, J.S. (2009) *Mostly Harmless Econometrics. An Empiricist's Companion*, Princeton University Press, Princeton, NJ.

Austin, P.C. (2008) A critical appraisal of propensity-score matching in the medical literature between 1996 and 2003. *Stat. Med.*, **27**, 2037–2049.

Becker, S.O. and Ichino, A. (2002) Estimation of average treatment effects based on propensity scores. *Stata Journal*, **2**, 358–377.

Berk, R.A. (2008) *Statistical Learning from a Regression Perspective*, Springer-Verlag, New York.

Busso, M., DiNardo, J., and McCrary, J. (2009a) Finite sample properties of semiparametric estimators of average treatment effects, Unpublished Manuscript. Retrieved from `http://emlab.berkeley.edu/ jmccrary/BDM_JBES.pdf` (accessed 18 January 2016).

Busso, M., DiNardo, J., and McCrary, J. (2009b) New evidence on the finite sample properties of propensity score matching and reweighting estimators, IZA Discussion Paper, No. 3998. Retrieved from `http://ftp.iza.org/dp3998.pdf` (accessed 23 December 2015).

Cochran, W.G. (1968) The effectiveness of adjustment by subclassification in removing bias in observational studies. *Biometrics*, **24**, 295–313.

Cochran, W.G. and Rubin, D.B. (1973) Controlling bias in observational studies: a review. *Sankhya, A*, **35**, 417–446.

Cook, T.D., Shadish, W.R., and Wong, V.C. (2008) Three conditions under which experiments and observational studies produce comparable causal estimates: new findings from within-study comparisons. *Journal of Policy Analysis and Management*, **27** (4), 724–750.

Cook, T.D. and Steiner, P.M. (2010) Case matching and the reduction of selection bias in quasi-experiments: the relative importance of the pretest as a covariate, unreliable measurement and mode of data analysis. *Psychological Methods*, **15** (1), 56–68.

Cox, D.R. (2006) *Principles of Statistical Inference*, Cambridge University Press, Cambridge.

Cox, D.R. and Hinkley, D.V. (1974) *Theoretical Statistics*, Chapman & Hall/CRC, Boca Raton, FL.

Freedman, D.A. (2008) On regression adjustments to experimental data. *Advances in Applied Mathematics*, **40**, 180–193.

Glazerman, S., Levy, D.M., and Myers, D. (2003) Nonexperimental versus experimental estimates of earnings impacts. *Annals of the American Academy*, **589**, 63–93.

Guo, S. and Fraser, M.W. (2010) *Propensity Score Analysis: Statistical Methods and Applications*, Sage Publications, Thousand Oaks, CA.

Hahn, J. (1998) On the role of the propensity score in efficient semiparametric estimation of average treatment effects. *Econometrica*, **66** (2), 315–331.

Hansen, B.B. (2004) Full matching in an observational study of coaching for the SAT. *Journal of the American Statistical Association*, **99**, 609–618.

Hansen, B.B. and Klopfer, S.O. (2006) Optimal full matching and related designs via network flows. *Journal of Computational and Graphical Statistics*, **15**, 609–627.

Heckman, J.J. (2005) The scientific model of causality. *Sociological Methodology*, **35** (1), 1–98.

Heckman, J.J., Ichimura, H., and Todd, P.E. (1997) Matching as an econometric evaluation estimator: evidence from evaluating a job training programme. *Review of Economic Studies*, **64**, 605–654.

Hill, J. (2008) Discussion of research using propensity-score matching: comments on 'A critical appraisal of propensity-score matching in the medical literature between 1996 and 2003' by Peter Austin. *Statistics in Medicine*, **27**, 2055–2061.

Hill, J. and Reiter, J.P. (2006) Interval estimation for treatment effects using propensity score matching. *Statistics in Medicine*, **25** (13), 2230–2256.

Hirano, K., Imbens, G.W., and Ridder, G. (2003) Efficient estimation of average treatment effects using the estimated propensity score. *Econometrica*, **71** (4), 1161–1189.

Ho, D.E., Imai, K., King, G., and Stuart, E.A. (2004) MatchIt: nonparametric preprocessing for parametric causal inference. *Journal of Statistical Software*, **42** (8), 1–28.

Holland, P.W. (1986) Statistics and causal inference. *Journal of the American Statistical Association*, **81**, 945–970.

Hong, G. (2010) Marginal mean weighting through stratification: adjustment for selection bias in multi-level data. *Journal of Educational and Behavioral Statistics*, **35** (5), 499–531.

Horvitz, D.G. and Thompson, D.J. (1952) A generalization of sampling without replacement from a finite universe. *Journal of the American Statistical Association*, **47**, 663–685.

Imai, K., King, G., and Stuart, E. (2008) Misunderstandings between experimentalists and observationalists about causal inference. *Journal of the Royal Statistical Society, Series A (Statistics in Society)*, **171**, 481–502.

Imbens, G.W. (2004) Nonparametric estimation of average treatment effects under exogeneity: a review. *Review of Economics and Statistics*, **86** (1), 4–29.

Kang, J. and Schafer, J.L. (2007) Demystifying double robustness: a comparison of alternative strategies for estimating population means from incomplete data. *Statistical Science*, **26**, 523–539.

Kish, L. (1987) *Statistical Design for Research*, John Wiley & Sons, Inc., Hoboken, NJ.

Kolar, A., Vehovar, V., and Steiner, P.M. (2014) Small samples and propensity score methods for estimating causal effects from observational study designs, Unpublished Manuscript.

Kosanke, J. and Bergstralh, E. (2004) Match cases to controls using variable optimal matching, `http://mayoresearch.mayo.edu/mayo/research/biostat/upload/vmatch.sas` and match 1 or more controls to cases using the greedy algorithm: `http://mayoresearch.mayo.edu/mayo/research/biostat/upload/gmatch.sas` (accessed 23 December 2015).

Leuven, E. and Sianesi, B. (2003) PSMATCH2: Stata module to perform full Mahalanobis and propensity score matching, common support graphing, and covariate imbalance testing, Statistical Software Components S432001.

Lohr, S.L. (1999) *Sampling: Design and Analysis*, Duxbury Press, Pacific Grove, CA.

Lumley, T. (2010) *Complex Surveys: A Guide to Analysis Using R*, John Wiley & Sons, Inc., Hoboken, NJ.

Lunceford, J.K. and Davidian, M. (2004) Stratification and weighting via propensity score in estimation of causal treatment effects: a comparative study. *Statistical Medicine*, **23**, 2937–2960.

McCaffrey, D.F., Ridgeway, G., and Morral, A.R. (2004) Propensity score estimation with boosted regression for evaluating causal effects in observational studies. *Psychological Methods*, **9**, 403–425.

Morgan, S.L. and Winship, C. (2007) *Counterfactuals and Causal Inference: Methods and Principles for Social Research*, Cambridge University Press, Cambridge.

Neyman, J. (1990) On the application of probability theory to agricultural experiments. Essay on principles. Section 9. Translated and edited by D. M. Dabrowska and T. P. Speed. *Statistical Science*, **5** (4), 465–472.

Parsons, L.S. (2001) Reducing bias in a propensity score matched-pair sample using greedy matching techniques, in *Proceedings of the 26th Annual SAS ® Users Group International Conference, Paper 214–26*, SAS Institute Inc., Cary, NC. Retrieved from `http://www2.sas.com/proceedings/sugi26/p214-26.pdf` (accessed 23 December 2015).

Pearl, J. (2009) *Causality: Models, Reasoning, and Inference*, 2nd edn, Cambridge University Press, Cambridge.

Pearl, J. (2010) The foundations of causal inference. *Sociological Methodology*, **40** (1), 75–149.

Pohl, S., Steiner, P.M., Eisermann, J., Soellner, R., and Cook, T.D. (2009) Unbiased causal inference from an observational study: results of a within-study comparison. *Educational Evaluation and Policy Analysis*, **31** (4), 463–479.

Robins, J. (1986) A new approach to causal inference in mortality studies with a sustained exposure period-application to control of the healthy worker survivor effect. *Math Modelling*, **7**, 1393–1512.

Robins, J.M. (1999a) Associations, causation, and marginal structural models. *Synthese*, **101**, 151–179.

Robins, J.M. (1999b) Marginal structural models versus structural nested models as tools for causal inference, in *Statistical Models in Epidemiology: The Environment and Clinical Trials* (eds E. Halloran and D. Berry), (eds) Springer-Verlag, New York, pp. 95–134.

Robins, J.M. and Rotnitzky, A. (1995) Semiparametric efficiency in multivariate regression models with missing data. *Journal of the American Statistical Association*, **90**, 122–129.

Robins, J.M., Rotnitzky, A., and Zhao, L.P. (1995) Analysis of semiparametric regression models for repeated outcomes in the presence of missing data. *Journal of the American Statistical Association*, **90**, 106–121.

Rosenbaum, P.R. (1987) Model-based direct adjustment. *Journal of the American Statistical Association*, **82**, 387–394.

Rosenbaum, P.R. (2002) *Observational Studies*, 2nd edn, Springer-Verlag, New York.

Rosenbaum, P.R. (2009) *Design of Observational Studies*, Springer-Verlag, New York.

Rosenbaum, P.R. and Rubin, D.B. (1983) The central role of the propensity score in observational studies for causal effects. *Biometrika*, **70** (1), 41–55.

Rosenbaum, P.R. and Rubin, D.B. (1984) Reducing bias in observational studies using subclassification on the propensity score. *Journal of the American Statistical Association*, **79**, 516–524.

Rubin, D.B. (1973) The use of matched sampling and regression adjustment to remove bias in observational studies. *Biometrics*, **29**, 185–203.

Rubin, D.B. (1974) Estimating causal effects of treatments in randomized and nonrandomized studies. *Journal of Educational Psychology*, **66**, 688–701.

Rubin, D.B. (1977) Assignment to treatment group on the basis of a covariate. *Journal of Educational Statistics*, **2**, 1–26.

Rubin, D.B. (1978) Bayesian inference for causal effects: the role of randomization. *Annals of Statistics*, **6**, 34–58.

Rubin, D.B. (1990a) Formal modes of statistical inference for causal effects. *Journal of Statistical Planning and Inference*, **25** (3), 279–292.

Rubin, D.B. (1990b) Neyman (1923) and causal inference in experiments and observational studies. Comment on Neyman, J. (1990), on the application of probability theory to agricultural experiments. *Statistical Science*, **5** (4), 472–480.

Rubin, D.B. (2001) Using propensity scores to help design observational studies: application to the tobacco litigation. *Health Services and Outcomes Research Methodology*, **2**, 169–188.

Rubin, D.B. (2006) *Matched Sampling for Causal Effects*, Cambridge University Press, Cambridge.

Rubin, D.B. (2007) The design versus the analysis of observational studies for causal effects: parallels with the design of randomized trials. *Statistics in Medicine*, **26** (1), 20–36.

Rubin, D.B. (2008) For objective causal inference. Design trumps analysis. *Annals of Applied Statistics*, **2**, 808–840.

Rubin, D.B. (2010) Reflections stimulated by the comments of Shadish (2010) and West and Thoemmes (2010). *Psychological Methods*, **15** (1), 38–46.

Rubin, D.B. and Thomas, N. (1996) Matching using estimated propensity scores: relating theory to practice. *Biometrics*, **52**, 249–264.

Rubin, D.B. and Thomas, N. (2000) Combining propensity score matching with additional adjustments for prognostic covariates. *Journal of the American Statistical Association*, **95**, 573–585.

Särndal, C.E. (1978) Design-based and model-based inference in survey sampling. *Scandinavian Journal of Statistics*, **5**, 27–52.

Schafer, J.L. and Kang, J. (2008) Average causal effects from non-randomized studies: a practical guide and simulated example. *Psychological Methods*, **13** (4), 279–313.

Schochet, P.Z. (2010) Is regression adjustment supported by the Neyman model for causal inference? *Journal of Statistical Planning and Inference*, **140**, 246–259.

Sekhon, J.S. (2011) Multivariate and propensity score matching software with automated balance optimization: the matching package for R. *Journal of Statistical Software*, **42** (7), Retrieved from `http://www.jstatsoft.org/v42/i07`.

Shadish, W.R., Clark, M.H., and Steiner, P.M. (2008) Can nonrandomized experiments yield accurate answers? A randomized experiment comparing random to nonrandom assignment. *Journal of the American Statistical Association*, **103**, 1334–1343.

Snowden, J.M., Rose, S., and Mortimer, K. (2011) Implementation of G-computation on a simulated data set: demonstration of a causal inference technique. *American Journal of Epidemiology*, **173**, 73–738.

Spanos, A. (1999) *Probability Theory and Statistical Inference: Econometric Modeling with Observational Data*, Cambridge University Press, Cambridge.

Spirtes, P., Glymour, C.N., and Scheines, R. (1993) *Causation, Prediction, and Search*, Springer-Verlag, New York.

Steiner, P.M. and Cook, D.L. (2013) Matching and propensity scores, in *The Oxford Handbook of Quantitative Methods*, vol. 1, Foundations (ed. T.D. Little), Oxford University Press, New York.

Steiner, P.M., Cook, T.D., and Shadish, W.R. (2011) On the importance of reliable covariate measurement in selection bias adjustments using propensity scores. *Journal of Educational and Behavioral Statistics*, **36** (2), 213–236.

Steiner, P.M., Cook, T.D., Shadish, W.R., and Clark, M.H. (2010) The importance of covariate selection in controlling for selection bias in observational studies. *Psychological Methods*, **15** (3), 250–267.

Steiner, P.M., Kim, Y., Hall, C., and Su, D. (2014) Graphical models for quasi-experimental designs. *Sociological Methods & Research*, Online First: `http://smr.sagepub.com/content/early/2015/05/13/0049124115582272.full.pdf+html` (accessed 18 January 2016).

Steyer, R. (2005) Analyzing individual and average causal effects via structural equation models. *Methodology*, **1**, 39–64.

Steyer, R., Gabler, S., von Davier, A.A., and Nachtigall, C. (2000a) Causal regression models II: unconfoundedness and causal unbiasedness. *Methods of Psychological Research Online*, **5** (3), 55–87.

Steyer, R., Gabler, S., von Davier, A.A., Nachtigall, C., and Buhl, T. (2000b) Causal regression models I: Individual and average causal effects. *Methods of Psychological Research Online*, **5** (2), 39–71.

Stuart, E.A. (2008) Developing practical recommendations for the use of propensity scores: discussion of 'a critical appraisal of propensity score matching in the medical literature between 1996 and 2003' by Peter Austin. *Statistics in Medicine*, **27** (12), 2062–2065.

Stuart, E.A. (2010) Matching methods for causal inference: a review and a look forward. *Statistical Sciences*, **25** (1), 1–21.

Thoemmes, F. (2012) Propensity score matching in SPSS, Unpublished Manuscript. Retrieved from `http://sourceforge.net/projects/psmspss/` & `http://arxiv.org/abs/1201.6385` (accessed 23 December 2015).

Waernbaum, I. (2010) Propensity score model specification for estimation of average treatment effects. *Journal of Statistical Planning and Inference*, **140** (7), 1948–1956.

15

ADJUSTMENT WHEN COVARIATES ARE FALLIBLE

STEFFI POHL

Department of Education and Psychology, Methods and Evaluation/Quality Management, Freie Universität Berlin, Berlin, Germany

MARIE-ANN SENGEWALD

Methodology and Evaluation Research, Institute of Psychology, Friedrich-Schiller-University Jena, Jena, Germany

ROLF STEYER

Methodology and Evaluation Research, Institute of Psychology, Friedrich-Schiller-University Jena, Jena, Germany

15.1 INTRODUCTION

Researchers in the social and behavioral sciences frequently investigate the effects of a particular treatment such as a training or therapy. For this aim, the performances of a treatment and a control group are compared with respect to an outcome measure of interest. Most of these studies rely on non-randomized designs in which participants either self-select into treatment or are assigned to treatment by a third person (e.g., administrators assign participants according to specific criteria that might be known or unknown). Because of this selection, systematic treatment group differences typically occur prior to treatment, resulting in initially incomparable groups. In these cases, the simple difference in the treatment and control group's average outcome usually is a biased estimator of the average treatment effect. A well-established strategy to avoid selection bias is covariate adjustment. Unbiased

Statistics and Causality: Methods for Applied Empirical Research, First Edition.
Wolfgang Wiedermann and Alexander von Eye.

treatment effects can be estimated via adjusting for all relevant covariates. Covariate adjusted estimation of treatment effects can be achieved using, for example, analysis of covariance (ANCOVA) or propensity score (PS) methods (Rosenbaum and Rubin, 1983), provided that certain assumptions hold. In empirical analyses, pretest measures are often the most important covariates as they are highly correlated to the outcome and often also to the treatment (e.g., Steiner *et al.*, 2010, Cook *et al.*, 2009).

While in empirical applications the choice of covariates is an important issue, the issue of measurement error in covariates is typically ignored. However, in social and behavioral sciences many covariates – such as competencies and personality characteristics – cannot be assessed without measurement error. As a consequence, in practice these constructs are usually assessed by multiple indicators allowing to model measurement error purged, latent variables. Measurement error of covariates may affect the estimate of the causal effect. While in the social sciences modeling of latent variables is state of the art, manifest covariates are usually considered in the adjustment model when estimating causal effects. Only few papers (e.g., Aiken and West, 1991, Cook *et al.*, 2009, Steiner *et al.*, 2011) focus on the effect of ignoring measurement error in covariates or on the development of adjustment methods based on latent covariates. It is to note that measurement error in covariates does not always bias causal effect estimates, and adjustment based on latent instead of on manifest covariates may in some cases even induce bias. The conditions under which covariate adjustment needs to be based on latent covariates or on manifest covariates have hardly been considered in methodological research. One of the few exceptions is Sengewald *et al.* 2016.

In this chapter, we discuss adjustment when covariates are not perfectly reliable. We start with reviewing a theoretical framework that applies to fallible and latent covariates. This framework allows for deriving conditions under which adjustment has to be based on the latent covariate and conditions under which adjustment has to be based on the fallible covariate. As most methodological research has considered the case in which adjustment has to be based on the latent covariates, we focus on this case in the following sections. We present analytic derivations, simulation studies, and empirical analyses on the biasing effect of measurement error in covariates for causal effect estimation. We also present and evaluate different approaches to adjust for latent covariates. Finally, we consider the role of further covariates for the biasing effect of measurement error in another covariate. For this, we present an empirical analysis and discuss results from simulation studies. We conclude discussing the implications for adjustment in empirical applications when covariates are fallible.

15.2 THEORETICAL FRAMEWORK

There are different theoretical frameworks for the analysis of causal effects. Two very influential frameworks, which have stimulated development of adjustment methods, are graphical modeling (e.g., Pearl, 1995, 2009) and the potential-outcome approach (e.g., Rubin, 1974, 2005). However, none of these frameworks comprises an explicit theory of latent variables. Steyer *et al.* (2014) and Mayer *et al.* (2014)

introduced a stochastic theory of causal effects (TCE) that extends and formalizes the potential-outcome framework and is compatible with the theory of latent variables presented by Steyer *et al.* (2015). In the following sections, we briefly outline the basic ideas of TCE for the definition and identification of the average causal effect of a treatment and derive conditions in which latent instead of observed (manifest) covariates need to be used for adjustment.

15.2.1 Definition of Causal Effects

We focus on defining the (total) causal effect of a dichotomous treatment variable X on an outcome variable Y, comparing two treatment conditions, say treatment (1) and control (0). For this setting, Rubin (1974) defines causal effects based on potential-outcome variables $Y_0(U)$ and $Y_1(U)$, the values of which denote the outcome of a specific person u under the control condition and the outcome of the same person u under treatment. In Rubin's definition, potential outcomes are assumed to be fixed numbers, implying that the outcome is fully determined by the person u and the treatment condition. In contrast, Steyer *et al.* (2014) pose the idea of a stochastic outcome not fully determined by the person and the treatment condition. Instead, the authors assume that every person has a distribution of outcomes under treatment and a possible different one under control. This distribution is due to (i) measurement error in the outcome variable and (ii) events and other variables occurring in between treatment and outcome, such as mediating variables or critical life events. As a consequence, instead of potential-outcome variables they consider true-outcome variables, the values of which are defined as the expectations of these conditional distributions of Y given person u and treatment x. In fact, this is the original approach of Neyman (1923/1990) and has also been used by Steyer *et al.* (2000a,b, 2002).

The basic idea in the definition of an atomic causal effect is to condition on (keep constant) all variables that are prior or simultaneous to X – except for X itself – and then see how Y depends on X. Variables that are prior or simultaneous to X except for X itself are also called potential confounders. Holding constant the person u, a value of the person variable U, in the definition of the true outcomes implies holding constant all attributes of the persons that can be represented as a mapping of the person variable U (e.g., perfectly reliable covariates such as gender). However, due to measurement error, fallible covariates prior to treatment are not mappings of U, that is, they are not constant when conditioning on a specific person u. Steyer and colleagues also consider fallible covariates as potential confounders because they may be correlated to X and Y. For instance, a fallible pretest Z of the outcome Y can determine treatment selection. Furthermore, the experience made through pretest assessment (e.g., having a low test score and thus feeling bad and less motivated) can affect the outcome (Steyer *et al.*, 2015). Thus, in order to condition on *all potential confounders*, we need to consider not only U but also fallible covariates *in the definition* of the true-outcome variables. Considering m fallible covariates $Z_1, \ldots, Z_m$, Steyer *et al.* (2014) define true-outcome variables by conditional expectations of Y in treatment ($X = 1$) and in control ($X = 0$), conditioning on U and the vector of all fallible covariates $Z_{all} := (Z_1, \dots, Z_m)$:

$$\tau_0 := E^{X=0}(Y|U, Z_{all}) \quad \text{and} \quad \tau_1 := E^{X=1}(Y|U, Z_{all}). \tag{15.1}$$

Note that in Equation (15.1) potential confounding due to perfectly reliable covariates, (measurement error purged) latent covariates, and potential confounding due to fallible covariates is considered. To emphasize, *for the definition* of causal effects it is essential to condition on all potential confounders. This does not imply that we have to condition on all potential confounders when identifying and estimating causal effects. Hence, we may distinguish between a potential confounder and an actual confounder.

The difference $\delta_{10} = \tau_1 - \tau_0$ between the true-outcome variables is called the atomic total effect variable. Due to the fundamental problem of causal inference (Holland, 1986, Rubin, 1974), its values, the atomic effects cannot be estimated in empirical applications. Instead, we often consider the average total treatment effect $ATE = E(\delta_{10}) = E(\tau_1 - \tau_0)$, that is, the expectation of the atomic total effect variable, integrating over the distribution of (U, Z_{all}).

15.2.2 Identification of Causal Effects

The *ATE* can be linked to quantities that are estimable in empirical applications and that are identical to the *ATE* under certain conditions. For example, the *conditional expectations* $E(Y|X = 0)$ and $E(Y|X = 1)$ are called *(causally) unbiased*, if

$$E(Y|X = 0) = E(\tau_0) \quad \text{and} \quad E(Y|X = 1) = E(\tau_1). \tag{15.2}$$

Hence, under this unbiasedness assumption $E(Y|X = 1) - E(Y|X = 0) = E(\tau_1) - E(\tau_0) = ATE$. In a randomized experiment treatment X and (U, Z_{all}), that is, X and all observed and unobserved covariates are stochastically independent, and this independence implies that the conditional expectations $E(Y|X = x)$ are unbiased (Eq. (15.2)).

In contrast, in a nonrandomized experiment the prima facie effect $E(Y|X = 1) - E(Y|X = 0)$ is typically biased due to systematic selection into treatment conditions. This systematic selection induces dependence of X and (U, Z_{all}). In those designs, *ATE* estimation requires finding a (possibly multivariate) covariate $Z^* = f(U, Z_{all})$ for which

$$E^{X=x}(Y|Z^*) = E(\tau_x|Z^*), \quad \text{for } x = 0, 1 \tag{15.3}$$

holds. This condition defines *unbiasedness of the conditional expectations* $E^{X=x}(Y|Z^*)$. Under this condition,[1]

$$E[E^{X=1}(Y|Z^*) - E^{X=0}(Y|Z^*)] = E(\delta_{10}) = ATE. \tag{15.4}$$

[1]If Equation (15.3) holds, $E[E^{X=1}(Y|Z^*) - E^{X=0}(Y|Z^*)] = E[E^{X=1}(Y|Z^*)] - E[E^{X=0}(Y|Z^*)] = E[E(\tau_1|Z^*)] - E[E(\tau_0|Z^*)] = E(\tau_1) - E(\tau_0) = E(\delta_{10})$. The expectations $E[E^{X=x}(Y|Z^*)], x = 0, 1,$ are called the Z^*- adjusted expectations of Y in treatment x.

Note that $E[E^{X=1}(Y|Z^*) - E^{X=0}(Y|Z^*)]$ integrates over the distribution of Z^*. Equations (15.3) and (15.4) follow from each of several causality conditions (see Steyer *et al.*, 2014). One causality condition is *Z_1^*-conditional independence of X and (U, Z_{all})*. For $Z_1^* = f_1(U, Z_{all})$ it is defined by

$$P(X = 1|U, Z_{all}) = P(X = 1|Z_1^*). \tag{15.5}$$

Equation (15.5) implies that Z_1^* comprises all covariates that are prior or simultaneous to X and determine treatment probability. Thus, in addition to the (possibly multivariate) covariate Z_1^*, there are no other attributes of the persons and no other fallible covariates that determine treatment probability. Equation (15.5) holds, for instance, when persons were randomly assigned to one of the treatment conditions with an assignment probability that *only* depends on Z_1^*. Equation (15.5) may also hold when treatment assignment is not randomized, but when we select the (multivariate) random variable Z_1^* such that this equation holds.

Another causality condition is *completeness of the conditional expectations* $E^{X=x}(Y|Z_2^*)$. For $Z_2^* = f_2(U, Z_{all})$, it is defined by

$$E^{X=x}(Y|U, Z_{all}) = E^{X=x}(Y|Z_2^*) \quad \text{for } x = 0, 1. \tag{15.6}$$

Equation (15.6) implies that Z_2^* comprises all covariates that are prior or simultaneous to X and that, together with X, determine the conditional expectations of the outcome variable. Note that Z_1^* is not necessarily identical to Z_2^*. Hence, the two causality conditions can be satisfied for different (multivariate) covariates Z_1^* and Z_2^*. Also note that, although $Z_1^* = f_1(U, Z_{all})$ and $Z_2^* = f_2(U, Z_{all})$, it is possible that Z_1^* and Z_2^* are mappings of U alone or mappings of Z_{all} alone, that is, Z_1^* and Z_2^* can only consist of perfectly reliable or (measurement error purged) latent covariates (mappings of U) or they can only consist of fallible covariates (mappings of Z_{all}). Most important to note, Equations (15.3) and (15.4) follow from each of the two causality conditions introduced above. (Note that there are more than these two causality conditions; see Steyer *et al.*, 2000b, 2014)

15.2.3 Adjusting for Latent or Fallible Covariates

In the previous section, we delineated under which conditions the *ATE* can be identified by estimable quantities. If there is measurement error in the covariates, the question is whether we should adjust for the latent or for the manifest (fallible) covariates (Sengewald et al., 2016). For simplicity, Sengewald and colleagues consider the case of just one fallible covariate $Z = \eta + \epsilon$. Four different scenarios can be distinguished: First, the treatment probability and the outcome variable depend on the latent variable η (scenario 1). This means that Equations (15.5) and (15.6) hold for $Z_1^* = f_1(U) = Z_2^* = f_2(U) = \eta$. This scenario is plausible for self-selection in to treatment, in which attributes of persons (latent variables) determine the treatment probability and the outcome variable is only determined by the latent covariate. In this case, in order to obtain an unbiased *ATE* estimate, the latent covariate needs to be used for adjustment. Second, the treatment probability and the outcome variable depend on the observed fallible covariate (i.e., Eqs (15.5) and (15.6) hold for $Z_1^* = f_1(Z_{all}) = Z_2^* = f_2(Z_{all}) = Z$ (scenario 2). The treatment probability may depend on the observed fallible score when, for instance, a teacher decides on participation of his or her students in basic mathematical tutoring based on the scores on a math test. The outcome variable may depend on the fallible covariate, when, due to being tired, the student performs low on the pretest and, because of this poor performance, is less motivated in the posttest after the treatment (outcome variable). Third, the treatment probability depends on the observed covariate (i.e., $Z_1^* = f_1(Z_{all}) = Z$ and the outcome variable is determined by the latent variable (i.e., $Z_2^* = f_2(U) = \eta$) (scenario 3) and fourth vice versa, that is $Z_1^* = f_1(U) = \eta$ and $Z_2^* = f_2(Z_{all}) = Z$ (scenario 4). In scenarios 3 and 4, the observed covariate and the latent variable of the covariate satisfy one of the two causality conditions (Eqs (15.5) or (15.6)) and either one suffices to identify the *ATE* via Equation (15.4) and to obtain an unbiased *ATE* estimate by estimating $E[E^{X=1}(Y|Z^*) - E^{X=0}(Y|Z^*)]$.

Note that only in scenario 1, it is *necessary* to use the latent covariate for adjustment. Using fallible covariates in this scenario could lead to a biased *ATE* estimate. However, while using the latent covariate would result in unbiased *ATE* estimates in most of the scenarios (scenarios 1, 3, and 4), its use is not always warranted. In scenario 2, adjusting for the latent instead of the fallible covariate could even induce bias. Thus, even when fallible covariates have been measured, we do not always have to (scenarios 3 and 4) or should (scenario 2) use latent variables in order to obtain unbiased *ATE* estimates.

Sengewald et al. (2016) delineated these conditions using only one covariate. When more than one covariate is considered – as it will be the case in most applications – the situation becomes more complex as also the correlations between the covariates need to be regarded and other covariates may compensate for effects of measurement error. This is addressed later on.

So far, methodological research on the impact of covariates' measurement error on causal effect estimates has only considered the case in which confounding occurs on the latent rather than the observed covariate (scenario 1). Although this scenario may be plausible in many applications, there may also be cases in which one of the other scenarios holds. In these cases, it suffices to use manifest covariates for adjustment. In settings where scenario 2 holds, adjusting for latent instead of manifest covariates may result in a biased *ATE* estimate. The impact of misspecification, when adjusting for a latent covariate in settings where confounding is due to the manifest covariate (scenario 2), has not, yet, been investigated. This may be an objective for future research. One empirical study presented later in the manuscript did, however, explicitly investigate whether scenario 1 holds. As most other previous research had only dealt with settings according to scenario 1, the research presented in the following sections is based on settings in which scenario 1 holds.

15.3 THE IMPACT OF MEASUREMENT ERROR IN COVARIATES ON CAUSAL EFFECT ESTIMATION

So far, in methodological research on fallible covariates it has implicitly been assumed that the treatment probability and the outcome variable depend on latent covariates (scenario 1). Only in this scenario the use of manifest instead of latent covariates can bias causal effect estimates. Hence, focusing on scenario 1, previous methodological research has tried to quantify the impact of using manifest instead of latent covariates on causal effect estimation. In the following sections, we review the respective research. First, we consider the implications of using the manifest instead of the latent variable of *one* fallible covariate (derived by analytic derivations). Then this effect is considered for a whole set of covariates in simulation studies. Finally, we consider an empirical study in which the impact of adjusting for latent instead of manifest covariates in real settings is investigated. In this study, it is also investigated whether in that application confounding is due to the latent covariate or due to the manifest covariate.

15.3.1 Theoretical Impact of One Fallible Covariate

Assuming scenario 1 holds, the impact of using a manifest instead of a latent covariate on the *ATE* estimate can be analytically derived for the case of one single fallible covariate Z (Aiken and West, 1991, Cohen *et al.*, 2003). We assume that the fallible covariate Z may be decomposed into a true-score variable η and measurement error ϵ, such that $Z = \eta + \epsilon$, and that the conditional expectation of the outcome variable Y given a dichotomous treatment variable X with values 0 and 1 and the latent covariate η is $E(Y|\eta, X) = \alpha_0 + \alpha_1\eta + \alpha_2 X$. The *ATE* of X is then the partial regression coefficient α_2:

$$\alpha_2 = \frac{\rho_{YX} - \rho_{Y\eta}\rho_{\eta X}}{1 - \rho_{\eta X}^2} * \frac{\sigma_Y}{\sigma_X} \tag{15.7}$$

with $\rho_{Y\eta}$, $\rho_{\eta X}$, and ρ_{YX} being the bivariate correlations between Y, η, and X, and σ_Y as well as σ_X being the standard deviations of Y and X, respectively (e.g., Aiken and West, 1991). Now consider that the fallible measure Z is used instead of the latent variable η in the outcome model. When $Rel(Z) < 1$ for all bivariate correlations of Z with another perfectly measured variable (e.g., X) $\rho_{ZX} = \rho_{\eta X} * \sqrt{Rel(Z)}$ holds. Falsely assuming $Rel(Z) = 1$ leads to an attenuation of the correlations involving Z because $\rho_{ZX} < \rho_{\eta X}$ and, therefore, to an attenuation in the estimation of α_2. According to Steiner *et al.* (2011), within this setting, the total bias reduction potential (100%) of one covariate is on average attenuated by $(1 - Rel_X(Z)) * 100\%$, with $Rel_X(Z)$ being the reliability coefficient of Z within the treatment and control group. Consequently, decreasing the reliability of Z by 0.1 removes on average 10% less bias in settings where unbiasedness of the conditional expectations can be achieved with $Z^* = \eta$ (see Eq. 15.3).

In more complex settings in which several covariates are necessary to achieve unbiasedness of the conditional expectations, the impact of using manifest instead of latent covariates on treatment effect estimates is more difficult to derive (e.g., Aiken and West, 1991). If just one of several covariates is measured with error, this distorts the partial regression coefficients of all other variables in the model (Cohen *et al.*, 2003). The effect of measurement error in covariates on causal effect estimates depends on (i) the reliabilities of all covariates and (ii) the intercorrelations among all variables in the model. Therefore, the impact of one fallible covariate within a set of covariates can be lower or higher than in the single covariate case. The effect of additional covariates on the bias introduced by using a manifest instead of the respective latent covariate is addressed later on in more detail.

15.3.2 Investigation of the Impact of Fallible Covariates in Simulation Studies

Steiner *et al.* (2011) and Cook *et al.* (2009) investigated the consequences of measurement error in covariates in simulation studies. In order to depict the complexity of real data, these authors conducted a simulation study that is based on the results of empirical studies (Shadish *et al.*, 2008, Pohl *et al.*, 2009). In these empirical studies, effects of educational trainings (i.e., a vocabulary/English and a math training), into which university students could self-select, were evaluated. Various covariates were collected prior to treatment; these may be grouped into five categories: demographic variables, proxy pretest measures, topic preference, prior academic achievement, and personality variables. In the analyses of the empirical data, Steiner *et al.* (2010) as well as Cook *et al.* (2009) showed that proxy pretests and topic preference were sufficient to obtain an unbiased causal effect estimate. In their simulations, Steiner *et al.* (2011) and Cook *et al.* (2009) used these data. They considered the original covariates as perfectly reliable and added measurement error to the covariates, resulting in covariate reliabilities ranging from 1.0 to 0.6. In the simulations measurement error was assumed to be unrelated to the treatment and to the outcome (scenario 1). The authors demonstrated that measurement error leads to bias of *ATE* estimates. The lower the reliability, the less selection bias is removed. Moreover, a reliable measurement of constructs is of particular importance for covariates that had a great potential for removing selection bias (e.g., proxy pretests and topic preference). The reliability of covariates with a low potential to reduce selection bias (e.g., psychological predispositions) had only a negligible effect on *ATE* estimates. Furthermore, the authors outlined that, in their example, the biasing effect of measurement error was lower in a set of covariates than for a single covariate.

15.3.3 Investigation of the Impact of Fallible Covariates in an Empirical Study

So far, methodological research on the impact of measurement error in covariates on causal effect estimates has only considered the case where confounding is due to the latent covariate (scenario 1). Whether in empirical settings, the treatment probability and the outcome variable really depend on the latent covariate or rather

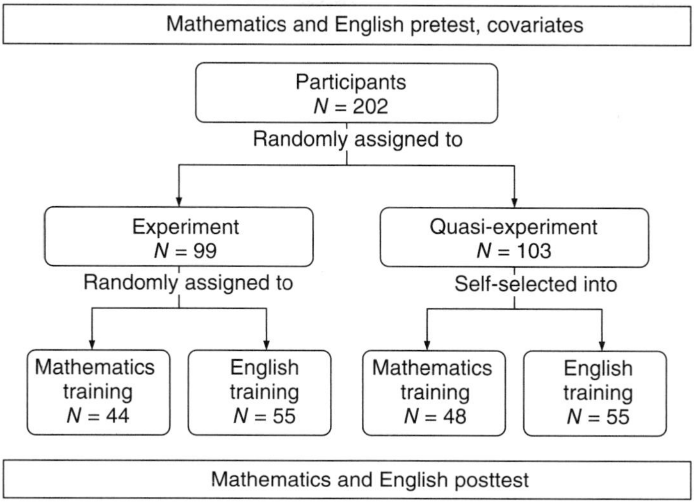

Figure 15.1 Design of the four-arm within-study comparison.
Reproduced from Pohl *et al.* (2009) with permission of SAGE Publishers.

on the manifest one has hardly been investigated. Neither has it been investigated whether the biasing effect of fallible covariates on *ATE* estimation is noteworthy in empirical applications. One reason for this is that an empirical evaluation of the impact of fallible covariates is usually difficult, because the true *ATE* is not available. However, the true *ATE* can be estimated in within-study comparisons (WSCs) that compare the effect of a nonrandomized experiment to the corresponding benchmark effect of a randomized trial within one study. As such they offer the opportunity for an empirical evaluation of the accuracy of causal effect estimates. A WSC design in which a full nonrandomized experiment is contrasted with a full experiment was proposed by Shadish *et al.* (2008) and also used by Pohl *et al.* (2009). Figure 15.1 illustrates the four-arm WSC design of Pohl *et al.* (2009). The overall sample consisted of 202 first-year students of education and psychology, who were randomly assigned either to a randomized experiment or a nonrandomized experiment. Students in the randomized experiment were then randomly assigned to one of two educational treatments – either English or mathematics training. Students in the nonrandomized experiment could choose which training they wanted to take. A set of covariates including proxy-pretest measures of the outcome, preference for English and math, demographic variables, prior academic achievement, and personality variables were assessed prior to treatment. Participants in the randomized and nonrandomized group attended the same pretest and training sessions. After the training, English and math skills were assessed by a posttest. In this design, two training effects can be investigated: First, the effect of English training with the math group being the reference group (control group without English training) and second, the effect of mathematics training with the English group being the reference group (control group without math training). The training effects were estimated in the randomized and in the nonrandomized experiment. Since all participating students were randomly assigned to the randomized and the nonrandomized experiment, the *ATE* estimated from the randomized experiment may serve as a benchmark for evaluating the nonadjusted (initial bias) and covariate-adjusted *ATE* estimates from the nonrandomized experiment.

Sengewald *et al.* (2016) used these data to investigate whether treatment and outcome really depend on the latent covariates (scenario 1). If confounding in the data occurs due to scenario 1, given that model assumptions are correct, adjusting for latent instead of manifest covariates should improve causal effect estimation. If scenario 2, that is, dependence of the treatment and the outcome on the manifest covariate, holds, using manifest covariates in the adjustment should lead to less biased effect estimates than using latent covariates. Similar effect estimates when using manifest or latent covariates would support the hypothesis that scenario 3 or 4 holds. The authors used proxy-pretest measures in English and math, since in previous analyses they have proven to be relevant covariates for adjustment (Steiner *et al.*, 2010, Cook *et al.*, 2009) and since these were assessed by a number of test items (indicators) allowing for modeling latent covariates. The manifest covariates in this study contained considerable measurement error (reliability estimates ranged from 0.70 to 0.75 for the English pretest and from 0.46 to 0.59 for the math pretest). Adjusting for the latent instead of the manifest covariates improved *ATE* estimation up to 22%. The results suggest that in the empirical application, the treatment probability and the outcome variable depend on the latent covariates (proxy pretest ability in English and math), that is that scenario 1 holds. Note that this may be different in other settings.

15.4 APPROACHES ACCOUNTING FOR LATENT COVARIATES

If the treatment probability and the outcome variable depend on the latent covariate instead of the manifest covariate, that is, if scenario 1 holds, the latent covariate needs to be used for adjustment. For both, PS methods and ANCOVA, new approaches, which allow adjustment for latent covariates, have been proposed. Both the PS approach of Raykov (2012) and the ANCOVA approach of Steyer and Partchev (2008) rely on the use of a measurement model for the latent covariates. The approaches are introduced using measurement models in the tradition of classical test theory (CTT; Lord *et al.*, 1968, Steyer *et al.*, 2015). They can, however, easily be extended to incorporate other measurement models such as those of item response theory (IRT; e.g., Embretson and Reise, 2000). For modeling the true-score variable of the fallible covariate, it is necessary that multiple manifest measures (indicators) V_i with $i = 1, \ldots, s$ of the latent covariate are available. These manifest measures are, for example, the observed scores on at least two (parallel) tests, items, or item parcels assessing the same construct. In CTT, each manifest indicator V_i is defined as the sum of a true-score variable η_i and an error variable ϵ_i. In the latent variable approach, based on the s manifest indicators, a measurement model is specified for the estimation of a latent variable η. Each true-score variable η_i is assumed to be a linear function of the latent variable η, that is $\eta_i = \nu_i + \lambda_i * \eta$, with ν_i and λ_i representing the indicator specific intercept and factor loading. A latent variable model is specified for all q latent covariates η_k, involving all s manifest indicators V_{ki} of each covariate η_k. This model is specified for the treatment group and the control group separately, assuming strong measurement invariance (i.e., equal factor loadings λ_{ki} and intercepts ν_{ki} between the groups, Millsap, 2011). This is the same for both methods, PS and ANCOVA.

15.4.1 Latent Covariates in Propensity Score Methods

Raykov (2012) distinguishes between p perfectly reliable covariates $W := (W_1, \ldots, W_p)$ and q fallible covariates $V := (V_1, \ldots, V_q)$. While in traditional PS analyses, the PS is usually estimated based on all observed covariates W and V, the modified PS (π) is defined using the vector of perfectly reliable covariates W and the underlying latent variables η_k of the q fallible covariates V_k with $k = 1, \ldots, q$:

$$\pi = P(X = 1 | W, \eta_1, \ldots, \eta_q). \tag{15.8}$$

Raykov (2012) outlines the estimation of the modified PS in two steps. First, a latent variable model is specified for all q latent covariates η_k. Raykov then uses factor score estimates as proxies for the individual values on the latent covariates η_k. These factor score estimates are then used in a second step for estimating the modified PS. Therefore, all perfectly measured observed covariates W and the estimated factor scores for all q latent variables η_k are incorporated as predictors in the PS model (e.g., in a logistic regression). In other words, Raykov does not use the latent covariates for the estimation of the PS but the estimated scores of the latent covariates, which are weighted sum scores of the manifest variables.

In PS analyses, if all confounding covariates are considered when estimating the PS, the PS balances all pretreatment group differences. Balance means that given a specific PS, the joint distribution of the covariates is the same in the treatment and control group (e.g., Rosenbaum and Rubin, 1983, 1984). The appropriateness of the estimated PS is usually checked via balance criteria (Rubin, 2001, Stuart and Rubin, 2007). For the modified PS, balance has to be checked on the observed error-free covariates W and all q latent variables η_k. Accordingly, in Raykov's approach balance can be checked on the observed error-free covariates W and the estimated factor scores of all q latent variables η_k.

Rubin (2001) furthermore argues that causal inference can only be drawn for the population described by the area of common support. Common support refers to the population for which every unit has the chance (i.e., a nonzero probability) to be in both the treatment and the control group. Cases with no common support are, for instance, persons with severe symptoms that always get into treatment and have no chance to be in the control group. Common support is indicated by the overlap of the PS distribution between treatment and control group and nonoverlapping persons are commonly discarded (e.g., Rubin, 2001).

The estimated PS is then used in a further step for estimating the *ATE*. This can be done via PS regression, inverse-propensity weighting, PS subclassification, or PS matching (e.g., Rosenbaum and Rubin, 1983, Schafer and Kang, 2008). In a simulation study, Raykov (2012) showed that when the treatment probability and the outcome variable depend on the latent covariates, that is, when scenario 1 holds, using the modified PS results in more accurate *ATE* estimates than a PS score that is based on manifest covariates. As factor score estimates are just proxies for the scores of the latent covariates, the *ATE* estimate using factor score estimates will be less accurate than the *ATE* estimate using the latent covariates. Furthermore, note that, as this is a three-step approach ((i) estimation of factor scores, (ii) estimation of the propensity score, and (iii) estimation of the *ATE*), no standard errors that incorporate all three steps are available, yet.

15.4.2 Latent Covariates in ANCOVA Models

In the analysis of covariance, covariates are directly incorporated in the outcome model. Steyer and Partchev (2008; with further developments from Kröhne, 2010, Mayer, 2015, and Mayer *et al.*, 2016) provide the software EffectLite to estimate these regression models within the framework of structural equation modeling. The version of EffectLite that relies on the latent variable software Mplus (Muthén and Muthén, 1998/2012) or LISREL (du Toit *et al.*, 2005) is available from `http://www.causal-effects.de/`. The first version of the R-package *EffectLiteR* (Mayer *et al.*, 2016), that relies on the R-package lavaan (Rosseel, 2012), is available from `https://github.com/amayer2010/EffectLiteR`.

The ANCOVA model is specified using all perfectly reliable covariates W as well as the underlying latent variables η_k of all q fallible covariates V_k with $k = 1, \ldots, q$:

$$E(Y|X, W, \eta_1, \ldots, \eta_q) = g_0(W, \eta_1, \ldots, \eta_q) + g_1(W, \eta_1, \ldots, \eta_q)X. \tag{15.9}$$

Instead of using factor score estimates as covariates, the latent variables are incorporated directly in the ANCOVA model. The ANCOVA model is specified as a multigroup model with the treatment variable defining the groups. The conditional treatment effect variable is represented by $g_1(W, \eta_1, \ldots, \eta_q) = E^{X=1}(Y|W, \eta_1, \ldots, \eta_q) - E^{X=0}(Y|W, \eta_1, \ldots, \eta_q)$. If unbiasedness of the conditional expectations (Eq. (15.3)) holds for $Z^* = (W, \eta_1, \ldots, \eta_q)$, then $E[g_1(W, \eta_1, \ldots, \eta_q)] = ATE$. The generalized ANCOVA model in EffectLite allows for modeling nonlinear relationships and covariate-treatment interactions as well as for heteroscedasticity across treatment groups (e.g., Kröhne, 2010, Steyer and Partchev, 2008). Furthermore, stochastic rather than fixed covariates can be modeled in EffectLite, which is important for obtaining correct standard error estimates (Kröhne, 2010). In applications using ANCOVA, effects are typically estimated for the entire target population by extrapolating also to regions of nonoverlap. This extrapolation is based on the assumptions on the functional form made in the model. In contrast to the PS-approach, conclusions using ANCOVA are usually drawn for the whole population of interest. For perfectly reliable covariates, Kröhne (2010) showed a good performance of the generalized ANCOVA model in EffectLite regarding point estimation and estimation of standard errors of the *ATE*.

15.4.3 Performance of the Approaches in an Empirical Study

Sengewald *et al.* (2016) also evaluated whether generalized ANCOVA and PS methods for latent variables improve *ATE* estimation in empirical settings. The authors used the WSC of Pohl *et al.* (2009) and estimated the *ATE* adjusting for the proxy-pretest in English and math. Instead of using the observed sum score of the pretests, the authors also adjusted for pretreatment differences by incorporating the latent variables for the pretests. For that the authors proposed measurement models on item parcels for the pretests that were assumed to be measurement invariant across the two treatment groups. The reliability of the test halves ranged from 0.70 to 0.75 for the proxy-pretest in English and from 0.46 to 0.59 for the proxy-pretest in math. Generalized ANCOVA showed a good performance when adjusting for the latent covariates; 22% more bias could be reduced when using latent instead of manifest pretest values. For PS methods, a direct comparison of PS methods using latent covariates (latent PS) and manifest covariates (manifest PS) is limited due to different PS models with differences in the area of common support and the achieved balance. While with generalized ANCOVA the *ATE* is estimated for the whole population, with PS methods conclusions are only drawn for the area of common support. It is to note that the area of common support, and thus the population for which inferences are drawn, differs between using manifest and latent covariates. This has to be considered in the interpretation of the results. As far as the results could be compared, latent PS methods seemed to reduce more bias than manifest PS.

Summarizing the results, approaches for modeling latent covariates in ANCOVA as well as in PS analyses show good performance in simulation studies as well as in an empirical application. Note that, all latent covariate models can only be applied if multiple indicators for each latent covariate are available. For situations where multiple indicators of fallibly measured covariates are not available, most recently alternative methods were developed. These rely on assumptions either about the structure and extent of measurement error (Kuroki and Pearl, 2014, McCaffrey *et al.*, 2013; Stuart, 2013; Yi *et al.*, 2012) or about the functional form between the extent of measurement error and bias (Lockwood and McCaffrey, 2013, 2014).

15.5 THE IMPACT OF ADDITIONAL COVARIATES ON THE BIASING EFFECT OF A FALLIBLE COVARIATE

So far, research on the impact of adjusting for manifest instead of latent covariates on causal effect estimation, pertaining to scenario 1, has mainly focused on the fallible covariate itself. Researchers dealt with the quantification of the effect when manifest instead of latent covariates are used for adjustment as well as the development and evaluation of methods incorporating latent covariates. However, additional covariates may have an impact on the bias induced by using a manifest instead of the respective latent covariate. This was already acknowledged by Aiken and West (1991) as well as Cohen *et al.* (2003). So far, systematic research on the effect of additional covariates in the context of accounting for fallible covariates is rare.

From research on *unobserved* covariates, it is known that additional covariates can reduce as well as amplify hidden bias and variance of effect estimates (e.g., Brookhart *et al.*, 2006, Kelcey, 2011, Pearl, 2010, Rubin and Thomas, 1996,

Stuart and Rubin, 2007). Covariate "measurement error can be treated as simply contributing to the problem of hidden bias" (West and Thoemmes, 2010, p. 34). That is, in settings where relevant covariates are fallible and the treatment probability as well as the outcome variable depend on the respective latent covariate (scenario 1), covariates are partly hidden (in the most extreme case they are fully hidden, i.e., unobserved). Thus, results from research on unobserved covariates may also apply to the problem of measurement error in relevant covariates.

To reduce hidden bias due to omitted (i.e., unobserved) relevant confounders, studies (e.g., Rubin and Thomas, 1996, Stuart and Rubin, 2007) recommend including covariates that determine treatment assignment or the outcome variable. However, additional covariates may also amplify bias. In addition to collider variables that have a complex relationship with (unobserved) covariates (e.g., Cole *et al.*, 2010; Pearl, 2013), recently a new class of bias amplifying variables (i.e., instrumental variables; Pearl, 2010, Pearl, 2013; Steiner and Yongnam, 2014) has been investigated. Instrumental variables are variables that are only related to the treatment variable but not directly to the outcome variable or to the omitted confounder. Adjusting for instrumental variables can amplify bias that exists due to omitting a relevant covariate.

Next to having an impact on bias of effect estimates, additional covariates may also have an impact on the variance of effect estimates. Brookhart *et al.* (2006) outline that even if all relevant covariates are controlled for and an unbiased ATE can be estimated, additional covariates that are only related to the treatment variable can increase variance of effect estimates and, thus, decrease efficiency. In contrast, using additional covariates that are only related to the outcome variable can decrease unexplained outcome variance and increase efficiency of effect estimates. Similar conclusions were drawn in the context of missing data problems. Collins *et al.* (2001) showed that an additional auxiliary variable, a variable that is not related to the missing process itself but to the outcome on which missing values occur, can increase efficiency of estimates and may compensate for bias due to missing values on the outcome variable.

If results from research on unobserved covariates also apply to the problem of unreliably measured covariates (i.e., the fallible covariate Z is a partly observation of the latent covariate η), an additional covariate W could reduce the bias occurring due to using the manifest covariate Z instead of the latent relevant covariate η for adjustment. Thereby, the relationship of the additional covariate W with the treatment variable X and the outcome variable Y can affect the accuracy of ATE estimation. If the additional covariate is related only to the treatment variable but not to the outcome variable, the additional covariate could even amplify the biasing effect of using the manifest instead of the latent covariate. In empirical analyses and simulation studies, the impact of using additional covariates when relevant covariates are fallible was investigated.

15.5.1 Investigation of the Impact of Additional Covariates in an Empirical Study

Using the WSC-data of Pohl *et al.* (2009) and Pohl and Sengewald (2014) investigated the impact of additional covariates in a model with a fallible relevant covariate.

The authors considered the causal effect of English training in the nonrandomized experiment and the proxy-pretest in English as a relevant fallible covariate. The English pretest was used as both, a manifest covariate (in form of the mean score across all items) and a latent covariate (using a measurement model) in the adjustment model. Three additional covariates that were relevant for adjustment and three covariates that were not relevant for adjustment were also included in the model. The relevant covariates were the attitude toward English (Like Eng), Grade in English (Grade Eng), and self-rated knowledge in English (knowledge Eng), while the nonrelevant covariates were positive affect, negative affect, and grade in biology (Grade biology). The correlations of the covariates with the outcome, the treatment, and the fallible pretest variable are given in Table 15.1. As can be seen in the table, the fallible covariate was highly correlated with the outcome and the treatment variable. Also, the three relevant covariates correlated with both outcome and treatment variable, while the correlations with the irrelevant covariates were negligible. Note that the relevant covariates were also highly correlated with the fallible pretest score in English, whereas the irrelevant covariates were not.

In six outcome models, the *ATE* was estimated using generalized ANCOVA. The outcome models differed in whether the English pretest was included as a manifest or a latent covariate and in which additional covariates were included in the model (none, only relevant ones, or only irrelevant ones). The impact of measurement error on *ATE* estimation was investigated by comparing the estimated bias of the *ATE* estimate between using the manifest or the latent English pretest (see Table 15.2).

Using a manifest instead of a latent English pretest score resulted in 14.29% more bias. Including three additional covariates reduced the biasing effect of measurement error in the English pretest only if the additional covariates were relevant. Note that

TABLE 15.1 Correlation Between the Covariates and the Outcome Variable *Y*, the Treatment Variable *X* and the Fallible Covariate *(Pretest Eng)*.

	Y	*X*	*Pretest Eng*
Fallible covariate			
Pretest Eng	0.77	0.24	1.00
Additional relevant covariates			
Like Eng	0.53	0.25	0.45
Grade Eng	0.41	0.16	0.37
Knowledge Eng	0.56	0.07	0.49
Additional irrelevant covariates			
Positive Affect	0.05	0.10	0.08
Negative Affect	0.10	0.06	0.12
Grade Biology	0.08	0.16	0.12

Abbreviations: Pretest Eng: pretest in English; Like Eng: attitude toward English; Grade Eng: grade in English; Knowledge Eng: self-rated knowledge in English; Positive Affect: positive affect; Negative Affect: negative affect; Grade Biology: grade in biology.

TABLE 15.2 Difference in the Estimated Bias Using Manifest or Latent English Pretest Scores and Including Additional Relevant or Irrelevant Covariates in the Model.

Pretest Eng	Additional Covariates	Bias (%)	Difference (%)
Manifest	None	34.07	14.29
Latent	None	19.78	
Manifest	Three relevant	32.97	12.09
Latent	Three relevant	20.88	
Manifest	Three irrelevant	35.16	14.28
Latent	Three irrelevant	20.88	

Note: Additional relevant covariates are attitude toward English, grade in English, and self-rated knowledge in English; additional irrelevant covariates are positive affect, negative affect, and grade in biology.

in these analyses, the irrelevant covariates were hardly correlated with the fallible pretest. Also note that, although the additional relevant covariates did compensate to some extent for the biasing effect of measurement error in the English pretest, they did not make up for the whole biasing effect. Still a considerable amount of bias remained. Thus, even in the presence of other relevant covariates, it was still necessary to account for the latent instead of the manifest English pretest. The empirical analyses clearly show what impact additional covariates may have in real data. They do, however, not allow to systematically vary certain parameters. In the data, additional covariates were mostly either correlated to the treatment variable, the outcome variable, and the fallible covariate, or they were not correlated with any of the three variables. In order to systematically study the impact of the different relationships of the additional covariate on effect estimates, simulation studies are needed.

15.5.2 Investigation of the Impact of Additional Covariates in Simulation Studies

In research on the impact of additional covariates in settings with an unobserved relevant covariate, it was assumed that the covariates are not correlated among each other but only to the treatment or to the outcome variable. However, the correlation between the unobserved (or partly observed) covariate and the additional covariate does play a great role for the impact of the additional covariate. Assume the extreme case that the unobserved relevant covariate and the additional covariate correlate to one, then the additional covariate could fully reduce the hidden bias. In fact, there is some evidence for that from simulation studies from Steiner *et al.* (2011) and Cook *et al.* (2009). The authors conducted their simulation studies based on real quasi-experimental data and investigated the effect of adding measurement error to several correlated covariates on *ATE* estimates. In the single covariate case, bias reduction was attenuated by 10% for a decrease in covariates reliability of 0.1. Considering measurement error on a whole set of covariates, attenuation of bias reduction reduced to 4–6% for a decrease in the covariates reliability of 0.1. Thus, the authors reported less bias due to measurement error if a set of covariates was used for adjustment compared to the single covariate case. Steiner and Cook (2013) conclude, that a set of (highly) correlated covariates may partially compensate for each other's unreliability depending on the covariate's correlation structure. However, so far this compensating effect of additional covariates on the effect of a fallible covariate was not systematically studied. In a simulation study, Sengewald and Pohl (2016) currently investigate the circumstances under which additional (irrelevant) covariates may reduce the biasing effect of using a manifest instead of a latent relevant covariate on *ATE* estimation. For this purpose, they systematically vary the correlational structure of the variables in the model as well as the amount of measurement error. The goal is to disentangle the bias-amplifying and bias-compensating potential of an additional (irrelevant) covariate on the biasing effect of using manifest instead of latent relevant covariates for adjustment. The results of this study may guide evaluators in the choice of covariates measured before treatment.

15.6 DISCUSSION

In order to appropriately perform adjustment when covariates are fallible, a researcher first needs to evaluate whether the treatment probability and the outcome variable depend on the manifest covariate (scenario 2), the latent covariate (scenario 1), or both (scenarios 3 and 4). Only if both, treatment and outcome depend on the latent covariate (scenario 1), it is necessary to use latent instead of manifest covariates for adjustment. If only one, treatment or outcome depends on the latent covariate and the other one on the manifest one (scenarios 3 and 4), a researcher can use either manifest or latent covariates for adjustment. While in these cases latent covariates may (scenarios 3 and 4) or have to (scenario 1) be used in order to obtain unbiased *ATE* estimates, adjustment using latent covariates is not always warranted. If both, treatment and outcome depend on the manifest covariate (scenario 2), adjusting for latent instead of manifest covariates can induce bias. In an application, a researcher needs to judge whether this may be the case. The treatment probability is most likely determined by the manifest covariate when selection into treatment depends on observed test scores, for instance, when a doctor decides on the participation of a patient in a special treatment based on results of a blood test. Treatment selection most likely depends on the latent covariate when participants self-select into treatment or when a teacher decides on the necessity of an extra training for the student based on his or her opinion of the student. Dependence of the outcome variable on the manifest covariate can, for instance, occur when the measurement error in the covariate (e.g., being more or less tired, resulting in a higher or lower pretest score) impacts the outcome variable (e.g., the motivation to perform well in the posttest). It may also be present when the same test instrument is used for both the covariate and the outcome variable and the measurement errors of both scores are correlated (e.g., due to method effects). This may especially occur for pretest measures as covariates. If none of these possible dependencies of the outcome variable on the manifest covariate are plausible, the outcome variable most likely depends on the latent covariate.

For the majority of applications, it is plausible that the outcome variable depends on latent instead of the fallible manifest covariates. In applications in which treatment selection is not based on observed pretest scores, it is also very plausible that the treatment variable depends on latent covariates instead of fallible manifest covariates.

Thus, we think that for most applications adjustment based on latent covariates is warranted – as it was the case in the considered empirical study. If the treatment probability and the outcome variable depend on the latent covariates, it is worthwhile to adjust for latent instead of manifest covariates. Thus, in practice, researchers should try to assess possibly fallible relevant covariates with multiple indicators (e.g., items) in order to be able to evaluate the amount of measurement error and to use latent variables for adjustment. If multiple indicators are assessed, a researcher can also specify two adjustment models – one with the manifest and one with the latent covariate – and compare the resulting *ATE* estimates, in order to evaluate the possible impact of model misspecification in terms of using manifest or latent covariates. For estimating treatment effects, a researcher can use ANCOVA as well as PS methods. The choice between the approaches may be guided by the questions (i) whether the researcher feels more confident with finding a correct model for the treatment selection (PS methods) or for the outcome (ANCOVA), (ii) in case of no perfect overlap, whether the researcher wants to draw inferences for the whole population relying on the assumption of extrapolation (ANCOVA) or only for the area of common support, that represents just a subset of persons (PS methods),[2] (iii) whether for reasons of prevention of manipulation, the outcome need not be available when finding the adjustment model (PS methods) or whether this is not an issue (ANCOVA), and (iv) whether the researcher is interested in the estimation of appropriate standard errors with latent covariates (ANCOVA) or not (PS methods). The proposed ANCOVA and PS method for modeling latent covariates also differ (v) in the way latent variables are used for adjustment. While in ANCOVA the latent covariate is directly used for adjustment, in the proposed PS method although the latent covariate is modeled, it is not directly used for adjustment. Instead, only manifest estimated factor scores, which are weighted sum scores of manifest variables, are included in the PS model. Thus, the ANCOVA model may better account for latent covariates than the PS model.

If it is not possible to use multiple indicators for the measurement of a covariate or to model the covariate as a latent variable, a researcher may consider to use additional covariates for adjustment. Of course, if additional covariates are relevant, they should be included in the adjustment model. But even if additional covariates are not relevant, it may be worthwhile to include them, if they are correlated with the fallible relevant covariates and with the outcome. Ideally, this is already considered when planning the study and deciding on the collected covariates. The empirical application showed that the biasing effect of using a manifest instead of a latent relevant covariate could only partly be compensated by other covariates. It is also to note that additional covariates also have the potential to amplify this biasing effect and increase variance of effect estimates. Thus, when possible, the first choice of a researcher should be to strive for collecting data on multiple indicators for fallible, possibly relevant covariates and to use latent variable modeling.

[2]The question of extrapolation or drawing inferences only for the area of common support is only relevant, when there is no perfect overlap, that is, when there are persons that have a probability of zero to get into the treatment group or the control group. Also note that is it possible to rely on extrapolation when using PS methods or to draw only inferences for the area of common support in ANCOVA analyses. This is, however, less commonly done in practice.

REFERENCES

Aiken, L.S. and West, S.G. (1991) *Multiple Regression: Testing and Interpreting Interactions*, Sage Publications, Newbury Park, CA.

Brookhart, M.A., Schneeweiss, S., Rothman, K.J., Glynn, R.J., Avorn, J., and Stürmer, T. (2006) Variable selection for propensity score models. *American Journal of Epidemiology*, **163** (12), 1149–1156.

Cohen, J., Cohen, P., West, S.G., and Aiken, L.S. (2003) *Applied Multiple Regression/ Correlation Analysis for the Behavioral Sciences*, Erlbaum, Hillsdale, NJ.

Cole, S.R., Platt, R.W., Schisterman, E.F., Chu, H., Westreich, D., Richardson, D., and Poole, C. (2010) Illustrating bias due to conditioning on a collider. *International Journal of Epidemiology*, **39** (2), 417–420.

Collins, L.M., Schafer, J.L., and Kam, C.M. (2001) A comparison of inclusive and restrictive strategies in modern missing data procedures. *Psychological Methods*, **6** (4), 330–351.

Cook, T.D., Steiner, P.M., and Pohl, S. (2009) How bias reduction is affected by covariate choice, unreliability, and mode of data analysis: Results from two types of within-study comparisons. *Multivariate Behavioral Research*, **44** (6), 828–847.

Embretson, S.E. and Reise, S.P. (2000) *Item Response Theory for Psychologists*, Erlbaum, London.

Holland, P.W. (1986) Statistics and causal inference. *Journal of the American Statistical Association*, **81** (396), 945–960.

Kelcey, B. (2011) Covariate selection in propensity scores using outcome proxies. *Multivariate Behavioral Research*, **46** (3), 453–476.

Kröhne, J.U. (2010) Estimation of average total effects in quasi-experimental designs: Nonlinear constraints in structural equation models, PhD thesis, Friedrich Schiller Universität Jena, Diss., 2010, Jena.

Kuroki, M. and Pearl, J. (2014) Measurement bias and effect restoration in causal inference. *Biometrika*, **101** (2), 423–437.

Lockwood, J. and McCaffrey, D. (2013) SIMEX for weighting and matching applications with error-prone covariates, *Paper presented at Society for Research on Educational Effectiveness Fall 2013 Conference, Washington, DC*. Retrieved from `https://www.sree.org/download/files/1414578570w0_conf_976.pdf`.

Lockwood, J. and McCaffrey, D.F. (2014) Correcting for test score measurement error in ANCOVA models for estimating treatment effects. *Journal of Educational and Behavioral Statistics*, **39** (1), 22–52.

Lord, F.M., Novick, M.R., and Birnbaum, A. (1968) *Statistical Theories of Mental Test Scores*, Addison-Wesley, Reading, MA.

Mayer, A. (2015) EffectLiteR, Unpublished R package. Retrieved from `https://github.com/amayer2010/effectliter`.

Mayer, A., Dietzfelbinger, L., Rosseel, Y., and Steyer, R. (2016) The EffectLiteR approach for analyzing average and conditional effects, Manuscript submitted for publication.

Mayer, A., Thoemmes, F., Rose, N., Steyer, R., and West, S.G. (2014) Theory and analysis of total, direct, and indirect causal effects. *Multivariate Behavioral Research*, **49** (5), 425–442.

McCaffrey, D.F., Lockwood, J., and Setodji, C.M. (2013) Inverse probability weighting with error-prone covariates. *Biometrika*, **100** (3), 671–680.

Millsap, R.E. (2011) *Statistical Approaches to Measurement Invariance*, Routledge, New York.

Muthén, L. and Muthén, B. (1998/2012) *Mplus User's Guide*, 7th edn [*Computer software manual*], Muthén & Muthén, Los Angeles, CA.

Neyman, J. (1923/1990) On the application of probability theory to agricultural experiments. Essay on principles. Section 9. *Statistical Science*, **5**, 465–480. Originally published 1923.

Pearl, J. (1995) Causal diagrams for empirical research. *Biometrika*, **82** (4), 669–688.

Pearl, J. (2009) *Causality: Models, Reasoning, and Inference*, Cambridge University Press, Cambridge.

Pearl, J. (2010) On a class of bias-amplifying covariates that endanger effect estimates, (Tech. Rep. R-356), University of California, Los Angeles, CA.

Pearl, J. (2013) Linear models: A useful "microscope" for causal analysis. *Journal of Causal Inference*, **1** (1), 155–170.

Pohl, S. and Sengewald, M.A. (2014) On the importance of adjustment for latent covariates, *Presentation at the Conference on Statistics and Causality*, Vienna, Austria.

Pohl, S., Steiner, P.M., Eisermann, J., Soellner, R., and Cook, T.D. (2009) Unbiased causal inference from an observational study: Results of a within-study comparison. *Educational Evaluation and Policy Analysis*, **31** (4), 463–479.

Raykov, T. (2012) Propensity score analysis with fallible covariates: A note on a latent variable modeling approach. *Educational and Psychological Measurement*, **72** (5), 715–733.

Rosenbaum, P.R. and Rubin, D.B. (1983) The central role of the propensity score in observational studies for causal effects. *Biometrika*, **70** (1), 41–55.

Rosenbaum, P.R. and Rubin, D.B. (1984) Reducing bias in observational studies using subclassification on the propensity score. *Journal of the American Statistical Association*, **79** (387), 516–524.

Rosseel, Y. (2012) lavaan: An R package for structural equation modeling. *Journal of Statistical Software*, **48** (2), 1–36, Retrieved from `http://www.jstatsoft.org/v48/i02/`.

Rubin, D.B. (1974) Estimating causal effects of treatments in randomized and nonrandomized studies. *Journal of Educational Psychology*, **66** (5), 688.

Rubin, D.B. (2001) Using propensity scores to help design observational studies: Application to the tobacco litigation. *Health Services and Outcomes Research Methodology*, **2**, 169–188.

Rubin, D.B. (2005) Causal inference using potential outcomes: Design, modeling, decisions. *Journal of the American Statistical Association*, **100**, 322–331.

Rubin, D.B. and Thomas, N. (1996) Matching using estimated propensity scores: Relating theory to practice. *Biometrics*, **52** (1), 249–264.

Schafer, J.L. and Kang, J. (2008) Average causal effects from nonrandomized studies: A practical guide and simulated example. *Psychological Methods*, **13** (4), 279–313.

Sengewald, M.-A. and Pohl, S. (2016) Compensating for attenuation bias: The impact of additional covariates. Manuscript in preparation.

Sengewald, M.-A., Pohl, S., and Steiner, P.M. (2016) On the importance of adjusting for latent covariates, Manuscript submitted for publication.

Shadish, W.R., Clark, M.H., and Steiner, P.M. (2008) Can nonrandomized experiments yield accurate answers? A randomized experiment comparing random and nonrandom assignments (with comments by Little/Long/Lin, Hill, and Rubin, and a rejoinder). *Journal of the American Statistical Association*, **103** (484), 1334–1356.

Steiner, P.M. and Cook, D.L. (2013) Matching and propensity scores, in *The Oxford Handbook of Quantitative Methods*, vol. 1, Foundations (ed. T.D. Little), Oxford University Press, New York.

Steiner, P.M., Cook, T.D., and Shadish, W.R. (2011) On the importance of reliable covariate measurement in selection bias adjustments using propensity scores. *Journal of Educational and Behavioral Statistics*, **36** (2), 213–236.

Steiner, P.M., Cook, T.D., Shadish, W.R., and Clark, M.H. (2010) The importance of covariate selection in controlling for selection bias in observational studies. *Psychological Methods*, **15** (3), 250–267.

Steiner, P.M. and Yongnam, K. (2014) On the bias-amplifying effect of near instruments in observational studies, *Paper presented at the Society for Research on Education Effectiveness Spring 2014 Conference*, Washington, DC. Retrieved from `https://www.sree.org/conferences/2014s/program/downloads/abstracts/1197.pdf`.

Steyer, R., Gabler, S., von Davier, A.A., and Nachtigall, C. (2000a) Causal regression models II: Unconfoundedness and causal unbiasedness. *Methods of Psychological Research Online*, **5** (3), 55–86.

Steyer, R., Gabler, S., von Davier, A.A., Nachtigall, C., and Buhl, T. (2000b) Causal regression models I: Individual and average causal effects. *Methods of Psychological Research Online*, **5** (2), 39–71.

Steyer, R., Mayer, A., and Fiege, C. (2014) Causal inference on total, direct, and indirect effects, in *Encyclopedia of Quality of Life and Well-Being Research* (ed. A.C. Michalos), Springer-Verlag, Dordrecht, The Netherlands, pp. 606–631.

Steyer, R., Mayer, A., Geiser, C., and Cole, D.A. (2015) A theory of states and traits-revised. *Annual Review of Clinical Psychology*, **11**, 71–98.

Steyer, R., Nachtigall, C., Wüthrich-Martone, O., and Kraus, K. (2002) Causal regression models III: Covariates, conditional, and unconditional average causal effects. *Methods of Psychological Research Online*, **7** (1), 41–68.

Steyer, R. and Partchev, I. (2008) *EffectLite: User's Manual*, Department of Methodology and Evaluation Research, Jena, Germany. Retrieved from `http://www.causal-effects.de/`.

Stuart, E.A. (2013) Strategies for dealing with covariate measurement error for propensity scores. *Paper presented at the Society for Research on Education Effectiveness Spring 2013 Conference*, Washington, DC Retrieved from `https://www.sree.org/conferences/2013s/program/downloads/abstracts/915_3.pdf`.

Stuart, E.A. and Rubin, D.B. (2007) Best practices in quasi-experimental designs: Matching methods for causal inference, in *Best Practices in Quantitative Methods* Chapter 11 (ed. J. Osborne), Sage Publications, Thousand Oaks, CA, pp. 155–176.

du Toit, S., du Toit, M., Mels, G., and Cheng, Y. (2005) *LISREL for Windows: PRELIS User's Guide*, Scientific Software International, Lincolnwood, IL.

West, S.G. and Thoemmes, F. (2010) Campbell's and Rubin's perspectives on causal inference. *Psychological Methods*, **15** (1), 18–37.

Yi, G., Ma, Y., and Carroll, R.J. (2012) A functional generalized method of moments approach for longitudinal studies with missing responses and covariate measurement error. *Biometrika*, **99** (1), 151–165.

16

LATENT CLASS ANALYSIS WITH CAUSAL INFERENCE: THE EFFECT OF ADOLESCENT DEPRESSION ON YOUNG ADULT SUBSTANCE USE PROFILE

STEPHANIE T. LANZA

Department of Biobehavioral Health and The Methodology Center, The College of Health and Human Development, The Pennsylvania State University, University Park, PA, USA

MEGAN S. SCHULER

Department of Health Care Policy, Harvard Medical School, Boston, MA, USA

BETHANY C. BRAY

The Methodology Center and The College of Health and Human Development, The Pennsylvania State University, University Park, PA, USA

16.1 INTRODUCTION

Latent class analysis (LCA) is an effective tool for identifying population subgroups characterized by particular patterns of responses on a set of observed variables. The resultant classes explain heterogeneity in individuals' responses to the set of observed variables. These population subgroups are latent, that is, otherwise unobservable in a population (Collins and Lanza, 2010). Prior studies have used LCA to characterize complex drug use behavior patterns (e.g. Kuramoto *et al.*, 2011, Lanza and Bray, 2010, Lanza *et al.*, 2010), sexual risk behavior patterns (e.g. Lanza and Collins, 2008, Vasilenko *et al.*, 2014), and early profiles of risk (e.g. Cooper and Lanza, 2014, Lanza and Rhoades, 2013).

Statistics and Causality: Methods for Applied Empirical Research, First Edition.
Wolfgang Wiedermann and Alexander von Eye.

Predicting latent class membership is often of interest; an LCA with covariates models the association between classes and predictors. This can be useful for describing individual characteristics that are associated with class membership (i.e., describing who is in each latent class), providing predictive or concurrent validation of the latent class variable by demonstrating that anticipated associations between correlates and class membership hold, or using longitudinal data to identify early factors that indicate likely class membership as individuals develop. This third motivation for estimating LCA with covariates can be useful for informing the use of intervention programs that target individuals who may, for example, develop problematic behavior profiles later in life.

Unless individuals are randomly assigned to levels of the predictor of latent class membership, causation may not be inferred from the estimated associations obtained from LCA with covariates. However, modern causal inference methods exist to adjust for potential confounding in observational data. Propensity score methods represent one general approach to drawing causal inferences from observational data. These methods model the selection process into levels of the treatment or exposure of interest in order to control for many potential confounders of the observed relation between the exposure and an outcome (Rosenbaum, 2002, Rosenbaum and Rubin, 1983). These propensity scores can be used to create weights that, when applied to an outcome analysis, such as a regression model estimating the effect of an exposure on an outcome, reweight the data in such a way that the exposure groups are balanced on a large set of potential confounders. This allows causation to be inferred. Propensity score analysis has been adopted quite widely by behavioral scientists in investigations involving manifest variables; however, only recently has this approach been integrated with latent variable modeling (see Butera *et al.*, 2014, Lanza *et al.*, 2013a, Schuler *et al.*, 2014). The purpose of this chapter is to describe this integration of propensity score analysis in LCA with covariates.

In the current investigation, we are using data from a National longitudinal study of US adolescents and young adults to investigate the association between adolescent depression risk and early adult substance use. Latent classes will be derived on the basis of multiple substance use behaviors that together describe a young adult's substance use profile. Adolescent risk for clinical depression will be examined as the predictor of interest; because depression risk is not randomized, propensity score weighting will be used to ensure the comparability of the at-risk and not at-risk groups. In addition, gender will be examined as a moderator of the causal effect.

We begin with a brief introduction to the latent class mathematical model, followed by a description of propensity score analysis for drawing causal inferences from observational data. We then move to our empirical demonstration of incorporating propensity score weighting with LCA with covariates in order to estimate the causal effect of adolescent depression risk on adult substance use class membership. Gender is considered as a moderator of the average causal effect (ACE) of adolescent depression risk on young adult substance use profile. We conclude with commentary on this integrated approach and recommendations for future research.

16.2 LATENT CLASS ANALYSIS

Suppose we observe M ($m = 1, 2, \ldots, M$) indicators of adult substance use for individual i ($i = 1, 2, \ldots, n$), and that each indicator has R_m ($r_m = 1, 2, \ldots, R_m$) response options. Let $\boldsymbol{u}$ represent a response pattern (i.e., a vector of possible responses to the observed indicators); similarly, let $\boldsymbol{U}$ represent the array of all possible $\boldsymbol{u}$s. Each response pattern $\boldsymbol{u}$ corresponds to a cell of the contingency table formed by cross-tabulating all of the observed indicators, and the length of the $\boldsymbol{U}$ array is equal to the number of cells in this table. A particular response pattern to all indicators of adult substance use given by individual i is denoted $\boldsymbol{u}_i$. Let us also establish an indicator function $I(u_m = r_m)$ that equals 1 when the response to indicator $m = r_m$ and equals 0 otherwise. The latent class model with K ($c = 1, 2, \ldots, K$) latent classes can be expressed as

$$P[\boldsymbol{U}_i = \boldsymbol{u}_i] = \sum_{c=1}^{K} \gamma_c \prod_{m=1}^{M} \prod_{r_m=1}^{R_m} \rho_{mr_m|c}^{I(u_m=r_m)}$$

where γ_c is the probability of membership in latent class c and $\rho_{mr_m|c}^{I(u_m=r_m)}$ is the probability of response r_m to indicator m, conditional on membership in latent class c. The γ parameters represent a vector of latent class membership probabilities that sum to 1. The ρ parameters represent a matrix of item response probabilities conditional on latent class membership. This model assumes conditional independence of the observed indicators given latent class; this implies that within each latent class, the M indicators are independent of one another.

16.2.1 LCA With Covariates

Covariates, which also might be referred to as predictors or exogenous variables, can be incorporated into the latent class model so that they may be used to predict latent class membership (Collins and Lanza, 2010, Dayton and Macready, 1988). Typically, covariates are added to the model via baseline-category multinomial logistic regression (e.g. Agresti, 2013); note that this reduces to binomial logistic regression when the number of classes equals 2. Variants of this prediction model are available, however, including adding covariates via binomial logistic regression, which collapses latent classes to assess the effect of a covariate on membership in one latent class versus the remaining latent classes (Lanza *et al.*, 2013a). As with other generalized linear models, covariates in LCA can be discrete, continuous, or higher order terms (e.g., interactions or powers). LCA with covariates is equivalent to standard multinomial logistic regression analysis, except that the categorical outcome is modeled as a latent variable, rather than manifest.

Suppose we observe P ($p = 1, 2, \ldots, P$) covariates of interest that we want to use to predict latent class membership. Let $\boldsymbol{x}$ represent a response pattern (i.e., a vector of possible responses to the covariates); similarly, let $\boldsymbol{X}$ represent the array of all possible $\boldsymbol{x}$s. A particular response pattern to all covariates of interest given by individual i is denoted $\boldsymbol{x}_i$ (i.e., $\boldsymbol{X}_i = \boldsymbol{x}_i = x_{i1}, x_{i2}, \ldots, x_{iP}$). The latent class model with K latent classes and P covariates of interest can be expressed as

$$P[\boldsymbol{U}_i = \boldsymbol{u}_i | \boldsymbol{X}_i = \boldsymbol{x}_i] = \sum_{c=1}^{K} \gamma_c(\boldsymbol{X}_i) \prod_{m=1}^{M} \prod_{r_m=1}^{R_m} \rho_{mr_m|c}^{I(u_m=r_m)}$$

where $\gamma_c(\boldsymbol{X}_i)$ is a baseline-category multinomial logistic regression model expressed as

$$\gamma_c(\boldsymbol{X}_i) = P[C = c | \boldsymbol{X}_i = \boldsymbol{x}_i] = \frac{\exp[\beta_{0c} + \beta_{1c} x_{i1} + \cdots + \beta_{Pc} x_{iP}]}{1 + \sum_{c'=1}^{K-1} \exp[\beta_{0c'} + \beta_{1c'} x_{i1} + \cdots + \beta_{Pc'} x_{iP}]}$$

for $c' = 1, 2, \ldots, K-1$ and latent class K designated as the reference class. Technical details on LCA with covariates can be found in a variety of resources (e.g. Collins and Lanza, 2010, Lanza *et al.*, 2007). Software options for LCA with covariates include PROC LCA (Lanza *et al.*, 2013b), Latent Gold (Vermunt and Magidson, 2005), and Mplus (Muthén and Muthén, 1998–2012).

16.2.1.1 ***Grouping Variables in LCA: Moderated Effects*** Moderation occurs when covariates have differential associations with latent class membership for different groups of individuals. For example, adolescent depression may be differentially associated with adult substance use profiles for males and females. Questions of moderation can be addressed directly in LCA by adding a grouping variable to the model that includes covariates. Logistic regression coefficients and corresponding confidence intervals are estimated within each group.

Grouping variables (e.g., gender) can be incorporated into the latent class model so that they may be used to test measurement invariance in the latent class structure across groups, whether the distribution of the latent classes is the same across groups, and whether the association between a covariate and the latent class variable is moderated by group membership. When a grouping variable is included in LCA, the γ, ρ, and β parameters may be allowed to vary across groups. The examination of moderation of the effect of a predictor on latent class membership, as we define it, requires that the measurement model be constrained equal across groups (i.e., the ρ parameters do not vary across groups). In this case, the associations of interest are the group-specific β parameters.

Suppose we observe a grouping variable where $G_i = g_i$ represents the group to which individual i belongs. The latent class model with K latent classes, P covariates of interest, and a grouping variable can be expressed as

$$P[\boldsymbol{U}_i = \boldsymbol{u}_i | \boldsymbol{X}_i = \boldsymbol{x}_i, G_i = g_i] = \sum_{c=1}^{K} \gamma_{c|g_i}(\boldsymbol{X}_i) \prod_{m=1}^{M} \prod_{r_m=1}^{R_m} \rho_{mr_m|c,g_i}^{I(u_m=r_m)}$$

where $\gamma_{c|g_i}(\boldsymbol{X}_i)$ is a baseline-category multinomial logistic regression model for group g_i expressed as

$$\gamma_{c|g_i}(\boldsymbol{X}_i) = P[C = c|\boldsymbol{X}_i = \boldsymbol{x}_i, G_i = g_i]$$
$$= \frac{\exp[\beta_{0c|g_i} + \beta_{1c|g_1}x_{i1} + \cdots + \beta_{Pc|g_i}x_{iP}]}{1 + \sum_{c'=1}^{K-1}\exp[\beta_{0c'|g_i} + \beta_{1c'|g_i}x_{i1} + \cdots + \beta_{Pc'|g_i}x_{iP}]}$$

for $c' = 1, 2, \ldots, K - 1$ and latent class K designated as the reference class. Because the γ, ρ, and β parameters are allowed to vary across groups, the γ parameters represent a vector of latent class membership probabilities that sum to 1 within each group; the ρ parameters represent a matrix of item response probabilities conditional on latent class membership and group membership; and the β parameters are baseline-category multinomial logistic regression coefficients expressing covariate effects on latent class membership conditional on group membership.

16.3 PROPENSITY SCORE ANALYSIS

Unlike in randomized studies in which the treatment assignment mechanism is fully known and controlled by the experimenters, the assignment mechanism to the exposure of interest in observational studies is not fully known. The propensity score model is a statistical model of this assignment mechanism; the estimated propensity score can then be used to balance exposure groups with regard to their likelihood of receiving the exposure. The probability of an individual receiving the exposure of interest (e.g., clinical depression risk versus no clinical depression risk) is modeled as a function of many variables believed to confound the association between the exposure and the outcome. Propensity score estimates, denoted $\hat{\pi}_i$ for individual i, are typically obtained by logistic regression of the exposure of interest, T_i, on a set of confounders, although more flexible alternatives such as generalized boosted regression (McCaffrey *et al.*, 2004) and classification and regression trees (Luellen *et al.*, 2005) have also been used.

The ultimate goal of propensity score methods is to equate the exposure groups with respect to the distributions of the covariates included in the propensity score estimates; this is known as achieving balance. Balance is desirable because if the distributions of the covariates (i.e., confounders) are equal across groups, the groups may be compared directly, similar to a randomized experiment. While randomization theoretically creates comparable treatment groups with respect to both measured and unmeasured variables, propensity score methods can only balance groups with respect to measured variables. Thus, it is important to include a comprehensive set of potential confounders in the propensity score model in order to best balance the groups.

In our application, being at risk for clinical depression during adolescence is the exposure of interest; our propensity score model will include a comprehensive set of variables that are potentially predictive of depression risk. The propensity score estimate is simply the predicted probability of clinical depression risk,

$$\hat{\pi}_i = \frac{\exp(\boldsymbol{X}_i^T \hat{\beta})}{1 + \exp(\boldsymbol{X}_i^T \hat{\beta})}$$

from the logistic regression model

$$\log\left(\frac{\pi_i}{1 - \pi_i}\right) = \boldsymbol{X}_i^T \beta$$

where $\boldsymbol{X}_i = [1, \text{confounders}]'$ and $\hat{\beta}$ are the estimated logistic regression coefficients.

Once propensity scores are obtained for all individuals in the sample, the degree of overlap in the estimated propensity scores between the exposure groups should be assessed. Poor propensity score overlap, where the range of propensity scores for individuals in one exposure group does not correspond to the range of propensity scores for individuals in another group, indicates that the groups are too dissimilar to warrant causal inferences. Propensity score distributions that show full overlap indicate that the exposure groups are comprised of a similar range of individuals who can potentially be equated using propensity score methods. There are no specific rules for what constitutes sufficient overlap; however, substantial overlap in the distributions is desirable (McCaffrey *et al.*, 2013).

The primary propensity score techniques for use in the final analysis include matching (Rosenbaum and Rubin, 1985), subclassification (Rosenbaum and Rubin, 1984), and inverse propensity weighting (Robins *et al.*, 1995); in this chapter, we focus on weighting.

16.3.1 Inverse Propensity Weights (IPWs)

Propensity scores can be used to create inverse propensity weights (IPWs) that are designed to balance groups with respect to the set of covariates used to estimate the propensity scores. The ACE is defined as the mean on the outcome if all individuals had one exposure level minus the mean on the outcome if all had another exposure level. When the ACE is the estimand of interest, the IPWs are created using the inverse probabilities of exposure received, where individuals in the exposure group (e.g., clinical depression risk) receive a weight of $w_i = \frac{1}{\hat{\pi}_i}$ and individuals in the control group (e.g., no clinical depression risk) receive a weight of $w_i = \frac{1}{(1-\hat{\pi}_i)}$. Balance between groups after weighting can be assessed by the standardized mean difference (SMD), that is, the standardized difference for a given covariate between the mean for the exposure group and the mean for the control group. SMD values close to 0 reflect that covariate means between the groups are similar; SMD values less than 0.20 in magnitude are typically considered indicative of good balance. The weights are then treated similar to survey weights in all subsequent analyses examining the effect of the exposure on an outcome.

16.4 EMPIRICAL DEMONSTRATION

16.4.1 The Causal Question: A Moderated Average Causal Effect

In our example, by estimating the ACE, we are seeking to answer the following specific research question: *What differences in adult substance use patterns (i.e., latent class membership) are expected if all individuals in the population had been at risk for clinical depression during adolescence, compared to if no individuals in the population had been at risk for clinical depression during adolescence?* Given the potential differential impact of adolescent depression for males and females, gender will be treated as moderator of the causal effect of adolescent depression risk on adult substance use latent class membership.

16.4.2 Participants

The data for this study were drawn from the National Longitudinal Study of Adolescent to Adult Health (Add Health; Harris *et al.*, 2009), a national survey of US youth followed from middle and high school at Wave 1 through early adulthood at Wave 4. Wave 1 of Add Health assessed participants in the 7th–12th grade (1994–1995). Three follow-up surveys were conducted: Wave 2 during 1995–1996 (participants in the 12th grade during Wave 1 were excluded); Wave 3 during 2001–2002; and Wave 4 during 2007–2008. This analysis was restricted to the 1642 individuals who were in grade 11 or 12 at Wave 2 when depression risk was assessed, and who had nonmissing data on at least one Wave 4 indicator of substance use.

16.4.3 Measures

16.4.3.1 Potential Confounders In order to control for the baseline differences between adolescents with and without depression risk at Wave 2, we used inverse propensity weighting to ensure that the two groups would be comparable on 37 covariates measured at Wave 1. Specifically, covariates included demographic factors (age, race/ethnicity), depression symptoms, early substance use behaviors, and factors relating to family, school, and neighborhood climate (see Table 16.1 for a complete list of potential confounders).

16.4.3.2 Exposure Depressive symptoms were measured at Wave 2 using a set of nine items that correspond to items in the Center for Epidemiological Studies Depression (CES-D) scale (Radloff, 1991). For each of the nine items, participants endorsed how frequently they experienced the symptom on a scale of 0 for "never or rarely" to 3 for "most or all of the time." Following prior research (Lehrer *et al.*, 2006, Roberts *et al.*, 1991), a dichotomous indicator reflecting high risk for clinical depression was created with a cut-point of 11 or greater for girls and 10 or greater for boys on a sum score of these nine items. At Wave 2, approximately 10% of males and 16% of females were at risk for clinical depression.

TABLE 16.1 Balance Table Showing Means/Proportions for Each Depression Risk Exposure Group and Standardized Mean Difference (Unweighted and Weighted) for All Potential Confounders.

	Unweighted			Propensity Score Weighted		
	Wave 2 Dep Risk	No Wave 2 Dep Risk	SMD	Wave 2 Dep Risk	No Wave 2 Dep Risk	SMD
Age	16.44	16.44	0.01	16.48	16.44	0.02
Female	63.40%	49.90%	**0.28**	60.20%	52.40%	0.16
White	67.60%	73.30%	0.13	69.50%	71.50%	0.04
Black	20.40%	17.40%	0.08	19.80%	18.50%	0.04
Other	14.40%	10.30%	0.12	12.30%	11.00%	0.04
Hispanic	15.30%	12.00%	0.10	15.70%	12.70%	0.08
Rural	29.00%	29.70%	0.02	31.00%	29.30%	0.04
Suburban	40.20%	37.50%	0.02	34.10%	37.50%	0.04
Urban	27.60%	29.90%	0.05	30.40%	30.20%	0.00
CESD-9 score ($Max = 27$)	10.18	5.35	**1.17**	7.50	6.14	0.19
Past month cigarette use	40.90%	29.70%	**0.24**	35.20%	31.30%	0.08
Lifetime alcohol use	72.80%	64.00%	0.19	67.20%	65.30%	0.04
Past year binge drink	41.20%	33.50%	0.16	35.80%	35.10%	0.02
Lifetime marijuana use	40.60%	32.00%	0.18	40.20%	34.00%	0.13
Lifetime illicit drug use	19.30%	12.40%	0.19	13.40%	14.10%	0.02
Lifetime regular cigarette use	31.90%	23.70%	0.09	26.90%	25.00%	0.04
Past year alcohol problems	16.20%	13.10%	0.14	15.00%	14.40%	0.02
Drunk in past year	43.10%	36.40%	0.10	38.20%	37.60%	0.01
Parent with alcohol problems	18.60%	16.20%	0.06	19.50%	16.30%	0.08

Cigarettes easily available in home	49.80%	31.20%	**0.39**	37.10%	33.40%	0.08
Alcohol easily available in home	37.00%	30.80%	0.13	33.30%	31.60%	0.04
Drugs easily available in home	6.50%	3.00%	0.16	3.40%	3.60%	0.01
# of 3 closest friends who smoke daily	1.23	0.90	**0.29**	1.11	0.96	0.07
# of 3 closest friends who drink monthly	1.50	1.37	0.11	1.37	1.39	0.01
# of 3 closest friends who use marijuana monthly	0.94	0.68	**0.24**	0.76	0.72	0.02
Used cigarettes before age 12	22.20%	16.00%	0.16	21.30%	16.70%	0.12
Used alcohol before age 12	15.40%	10.60%	0.14	11.90%	11.20%	0.02
Family bonding (*Max* = 5)	3.44	3.69	**0.54**	3.62	3.66	0.05
Family receives public assistance	9.70%	7.80%	0.07	9.40%	8.10%	0.05
Mom works 30+ h/wk	70.90%	72.80%	0.04	71.70%	72.50%	0.02
Maternal birth age	25.82	25.60	0.05	25.54	25.6	0.01
School alienation (*Max* = 5)	2.57	2.24	**0.35**	2.43	2.30	0.07
School rejection (*Max* = 5)	2.18	1.79	**0.56**	1.91	1.85	0.05
Feel that friends care (*Max* = 5)	1.92	1.73	**0.24**	1.77	1.76	0.00
Impulsivity (*Max* = 5)	2.22	2.18	0.08	2.23	2.19	0.04
Aggression problems	37.80%	28.40%	**0.20**	32.70%	30.40%	0.03
Neighborhood attitude (*Max* = 5)	3.04	3.13	0.09	3.02	3.12	0.05
Neighborhood happiness (*Max* = 5)	3.14	3.20	0.12	3.15	3.19	0.03
School prejudice (*Max* = 5)	2.60	2.73	0.11	2.73	2.72	0.01
School safety (*Max* = 5)	2.40	2.17	**0.22**	2.38	2.21	0.08

Note: SMD = Standardized mean difference in covariate between exposed and control group. > 0.20 are bolded.

16.4.3.3 ***Outcome*** Substance use latent class membership in early adulthood, assessed at Wave 4, was measured with five indicators. Past year alcohol use was categorized as none (28%), less than weekly (40%), or weekly or more (32%). Past year binge drinking (5 or more drinks in a sitting) was categorized as none (51%) or any (49%). Past-month cigarette use was categorized as none (64%), less than daily (15%), or daily (21%).

Past year marijuana use was categorized as none (79%) or any (21%). Past year illicit drug use was categorized as none (91%) or any (9%).

16.4.3.4 ***Moderator*** To examine a moderated causal effect, the final LCA model with depression risk as a predictor included gender as a grouping variable so that the gender-specific ACE of depression risk on adult substance use latent class membership could be estimated.

16.4.4 Analytic Strategy for LCA With Causal Inference

First, a propensity score model was estimated using logistic regression in which covariates from Wave 1 were used to predict depression risk at Wave 2. We calculated IPWs from the predicted propensity scores as described above. Overlap was examined, and balance between the two depression risk groups was assessed before and after applying IPW by the SMD, with values 0.20 in magnitude considered to be indicative of good balance.

After ensuring that there was sufficient overlap between exposure groups and that the IPWs yielded well-balanced exposure groups, we next fit the latent class measurement model. We considered LCA models with one through six classes; latent class model selection was performed with IPWs and treating gender as a grouping variable. Model selection was performed both allowing the classes to be freely estimated across genders and constraining the classes to be equal across genders. The final model was selected based on fit statistics (BIC, AIC, adjusted BIC, G^2, and entropy), as well as interpretability of classes (see Collins and Lanza, 2010, for a discussion of model selection in LCA).

To estimate the causal effect of adolescent depression risk on adult substance use class membership, we implemented an inverse propensity weighted latent class regression model. This model included Wave 2 depression risk status as a predictor of Wave 4 substance use latent class using the LCA model identified during the model selection step. As before, the outcome analysis was weighted using IPW and gender was treated as a grouping variable.

SAS PROC GENMOD was used to estimate the propensity scores and PROC LCA (Lanza *et al.*, 2007, Lanza *et al.*, 2013b) was used to fit the outcome models.

16.4.5 Results From Empirical Demonstration

Figure 16.1 shows a boxplot for the distribution of propensity scores in each exposure group. As both distributions span essentially the same range of propensity scores, we determined that overlap was sufficient to continue with a causal analysis. As

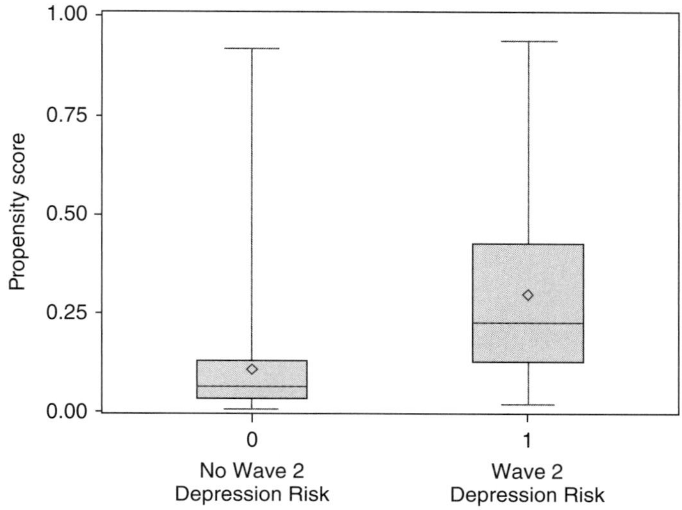

Figure 16.1 Boxplots showing overlap of propensity score distributions for adolescent depression risk groups.

Table 16.1 shows, inverse propensity weighting successfully balanced the two groups (individuals at risk for clinical depression at Wave 2 and those not at risk) with respect to all potential confounders. Prior to weighting, notable differences were observed between groups. Those at risk for depression were more likely to be female, had higher CESD-9 scores at Wave 1, showed elevated substance use on nearly every indicator, had more friends who regularly used substances, and reported more adverse family and social environments. Inverse propensity weighting balanced the groups such that the SMD for all variables was less than 0.10, except gender and CESD-9, which had SMD less than 0.20. In the LCA outcome model, gender was treated as a grouping variable and CESD-9 was included as a covariate to further adjust (known as a doubly robust approach, see Kang and Schafer, 2007). In propensity score weighted analyses, the mean age at Wave 1 was 16.5, 56% were females, 71% identified as White, 19% as Black, 14% as Hispanic, and 12% as another race/ethnicity. The mean CESD-9 score was 6.7 (max of 27), 33% reported cigarette use in the past month, 35% reported past year binge drinking, 37% reported lifetime marijuana use, and 14% reported lifetime illicit drug use.

Five indicators of substance use at Wave 4 were used to define latent classes of adult substance use: past year alcohol use, past year binge drinking, past month cigarette use, past month marijuana use, and past year illicit drug use. When determining the optimal number of classes for substance use at Wave 4, we considered LCA models in which the measurement model was both freely estimated and constrained across genders and determined that a constrained 4-class model was optimal (see Table 16.2). We describe the four classes as follows: Low Use (males: 37%, females: 44%), Primarily Alcohol Use (males: 39%, females: 40%), Polysubstance Use without Binge Drinking (males: 4%, females: 7%), and Polysubstance Use (males: 20%,

TABLE 16.2 Model Fit Statistics (Weighted) for Models With Measurement Parameters Freely Estimated and Constrained to Be Equal Across Gender.

Method	#	df	G^2	AIC	BIC	Adj BIC	Entropy
LCA	1	129	1678.53	1706.53	1782.19	1737.71	1.00
model	2	113	508.88	568.88	730.99	635.69	0.90
freely	3	97	266.95	358.95	607.52	461.39	0.87
estimated	4	81	152.21	276.21	611.23	414.27	0.81
across	5	65	110.99	266.99	688.48	440.68	0.75
gender	6	49	74.19	262.19	770.14	471.51	0.77
LCA	1	136	1760.29	1774.29	1812.12	1789.88	1.00
model	2	127	589.42	621.42	707.87	657.04	0.90
constrained	3	118	348.39	398.39	533.48	454.06	0.87
across	4	109	242.64	310.64	494.37	386.35	0.79
gender	5	100	187.44	273.44	505.8	369.19	0.74
	6	91	162.41	266.41	547.41	382.21	0.75

Note: # = Number of classes.

females: 9%). The low use class is defined by Low Use across the five indicators; the Primarily Alcohol Use class is defined by occasional or weekly alcohol use with a high probability of binge drinking (84%); the Polysubstance Use without Binge Drinking class is defined by daily cigarette smoking (66%), regular marijuana use (64%), some illicit drug use (37%) but a near-zero probability of binge drinking; and the Polysubstance Use class is defined by regular alcohol, cigarette, marijuana (85%), and illicit drug use (45%) (see Table 16.3).

Table 16.4 presents the results from estimating the ACE of high adolescent depression risk on young adult substance use class. These estimates were obtained from an inverse propensity weighted LCA with covariates model, in which Wave 2 depression risk predicted substance use class. Based on the previously described results from our LCA model fitting, the LCA measurement model was specified as a constrained 4-class model. This model simultaneously estimated the latent class model and the association of depression risk and latent class membership. Because Wave 1 depression was the strongest predictor of Wave 2 depression, we estimated the same model but also including a single confounder, Wave 1 CESD-9 score, to control for this variable in a doubly robust manner. For comparison purposes, Table 16.4 presents results from an unweighted model, a weighted model, and the doubly robust model.

In the unweighted model, adolescent depression risk was significantly associated with membership in the Polysubstance Use without Binge Drinking class relative to the Low Use class for both males (OR = 3.29, 95% CI = $[1.13, 9.54]$) and females (OR = 5.50, 95% CI = $[2.43, 12.48]$). For males, both the standard inverse propensity weighted model and the doubly robust model (IPW + W1 depression) indicate that adolescent depression risk was not causally related to young adult substance use profile. However, for females, even after adjusting for 37 potential confounders, the causal effect of adolescent depression risk on young adult substance use profile was significant, with elevated odds of membership in the Polysubstance Use without

TABLE 16.3 The Four-Class Model of Young Adult Substance Use With Measurement Constrained to Be Equal Across Gender.

Substance	Response	Low Use M: 37% F: 44%	Primarily Alcohol M: 39% F: 40%	Poly Use, No Binge M: 4% F: 7%	Poly Use M: 20% F: 9%
Alcohol (past year)	No use	**66.1%**	0.0%	**70.4%**	0.0%
	< weekly use	31.2%	47.8%	29.1%	30.0%
	≥ weekly use	2.7%	**52.2%**	0.6%	**70.0%**
Binge drinking (past year)	None	**100.0%**	15.6%	**99.8%**	5.4%
	Any	0.0%	**84.4%**	0.2%	**94.6%**
Cigarettes (past month)	No use	**74.7%**	**72.4%**	10.1%	20.6%
	< daily use	8.0%	13.0%	23.6%	36.7%
	≥ daily use	17.4%	14.6%	**66.3%**	42.7%
Marijuana (past month)	No use	**96.3%**	**88.9%**	36.4%	15.1%
	Any use	3.8%	11.1%	**63.6%**	**84.9%**
Illicit drugs (past year)	No use	**99.2%**	**95.0%**	**62.9%**	**55.2%**
	Any use	0.8%	5.0%	37.1%	44.8%

Note: M = male, F = female; item response probabilities greater than 50% marked in bold to facilitate interpretation.

TABLE 16.4 Average Causal Effect of Adolescent Depression Risk on Adult Substance Use Class.

Causal Estimand	*Primarily Alcohol* OR	95% CI	*Poly Use, No Binge* OR	95% CI	*Poly Use* OR	95% CI
ACE: Males						
Unweighted	0.9	[0.42–1.94]	3.29[1]	**[1.13–9.54]**	1.57	[0.71–3.46]
IPW	0.62	[0.19–2.07]	1	[0.23–4.26]	0.85	[0.29–2.53]
IPW + W1 depression	0.63	[0.30–1.33]	0.78	[0.24–2.45]	0.70	[0.31–1.61]
ACE: Females						
Unweighted	1.01	[0.59–1.74]	5.50[1]	**[2.43–12.48]**	1.58	[0.67–3.75]
IPW	1.27	[0.55–2.91]	5.52[1]	**[1.61–18.92]**	1.89	[0.51–6.98]
IPW + W1 depression	1.21	[0.73–2.03]	3.62[1]	**[1.56–8.38]**	1.65	[0.65–4.20]

[1] $p < 0.05$; significant effects are marked in bold. The reference class for the ORs is the Low Use class.

Binge Drinking class relative to the Low Use class. This was significant using both the standard inverse propensity weighted model (OR = 5.52, 95% CI = $[1.61, 18.92]$) and the doubly robust model that also adjusts for Wave 1 depression (OR = 3.62, 95% CI = $[1.56, 8.38]$). Thus, our results are consistent with a causal effect of adolescent depression risk status on young adult substance use patterns for females but not males, indicating that gender moderates this causal effect.

16.5 DISCUSSION

Modern causal inference methods have great potential to advance behavioral research. There is a preponderance of behavioral data from observational studies on a variety of behavioral topics, including alcohol, tobacco and drug use, delinquent behavior, eating behavior, and physical activity. By applying causal inference methods to these data sets, a wealth of new scientific knowledge can be gained about the antecedents and consequences of behavior. Such methods, including inverse propensity weighting and propensity score matching, are now being implemented fairly widely in behavioral research; however, a substantial opportunity remains to integrate these methods into latent variable models. By integrating IPW with LCA, for instance, researchers can estimate causal effects of exposures on multidimensional outcomes such as complex substance use behavior profiles.

As our applied example highlights, applying IPW to LCA with covariates is not as straightforward as simply assigning individuals to latent class membership and then conducting propensity score weighting as usual, treating the outcome as having a multinomial distribution. As described in the latent class literature, assigning class membership to individuals and then treating class membership as an "observed" variable results in misclassification of individuals with regard to class, and ultimately attenuates effect estimates. Instead, the latent class measurement model and the effect of the predictor on the latent class outcome should be estimated simultaneously using software for LCA (e.g., PROC LCA, Lanza *et al.*, 2013b; Latent Gold, Vermunt and Magidson, 2005; Mplus, 1998–2012). In other words, individuals are never actually classified on the latent class outcome. A key advantage of this simultaneous estimation is the fact that measurement error with regard to the latent class is properly modeled and thus does not bias the effect estimated.

There is still much methodological work to be done on integrating propensity score methods with latent variable analysis. The use of any weighting approach in LCA can be challenging because, as the weights modify the information in the data, the actual number and meaning of the latent classes themselves can be altered relative to the unweighted model. This issues is discussed in more detail by Lanza *et al.* (2013a). Another area for future work is to address the subtle complexities that arise with respect to estimating ACEs across levels of a moderator. Due to the nature of the outcome variable in the current study (a latent class variable), and the fact that we wished to impose measurement invariance of the latent class variable across gender groups, we confirmed that balance on all potential confounders was achieved overall; however it does not necessarily follow that sufficient balance was achieved within each group. Future research is necessary to clarify the balance requirements for moderated causal effects in the context of both observed and latent outcomes. Finally, our application included a binary exposure, yet recent methodological development has focused on extending propensity score methods to multiple treatment groups (McCaffrey *et al.*, 2013) and continuous exposure variables (Zhu *et al.*, 2014). Future work is necessary to integrate multiple-group propensity score methods and continuous exposure variables into the latent variable context.

16.5.1 Limitations

There are several limitations of this investigation worth noting. First, as with many behavioral studies, all assessments were self-reports. Second, as with any propensity score approach to causal inference, a fundamental assumption is that all potential confounders have been measured and included in the selection model. Omission of key confounders can bias the estimated ACEs. Fortunately, to the extent that any unobserved confounders are correlated with observed ones, this bias is mitigated. Third, there is potential for posttreatment confounding in the causal effect. That is, during the approximately 10-year time span between the exposure (adolescent depression risk) and the outcome (young adult substance use behavior profile), other factors not accounted for in this study may have contributed to any causal effect identified here. The degree to which adolescent depression symptomology is stable into emerging adulthood, however, may lessen this particular concern. Finally, statistical power and sample size requirements for estimating causal effects in LCA is an important area for future research. In this empirical study, the smallest latent class for both males and females was Polysubstance Use without Binge Drinking. Based on the overall sample size of $N = 1642$, approximately 29 males and 64 females are expected to be in this class. These groups were of sufficient size to detect significant causal effects of adolescent depression risk, most likely because the effect sizes were quite large (ORs > 3.0).

ACKNOWLEDGMENTS

This work was supported by grants P50 DA010075 and T32 DA017629 from the National Institute on Drug Abuse (NIDA). The content is solely the responsibility of the authors and does not necessarily represent the official views of NIDA or the National Institutes of Health.

REFERENCES

Agresti, A. (2013) *Categorical Data Analysis*, 3rd edn, John Wiley & Sons, Inc., New York.

Butera, N.M., Lanza, S.T., and Coffman, D.L. (2014) A framework for estimating causal effects in latent class analysis: is there a causal link between early sex and subsequent profiles of delinquency? *Prevention Science*, **15** (3), 397–407.

Collins, L.M. and Lanza, S.T. (2010) *Latent Class and Latent Transition Analysis: With Applications in the Social, Behavioral, and Health Sciences*, John Wiley & Sons, Inc., New York.

Cooper, B.R. and Lanza, S.T. (2014) Who benefits most from head start? Using latent class moderation to examine differential treatment effects. *Child Development*, **85**, 2317–2338, doi: 10.1111/cdev.12278. PMCID: PMC4236237.

Dayton, C.M. and Macready, G.B. (1988) Concomitant-variable latent-class models. *Journal of the American Statistical Association*, **83** (401), 173–178.

Harris, K.M., Halpern, C.T., Whitsel, E., Hussey, J., Tabor, J., Entzel, P., and Udry, J.R. (2009) The national longitudinal study of adolescent to adult health: Research design [www document], URL: `http://www.cpc.unc.edu/projects/addhealth/design` (accessed 24 December 2015).

Kang, J.D. and Schafer, J.L. (2007) Demystifying double robustness: a comparison of alternative strategies for estimating a population mean from incomplete data. *Statistical science*, **22** (4), 523–539.

Kuramoto, S., Bohnert, A., and Latkin, C. (2011) Understanding subtypes of inner-city drug users with a latent class approach. *Drug and Alcohol Dependence*, **118** (2), 237–243.

Lanza, S.T. and Bray, B.C. (2010) Transitions in drug use among high-risk women: an application of latent class and latent transition analysis. *Advances and Applications in Statistical Sciences*, **3** (2), 203–235.

Lanza, S.T., Coffman, D.L., and Xu, S. (2013a) Causal inference in latent class analysis. *Structural Equation Modeling: A Multidisciplinary Journal*, **20** (3), 361–383.

Lanza, S.T., Dziak, J.J., Wagner, A., and Collins, L.M. (2013b) PROC LCA & PROC LTA users' guide (version 1.3.0), The Methodology Center, Penn State, University Park. Retrieved from `http://methodology.psu.edu` (accessed 24 December 2015).

Lanza, S.T. and Collins, L.M. (2008) A new SAS procedure for latent transition analysis: transitions in dating and sexual risk behavior. *Developmental Psychology*, **44** (2), 446–456.

Lanza, S.T., Collins, L.M., Lemmon, D.R., and Schafer, J.L. (2007) PROC LCA: a SAS procedure for latent class analysis. *Structural Equation Modeling*, **14** (4), 671–694.

Lanza, S.T., Patrick, M.E., and Maggs, J.L. (2010) Latent transition analysis: benefits of a latent variable approach to modeling transitions in substance use. *Journal of Drug Issues*, **40** (1), 93–120.

Lanza, S.T. and Rhoades, B.L. (2013) Latent class analysis: an alternative perspective on subgroup analysis in prevention and treatment. *Prevention Science*, **14** (2), 157–168.

Lehrer, J.A., Shrier, L.A., Gortmaker, S., and Buka, S. (2006) Depressive symptoms as a longitudinal predictor of sexual risk behaviors among us middle and high school students. *Pediatrics*, **118** (1), 189–200.

Luellen, J.K., Shadish, W.R., and Clark, M. (2005) Propensity scores an introduction and experimental test. *Evaluation Review*, **29** (6), 530–558.

McCaffrey, D.F., Griffin, B.A., Almirall, D., Slaughter, M.E., Ramchand, R., and Burgette, L.F. (2013) A tutorial on propensity score estimation for multiple treatments using generalized boosted models. *Statistics in Medicine*, **32** (19), 3388–3414.

McCaffrey, D.F., Ridgeway, G., and Morral, A.R. (2004) Propensity score estimation with boosted regression for evaluating causal effects in observational studies. *Psychological Methods*, **9** (4), 403–425.

Muthén, L.K. and Muthén, B.O. (1998–2012) *Mplus User's Guide*, 7th edn, Muthén & Muthén, Los Angeles, CA.

Radloff, L.S. (1991) The use of the center for epidemiologic studies depression scale in adolescents and young adults. *Journal of Youth and Adolescence*, **20** (2), 149–166.

Roberts, R.E., Lewinsohn, P.M., and Seeley, J.R. (1991) Screening for adolescent depression: a comparison of depression scales. *Journal of the American Academy of Child & Adolescent Psychiatry*, **30** (1), 58–66.

Robins, J.M., Rotnitzky, A., and Zhao, L.P. (1995) Analysis of semiparametric regression models for repeated outcomes in the presence of missing data. *Journal of the American Statistical Association*, **90** (429), 106–121.

Rosenbaum, P.R. (2002) *Observational Studies*, 2nd edn, Springer-Verlag, New York.

Rosenbaum, P.R. and Rubin, D.B. (1983) The central role of the propensity score in observational studies for causal effects. *Biometrika*, **70** (1), 41–55.

Rosenbaum, P.R. and Rubin, D.B. (1984) Reducing bias in observational studies using subclassification on the propensity score. *Journal of the American Statistical Association*, **79** (387), 516–524.

Rosenbaum, P.R. and Rubin, D.B. (1985) Constructing a control group using multivariate matched sampling methods that incorporate the propensity score. *American Statistician*, **39** (1), 33–38.

Schuler, M.S., Leoutsakos, J.M.S., and Stuart, E.A. (2014) Addressing confounding when estimating the effects of latent classes on a distal outcome. *Health Services and Outcomes Research Methodology*, **14** (4), 232–254.

Vasilenko, S., Kugler, K., Butera, N., and Lanza, S. (2014) Patterns of adolescent sexual behavior predicting young adult sexually transmitted infections: a latent class analysis approach. *Archives of Sexual Behavior*, **44**, 705–715, doi: 10.1007/s10508-014-0258-6. PMCID: PMC4107199.

Vermunt, J.K. and Magidson, J. (2005) *Technical Guide for Latent GOLD 4.0: Basic and Advanced*, Statistical Innovations Inc., Belmont, MA.

Zhu, Y., Coffman, D.L., and Ghosh, D. (2014) A boosting algorithm for estimating generalized propensity scores with continuous treatments. *Journal of Causal Inference*, **3**, 25–40.

APPENDIX

```
*********************************************************
* ESTIMATE PROPENSITY SCORE MODEL: WAVE 2 DEPRESSION
*********************************************************;
proc logistic data=addhlth.ps_complete descending;
 model w2dep_sbin = sex age black other hisp w1dep_short w1cigpm
                    w1alc w1bingepy w1mar w1drug w1drunkpy
                    w1f_drink pc49e_23 h1to50 h1to51 h1to52 h1to9
                    h1to29 h1to33 avebond welfare mom_work
                    age_birth alienation rejection fcare
                    impulsivity aggression avggrade neiatt neihap
                    rural suburb urban h1ed21 h1ed24;
 output out=addhlth.ps_complete prob = pscore;
run;
*********************************************************
*CALCULATE ATE PROPENSITY SCORE WEIGHTS
*********************************************************;
data addhlth.ps_complete;
 set addhlth.ps_complete;
 *For TX=1;
        if w2dep_sbin=1 then w_ate = 1/pscore;
 *For TX=0;
        if w2dep_sbin=0 then w_ate = 1/(1-pscore);
run;
*********************************************************
*TRIM PROPENSITY SCORE WEIGHTS
*********************************************************;
data addhlth.ps_complete;
 set addhlth.ps_complete;
 w_ate1=.;
 if w_ate<=20 then w_ate1=w_ate;
 if w_ate>20 then w_ate1=20;
run;
*********************************************************
*ASSESS PROPENSITY SCORE BALANCE USING STANDARDIZED DIFFS
*********************************************************;
*MACRO FOR COMPUTING STANDARDIZED DIFFERENCES FOR CONTINUOUS
*VARIABLES; RUN TWICE: WITH AND WITHOUT WEIGHT;
proc means mean stddev data=addhlth.ps_complete noprint;
 weight w_ate1;
 var &var;
 by w2dep_sbin;
 output out=outmean (keep=w2dep_sbin mean stddev) mean=mean
 stddev=stddev;
run;
*Calculate mean, SD for Tx=0 group;
data w2dep_sbin_0;
 set outmean;
 if w2dep_sbin = 0;
 mean_0 = mean;
 s_0 = stddev;
 keep mean_0 s_0;
run;
```

```
*Calculate mean, SD for Tx=1 group;
data w2dep_sbin_1;
 set outmean;
 if w2dep_sbin = 1;
 mean_1 = mean;
 s_1 = stddev;
 keep mean_1 s_1;
run;
*Using these 2 datasets, calculate SMD;
data newcont;
 length label $ 25;
 merge w2dep_sbin_0 w2dep_sbin_1 ;
 d1 = (mean_1 - mean_0) / sqrt ((s_1*s_1 + s_0*s_0)/2);
 d1 = round (abs (d1), 0.001);
 label = &label;
 keep d1 label;
run;
proc append data = newcont base=standiff force;
run;
*MACRO FOR COMPUTING STANDARDIZED DIFFERENCES FOR
*BINARY VARIABLES;
options spool;
proc means mean data=addhlth.ps_complete noprint;
 weight w_ate1;
 var &var;
 by w2dep_sbin;
 output out=outmean (keep=w2dep_sbin mean) mean= mean;
run;
*Calculate mean, SD for Tx=0 group;
data w2dep_sbin_0;
 set outmean;
 if w2dep_sbin = 0;
 mean_0 = mean;
 keep mean_0;
run;
*Calculate mean, SD for Tx=1 group;
data w2dep_sbin_1;
 set outmean;
 if w2dep_sbin = 1;
 mean_1 = mean;
 keep mean_1;
run;
*Using these 2 datasets, calculate SMD;
data newcont;
 length label $ 25;
 merge w2dep_sbin_0 w2dep_sbin_1 ;
 d1 = (mean_1 - mean_0) / sqrt ((mean_1*(1-mean_1)
      + mean_0*(1- mean_0))/2);
 d1 = round (abs (d1), 0.001);
 label = &label;
 keep d1 label;
run;
proc append data = newcont base=standiff force;
run;
```

```
*********************************************************
* FIT LCA MODEL TO DETERMINE # OF CLASSES
* NO COVARIATES INCLUDED; PS WEIGHTS INCLUDED
*********************************************************;
* MALE / FEMALE GROUPS FREELY ESTIMATED;
proc lca data=addhlth.ps_complete;
  weight w_ate1;
  nclass 4;
  items w4cigpm w4alcpy w4bingepy w4marpy w4drugpy;
  categories 3 3 2 2 2;
  nstarts 1000;
  groups sex_grp;
  groupnames male female;
  rho prior=1;
run;
*MEASUREMENT INVARIANCE WITH REGARD TO MALE / FEMALE GROUPS;
proc lca data=addhlth.ps_complete;
 weight w_ate1;
 nclass 4;
 items w4cigpm w4alcpy w4bingepy w4marpy w4drugpy;
 categories 3 3 2 2 2;
 nstarts 1000;
 groups sex_grp;
 groupnames male female;
 measurement groups;
 rho prior=1;
run;
*********************************************************
* ESTIMATE LATENT CLASS REGRESSION MODEL (LCA W COVARIATES)
* INCLUDE COVARIATES AND PS WEIGHTS
*********************************************************;
proc lca data=addhlth.ps_complete;
 *weight w_ate1;
 nclass 4;
 items w4cigpm w4alcpy w4bingepy w4marpy w4drugpy;
 categories 3 3 2 2 2;
 covariate w2dep_sbin;
 groups sex_grp;
 groupnames male female;
 measurement groups;
 nstarts 1000;
 rho prior=1;
 beta prior=1;
run;
```

PART V

DESIGNS FOR CAUSAL INFERENCE

17

CAN WE ESTABLISH CAUSALITY WITH STATISTICAL ANALYSES? THE EXAMPLE OF EPIDEMIOLOGY

ULRICH FRICK

Department of Applied Psychology, HSD University of Applied Sciences,Cologne, Germany; Swiss Research Institute on Public Health and Addiction, University of Zurich, Zurich, Switzerland; Psychiatric University Hospital, University of Regensburg, Regensburg, Germany

JÜRGEN REHM

Social and Epidemiological Research (SER) Department, Centre for Addiction and Mental Health, Toronto, Canada; Addiction Policy, Dalla Lana School of Public Health, University of Toronto, Toronto, Canada; Department of Psychiatry, Faculty of Medicine, University of Toronto, Toronto, Canada; PAHO/WHO Collaborating Centre for Mental Health & Addiction, Toronto, Canada; Institute of Medical Science, University of Toronto, Toronto, Canada; Epidemiological Research Unit, Technische Universität Dresden, Klinische Psychologie & Psychotherapie, Dresden, Germany

17.1 WHY A CHAPTER ON DESIGN?

This book provides an overview of novel methods of analysis of hypotheses that are compatible with theories of causality in modern empirical research. The other contributions in this volume mainly deal with statistical tools to establish causal relations based on given data configurations. This contribution, by contrast, focuses on design (e.g., Were the data derived from an experiment with randomization or from an observational study?), that is, on the question of data generation rather than statistical analysis. Epidemiology as a discipline where observational studies are not

Statistics and Causality: Methods for Applied Empirical Research, First Edition.
Wolfgang Wiedermann and Alexander von Eye.

uncommon will serve as a basis for theory and empirical examples. A second focus of this contribution will be on measurement. At least in systems of philosophy of science implying the existence of a real world, which can be understood—such as critical realism or critical rationalism (Popper, 1934)—data are seen as representations of empirical reality with numbers (Coombs, 1964, Osborne, 1976). This means that establishing causality can be impacted by the quality of measurement, as operationalized by the classic criteria of validity and reliability (Gadenne, 1976).

This contribution starts with the classic epidemiological definition of causality by Sir Bradford Hill (1965) and its evolution to, in Epidemiology, the currently accepted theory of causality by Rothman (1986, 1988). Throughout the chapter, we give examples to illustrate the conceptual thinking and its implementation in design and analysis. Special emphasis is given to the two classical designs of epidemiology, case-control and cohort studies (Rothman *et al.*, 2008). We will deal with both designs to illustrate the theoretical underlying principles knowing that in current epidemiology most of the designs are actually variants of the pure forms (such as case-crossover, Maclure, 1991; retrospective cohort Euser *et al.*, 2009; or retrospective case-control-cohort designs Petri and Allbritton, 1993; just to name a few).

The core of our contribution will be an in-depth analysis of key elements necessary to strengthen causality such as measurement, use of pseudoexperimental techniques, Mendelian randomization, experimentation within observational pathways where possible, inclusion of other elements allowing more control such as multiple nonexperimental control groups, and statistical techniques in line with epidemiological theory of causality. We conclude stating that the majority of thoughts and resources in empirical research should be put into implementation of coherent design, measurement, and analysis strategies.

17.2 THE EPIDEMIOLOGICAL THEORY OF CAUSALITY

Epidemiology is a modern science, which was mainly established in the 20th century, and which has been plagued by the impossibility of applying the gold standard principles of natural sciences, that is, the experiment with the possibility of randomization. Consider the example of smoking and lung cancer: while in principle it would be possible to randomize people into two groups who smoke or abstain based on experimental manipulation, this procedure fails for at least three reasons: (i) it is unethical given current standards (World Health Organization, 2005) to expose people to potential harmful substances in order to find out causal relations between these substances and the respective disease endpoints (= main dependent variable in epidemiology); (ii) the implementation of such an experimental manipulation is almost impossible as it would require very high incentives and/or punishments to guarantee compliant behavior in the sense of the experimental manipulation; (iii) the time lag between cause and effect, that is, exposure and disease outcome is long (at least 25 years in this case Loeb *et al.*, 1984), so that even if an initial randomization was successful, compliance over time and exclusion of potential confounding over decades seems virtually impossible.

Epidemiological research thus had to invent other procedures to demonstrate causality. Ironically, the theoretical foundation of these designs and techniques came after their application, in part as the result of the discussion whether some of the results could be interpreted as causal or not. Consider again the example of smoking and lung cancer: there had been some speculation about an effect of cigarette smoking on lung cancer in the first half of the 20th century, for example, in Nazi-Germany (Proctor, 1996). The work of Doll and Hill (1950, 1954, 1956, 1964) on British doctors is generally considered as a breakthrough to "prove" causality; they used a cohort of British doctors, assessed exposure at the beginning and lung cancer after 3, 10, 20 years, and so on. Of course, these results—at the time new and shocking—were strongly contested, especially in the early years when there were not sufficient numbers of cases, and the uncertainty was large. The most notable critique came from the eminent Sir Ronald Fisher, one of the most important and influential statisticians of the 20th century. In short, Fisher postulated a potential third cause leading to both smoking and lung cancer, that is, certain genetic yet unknown factors were postulated by him to cause both smoking behavior and lung cancer (Fisher, 1959).

We deal with this historical controversy to illustrate two things: (i) classical cohort studies can never be fully conclusive and (ii) the theoretical foundation for causality in epidemiology was derived from empirical study results and controversies there off. In fact, one of the authors of the classic papers, the later knighted Sir Bradford Hill, has formulated some of the causal criteria in response to this controversy (Hill, 1965):

(1) Strength: strong associations are more likely to be causal, because if this association could be explained by another factor, this alternative factor must have an even stronger association to the outcome, and therefore has little chance to remain overlooked (see Rothman and Greenland, 2005, p. S148 explaining Hill's original reasoning);
(2) Consistency: the effect can be replicated in independent studies in different populations under different circumstances;
(3) Specificity: a cause leads to a specific single effect, not multiple effects;
(4) Temporality: the cause should precede the effect;
(5) Biological gradient: this criterion often is also called a dose-response relationship requiring a monotonic function between a quantifiable exposure and the intensity of the effect;
(6) Plausibility: biological pathways can be described to explain the effect;
(7) Coherence: a cause should not contradict to the current state of knowledge in the respective field;
(8) Experiment: removal of a specific antecedent condition in an experimental intervention that leads to a reduction of the outcome could serve as a proof for its causality;
(9) Analogy: similar evidence exists from related research topics.

This list was the accepted standard in epidemiological reasoning and was almost used as "proof" until Rothman (1986) not only criticized all of the assumptions as nonconclusive (see also Charlton, 1996), but replaced them by a deterministic theory about causality for epidemiological research (Rothman, 1988). Major arguments against the unreflected use of Hill's "viewpoints" (as Hill himself named them) were as follows: none of the criteria even could serve as a necessary or sufficient condition for causality; for each of the criteria examples Rothman *et al.* (2008) gave counterexamples, where the criterion was not fulfilled even though the respective relationship was obviously causal (Rothman and Greenland, 2005). Nevertheless, the Hill criteria still continue to be used in a heuristic manner, which is reflected in the almost 5000 citations since the year 2000 (Google Scholar search from 02/05/2015). Apparently, those criteria relying on a probabilistic regularity view of causality (strength, specificity, consistency, biological gradient; see Thygesen *et al.*, 2005) have been mentioned in epidemiological studies more frequently than criteria stressing a generative view on causality (like coherence, plausibility, and analogy; see also Weed and Gorelic, 1996, Weed, 1997). Within the context of this book, the criterion of temporality is a pivotal component of the theory of Granger causality. Most chapters are in line with the pragmatic statistical view on causality given by Cox (1992), that is, causality as statistical association that cannot be explained by confounding variables.

While the Hill criteria give some useful heuristics for determination of causality, the current system of epidemiology is very much determined by the thinking of Rothman. His theory of causality basically stipulates the following: epidemiological outcomes (most often the new occurrence = "incidence" of a disease or death) are determined by the complex interaction of multiple causes. Think of the following constellation: a blood alcohol level of 0.18% plus driving a car over 60 minutes and with a speed of 130 km/hour on an unknown and icy road may lead to a traffic accident, and if this constellation always leads to a traffic accident, then this combination is causal. In Rothman's thinking, causality is given when we find the minimal number of necessary and sufficient preconditions for this accident. We do not know whether the conditions listed above are really necessary and sufficient for this accident, but we use them as a thought experiment to derive the best strategies to empirically assess causality in epidemiology. If all of these conditions together are necessary and sufficient, it follows that removing one would not lead to the accident. In order to test whether alcohol is causal for traffic accidents, or more precisely: whether a BAC of 0.18 is causal, all we have to do is to remove alcohol from the constellation of antecedent conditions. Unfortunately, for most outcomes, the exact constellation of antecedent conditions, which are necessary and sufficient, is not known, thus implying the existence of unknown conditions. This leads to a probabilistic reasoning. Removal of one condition will thus only reduce the risk of accidents without any certainty that no outcome occurs.

What does that mean for epidemiological design? In terms of a longitudinal study, we would strive to create constellations, where many of the necessary preconditions are present with and without the exposure in question being present. We would wait for a sufficient number of outcomes to allow for enough statistical power to compare the risks of the exposed and the unexposed group. Often, we would estimate a relative risk of exposed to unexposed people, where a significant increase in frequency of outcomes in the exposed group could be regarded as an indication of a causal relation. Of course, the whole reasoning hinges on a ceteris paribus assumption that there are no other impacts both on exposure and outcome that could explain the relative risk. In natural sciences, experimentation could help insure that the above conditions are true, that is, that all potential influencing factors are randomly distributed over conditions. In epidemiology, we have to try to control for known alternative risk factors by other means.

The classic research strategies to control in epidemiological research are the cohort study and the case-control study. We will explain these studies in principle to introduce different design elements to strengthen interpretations of causality.

17.3 COHORT AND CASE-CONTROL STUDIES

Cohort Studies The design of a cohort study seems simple and straightforward: Define a group of subjects that does not display the disease under study at the starting point of your study, measure subjects' status on the relevant exposure variables, and follow their status on the outcome variable.

Cohort studies in most cases are not designed to assess the impact of a single risk factor on a single outcome. They usually combine multiple domains of risk factors and people with different constellations of these risk factors are followed prospectively in defined time intervals for a longer period. Study endpoints comprise death or the incidence of specific diseases but need not be restricted to that. It would be also possible to monitor the course of some health-related variables like physiological measures, subjective health status, or social characteristics (employment periods, living situation, marital status, birth of children, etc.).

Following the argumentation in Rothman *et al.* (2008) or other epidemiological textbooks, cohort studies do not need to fulfill the criterion of representativeness for some population as a prerequisite for causality. It is only necessary that the causal factor under study and the outcome variable are represented in the cohort with variation (e.g., exposed and not exposed subjects leading to the study endpoint or not).

Subjects participating in a cohort study qualify for enrollment by a certain attribute they have in common. Most often this attribute is selected to allow for high probability of easy follow-up (i.e., health professionals; nurses; people from the same jurisdiction, where mortality can be followed up via register and the like). It is important that cohorts can be followed fully, that is, the dropout over time is minimalized (see below as well). Dropout in this case can be passive via register data such as mortality registers.

Cohort studies are costly, especially when the disease under study is of low incidence. Their sample size for the latter reason usually exceeds that of a clinical trial by far in order to reach enough statistical power. Cohort studies are time consuming, as some diseases occur only with considerable time lag after exposure. This is the reason, why for rare diseases and large time lags, case-cohort studies usually are selected as the preferred design (Rothman *et al.*, 2008). Also, cohort studies bear the inherent risk of omitting important variables because the relevant predictor and confounding variables may not be known at the starting point of a cohort study. However, cohort designs allow for statistical control of omitted variables under certain circumstances (Rehm *et al.*, 1992). Therefore, sometimes, analyses of cohort studies after the statistical analyses of the main relationships can be compared to a scientific trawl net with respect to their prospects of identifying causal relationships.

Their strengths are the measurement of exposure, where the outcome is not known and thus cannot influence on reporting, the possibility of measuring exposure over time, the possibility to check for interactions between multiple risk factors to cause disease onset, and a time horizon long enough to detect subtle, cumulative effects over longer periods. The sequence of exposure-effect can be determined with better validity than in retrospective studies because exposure is ascertained before outcome is ascertained. Cohort studies are quite flexible: once having started observation of exposure, a completely new study endpoint could be introduced even years after the initial enrollment of subjects.

A very important risk for the interpretation of results from cohort studies is given by attrition during follow-up measurements. Informative censoring (selective withdrawal from the study) would invalidate conclusions drawn from this biased sample. As cohort studies often measure a multitude of supposed risk factors and/or confounding variables, they pose a high cognitive burden on the respondents and may pose a heavy burden on respondents. Shortening the length of interviews has been shown to improve response rates (Edwards *et al.*, 2002). Consequently, researchers have developed methods to systematically omit parts of the complete measurement program for defined (and randomly chosen) subsamples of the total cohort. "Partial Questionnaire Design" (Wacholder, 1995) or "Multiple Matrix Sampling" (Gonzalez and Eltinge, 2007) denote the same idea: variables Z (not asked in subsample A) and variables X (not asked in subsample B) are both measured in a third subsample C. Their missing values in subsamples A and B can be considered missing completely at random (MCAR), as their origin stems from a randomization procedure. Therefore, multiple imputation methods (Little, 1992) can be used when estimating complex statistical models (see also Smits and Vorst, 2007 for an illustration and Chipperfield and Steel, 2011 for a simulation study how to balance efficiency of estimators with burden for the respondents).

Case-Control Studies A major drawback of cohort studies is given by the time lag between exposure and onset of the study endpoint. Even if this "incubation period" (not necessarily used in the strict biological meaning of this term) seems quite short, there are situations where you might have no time left to wait prospectively until the occurrence of your study endpoint. Imagine the epidemic outbreak of the hemolytic-uremic syndrome (HUS) in Western Europe starting on May 21, 2011, with over 3400 EHEC infections, of which 796 were of serious character (HUS) leading to over 50 deaths (Robert Koch Institut, Epidemiologisches Bulletin, 8. August 2011, Nr. 31). Though it was clear that the toxic agents of the enterohemorrhagic disease was an infection with Shiga-toxin-producing *Escherichia coli*

O104:H4 (short: EHEC) bacterium, this was not a satisfactory result for the German health authorities, because the full causal chain was not identified: Where did the bacteria stem from and how were they transmitted to the rapidly growing number of patients? A well-known pathway of transmission is via raw nutritional products. In this situation, a case-control study was performed, asking the patients of three specialized hospitals in Bremen, Bremerhaven, and Lübeck on their intake of raw vegetables during the 14 days before the onset of illness. As controls, subjects of same sex and age living in a patient's neighborhood were assessed. Recall of food intake was asked in personal interviews and pointed at tomatoes, cucumbers, and green salad as the potential sources of the outbreak (Appel *et al.*, 2011). Authorities therefore warned to consume these products. But the exact species of *E. coli* could not be found via polymerase chain reaction (PCR) tests in samples of various suspected horticultural farms (e.g., Spanish cucumbers formerly alleged as source). Additional information from a growing number of cases led to identification of 41 restaurants where the cases had dinner prior to their EHEC/HUS infection. Composition of salads offered during the time prior to the outbreak then could also be reconstructed from restaurants' recipes. A new case-control study (Buchholz *et al.*, 2011) used not only a listing of nutritional components as answering format but also offered photographs of various salad compositions to the patients, from which they identified their ordering (information given during a session of the BfR-commission on risk research, autumn 2011, member Ulrich Frick). This procedure resulted in the conclusion that sprouts (fenugreek; trigonella foenum-graecum) were most likely ($OR = 14.23$; C.I. 2.55 to ∞) to have caused this epidemic, though a final proof based on laboratory PCR tests could not be found. After a second outbreak of EHEC/HUS in France, sprouts drawn from seeds that had been imported from Egypt could be identified via trace-back trace-forward investigations on distributors of fenugreek seeds as the probable pathway of the epidemics in Western Europe (Appel *et al.*, 2011). At the end of July, the official end of the EHEC epidemic in Germany could be declared.

This case history demonstrates typical characteristics of the strengths and potential weaknesses of case-control studies. Case-control studies offer a rapid assessment of exposures potentially leading to a defined outcome. An outbreak similar to the 2011 EHEC crisis requires research results as fast as possible at the cost of a potentially false identification of causal factors (e.g., Spanish cucumbers). Choosing adequate controls is a pivotal topic of case-control studies. Cases and controls can additionally be "matched" on potential confounders such as age, sex, enabling better estimation of the main causal relation.

In this example, neither general population nor neighborhood controls were the best choice from that moment, it had become clear that having dinner in a series of restaurants is a criterion to define the so-called "source population." People cooking at home, for instance, were not at risk and therefore not the best choice to supply controls for the guests of the affected restaurants. Asking retrospectively for prior exposure to known or alleged causal factors bears the considerable risk of false memory of respondents, omission of relevant factors, or other biases. A "recall bias" in our example was the human memory, which tended to omit topics of the ordered salads such as sprouts, even if they had been listed among salad ingredients. Recognition from photographs in this case was the superior method of data gathering.

Taking these arguments together, the most important elements of planning a case-control study can be seen in (Wacholder, 1995):

- Clear and reliable identification of cases
- Selection of adequate controls
- High quality of the exposure measurement.

The above-mentioned study is not very typical for a case-control study, as it involved finding the causally relevant exposure. In the classic case-control study, an exposure is theoretically derived and tested (e.g., alcohol drinking and lung cancer). All the main points about the study listed above still apply. While cohort studies are convenient to study many diseases within one study, case-control studies are better suitable to study many exposures for a given outcome within that study. They are usually less expensive than a long-term cohort study that requires a long period of fieldwork.

17.4 IMPROVING CONTROL IN EPIDEMIOLOGICAL RESEARCH

17.4.1 Measurement

We will take for granted that most readers of this book are familiar with the concept of measurement error, as it has been described in numerous textbooks on psychometrics, namely within the concept of classical test theory (e.g., Lord *et al.*, 1968, Nunnally Jr, 1970). Establishing any statistical association between a predictor variable and a dependent variable becomes more difficult, the more the "signal" (or "true score") in a variable is covered by "noise" (or "measurement error"). Beyond this lowering of statistical power, a second potential consequence of low reliability has been described. In the case of multiple predictor variables with substantial correlations between them, a causal predictor might be rejected in favor of a noncausal variable, which itself only spuriously correlates with the dependent variable. Effects such as this have been extensively discussed by Fuller (1987) for the case of linear relationships and, for example, by Zidek *et al.* (1996) for nonlinear models. Note that these models implicitly rely on causality being strongly related to the strength of the association (see above, p.409).

In epidemiological studies, not only random error could hamper detection of causal relationships but also systematic error in the measurement process could completely mislead the conclusions to be drawn from an observational study. Let us consider an example from the field of lung cancer: during the 1980s, there was a debate in oncology whether feathers of pet birds exposed their owners to an increased risk of lung cancer. It was speculated that microscopic fibers of feathers could be inhaled and then have the same effect as asbestos fibers. To test this hypothesis, the former Federal Health Office of Germany started a case-control study investigating this suspicion.

The interview was conducted by trained staff of the Federal Health Office. Controls were randomly selected from the population registry and matched for sex and age and place of residence (cases and controls stemmed all from West Berlin). The interviewers asked each respondent questions concerning exposure to pet birds during childhood as well as for having owned pet birds in adulthood, and checked a list of alternative exposure variables known to cause lung cancer (active and passive exposure to tobacco smoke, etc.). The statistical results were unequivocal (Kohlmeier *et al.*, 1992): after adjusting for active smoking history, passive exposure to tobacco smoke, professions with occupational exposure to pulmonary carcinogens, and nutritional habits (carrots consumption), there remained a significant independent risk for the duration of years subjects had lived in a household together with pet birds. During a lively discussion in the publishing British Medical Journal, the authors were able to respond to and reject various arguments that had been risen as potential design errors of this study (Kohlmeier *et al.*, 1993). At a first glance, the study seemed to have successfully demonstrated a new causal relation for incidence of lung cancer.

But other and larger studies in Sweden (Modigh *et al.*, 1996), Missouri (Alavanja *et al.*, 1996), and New York (Morabia *et al.*, 1998) failed to arrive at the same conclusions. For the population in Germany, Jöckel and colleagues used the same interview as Kohlmeier and colleagues. They found an adjusted odds ratio of 0.85 (95% CI: 0.53–1.35) for having ever kept pet birds (Jöckel *et al.*, 2002). An age gradient (younger age at onset associated with higher risk) was interpreted as an age-related recall bias of younger patients: they remembered better their pet birds due to the shorter time distance to former exposure.

We would like to add another potential source of bias that might have caused the differences between these studies. The Kohlmeier study used staff members of the Federal Health Office as interviewers. They all had known the status of their interview partners (case or control could be concluded from the place of the interview: hospital or living home). And all interviewers had also known the reason for this case-control study and its main hypothesis: *Do inhaled fibers of pet birds' feathers cause cancer?* We know this from Jürgen Rehm's involvement in this study. An "ambitious interviewer" effect (scrutinizing a history of pet birds more intensive in patients) thus seems not implausible. While the status of respondents (as cases versus controls) could not be masked in this study, it certainly would have been advantageous and feasible to preclude interviewers from knowledge of scope and main hypotheses of this study. This would have prevented a potentially selective alertness on certain topics during the interview.

Biases (= systematic errors) in epidemiological studies do not only occur during sampling of subjects but can also stem from the data gathering process itself (Choi and Pak, 2005). Interviews and also self-administered questionnaires are (real or virtual) conversations and therefore subject to all those effects that have been described in survey research from a social cognition perspective (Hippler *et al.*, 1987, Sudman *et al.*, 1996). Cognitive processes during the question–answer process can also cause bias via nonresponse to ill-designed questionnaires or interviews (De Leeuw, 2001).

We see it as justified that the seminal textbook by Biemer and colleagues (2011) contains twice as many pages (over 480) on errors that can occur during the data gathering process, compared to 200 pages dealing with statistical handling of measurement error after completion of the field work. Construction of a questionnaire and/or interview as a design topic worth serious effort to control on potential biases is too often neglected not only in epidemiology.

"Recall bias" (intentional or unintentional differential recall of information about exposure and/or outcome) can not only be provoked by memory failure (Strube, 1987, Schwarz and Sudman, 1994) and differential affectedness by the relevant details (Coughlin, 1990). Also characteristics of the interview situation itself are used by the interviewees when formulating an answer for the interviewer: report of symptoms (allegedly connected to exposure to electromagnetic fields) was dependent on the introduction of the purpose of the study in one of our previous investigations (Frick *et al.*, 2004). By introducing different anchoring stimuli with high or low physical thread, respondents' reports on their health status (complaint list for symptoms present during 30 days prior to the interview) could be changed markedly (Frick *et al.*, 2002).

Because answers often require a valuative judgment by the interviewee rather than a recall of a stable personality trait, many of those answers do not only reflect personal habits but also responses to characteristics of the interview situation. For example, music festival visitors include into their judgment on the potential harm that might result from various psychotropic substances their subjective knowledge on the prevalence of drug consumption in their reference group (=festival visitors) at the time of the interview. If cognitive availability of high prevalence is promoted by asking a respective question before the harm perception, more frequently used drugs are more strongly trivialized by respondents with respect to potential harm (Wiedermann *et al.*, 2014).

In sum, modern observational research has to take into account the current knowledge on social cognition as applied to questionnaire construction.

17.4.2 Mendelian Randomization

One reasoned solution to strengthen causal interpretation in epidemiological research has been via the use of "Mendelian randomization" studies. The concept was introduced by Katan (2004) already in the 1980s, although empirical studies started to appear only after the turn of the century and became more numerous after Davey Smith and Ebrahim's seminal paper in 2003 (Davey Smith and Ebrahim, 2003). The original illustration tried to solve the question whether the association between low serum cholesterol levels and cancer was causal or not. One hypothesis would stipulate that low cholesterol leads to cancer. Alternatively, cancer may cause decreases in cholesterol. As cancer takes decades to develop temporal sequencing is difficult. Moreover, there exists a number of empirical studies that point into different directions, partly because of confounding factors (such as diet and smoking) impacting both serum cholesterol and cancer. The solution to disentangling the direction of causality involved using the genetic constellation. The apolipoprotein E (ApoE) gene

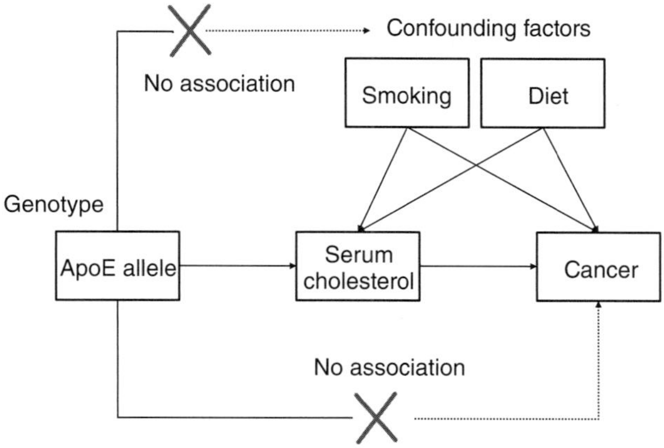

Figure 17.1 Principle of Mendelian randomization: example of clarifying the causal direction between low cholesterol and cancer.

affects levels of serum cholesterol, with ApoE2 being associated with lower levels. Clearly, the genetic constellation was formed during meiosis independent of and prior to serum cholesterol level measured and incidence of cancer. Moreover, during meiosis genes are randomly assigned (hence the name "Mendelian randomization"). Finally, there is no direct link between the ApoE gene and cancer.

These assumptions together allow the following empirical test: if the variants of the ApoE are distributed equally between cancer and noncancer cases, then the relationship between serum cholesterol and cancer cannot be due to a causal impact of cholesterol. On the other hand, if there is proportionally more ApoE2 alleles among cancer patients compared to noncancer patients, this is an indication of a causal pathway leading from ApoE2 → low cholesterol → cancer (Fig. 17.1).

The above reasoning seems to point to an easy way to prove causality in observational epidemiology (Sheehan *et al.*, 2008, Davey Smith and Ebrahim, 2003, Davey Smith *et al.*, 2005). Unfortunately, like any other methods, Mendelian randomization relies on a number of crucial assumptions. We will use the recent controversy about the paper of Holmes and colleagues (2014) on the impact of alcohol on ischemic heart disease to illustrate. Background is the so-called "cardioprotective" effect, that is, the postulate that on average low to moderate drinking without any heavy drinking occasions leads to a lowered risk of ischemic heart disease (Roerecke and Rehm, 2010, 2012). There had been many criticisms regarding the cardioprotective effect even though none of the postulated confounding up to now could be corroborated empirically.

Holmes and colleagues used the Mendelian randomization approach based on the rs1229984 A variant in the alcohol dehydrogenase-1b (ADH1B) gene, which impacts alcohol metabolism. The ADH1B rs1229984 A variant leads to a slightly unpleasant flushing reaction and thus to an overall lower level of alcohol intake. The reasoning of Holmes and colleagues was the following: within the light drinkers carriers of ADH1B rs1229984 A variant would drink less and thus, if there was a protective effect, they should have a higher risk of IHD (Roerecke and Rehm, 2015). However, an empirical meta-analysis of 56 studies showed the opposite: within light drinkers carriers of the ADH1B rs1229984 A variant showed less risk for IHD. Can these results be interpreted as a final proof that the cardioprotective effect is a spurious association without any causal impact of alcohol? The debate was animated with more than 15 rapid responses on the BMJ website (see also Roerecke and Rehm, 2015; Chikritzhs *et al.*, 2015; Rothman in: `http://www.bu.edu/alcohol-forum/critique-143-a-mendelian-randomization-assessment-of-alcohol-and-cardiovascular-disease-20-july-2014/`). The conclusion was that a causal relation could not be established in this case for the following reasons: (i) all of the effect of the ADH1B rs1229984 A variant would have to be mediated by average amount of drinking, which is not the case (Glymour, 2014); (ii) ADH1B rs1229984 A variant was indiscriminately associated with all alcohol indicators including heavy drinking occasions, which would predict the opposite of the cardioprotective effect, that is, light drinkers with occasional heavy drinking do not show any cardioprotective effect (Roerecke and Rehm, 2010); (iii) the results of Holmes and colleagues were also inconsistent with other biomarkers related to the risk of IHD such as HDL level. In sum, the assumptions of the Mendelian randomization were violated and thus no firm conclusions could be drawn (Fig. 17.2).

Mendelian randomization offers a way to potentially increase control and thus our belief in causal relations. Unfortunately, Mendelian randomization studies offer no panacea to the problem, although many risk factors both behavioral (smoking, drinking, and nutritional habits) and physiological (blood pressure, cholesterol, and glucose) have been shown to be genetically impacted. The assumptions of no other effects outside of a postulated pathway often do not hold true.

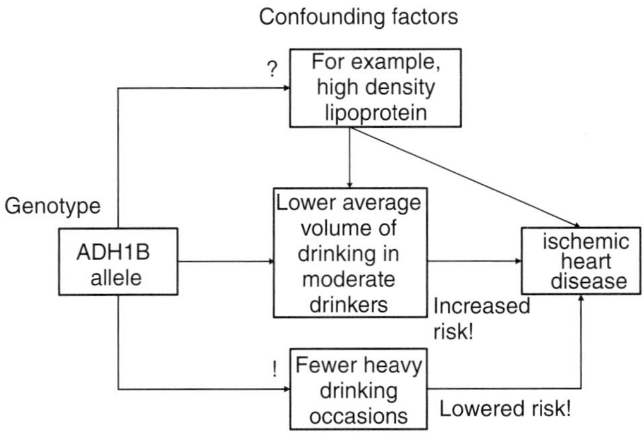

Figure 17.2 Principle of Mendelian randomization: example of clarifying the causal direction between low average volume of drinking and ischemic heart disease.

17.4.3 Surrogate Endpoints (Experimental)

Does alcohol consumption increase the risk of HIV? Again, there is a clear association between level of drinking and HIV as well as between other alcohol consumption dimensions (frequency of heavy drinking occasions, drinking before sex, etc.) and this endpoint (Scott-Sheldon *et al.*, 2013, Baliunas *et al.*, 2010, Shuper *et al.*, 2009, Fisher *et al.*, 2007). While some people interpret this association as causal, alternative explanations have been proposed, such as personality traits explaining both alcohol drinking and unprotected sex, most notable elevated risk taking (Shuper *et al.*, 2010). A randomized experiment seems hardly feasible for ethical and practical reasons. However, experimentation is possible if another endpoint instead of HIV status is used, intention to unsafe sex. This surrogate endpoint has been linked to the actual behavior in meta-analyses with a high effect size (Sheeran and Orbell, 1998, Sheeran *et al.*, 1999). In other words, intention to unsafe sex can serve as a surrogate for unsafe sex itself, which is the main pathway to HIV infection. The reasoning with surrogate endpoints has been applied for a variety of epidemiological areas: using elevated blood glucose over time as a surrogate endpoint for diabetes, serum cholesterol levels as a surrogate for IHD, or intestinal polyps as a surrogate for colon cancer. A surrogate endpoint needs to fulfill two criteria: (i) it should be easily assessed in a reliable manner, (ii) it should have close and quantifiable relationship with the final endpoint (for general considerations see Weir and Walley, 2006; for statistical considerations see Atkinson *et al.*, 2001).

How did the intention to have unsafe sex allow the testing of a causal hypothesis? While alcohol consumption cannot be experimentally manipulated over a long run (see above), short time manipulation is feasible. Up to a blood alcohol content of 0.1% (this is a higher BAC than stipulated in usual laws of alcohol and traffic participation), the subjects are not able to distinguish whether or not they have received alcohol. The manipulation is achieved by using mixed drinks (e.g., orange juice with rum or rum flavor in an environment where other clues point to an alcoholic drink—spilling of rum on the serving tablet or on the outside of the glass). Furthermore, in many of those experiments subjects were sexually aroused or not aroused as a second experimental condition (showing explicit photos or porn movies). The dependent variable then consisted of intention for unsafe sex usually on a Likert scale. For instance, explicit photos of attractive partners are shown with questions about willingness to engage in sexual interaction, followed by questions whether such interactions could be performed in an unsafe manner (anal intercourse without using a condom). The hypothesis to be tested would state that consumption of alcohol leads to a stronger intention for unsafe sex. A recent meta-analysis showed a clear dose–response relationship of all the experiments to date: the higher the BAC level, the higher the willingness to engage in unsafe sex (Rehm *et al.*, 2012).

The interpretation is clear: alcohol as the manipulated variable causally impacts the intention for unsafe sex. Alternative interpretations such as personality variables can be excluded, as potential confounding variables should be equally distributed if the randomization process was effective. For the causal interpretation it is inconsequential whether the interaction effect between BAC and arousal was significant or not. Unfortunately, establishing causality was only successful for the surrogate endpoint and it still can be argued that the real endpoint might be impacted by other determining factors on the pathway between intention and behavior. For instance, it has been argued that the experimental situation was seen as meaningless in terms of consequences and in such a situation alcohol may provoke extreme statements. Again, like with Mendelian randomization, experiments with surrogate endpoints may contribute to increase control and thus our belief in causality, but there exists no final proof.

17.4.4 Other Design Measures to Increase Control

In this section, we suggest several techniques to increase our confidence in causality in nonexperimental designs. Theoretically, the argumentation follows Rehm and Strack (1994), but contentwise we use examples from recent epidemiology:

The first design to be discussed is the **case-crossover design**, which is a method for studying transient effects on the risk of acute events (Maclure, 1991, Maclure *et al.*, 2000). The idea of control is the following: the best control from a theoretical perspective is an identical twin in the same environment, who is just experimentally matched to a different exposure. In the case-crossover design, not an identical twin is used but the same person. Consider the question whether alcohol has a causal effect in triggering myocardial infarction (Mostofsky *et al.*, 2015, Gerlich *et al.*, 2009). One way to study this question would be a case-control design, but the question would be how to find good controls. Hospital controls are problematic, as alcohol has been related to more than 200 disease and injury conditions (Rehm *et al.*, 2009), and it is hard to find a ward in the hospital, which is not affected by alcohol consumption. Population controls may be the answer, but then how to select good population controls, and most importantly, how to check for the triggering effect? This led to a case-crossover design. It basically starts with cases of myocardial infarction. These cases will be asked whether they consumed alcohol in temporal proximity to the infarction. This will give a rate of all cases with infarction who consumed alcohol. However, what would be the comparison group to judge if this rate is elevated or not. The answer is to ask whether the people with myocardial infarction consumed alcohol exactly 1 week before the infarction at the same time. This gives a two-by-two table, with infarction (yes/no—the no condition would be 1 week prior), and exposure at both time points. From this table, risks of myocardial infarction following alcohol consumption can be estimated, including the question, whether such risks are significantly elevated (Marshall and Jackson, 1993).

What are the caveats? Basically, the human brain including, but not limited to, memory effects (e.g., Gmel and Daeppen, 2007). First, it is easier to remember an event yesterday than a week ago. Second, it may be more salient to remember anything related to a salient event such as a myocardial infarction. And third, people try to make sense of an event, and may try to explain their serious condition by things, and alcohol may become salient there. In sum, it cannot be excluded that case-crossover designs exaggerate association and these association thus may appear causal even if they are not. Although there are a number of remedies such as the usual frequency method (Ye *et al.*, 2013), the main problem of biased memory and/or reporting cannot be fully solved.

There are many designs available to strengthen conclusions concerning causality, and this chapter does not allow sufficient room to list them all. They all follow the principles of control (Rehm and Strack, 1994), and the seminal works of Campbell and colleagues have given good examples how to make best use of designs and what analysis techniques are the most adequate (Campbell and Stanley, 1963, Cook and Campbell, 1979, Shadish *et al.*, 2002). In short the following points are stressed:

- In all situations, try to find better ways to control. Thus, to strengthen the causal interpretation of a simple before–after comparison, the introduction of a control group helps even if randomization is impossible. If the control group is not fully comparable, then adding multiple control groups may help.
- Make use of natural variation to test causality. The famous examples of alcohol and liver cirrhosis in wartime Paris (as described in Zatonski *et al.*, 2010) or whether raising the drinking age to 21 years reduced highway fatalities (Voas *et al.*, 2003) both made use of natural variation (in the Paris case the occupation of Paris by the Germans and the confiscation of wine, which saved many French liver deaths; the drinking age of the specific political situation of the state responsibility that led to introductions of the drinking age at multiple times where interrupted time series could be created, where one state served as a control to others).

This does not mean that such analyses cannot be challenged (e.g., Fillmore *et al.*, 2002, for the Paris example), but still, the strength of causality is based on the degree of control, and the more control there is, the harder it becomes to find convincing alternative explanations.

17.4.5 Methods of Analysis

Even in focused research questions, multivariate modeling techniques are used in the analyses of observational studies to control for confounding. As indicated above the ideal mode in epidemiology would be to have different exposures in identical twins. This ideal is not very feasible, so other forms of control have been introduced from matching (see above, section 17.3, p.412) to propensity score analysis (82).

Again, we introduce a concrete example to clarify these ideas: whether or not an intervention of the police (arresting the offender for a short period) after an incident of wife battering causally could prevent a reevent during a follow-up period was a lively debated research question in the United States through the 1980s. Experimental studies aiming to answer this question by randomly assigning offenders to short-time arrests or not yielded indecisive results: one study displaying preventive effects in Minneapolis (Minneapolis Domestic Violence Project, Sherman and Berk, 1984) could be criticized for serious protocol violations pointing at the fact that randomization proved hardly feasible. Police officers too often deliberately decided on arresting the offender beyond the study protocol by their own responsibility.

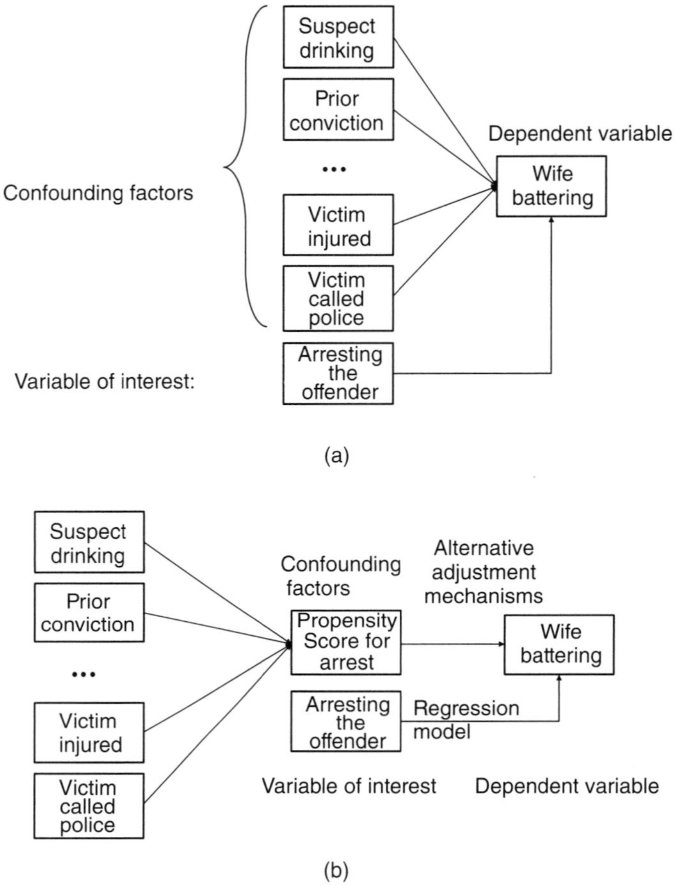

Figure 17.3 Principle of propensity score. (a) Risk of spousal violence from a traditional multivariate modeling perspective without PS. (b) Impact of arresting the offender calculated using PS.

A replication study in Omaha, Nebraska, failed to show any differences between the two randomized intervention groups (Dunford *et al.*, 1990). In this situation, Berk and Newton (1985) tried to use observational data to corroborate the conclusion of the Minneapolis Domestic Violence Project and published an early example for the use of "propensity scores" that had been proposed 2 years before by Rosenbaum and Rubin (1983).

While Figure 17.3a gives a visualization of how usual multivariate regression techniques would calculate a net effect for the impact of "Arresting the offender" on "Wife battering" (during a 6-month follow-up period), Figure 17.3b illustrates the central idea of using a "propensity score." The research question for Figure 17.3b is not to give a full picture of risk factors for wife battering but to specifically use all information available to model a score to predict an offender's chance of being arrested after an initial incident of domestic violence. This is the first step. A second step then is to use this "propensity" (to receive the "treatment") as the condensed confounding variable when evaluating the potentially causal impact of arrests on subsequent re-incidents.

If the prediction of the treatment variable is nearly perfect, then the decision on arresting an offender between two subjects with identical probability (= propensity score) can be regarded as simulating a randomization in which all these variables are controlled (d'Agostino, 1998). This is implied when the propensity score method is described as mimicking a randomized trial. Step 1 in Figure 17.3b is usually performed by discriminant analysis or (more frequently used and without assuming multivariate normal distribution among predictor variables) by logistic regression. An adjustment for multiple testing is not necessary during the scoring in step 1, and nonlinear effects of arrest-predicting variables can easily be modeled by including quadratic and/or cubic transformations of continuous variables (e.g., age). The score itself can be used in different ways of confounder adjustment (d'Agostino, 1998): (i) treated and untreated (here, arrested or not arrested offenders) subjects can be identified from the manifest, directly observed classification variable and matched in pairs according to comparable propensity scores. Statistical analysis is then performed by the usual conditional logistic regression model for case-control studies; (ii) the analysis of the outcome variable regressed on the potential causal variable can be performed in strata that were designed according to ranges of the propensity variable (e.g., quintiles); (iii) the variable of interest may be used as predictor variable for the outcome while simultaneously adjusting for the propensity score as a second predictor (traditional multivariate approach). A fourth method using the propensity score is called Inverse Probability of Treatment Weighting (IPTW, see Austin, 2011). Each sampling unit of the regression model in step 2 contributes only with his/her inverse probability of receiving the treatment. Weighting thus creates a distribution of measured covariates among the sample that is independent of treatment assignment.

The aforementioned historical example used a logistic regression approach to construct a propensity score for being arrested after an incident of domestic violence, resulting in over 96% of correctly classified subjects by using 14 variables ($n =$ 783). When this variable is added to a second logistic regression model on repeated domestic violence during the 6-month follow-up period, the assignment mechanism to police arrest could be regarded strongly ignorable (see also Rubin, 1991, for this notion). In the second step, Berk and Newton in a first attempt regressed the probability of wife battering separately for arrested and not arrested men using the propensity score as a continuous predictor variable. They found that only for the not arrested group a clear increase of the risk for wife battering with increasing values of the propensity score. The arrested group displayed virtually no association between the propensity score and the outcome. As a second modeling strategy, the proportional hazard model of Cox was used to take into account the time until the re-incident of familial violence. Both the propensity score and an interaction between propensity and factual arrest proved significant, thus pointing at a causal impact of the arrest on preventing repeated wife battering in a subgroup of initial offenders that showed only low probability to be arrested by the police after a first offense.

Propensity score methods became (with a considerable time lag to their theoretical foundation) quite popular in medicine after the year 2000 (Stürmer *et al.*, 2006). Though they are theoretically convincing and easy to perform, they cannot be regarded as panacea to the problem of establishing causality in observational studies. Their practical worth hinges on the possibility of prediction. For many diseases, such a prediction can only be done with lots of uncertainty. For instance, Bofetta found that known risk factors explained 35% of all cancers in smokers and 15% of all cancers in nonsmokers (Boffetta *et al.*, 2009). Similar numbers of explained variance are true for most disease and mortality outcomes. In this situation, differences between the results of different adjustment techniques often seem neglectable. Neither the review of Stürmer *et al.* (2006) nor the systematic review by Shah *et al.* (2005) could find dramatic differences when comparing studies using both methods of Figure 17.3a and b. Propensity score methods tended to estimate slightly smaller effect sizes for allegedly causal factors than traditional regression techniques.

Propensity scores can be seen as one form of matching (i.e., artificially creating twins), as per Rubin's theory of causality (Rubin, 1973). Another way, as we have seen above in the treatment of case-control studies, is given by finding matches as controls. The selection procedure is determined by key confounding variables as sex, age, and socioeconomic status. There is guidance on how to best select the number of matches to achieve efficiency (Austin, 2010, Kupper *et al.*, 1981, Stuart, 2010, e.g.,). In any form of matching or statistical control in general, the principle is, that only potential confounders should be used for matching, but not those variables that are on the causal pathway from exposure to outcome, or which are effects of the dependent (outcome) variable (Ho *et al.*, 2007). Clearly, all statistical techniques have to be used in a way congruent with the underlying theory or knowledge. Trying to optimize explained variance by introducing nonconfounders can result in biased estimates on the causal relation.

17.5 CONCLUSION: CONTROL IN EPIDEMIOLOGICAL RESEARCH CAN BE IMPROVED

We have illustrated how various measures can increase control in epidemiological studies and thus improve our understanding of causal relations (Rehm and Strack, 1994). Specifically, the following elements should be considered:

- Study design
- Measurement of dependent and independent variables
- Appropriate statistical techniques
- Accurate interpretation of results.

These four elements are also the core of the STRengthening Analytical Thinking for Observational Studies (STRATOS) initiative (Sauerbrei *et al.*, 2014). The points above also put a huge emphasis on the planning phase of any study to establish causal relations. First and foremost possibilities to establish control mechanisms outside of ex-post statistical control of observational data should be considered. It may be that there is genetic variation with impact on the independent variable of the causal relation and thus allowing a Mendelian randomization experiment. There may be surrogate endpoints close enough to the outcome under consideration to allow an experiment even when at first sight such a design looked impossible. Finally, there are a number of important design considerations in observational studies, such as better matching or inclusion of an additional control group. No matter what final design has been decided on, the next question should be about the best possible measurement of both dependent and independent variables. As shown above, formulation of questions or other assessment techniques should be constructed with the same attention to empirical research on social cognition and other relevant research. Finally, and only after the above steps have been taken, statistical techniques should be considered commensurate with the design and measurement (assumptions, scaling level, and possibility to maximize control). Overall, in our experience, the overwhelming effort should be spent in discussion of design alternatives, as even highly sophisticated statistical modeling cannot compensate weaknesses of the data generating process.

REFERENCES

Alavanja, M.C., Brownson, R.C., Berger, E., Lubin, J., and Modigh, C. (1996) Avian exposure and risk of lung cancer in women in Missouri: population based case-control study. *BMJ*, **313** (7067), 1233–1235. Using Smart Source Parsing Nov. 16.

Appel, B., Böl, G.F., Greiner, M., Lahrssen-Wiederholt, M., and Hensel, A. (2011) EHEC outbreak 2011. Investigation of the outbreak along the food chain. BfR Wissenschaft, Vol. 03/2012.

Atkinson, A.J., Colburn, W.A., DeGruttola, V.G., DeMets, D.L., Downing, G.J., Hoth, D.F., Oates, J.A., Peck, C.C., Schooley, R.T., and Spilker, B.A. (2001) Biomarkers and surrogate endpoints: preferred definitions and conceptual framework. *Clinical Pharmacology & Therapeutics*, **69** (3), 89–95.

Austin, P.C. (2010) Statistical criteria for selecting the optimal number of untreated subjects matched to each treated subject when using many-to-one matching on the propensity score. *American Journal of Epidemiology*, **172** (9), 1092–1097.

Austin, P.C. (2011) An introduction to propensity score methods for reducing the effects of confounding in observational studies. *Multivariate Behavioral Research*, **46** (3), 399–424.

Baliunas, D., Rehm, J., Irving, H., and Shuper, P. (2010) Alcohol consumption and risk of incident human immunodeficiency virus infection: a meta-analysis. *International Journal of Public Health*, **55** (3), 159–166.

Berk, R.A. and Newton, P.J. (1985) Does arrest really deter wife battery? An effort to replicate the findings of the Minneapolis spouse abuse experiment. *American Sociological Review*, **50** (2), 253–262.

Biemer, P.P., Groves, R.M., Lyberg, L.E., Mathiowetz, N.A., and Sudman, S. (2011) *Measurement Errors in Surveys*, vol. **173**, John Wiley & Sons, Inc. New York.

Boffetta, P., Tubiana, M., Hill, C., Boniol, M., Aurengo, A., Masse, R., Valleron, A.J., Monier, R., Boyle, P., and Autier, P. (2009) The causes of cancer in France. *Annals of Oncology*, **20** (3), 550–555.

Buchholz, U., Bernard, H., Werber, D., Böhmer, M.M., Remschmidt, C., Wilking, H., Deleré, Y., an der Heiden, M., Adlhoch, C., Dreesman, J., Ehlers, J., Ethelberg, S., Faber, M., Frank, C., Fricke, G., Greiner, M., Höhle, M., Ivarsson, S., Jark, U., Kirchner, M., Koch, J., Krause, G., Luber, P., Rosner, B., Stark, K., and Kühne, M. (2011) German outbreak of *Escherichia coli* O104: H4 associated with sprouts. *New England Journal of Medicine*, **365** (19), 1763–1770.

Campbell, D.T. and Stanley, J.C. (1963) *Experimental and Quasi-Experimental Designs for Generalized Causal Inference*, Houghton Mifflin, Boston, MA.

Charlton, B.G. (1996) Attribution of causation in epidemiology: chain or mosaic? *Journal of Clinical Epidemiology*, **49** (1), 105–107.

Chikritzhs, T., Stockwell, T., Naimi, T., Andreasson, S., Dangardt, F., and Liang, W. (2015) Has the leaning tower of presumed health benefits from 'moderate' alcohol use finally collapsed? *Addiction*, **110** (5), 726–727.

Chipperfield, J.O. and Steel, D.G. (2011) Efficiency of split questionnaire surveys. *Journal of Statistical Planning and Inference*, **141** (5), 1925–1932.

Choi, B.C. and Pak, A.W. (2005) A catalog of biases in questionnaires. *Preventing Chronic Disease*, **2** (1), A13.

Cook, T.D. and Campbell, D.T. (1979) *Quasi-Experimentation: Design & Analysis Issues for Field Settings*, Houghton Mifflin, Boston, MA.

Coombs, C. (1964) *A Theory of Data*, John Wiley & Sons, Inc., New York.

Coughlin, S.S. (1990) Recall bias in epidemiologic studies. *Journal of Clinical Epidemiology*, **43** (1), 87–91.

Cox, D.R. (1992) Causality: some statistical aspects. *Journal of the Royal Statistical Society. Series A (Statistics in Society)*, **155** (2), 291–301.

d'Agostino, R.B. (1998) Tutorial in biostatistics: propensity score methods for bias reduction in the comparison of a treatment to a non-randomized control group. *Statistics in Medicine*, **17** (19), 2265–2281.

Davey Smith, G. and Ebrahim, S. (2003) 'Mendelian randomization': can genetic epidemiology contribute to understanding environmental determinants of disease? *International Journal of Epidemiology*, **32** (1), 1–22.

Davey Smith, G., Ebrahim, S., Lewis, S., Hansell, A.L., Palmer, L.J., and Burton, P.R. (2005) Genetic epidemiology and public health: hope, hype, and future prospects. *Lancet*, **366** (9495), 1484–1498.

De Leeuw, E.D. (2001) Reducing missing data in surveys: an overview of methods. *Quality and Quantity*, **35** (2), 147–160.

Doll, R. and Hill, A.B. (1950) Smoking and carcinoma of the lung. *BMJ*, **2** (4682), 739–748.

Doll, R. and Hill, A.B. (1954) The mortality of doctors in relation to their smoking habits. *BMJ*, **1** (4877), 1451–1455.

Doll, R. and Hill, A.B. (1956) Lung cancer and other causes of death in relation to smoking. *BMJ*, **2** (5001), 1071.

Doll, R. and Hill, A. (1964) Mortality in relation to smoking: ten years' observations of British doctors. *BMJ*, **1** (5395), 1399–1410.

Dunford, F.W., Huizinga, D., and Elliott, D.S. (1990) The role of arrest in domestic assault: the Omaha police experiment. *Criminology*, **28** (2), 183–206.

Edwards, P., Roberts, I., Clarke, M., DiGuiseppi, C., Pratap, S., Wentz, R., and Kwan, I. (2002) Increasing response rates to postal questionnaires: systematic review. *BMJ*, **324** (7347), 1183.

Euser, A.M., Zoccali, C., Jager, K.J., and Dekker, F.W. (2009) Cohort studies: prospective versus retrospective. *Nephron Clinical Practice*, **113** (3), c214–c217.

Fillmore, K.M., Roizen, R., Farrell, M., Kerr, W., and Lemmens, P. (2002) Wartime Paris, cirrhosis mortality, and the ceteris paribus assumption. *Journal of Studies on Alcohol and Drugs*, **63** (4), 436–446.

Fisher, S.R.A. (1959) *Smoking: The Cancer Controversy: Some Attempts to Assess the Evidence*, Oliver and Boyd, London.

Fisher, J.C., Bang, H., and Kapiga, S.H. (2007) The association between HIV infection and alcohol use: a systematic review and meta-analysis of African studies. *Sexually Transmitted Diseases*, **34** (11), 856–863.

Frick, U., Meyer, M., Hauser, S., and Eichhammer, P. (2004) Machbarkeitsstudie: Verifizierung der Beschwerden "Elektrosensibler" vor und nach einer Sanierung. *Bundesministerium für Umwelt, Naturschutz und Reaktorsicherheit (Hrsg). Bonn.*

Frick, U., Rehm, J., and Eichhammer, P. (2002) Risk perception, somatization, and self report of complaints related to electromagnetic fields—a randomized survey study. *International Journal of Hygiene and Environmental Health*, **205** (5), 353–360.

Fuller, W.A. (1987) *Measurement Error Models*, vol. **305**, John Wiley & Sons, Inc.

Gadenne, V. (1976) *Die Gültigkeit psychologischer Untersuchungen*, Kohlhammer, Stuttgart.

Gerlich, M.G., Kramer, A., Gmel, G., Maggiorini, M., Luscher, T., Rickli, H., Kleger, G.R., and Rehm, J. (2009) Patterns of alcohol consumption and acute myocardial infarction: a case-crossover analysis. *European Addiction Research*, **15** (3), 143–149.

Glymour, M.M. (2014) Alcohol and cardiovascular disease. *BMJ*, **349**, g4334.

Gmel, G. and Daeppen, J.B. (2007) Recall bias for seven-day recall measurement of alcohol consumption among emergency department patients: implications for case-crossover designs. *Journal of Studies on Alcohol and Drugs*, **68** (2), 303–310.

Gonzalez, J.M. and Eltinge, J.L. (2007) Multiple matrix sampling: a review, in Proceedings of the Section on Survey Research Methods, American Statistical Association, 3069–3075.

Hill, A.B. (1965) The environment and disease: association or causation? *Proceedings of the Royal Society of Medicine*, **58**, 295–300.

Hippler, H.J., Schwarz, N., and Sudman, S. (1987) *Social Information Processing and Survey Methodology*, Springer Science, New York. Recent research in psychology.

Ho, D.E., Imai, K., King, G., and Stuart, E.A. (2007) Matching as nonparametric preprocessing for reducing model dependence in parametric causal inference. *Political Analysis*, **15** (3), 199–236.

Holmes, M.V., Dale, C.E., Zuccolo, L., Silverwood, R.J., Guo, Y., Ye, Z., Prieto-Merino, D., Dehghan, A., Trompet, S., and Wong, A. (2014) Association between alcohol and cardiovascular disease: Mendelian randomisation analysis based on individual participant data. *BMJ*, **349**, g4164.

Jöckel, K.H., Pohlabeln, H., Bromen, K., Ahrens, W., and Jahn, I. (2002) Pet birds and risk of lung cancer in North-Western Germany. *Lung Cancer*, **37** (1), 29–34.

Katan, M.B. (2004) Apolipoprotein E isoforms, serum cholesterol, and cancer. *International Journal of Epidemiology*, **33** (1), 9.

Kohlmeier, L., Arminger, G., Bartolomeycik, S., Bellach, B., Rehm, J., and Thamm, M. (1992) Pet birds as an independent risk factor for lung cancer: case-control study. *BMJ*, **305** (6860), 986–989.

Kohlmeier, L., Bellach, B., and Thamm, M. (1993) Pet birds and lung cancer. *BMJ*, **306** (6869), 60.

Kupper, L.L., Karon, J.M., Kleinbaum, D.G., Morgenstern, H., and Lewis, D.K. (1981) Matching in epidemiologic studies: validity and efficiency considerations. *Biometrics*, **37** (2), 271–291.

Little, R.J. (1992) Regression with missing X's: a review. *Journal of the American Statistical Association*, **87** (420), 1227–1237.

Loeb, L.A., Emster, V.L., Warner, K.E., Abbotts, J., and Laszlo, J. (1984) Smoking and lung cancer: an overview. *Cancer Research*, **44** (12 Part 1), 5940–5958.

Lord, F.M., Novick, M.R., and Birnbaum, A. (1968) *Statistical Theories of Mental Test Scores*, Addison-Wesley, Oxford.

Maclure, M. (1991) The case-crossover design: a method for studying transient effects on the risk of acute events. *American Journal of Epidemiology*, **133** (2), 144–153.

Maclure, M. and Mittleman, M.A. (2000) Should we use a case-crossover design? *Annual Review of Public Health*, **21** (1), 193–221.

Marshall, R.J. and Jackson, R.T. (1993) Analysis of case-crossover designs. *Statistics in Medicine*, **12** (24), 2333–2341.

Modigh, C., Axelsson, G., Alavanja, M., Andersson, L., and Rylander, R. (1996) Pet birds and risk of lung cancer in Sweden: a case-control study. *BMJ*, **313** (7067), 1236–1238.

Morabia, A., Stellman, S., Lumey, L., and Wynder, E. (1998) Parakeets, canaries, finches, parrots and lung cancer: no association. *British Journal of Cancer*, **77** (3), 501–504.

Mostofsky, E., van der Bom, J.G., Mukamal, K.J., Maclure, M., Tofler, G.H., Muller, J.E., and Mittleman, M.A. (2015) Risk of myocardial infarction immediately after alcohol consumption. *Epidemiology*, **26** (2), 143–150.

Nunnally, J.C. Jr. (1970) *Introduction to Psychological Measurement*, McGraw-Hill, New York.

Osborne, D.K. (1976) Unified theory of derived measurement. *Synthese*, **33** (1), 455–481.

Petri, M. and Allbritton, J. (1993) Fetal outcome of lupus pregnancy: a retrospective case-control study of the Hopkins lupus Cohort. *Obstetrical & Gynecological Survey*, **48** (11), 717–718.

Popper, K. (1934) *Logik der Forschung. Zur Erkenntnistheorie der modernen Naturwissenschaften*, Mohr Siebeck, Tübingen.

Proctor, R.N. (1996) The anti-tobacco campaign of the Nazis: a little known aspect of public health in Germany, 1933-45. *BMJ*, **313** (7070), 1450–1453.

Rehm, J. and Strack, F. (1994) Kontrolltechniken. In: Herrmann T, Tack W, Bierbaumer N, et al., (Eds.), *Enzyklopädie der Psychologie / Methodologische Grundlagen der Psychologie / Forschungsmethoden der Psychologie.* Vol Band 1. Göttingen: Hogrefe; 1994: 508–555.

Rehm, J., Arminger, G., and Kohlmeier, L. (1992) Using follow-up data to avoid omitted variable bias: an application to cardiovascular epidemiology. *Statistics in Medicine*, **11** (9), 1195–1208.

Rehm, J., Mathers, C., Popova, S., Thavorncharoensap, M., Teerawattananon, Y., and Patra, J. (2009) Global burden of disease and injury and economic cost attributable to alcohol use and alcohol-use disorders. *Lancet*, **373** (9682), 2223–2233.

Rehm, J., Shield, K.D., Joharchi, N., and Shuper, P.A. (2012) Alcohol consumption and the intention to engage in unprotected sex: systematic review and meta-analysis of experimental studies. *Addiction*, **107** (1), 51–59.

Roerecke, M. and Rehm, J. (2010) Irregular heavy drinking occasions and risk of ischemic heart disease: a systematic review and meta-analysis. *American Journal of Epidemiology*, **171** (6), 633–644.

Roerecke, M. and Rehm, J. (2012) The cardioprotective association of average alcohol consumption and ischaemic heart disease: a systematic review and meta-analysis. *Addiction*, **107** (7), 1246–1260.

Roerecke, M. and Rehm, J. (2015) Alcohol and ischaemic heart disease risk—finally moving beyond interpretation of observational epidemiology. *Addiction*, **110** (5), 723–725.

Rosenbaum, P.R. and Rubin, D.B. (1983) The central role of the propensity score in observational studies for causal effects. *Biometrika*, **70** (1), 41–55.

Rothman, K. (1986) *Modern Epidemiology*, 1st edn, Little Brown, Boston, MA.

Rothman, K. (1988) *Causal Inference*, Epidemiology Resources, Boston, MA.

Rothman, K. and Greenland, S. (2005) Causation and causal inference in epidemiology. *American Journal of Public Health*, **95** (S1), S144–S150.

Rothman, K., Greenland, S., and Lash, T. (2008) *Modern Epidemiology*, 3rd edn, Lippincott Williams & Wilkins, Philadelphia, PA.

Rubin, D.B. (1973) Matching to remove bias in observational studies. *Biometrics*, **29** (1), 159–183.

Rubin, D.B. (1991) Practical implications of modes of statistical inference for causal effects and the critical role of the assignment mechanism. *Biometrics*, **47** (4), 1213–1234.

Sauerbrei, W., Abrahamowicz, M., Altman, D.G., Cessie, S., and Carpenter, J. (2014) Strengthening analytical thinking for observational studies: the STRATOS initiative. *Statistics in Medicine*, **33** (30), 5413–5432.

Schwarz, N. and Sudman, S. (1994) *Autobiographical Memory and the Validity of Retrospective Reports*, Springer Science & Business Media, New York.

Scott-Sheldon, L.A., Walstrom, P., Carey, K.B., Johnson, B.T., Carey, M.P., and Team, M.R. (2013) Alcohol use and sexual risk behaviors among individuals infected with HIV: a systematic review and meta-analysis 2012 to early 2013. *Current HIV/AIDS Reports*, **10** (4), 314–323.

Shadish, W.R., Cook, T.D., and Campbell, D.T. (2002) *Experimental and Quasi-Experimental Designs for Generalized Causal Inference*, Houghton Mifflin Company, Boston, MA.

Shah, B.R., Laupacis, A., Hux, J.E., and Austin, P.C. (2005) Propensity score methods gave similar results to traditional regression modeling in observational studies: a systematic review. *Journal of Clinical Epidemiology*, **58** (6), 550–559.

Sheehan, N.A., Didelez, V., Burton, P.R., and Tobin, M.D. (2008) Mendelian randomisation and causal inference in observational epidemiology. *PLoS Medicine*, **5** (8), e177.

Sheeran, P., Abraham, C., and Orbell, S. (1999) Psychosocial correlates of heterosexual condom use: a meta-analysis. *Psychological Bulletin*, **125** (1), 90–132.

Sheeran, P. and Orbell, S. (1998) Do intentions predict condom use? Meta-analysis and examination of six moderator variables. *British Journal of Social Psychology*, **37** (2), 231–250.

Sherman, L.W. and Berk, R.A. (1984) The specific deterrent effects of arrest for domestic assault. *American Sociological Review*, **49** (2), 261–272.

Shuper, P.A., Joharchi, N., Irving, H., and Rehm, J. (2009) Alcohol as a correlate of unprotected sexual behavior among people living with HIV/AIDS: review and meta-analysis. *AIDS and Behavior*, **13** (6), 1021–1036.

Shuper, P.A., Neuman, M., Kanteres, F., Baliunas, D., Joharchi, N., and Rehm, J. (2010) Causal considerations on alcohol and HIV/AIDS—a systematic review. *Alcohol and Alcoholism*, **45** (2), 159–166.

Smits, N. and Vorst, H.C. (2007) Reducing the length of questionnaires through structurally incomplete designs: an illustration. *Learning and Individual Differences*, **17** (1), 25–34.

Strube, G. (1987) *Answering Survey Questions: The Role of Memory*, In H.J. Hippler, N. Schwarz, and S. Sudman (Eds.), *Social Information Processing and Survey Methodology* (pp. 86–101). New York: Springer.

Stuart, E.A. (2010) Matching methods for causal inference: a review and a look forward. *Statistical Science: A Review Journal of the Institute of Mathematical Statistics*, **25** (1), 1–21.

Stürmer, T., Joshi, M., Glynn, R.J., Avorn, J., Rothman, K.J., and Schneeweiss, S. (2006) A review of the application of propensity score methods yielded increasing use, advantages in specific settings, but not substantially different estimates compared with conventional multivariable methods. *Journal of Clinical Epidemiology*, **59** (5), 437.e1–437.e24.

Sudman, S., Bradburn, N.M., and Schwarz, N. (1996) *Thinking About Answers*, Josey-Bass, San Francisco, CA.

Thygesen, L.C., Andersen, G.S., and Andersen, H. (2005) A philosophical analysis of the Hill criteria. *Journal of Epidemiology and Community Health*, **59** (6), 512–516.

Voas, R.B., Tippetts, A.S., and Fell, J.C. (2003) Assessing the effectiveness of minimum legal drinking age and zero tolerance laws in the United States. *Accident Analysis and Prevention*, **35** (4), 579–587.

Wacholder, S. (1995) Design issues in case-control studies. *Statistical Methods in Medical Research*, **4** (4), 293–309.

Weed, D.L. (1997) On the use of causal criteria. *International Journal of Epidemiology*, **26** (6), 1137–1141.

Weed, D.L. and Gorelic, L.S. (1996) The practice of causal inference in cancer epidemiology. *Cancer Epidemiology Biomarkers & Prevention*, **5** (4), 303–311.

Weir, C.J. and Walley, R.J. (2006) Statistical evaluation of biomarkers as surrogate endpoints: a literature review. *Statistics in Medicine*, **25** (2), 183–203.

Wiedermann, W., Niggli, J., and Frick, U. (2014) The Lemming-effect: harm perception of psychotropic substances among music festival visitors. *Health, Risk & Society*, **16** (4), 323–338.

World Health Organization (2005) *Handbook for Good Clinical Research Practice (GCP): Guidance for Implementation*, WHO, Geneva.

Ye, Y., Bond, J., Cherpitel, C.J., Stockwell, T., Macdonald, S., and Rehm, J. (2013) Risk of injury due to alcohol: evaluating potential bias using the case-crossover usual-frequency method. *Epidemiology (Cambridge, Mass.)*, **24** (2), 240–243.

Zatoński, W.A., Sulkowska, U., Mańczuk, M., Rehm, J., Boffetta, P., Lowenfels, A.B., and La Vecchia, C. (2010) Liver cirrhosis mortality in Europe, with special attention to Central and Eastern Europe. *European Addiction Research*, **16** (4), 193–201.

Zidek, J.V., Wong, H., Le, N., and Burnett, R. (1996) Causality, measurement error and multicollinearity in epidemiology. *Environmetrics*, **7** (4), 441–451.

INDEX

Statistics and Causality: Methods for Applied Empirical Research, First Edition.
Wolfgang Wiedermann and Alexander von Eye.

WILEY SERIES IN PROBABILITY AND STATISTICS

Established by *Walter A. Shewhart and Samuel S. Wilks*

The Wiley Series in Probability and Statistics is well established and authoritative. It covers many topics of current research interest in both pure and applied statistics and probability theory. Written by leading statisticians and institutions, the titles span both state-of-the-art developments in the field and classical methods.
Reflecting the wide range of current research in statistics, the series encompasses applied, methodological and theoretical statistics, ranging from applications and new techniques made possible by advances in computerized practice to rigorous treatment of theoretical approaches. This series provides essential and invaluable reading for all statisticians, whether in aca–demia, industry, government, or research.

† ABRAHAM and LEDOLTER · Statistical Methods for Forecasting
AGRESTI · Analysis of Ordinal Categorical Data, Second Edition
AGRESTI · An Introduction to Categorical Data Analysis, Second Edition
AGRESTI · Categorical Data Analysis, Third Edition
AGRESTI · Foundations of Linear and Generalized Linear Models
ALSTON, MENGERSEN and PETTITT (editors) · Case Studies in Bayesian Statistical Modelling and Analysis
ALTMAN, GILL, and McDONALD · Numerical Issues in Statistical Computing for the Social Scientist
AMARATUNGA and CABRERA · Exploration and Analysis of DNA Microarray and Protein Array Data
AMARATUNGA, CABRERA, and SHKEDY · Exploration and Analysis of DNA Microarray and Other High-Dimensional Data, Second Edition
ANDEĚL · Mathematics of Chance
ANDERSON · An Introduction to Multivariate Statistical Analysis, Third Edition
* ANDERSON · The Statistical Analysis of Time Series
ANDERSON, AUQUIER, HAUCK, OAKES, VANDAELE, and WEISBERG · Statistical Methods for Comparative Studies
ANDERSON and LOYNES · The Teaching of Practical Statistics
ARMITAGE and DAVID (editors) · Advances in Biometry
ARNOLD, BALAKRISHNAN, and NAGARAJA · Records
* ARTHANARI and DODGE · Mathematical Programming in Statistics
AUGUSTIN, COOLEN, DE COOMAN and TROFFAES (editors) · Introduction to Imprecise Probabilities
* BAILEY · The Elements of Stochastic Processes with Applications to the Natural Sciences

BAJORSKI · Statistics for Imaging, Optics, and Photonics

BALAKRISHNAN and KOUTRAS · Runs and Scans with Applications

BALAKRISHNAN and NG · Precedence-Type Tests and Applications

BALI, ENGLE, and MURRAY · Empirical Asset Pricing: The Cross-Section of Stock Returns

BARNETT · Comparative Statistical Inference, Third Edition

BARNETT · Environmental Statistics

BARNETT and LEWIS · Outliers in Statistical Data, Third Edition

BARTHOLOMEW, KNOTT, and MOUSTAKI · Latent Variable Models and Factor Analysis: A Unified Approach, Third Edition

BARTOSZYNSKI and NIEWIADOMSKA-BUGAJ · Probability and Statistical Inference, Second Edition

BASILEVSKY · Statistical Factor Analysis and Related Methods: Theory and Applications

BATES and WATTS · Nonlinear Regression Analysis and Its Applications

BECHHOFER, SANTNER, and GOLDSMAN · Design and Analysis of Experiments for Statistical Selection, Screening, and Multiple Comparisons

BEH and LOMBARDO · Correspondence Analysis: Theory, Practice and New Strategies

BEIRLANT, GOEGEBEUR, SEGERS, TEUGELS, and DE WAAL · Statistics of Extremes: Theory and Applications

BELSLEY · Conditioning Diagnostics: Collinearity and Weak Data in Regression

†BELSLEY, KUH, and WELSCH · Regression Diagnostics: Identifying Influential Data and Sources of Collinearity

BENDAT and PIERSOL · Random Data: Analysis and Measurement Procedures, Fourth Edition

BERNARDO and SMITH · Bayesian Theory

BHAT and MILLER · Elements of Applied Stochastic Processes, Third Edition

BHATTACHARYA and WAYMIRE · Stochastic Processes with Applications

BIEMER, GROVES, LYBERG, MATHIOWETZ, and SUDMAN · Measurement Errors in Surveys

BILLINGSLEY · Convergence of Probability Measures, Second Edition

BILLINGSLEY · Probability and Measure, Anniversary Edition

BIRKES and DODGE · Alternative Methods of Regression

BISGAARD and KULAHCI · Time Series Analysis and Forecasting by Example

BISWAS, DATTA, FINE, and SEGAL · Statistical Advances in the Biomedical Sciences: Clinical Trials, Epidemiology, Survival Analysis, and Bioinformatics

BLISCHKE and MURTHY (editors) · Case Studies in Reliability and Maintenance

BLISCHKE and MURTHY · Reliability: Modeling, Prediction, and Optimization

BLOOMFIELD · Fourier Analysis of Time Series: An Introduction, Second Edition

BOLLEN · Structural Equations with Latent Variables

BOLLEN and CURRAN · Latent Curve Models: A Structural Equation Perspective

BONNINI, CORAIN, MAROZZI and SALMASO · Nonparametric Hypothesis Testing: Rank and Permutation Methods with Applications in R

BOROVKOV · Ergodicity and Stability of Stochastic Processes

BOSQ and BLANKE · Inference and Prediction in Large Dimensions

BOULEAU · Numerical Methods for Stochastic Processes

* BOX and TIAO Bayesian Inference in Statistical Analysis

BOX · Improving Almost Anything, Revised Edition

* BOX and DRAPER · Evolutionary Operation: A Statistical Method for Process Improvement

BOX and DRAPER · Response Surfaces, Mixtures, and Ridge Analyses, Second Edition

BOX, HUNTER, and HUNTER · Statistics for Experimenters: Design, Innovation, and Discovery, Second Editon

BOX, JENKINS, REINSEL, and LJUNG · Time Series Analysis: Forecasting and Control, Fifth Edition

BOX, LUCEÑO, and Paniagua-QuiÑones · Statistical Control by Monitoring and Adjustment, Second Edition

* BROWN and HOLLANDER · Statistics: A Biomedical Introduction

CAIROLI and DALANG · Sequential Stochastic Optimization

CASTILLO, HADI, BALAKRISHNAN, and SARABIA · Extreme Value and Related Models with Applications in Engineering and Science

CHAN Time · Series: Applications to Finance with R and S-Plus®, Second Edition

CHARALAMBIDES · Combinatorial Methods in Discrete Distributions

CHATTERJEE and HADI · Regression Analysis by Example, Fourth Edition

CHATTERJEE and HADI · Sensitivity Analysis in Linear Regression

CHEN · The Fitness of Information: Quantitative Assessments of Critical Evidence

CHERNICK · Bootstrap Methods: A Guide for Practitioners and Researchers, Second Edition

CHERNICK and FRIIS · Introductory Biostatistics for the Health Sciences

CHILÉS and DELFINER · Geostatistics: Modeling Spatial Uncertainty, Second Edition

CHIU, STOYAN, KENDALL and MECKE · Stochastic Geometry and Its Applications, Third Edition

CHOW and LIU · Design and Analysis of Clinical Trials: Concepts and Methodologies, Third Edition

CLARKE · Linear Models: The Theory and Application of Analysis of Variance

CLARKE and DISNEY · Probability and Random Processes: A First Course with Applications, Second Edition

* COCHRAN and COX · Experimental Designs, Second Edition

COLLINS and LANZA · Latent Class and Latent Transition Analysis: With Applications in the Social, Behavioral, and Health Sciences

CONGDON · Applied Bayesian Modelling, Second Edition

CONGDON · Bayesian Models for Categorical Data

CONGDON · Bayesian Statistical Modelling, Second Edition

CONOVER · Practical Nonparametric Statistics, Third Edition

COOK · Regression Graphics

COOK and WEISBERG · An Introduction to Regression Graphics

COOK and WEISBERG · Applied Regression Including Computing and Graphics

CORNELL · A Primer on Experiments with Mixtures

CORNELL · Experiments with Mixtures, Designs, Models, and the Analysis of Mixture Data, Third Edition

COX · A Handbook of Introductory Statistical Methods

CRESSIE · Statistics for Spatial Data, Revised Edition

CRESSIE and WIKLE · Statistics for Spatio-Temporal Data

CSÖRGO' and HORV ATH · Limit Theorems in Change Point Analysis

DAGPUNAR · Simulation and Monte Carlo: With Applications in Finance and MCMC

DANIEL · Applications of Statistics to Industrial Experimentation

DANIEL · Biostatistics: A Foundation for Analysis in the Health Sciences, Eighth Edition

* DANIEL · Fitting Equations to Data: Computer Analysis of Multifactor Data, Second Edition

DASU and JOHNSON · Exploratory Data Mining and Data Cleaning

DAVID and NAGARAJA · Order Statistics, Third Edition

DAVINO, FURNO and VISTOCCO · Quantile Regression: Theory and Applications

* DEGROOT, FIENBERG, and KADANE · Statistics and the Law

DEL CASTILLO · Statistical Process Adjustment for Quality Control

DEMARIS · Regression with Social Data: Modeling Continuous and Limited Response Variables

DEMIDENKO · Mixed Models: Theory and Applications with R, Second Edition

DENISON, HOLMES, MALLICK, and SMITH · Bayesian Methods for Nonlinear Classification and Regression

DETTE and STUDDEN · The Theory of Canonical Moments with Applications in Statistics, Probability, and Analysis

DEY and MUKERJEE · Fractional Factorial Plans

DILLON and GOLDSTEIN · Multivariate Analysis: Methods and Applications

* DODGE and ROMIG · Sampling Inspection Tables, Second Edition

* DOOB · Stochastic Processes

DOWDY, WEARDEN, and CHILKO · Statistics for Research, Third Edition

DRAPER and SMITH · Applied Regression Analysis, Third Edition

DRYDEN and MARDIA · Statistical Shape Analysis

DUDEWICZ and MISHRA · Modern Mathematical Statistics

DUNN and CLARK · Basic Statistics: A Primer for the Biomedical Sciences, Fourth Edition

DUPUIS and ELLIS · A Weak Convergence Approach to the Theory of Large Deviations

EDLER and KITSOS · Recent Advances in Quantitative Methods in Cancer and Human Health Risk Assessment

* ELANDT-JOHNSON and JOHNSON · Survival Models and Data Analysis

ENDERS · Applied Econometric Time Series, Third Edition

† ETHIER and KURTZ Markov · Processes: Characterization and Convergence

EVANS, HASTINGS, and PEACOCK · Statistical Distributions, Third Edition

EVERITT, LANDAU, LEESE, and STAHL · Cluster Analysis, Fifth Edition

FEDERER and KING · Variations on Split Plot and Split Block Experiment Designs

FELLER · An Introduction to Probability Theory and Its Applications, Volume I, Third Edition, Revised; Volume II, Second Edition

FITZMAURICE, LAIRD, and WARE · Applied Longitudinal Analysis, Second Edition

* FLEISS · The Design and Analysis of Clinical Experiments

FLEISS · Statistical Methods for Rates and Proportions, Third Edition

† FLEMING and HARRINGTON · Counting Processes and Survival Analysis

FUJIKOSHI, ULYANOV, and SHIMIZU · Multivariate Statistics: High-Dimensional and Large-Sample Approximations

FULLER · Introduction to Statistical Time Series, Second Edition

† FULLER · Measurement Error Models

GALLANT · Nonlinear Statistical Models

GEISSER · Modes of Parametric Statistical Inference

GELMAN and MENG · Applied Bayesian Modeling and Causal Inference from ncomplete-Data Perspectives

GEWEKE · Contemporary Bayesian Econometrics and Statistics

GHOSH, MUKHOPADHYAY, and SEN · Sequential Estimation

GIESBRECHT and GUMPERTZ · Planning, Construction, and Statistical Analysis of Comparative Experiments

GIFI · Nonlinear Multivariate Analysis

GIVENS and HOETING · Computational Statistics

GLASSERMAN and YAO · Monotone Structure in Discrete-Event Systems

GNANADESIKAN · Methods for Statistical Data Analysis of Multivariate Observations, Second Edition

GOLDSTEIN · Multilevel Statistical Models, Fourth Edition

GOLDSTEIN and LEWIS · Assessment: Problems, Development, and Statistical Issues

GOLDSTEIN and WOOFF · Bayes Linear Statistics

GRAHAM Markov · Chains: Analytic and Monte Carlo Computations

GREENWOOD and NIKULIN · A Guide to Chi-Squared Testing

GROSS, SHORTLE, THOMPSON, and HARRIS · Fundamentals of Queueing Theory, Fourth Edition

GROSS, SHORTLE, THOMPSON, and HARRIS · Solutions Manual to Accompany Fundamentals of Queueing Theory, Fourth Edition

* HAHN and SHAPIRO · Statistical Models in Engineering

HAHN and MEEKER · Statistical Intervals: A Guide for Practitioners

HALD · A History of Probability and Statistics and their Applications Before 1750

† HAMPEL · Robust Statistics: The Approach Based on Influence Functions

HARTUNG, KNAPP, and SINHA · Statistical Meta-Analysis with Applications

HEIBERGER · Computation for the Analysis of Designed Experiments

HEDAYAT and SINHA · Design and Inference in Finite Population Sampling

HEDEKER and GIBBONS · Longitudinal Data Analysis

HELLER MACSYMA · for Statisticians

HERITIER, CANTONI, COPT, and VICTORIA-FESER · Robust Methods in Biostatistics

HINKELMANN and KEMPTHORNE · Design and Analysis of Experiments, Volume 1: Introduction to Experimental Design, Second Edition

HINKELMANN and KEMPTHORNE · Design and Analysis of Experiments, Volume 2: Advanced Experimental Design

HINKELMANN (editor) · Design and Analysis of Experiments, Volume 3: Special Designs and Applications

HOAGLIN, MOSTELLER, and TUKEY · Fundamentals of Exploratory Analysis of Variance

* HOAGLIN, MOSTELLER, and TUKEY · Exploring Data Tables, Trends and Shapes

* HOAGLIN, MOSTELLER, and TUKEY · Understanding Robust and Exploratory Data Analysis

HOCHBERG and TAMHANE · Multiple Comparison Procedures

HOCKING · Methods and Applications of Linear Models: Regression and the Analysis of Variance, Third Edition

HOEL · Introduction to Mathematical Statistics, Fifth Edition

HOGG and KLUGMAN · Loss Distributions

HOLLANDER, WOLFE, and CHICKEN · Nonparametric Statistical Methods, Third Edition

HOSMER and LEMESHOW · Applied Logistic Regression, Second Edition

HOSMER, LEMESHOW, and MAY · Applied Survival Analysis: Regression Modeling of Time-to-Event Data, Second Edition

HUBER · Data Analysis: What Can Be Learned From the Past 50 Years

HUBER · Robust Statistics

† HUBER · and Ronchetti Robust Statistics, Second Edition

HUBERTY · Applied Discriminant Analysis, Second Edition

HUBERTY and OLEJNIK · Applied MANOVA and Discriminant Analysis, Second Edition

HUITEMA · The Analysis of Covariance and Alternatives: Statistical Methods for Experiments, Quasi-Experiments, and Single-Case Studies, Second Edition

HUNT and KENNEDY · Financial Derivatives in Theory and Practice, Revised Edition

HURD and MIAMEE · Periodically Correlated Random Sequences: Spectral Theory and Practice

HUSKOVA, BERAN, and DUPAC · Collected Works of Jaroslav Hajek–with Commentary

HUZURBAZAR · Flowgraph Models for Multistate Time-to-Event Data

JACKMAN · Bayesian Analysis for the Social Sciences

† JACKSON · A User's Guide to Principle Components

JOHN · Statistical Methods in Engineering and Quality Assurance

JOHNSON · Multivariate Statistical Simulation

JOHNSON and BALAKRISHNAN · Advances in the Theory and Practice of Statistics: A Volume in Honor of Samuel Kotz

JOHNSON, KEMP, and KOTZ · Univariate Discrete Distributions, Third Edition

JOHNSON and KOTZ (editors) · Leading Personalities in Statistical Sciences: From the Seventeenth Century to the Present

JOHNSON, KOTZ, and BALAKRISHNAN · Continuous Univariate Distributions, Volume 1, Second Edition

JOHNSON, KOTZ, and BALAKRISHNAN · Continuous Univariate Distributions, Volume 2, Second Edition

JOHNSON, KOTZ, and BALAKRISHNAN · Discrete Multivariate Distributions

JUDGE, GRIFFITHS, HILL, LÜTKEPOHL, and LEE · The Theory and Practice of Econometrics, Second Edition

JUREK and MASON · Operator-Limit Distributions in Probability Theory

KADANE · Bayesian Methods and Ethics in a Clinical Trial Design

KADANE and SCHUM · A Probabilistic Analysis of the Sacco and Vanzetti Evidence

KALBFLEISCH and PRENTICE · The Statistical Analysis of Failure Time Data, Second Edition

KARIYA and KURATA · Generalized Least Squares

KASS and VOS · Geometrical Foundations of Asymptotic Inference

† KAUFMAN and ROUSSEEUW · Finding Groups in Data: An Introduction to Cluster Analysis

KEDEM and FOKIANOS · Regression Models for Time Series Analysis

KENDALL, BARDEN, CARNE, and LE · Shape and Shape Theory

KHURI · Advanced Calculus with Applications in Statistics, Second Edition

KHURI, MATHEW, and SINHA · Statistical Tests for Mixed Linear Models

*KISH · Statistical Design for Research

KLEIBER and KOTZ · Statistical Size Distributions in Economics and Actuarial Sciences

KLEMELÄ · Smoothing of Multivariate Data: Density Estimation and Visualization

KLUGMAN, PANJER, and WILLMOT · Loss Models: From Data to Decisions, Third Edition

KLUGMAN, PANJER, and WILLMOT · Loss Models: Further Topics

KLUGMAN, PANJER, and WILLMOT · Solutions Manual to Accompany Loss Models: From Data to Decisions, Third Edition

KOSKI and NOBLE · Bayesian Networks: An Introduction

KOTZ, BALAKRISHNAN, and JOHNSON · Continuous Multivariate Distributions, Volume 1, Second Edition

KOTZ and JOHNSON (editors) · Encyclopedia of Statistical Sciences: Volumes 1 to 9 with Index

KOTZ and JOHNSON (editors) · Encyclopedia of Statistical Sciences: Supplement Volume

KOTZ, READ, and BANKS (editors) · Encyclopedia of Statistical Sciences: Update Volume 1

KOTZ, READ, and BANKS (editors) · Encyclopedia of Statistical Sciences: Update Volume 2

KOWALSKI and TU · Modern Applied U-Statistics

KRISHNAMOORTHY and MATHEW · Statistical Tolerance Regions: Theory, Applications, and Computation

KROESE, TAIMRE, and BOTEV · Handbook of Monte Carlo Methods

KROONENBERG · Applied Multiway Data Analysis

KULINSKAYA, MORGENTHALER, and STAUDTE · Meta Analysis: A Guide to Calibrating and Combining Statistical Evidence

KULKARNI and HARMAN · An Elementary Introduction to Statistical Learning Theory

KUROWICKA and COOKE · Uncertainty Analysis with High Dimensional Dependence Modelling

KVAM and VIDAKOVIC · Nonparametric Statistics with Applications to Science and Engineering

LACHIN · Biostatistical Methods: The Assessment of Relative Risks, Second Edition

LAD · Operational Subjective Statistical Methods: A Mathematical, Philosophical, and Historical Introduction

LAMPERTI · Probability: A Survey of the Mathematical Theory, Second Edition

LAWLESS · Statistical Models and Methods for Lifetime Data, Second Edition

LAWSON · Statistical Methods in Spatial Epidemiology, Second Edition

LE · Applied Categorical Data Analysis, Second Edition

LE · Applied Survival Analysis

LEE · Structural Equation Modeling: A Bayesian Approach

LEE and WANG · Statistical Methods for Survival Data Analysis, Fourth Edition

LePAGE and BILLARD · Exploring the Limits of Bootstrap

LESSLER and KALSBEEK · Nonsampling Errors in Surveys

LEYLAND and GOLDSTEIN (editors) · Multilevel Modelling of Health Statistics

LIAO · Statistical Group Comparison

LIN · Introductory Stochastic Analysis for Finance and Insurance

LINDLEY · Understanding Uncertainty, Revised Edition

LITTLE and RUBIN · Statistical Analysis with Missing Data, Second Edition

LLOYD · The Statistical Analysis of Categorical Data

LOWEN and TEICH · Fractal-Based Point Processes

MAGNUS and NEUDECKER · Matrix Differential Calculus with Applications in Statistics and Econometrics, Revised Edition

MALLER and ZHOU · Survival Analysis with Long Term Survivors

MARCHETTE · Random Graphs for Statistical Pattern Recognition

MARDIA and JUPP · Directional Statistics

MARKOVICH · Nonparametric Analysis of Univariate Heavy-Tailed Data: Research and Practice

MARONNA, MARTIN and YOHAI · Robust Statistics: Theory and Methods

MASON, GUNST, and HESS · Statistical Design and Analysis of Experiments with Applications to Engineering and Science, Second Edition

McCULLOCH, SEARLE, and NEUHAUS · Generalized, Linear, and Mixed Models, Second Edition

McFADDEN · Management of Data in Clinical Trials, Second Edition

* McLACHLAN Discriminant Analysis and Statistical Pattern Recognition

McLACHLAN, DO, and AMBROISE · Analyzing Microarray Gene Expression Data

McLACHLAN and KRISHNAN · The EM Algorithm and Extensions, Second Edition

McLACHLAN and PEEL · Finite Mixture Models

McNEIL · Epidemiological Research Methods

MEEKER and ESCOBAR · Statistical Methods for Reliability Data

MEERSCHAERT and SCHEFFLER · Limit Distributions for Sums of Independent Random Vectors: Heavy Tails in Theory and Practice

MENGERSEN, ROBERT, and TITTERINGTON · Mixtures: Estimation and Applications

MICKEY, DUNN, and CLARK · Applied Statistics: Analysis of Variance and Regression, Third Edition

*MILLER · Survival Analysis, Second Edition

MONTGOMERY, JENNINGS, and KULAHCI · Introduction to Time Series Analysis and Forecasting, Second Edition

MONTGOMERY, PECK, and VINING · Introduction to Linear Regression Analysis, Fifth Edition

MORGENTHALER and TUKEY · Configural Polysampling: A Route to Practical Robustness

MUIRHEAD · Aspects of Multivariate Statistical Theory

MULLER and STOYAN Comparison Methods for Stochastic Models and Risks

MURTHY, XIE, and JIANG · Weibull Models

MYERS, MONTGOMERY, and ANDERSON-COOK · Response Surface Methodology: Process and Product Optimization Using Designed Experiments, Third Edition

MYERS, MONTGOMERY, VINING, and ROBINSON · Generalized Linear Models. With Applications in Engineering and the Sciences, Second Edition

NATVIG · Multistate Systems Reliability Theory With Applications

† NELSON · Accelerated Testing, Statistical Models, Test Plans, and Data Analyses

† NELSON · Applied Life Data Analysis

NEWMAN · Biostatistical Methods in Epidemiology

NG, TAIN, and TANG · Dirichlet Theory: Theory, Methods and Applications

OKABE, BOOTS, SUGIHARA, and CHIU · Spatial Tesselations: Concepts and

Applications of Voronoi Diagrams, Second Edition

OLIVER and SMITH · Influence Diagrams, Belief Nets and Decision Analysis

PALTA · Quantitative Methods in Population Health: Extensions of Ordinary Regressions

PANJER · Operational Risk: Modeling and Analytics

PANKRATZ · Forecasting with Dynamic Regression Models

PANKRATZ · Forecasting with Univariate Box-Jenkins Models: Concepts and Cases

PARDOUX · Markov Processes and Applications: Algorithms, Networks, Genome and Finance

PARMIGIANI and INOUE · Decision Theory: Principles and Approaches

*PARZEN · Modern Probability Theory and Its Applications

PEÑA, TIAO, and TSAY · A Course in Time Series Analysis

Pesarin and Salmaso · Permutation Tests for Complex Data: Applications and Software

PIANTADOSI · Clinical Trials: A Methodologic Perspective, Second Edition

POURAHMADI · Foundations of Time Series Analysis and Prediction Theory

POURAHMADI · High-Dimensional Covariance Estimation

POWELL · Approximate Dynamic Programming: Solving the Curses of Dimensionality, Second Edition

POWELL and RYZHOV · Optimal Learning

PRESS · Subjective and Objective Bayesian Statistics, Second Edition
PRESS and TANUR · The Subjectivity of Scientists and the Bayesian Approach
PURI, VILAPLANA, and WERTZ · New Perspectives in Theoretical and Applied Statistics
† PUTERMAN · Markov Decision Processes: Discrete Stochastic Dynamic Programming
QIU · Image Processing and Jump Regression Analysis
*RAO · Linear Statistical Inference and Its Applications, Second Edition
RAO · Statistical Inference for Fractional Diffusion Processes
RAUSAND and HØYLAND · System Reliability Theory: Models, Statistical Methods, and Applications, Second Edition
RAYNER, THAS, and BEST · Smooth Tests of Goodnes of Fit: Using R, Second Edition
RENCHER and SCHAALJE · Linear Models in Statistics, Second Edition
RENCHER and CHRISTENSEN · Methods of Multivariate Analysis, Third Edition
RENCHER · Multivariate Statistical Inference with Applications
RIGDON and BASU · Statistical Methods for the Reliability of Repairable Systems
* RIPLEY · Spatial Statistics
* RIPLEY · Stochastic Simulation
ROHATGI and SALEH · An Introduction to Probability and Statistics, Third Edition
ROLSKI, SCHMIDLI, SCHMIDT, and TEUGELS · Stochastic Processes for Insurance and Finance
ROSENBERGER and LACHIN · Randomization in Clinical Trials: Theory and Practice
ROSSI, ALLENBY, and McCULLOCH · Bayesian Statistics and Marketing
† ROUSSEEUW and LEROY · Robust Regression and Outlier Detection
ROYSTON and SAUERBREI · Multivariate Model Building: A Pragmatic Approach to Regression Analysis Based on Fractional Polynomials for Modeling Continuous Variables
* RUBIN · Multiple Imputation for Nonresponse in Surveys
RUBINSTEIN and KROESE · Simulation and the Monte Carlo Method, Second Edition
RUBINSTEIN and MELAMED · Modern Simulation and Modeling
RUBINSTEIN, RIDDER, and VAISMAN · Fast Sequential Monte Carlo Methods for Counting and Optimization
RYAN · Modern Engineering Statistics
RYAN · Modern Experimental Design
RYAN · Modern Regression Methods, Second Edition
RYAN · Sample Size Determination and Power
RYAN · Statistical Methods for Quality Improvement, Third Edition
SALEH · Theory of Preliminary Test and Stein-Type Estimation with Applications
SALTELLI, CHAN, and SCOTT (editors) · Sensitivity Analysis
SCHERER · Batch Effects and Noise in Microarray Experiments: Sources and Solutions
*SCHEFFE · The Analysis of Variance
SCHIMEK · Smoothing and Regression: Approaches, Computation, and Application
SCHOTT · Matrix Analysis for Statistics, Second Edition
SCHOUTENS LEVY · Processes in Finance: Pricing Financial Derivatives
SCOTT · Multivariate Density Estimation

SCOTT · Multivariate Density Estimation: Theory, Practice, and Visualization

*SEARLE · Linear Models

† SEARLE · Linear Models for Unbalanced Data

† SEARLE · Matrix Algebra Useful for Statistics

† SEARLE, CASELLA, and McCULLOCH · Variance Components

SEARLE and WILLETT · Matrix Algebra for Applied Economics

SEBER · A Matrix Handbook For Statisticians

† SEBER · Multivariate Observations

SEBER and LEE · Linear Regression Analysis, Second Edition

† SEBER and WILD · Nonlinear Regression

SENNOTT · Stochastic Dynamic Programming and the Control of Queueing Systems

*SERFLING · Approximation Theorems of Mathematical Statistics

SHAFER and VOVK · Probability and Finance: It's Only a Game!

SHERMAN · Spatial Statistics and Spatio-Temporal Data: Covariance Functions and Directional Properties

SILVAPULLE and SEN · Constrained Statistical Inference: Inequality, Order, and Shape Restrictions

SINGPURWALLA · Reliability and Risk: A Bayesian Perspective

SMALL and McLEISH · Hilbert Space Methods in Probability and Statistical Inference

SRIVASTAVA · Methods of Multivariate Statistics

STAPLETON · Linear Statistical Models, Second Edition

STAPLETON · Models for Probability and Statistical Inference: Theory and Applications

STAUDTE and SHEATHER · Robust Estimation and Testing

STOYAN · Counterexamples in Probability, Second Edition

STOYAN and STOYAN · Fractals, Random Shapes and Point Fields: Methods of Geometrical Statistics

STREET and BURGESS · The Construction of Optimal Stated Choice Experiments: Theory and Methods

STYAN · The Collected Papers of T. W. Anderson: 1943–1985

SUTTON, ABRAMS, JONES, SHELDON, and SONG · Methods for Meta-Analysis in Medical Research

TAKEZAWA · Introduction to Nonparametric Regression

TAMHANE · Statistical Analysis of Designed Experiments: Theory and Applications

TANAKA · Time Series Analysis: Nonstationary and Noninvertible Distribution Theory

THOMPSON · Empirical Model Building: Data, Models, and Reality, Second Edition

THOMPSON · Sampling, Third Edition

THOMPSON · Simulation: A Modeler's Approach

THOMPSON and SEBER · Adaptive Sampling

THOMPSON, WILLIAMS, and FINDLAY · Models for Investors in Real World Markets

TIERNEY LISP-STAT: · An Object-Oriented Environment for Statistical Computing and Dynamic Graphics

TROFFAES and DE COOMAN · Lower Previsions

TSAY · Analysis of Financial Time Series, Third Edition
TSAY · An Introduction to Analysis of Financial Data with R
TSAY · Multivariate Time Series Analysis: With R and Financial Applications
UPTON and FINGLETON · Spatial Data Analysis by Example, Volume II: Categorical and Directional Data
† VAN BELLE · Statistical Rules of Thumb, Second Edition
VAN BELLE, FISHER, HEAGERTY, and LUMLEY · Biostatistics: A Methodology for the Health Sciences, Second Edition
VESTRUP · The Theory of Measures and Integration
VIDAKOVIC · Statistical Modeling by Wavelets
VIERTL · Statistical Methods for Fuzzy Data
VINOD and REAGLE · Preparing for the Worst: Incorporating Downside Risk in Stock Market Investments
WALLER and GOTWAY · Applied Spatial Statistics for Public Health Data
WEISBERG · Applied Linear Regression, Fourth Edition
WEISBERG · Bias and Causation: Models and Judgment for Valid Comparisons
WELSH · Aspects of Statistical Inference
WESTFALL and YOUNG · Resampling-Based Multiple Testing: Examples and Methods for p-Value Adjustment
*WHITTAKER · Graphical Models in Applied Multivariate Statistics
WINKER · Optimization Heuristics in Economics: Applications of Threshold Accepting
WOODWORTH · Biostatistics: A Bayesian Introduction
WOOLSON and CLARKE · Statistical Methods for the Analysis of Biomedical Data, Second Edition
WU and HAMADA · Experiments: Planning, Analysis, and Parameter Design Optimization, Second Edition
WU and ZHANG · Nonparametric Regression Methods for Longitudinal Data Analysis
YAKIR · Extremes in Random Fields
YIN · Clinical Trial Design: Bayesian and Frequentist Adaptive Methods
YOUNG, VALERO-MORA, and FRIENDLY · Visual Statistics: Seeing Data with Dynamic Interactive Graphics
ZACKS · Examples and Problems in Mathematical Statistics
ZACKS · Stage-Wise Adaptive Designs
*ZELLNER · An Introduction to Bayesian Inference in Econometrics
ZELTERMAN · Discrete Distributions–Applications in the Health Sciences
ZHOU, OBUCHOWSKI, and McCLISH · Statistical Methods in Diagnostic Medicine, Second Edition